ARID LAND IRRIGATION IN DEVELOPING COUNTRIES

ENVIRONMENTAL PROBLEMS AND EFFECTS

The International Symposium on which this volume is based and the subsequent work entailed was made possible by a grant from the United Nations Environment Programme, administered through UNESCO and the International Council of Scientific Unions, and in cooperation with the UN Food and Agriculture Organization, World Health Organization, Union of International Engineering Organizations and other government and non-governmental organizations.

ARID LAND IRRIGATION IN DEVELOPING COUNTRIES:

ENVIRONMENTAL PROBLEMS AND EFFECTS

Based on the International Symposium,
16–21 February 1976, Alexandria, Egypt

Editor

E. BARTON WORTHINGTON
President, Scientific Committee on Water Research, International Council of Scientific Unions

PERGAMON PRESS

OXFORD   NEW YORK   SYDNEY   TORONTO   PARIS   FRANKFURT

UK	Pergamon Press Ltd., Headington Hill Hall, Oxford OX3 0BW, England
USA	Pergamon Press Inc., Maxwell House, Fairview Park, Elmsford, New York 10523, USA
CANADA	Pergamon of Canada Ltd., 75 The East Mall, Toronto, Ontario, Canada
AUSTRALIA	Pergamon Press (Aust.) Pty. Ltd., 19a Boundary Street, Rushcutters Bay, NSW 2011, Australia
FRANCE	Pergamon Press SARL, 24 rue des Ecoles. 75240 Paris, Cedex 05, France
WEST GERMANY	Pergamon Press GmbH, 6242 Kronberg-Taunus, Pferdstrasse 1, Frankfurt-am-Main, West Germany

Library of Congress Cataloging in Publication Data

Main entry under title:

Arid land irrigation in developing countries.

   Includes index.
    1. Irrigation—Environmental aspects—Congresses.
2. Underdeveloped areas—Irrigation—Congresses.
3. Arid regions—Congresses. I. Worthington, Edgar
Barton, 1905–
S612.2.A74   333.7′3   76-58394
ISBN 0-08-021588-2

Printed in Great Britain by Butler & Tanner Ltd., Frome and London

CONTENTS

PREFACE

Where big irrigation schemes are concerned there has been some tendency in recent years to emphasize the detrimental environmental effects and to underplay the benefits which such schemes confer on mankind. The increase of diseases, reduction of fisheries, change in the erosion pattern, salinization of soils—however well predicted and taken into account—have received publicity, while the main purpose of the schemes—bringing water and a better standard of life to millions of people—is apt to be taken for granted. In this book emphasis is placed on how to get the best out of the potential benefits of irrigation and to ensure their permanence, as well as how to overcome the detrimental effects on the environment and on man himself.

The book is based on an international symposium which was held from 16 to 21 February 1976, in Alexandria, at the invitation of the Egyptian Government. The purpose was to bring all main scientific disciplines to bear on irrigation in arid lands, especially on the environmental problems and with the emphasis on developing countries. Some 30 papers were invited and were printed in advance by the Egyptian Government. Their authors provided a panel of consultants at each session and it was possible to focus mainly on discussion rather than presentation of papers. Registered participants numbered more than 200, and a number of others attended some of the sessions. They came from 28 different countries and 9 international organizations, the expenses for some of those from developing countries being defrayed by UNEP.

The interest of the Scientific Committee on Water Research (COWAR) of the International Council of Scientific Unions (ICSU) in irrigation, and also that of the Scientific Committee on Problems of the Environment (SCOPE), was a natural sequence to activity from 1970 onwards on man-made lakes. An international symposium organized by COWAR on that subject, held in 1971 at Knoxville, Tennessee, resulted in a substantial volume of proceedings and also in a report of SCOPE which drew heavily on the symposium but did not limit itself to what was discussed there. This method of taking maximum advantage of a large international meeting proved successful and has been followed in the present case. COWAR organized the symposium, and a working group of SCOPE sat for an intensive week immediately following it in order to produce a report which is published as Section I, and is also to be issued separately by the Man and Biosphere Programme (MAB).

The symposium was opened by Professor Dr. Abdel Moneim Abul Azm, President of the Academy of Scientific Research and Technology, and other speakers at the preliminary session were, in order of presentation, Dr. E. B. Worthington, President of COWAR; Dr. Kamal Raheem, representing the Director General of UNESCO; His Excellency Abdel Tawab Hudeib, Governor of Alexandria; His Excellency Engineer Abdel Azim Abul Ata, Minister of Irrigation, who spoke on behalf of His Excellency the Prime Minister of Egypt,

Mamdouh Salem, prevented from attending at the last moment. The sessions occupied four days divided by a day's excursion to examine some problems of newly reclaimed areas of irrigation within reach of Alexandria. Arrangements were in the hands of Dr. Gamel Abdel Samie, Vice President of the Egyptian Academy, and Dr. Fred Fournier of UNESCO and the International Association of Scientific Hydrology (IAHS). The symposium expressed much appreciation for all the facilities and hospitality which were provided.

Each session was arranged by an appropriate organization: the introductory session and session 2 on case studies by COWAR, session 3 on hydrological processes by IAHS, session 4 on land use and soil problems by the International Soil Science Society (ISSS), session 5 on biological balances by the International Union of Biological Science (IUBS), session 6 on the efficiency of irrigation by the International Commission on Irrigation and Drainage (ICID), and session 7 on health by the World Health Organization (WHO). The last session, number 8, initiated the synthesis of all these subjects into a whole.

The timing of the symposium was arranged to feed into other activities in this same general field. It was coordinated with the MAB series of meetings on irrigation of which the regional meeting for the Middle East was held following it in Alexandria. Our conclusions are expected to provide an input to the UN Water Conference of March 1977 and the Desertification Conference later that year. However, it was not talk at Conferences at which the symposium aimed so much as action towards improved systems of irrigation, in balance and in partnership with nature.

While this book results from the symposium, its content has been edited somewhat freely so that it will stand as a general account of the interest and activities of the various scientific disciplines which are concerned in arid land irrigation. Thus in the discussions and in the papers themselves some passages which were specifically related to the symposium as such have been omitted, while the parts of lasting scientific value are retained. Speeches of welcome at the opening session are not included, much as they were appreciated at the time. Contributions to the discussions were written down by the speakers themselves and appear in summarized form at the end of each Section. With the exception of their first mention, all organizations are referred to by initials only, for which there is a Glossary.

The papers, though all in either English or French, are by specialists in a number of countries, some of which continue to use non-metric systems for length, area and weight. While comprehensive conversion to metric would have the advantage of direct comparability, it would have disadvantages in some countries where the book is used. For example, the acre and the feddan have more agricultural meaning in countries where they are used than hectares. Therefore in the editing they have, with few exceptions, been left as originally presented.

There is some repetition among the papers, for problems such as the accumulation of salts in the soil, the need for drainage, training of farmers, recur frequently. Repetition is inevitable, and perhaps it is no bad thing, for such problems are so important that the lessons learned from many experiences need to be pressed home.

This example of the coordination of many viewpoints on irrigation, a branch of development which is certain to increase enormously in future years, is a last undertaking of COWAR as it was set up by ICSU in 1964. Experience has shown clearly that the sciences comprised in ICSU are not in themselves sufficient for the overall view which is essential to appreciate the influences of development on the environment and vice versa. In regard to water several organizations of the ICSU family are acknowledged leaders in their subjects, for example, IAHS and the International Association of Hydrology (IAH), the International Limnological Society (SIL) and the International Association of Water Pollution Research (IAWPR) could hardly have done more. However, the relationships of such branches of water science to engineering, hydraulics, medicine and sociology, which are not a part of ICSU, often becomes of overriding importance. Hence ICSU has decided, at its 16th Assembly in October 1976 at Washington, DC, to broaden the scope of its water committee as a coordinating body, and, with new terms of reference, to work towards a closer association with other international organizations concerned.

I express thanks to all the organizations and individuals who made the Symposium of February 1976 at Alexandria a success; and also those who have assisted me in the preparation of this book, including Dr. Fred Fournier for checking the parts in French, Dr. Rosemary Lowe-McConnell for help with the English text, Miss Sue Darell-Brown for administration and Mrs. Toni Sycamore for typing.

E. Barton Worthington

Sussex,
June 1976

SECTION I

The Main Effects and Problems of Irrigation

Prepared by a joint working group, convened by Gilbert White, President of SCOPE

This first section results from a working group of fourteen people who had participated in the International Symposium on Arid Lands Irrigation at Alexandria, February 16–20, 1976, and stayed for a further 5 days in order to assess its discussions and conclusions, together with information and opinion, not all of which was actually presented at the Symposium. Drafts by each member of the working group were subsequently brought together and edited by Gilbert White into this relatively short report, which, perhaps after some minor modifications, is expected to be published under a separate cover in the series of MAB and SCOPE reports. The members of the working group were:

Alexis Coumbaras, Faculty of Medicine, University of Paris
Jean de Forges, École National Supérieure d'Horticulture, Consultant, UNESCO
Jacques Deom, World Health Organization
Malin Falkenmark, Swedish National Committee for SCOPE
Frederic Fournier, Consultant, UNESCO
Milos Holý, International Commission for Irrigation and Drainage, Technical University, Prague
Clyde E. Houston, Food and Agriculture Organization
Mohamed Kassas, University of Cairo and Vice President, SCOPE
Gyorgy Kovacs, National Water Authority, Hungary
Letitia Eva Obeng, United Nations Environmental Programme
G. Boris Rozanov, Moscow State University
Gilbert F. White, International Geographical Union, SCOPE and University of Colorado, Chairman
Nathaniel Wollman, University of New Mexico
E. B. Worthington, President, Committee on Water Research

1. BENEFITS OF ARID LANDS IRRIGATION

In the face of consensus that supplying food for a prospective world population will require by 1985 a giant increase in cereals production over 1970, major revisions in systems for food production, storage, transport, distribution and consumption are a necessity. Production improvement can be achieved by increasing land productivity on existing lands or by expanding the total area of arable land utilized. Both approaches probably must be adopted simultaneously.

Irrigation is one means of improving the total volume or reliability of agricultural production by managing water for the crop. Water control is significant in affecting the returns from additional inputs of fertilizers, seed, diversification, mechanization and other management efforts.

It is estimated that about 13 percent of the world's arable lands are irrigated, and that they use about 1400 billion cubic meters of water per annum. The irrigated harvested area in developing countries is estimated to increase at about 2.9 percent per annum while nonirrigated arable land has been increasing at a rate of only 0.7 percent per annum (Holý 1971). To the extent that food problems are to be met by enlarged production the enhancement of irrigation is important. As a minimum, it is essential to maintain the present productivity of irrigated land.

Further investment in irrigation must be examined in the perspective of its full environmental effects if it is to develop in a stable and fruitful fashion. In arid lands of developing countries the need for such perspective is especially acute because of the paucity of alternative resources and the urgency that small stocks of available capital be spent wisely.

Irrigation in perspective

When people first moved from caves to open land in the valleys they learned that slight modifications in the natural environment produced increased returns in food production. By scratching the ground surface prior to planting they were able to obtain a a more even distribution in production than by flinging the seed upon a barren hard surface. When the rains came, much soil was washed away from the disturbed land and deposited in the flat bottoms downstream. These depositions were found to be fertile and the soil relatively easy to work for planting purposes. When the rains did not come, crops withered and died, except next to flowing streams where water diverted to fields produced the same results as though the rains had continued.

Man manipulated his environment since his beginning. For the most part, this provided support for enlarging population. In later years, some environmental manipulation has been on the negative side. Much of this can be attributed to drastic population increases which generated pressures for living space and for more food. These mounted during the past century, culminating in the present food crises.

At present, the battle to provide sufficient food to the world population is at a crucial stage. In some areas the human race is producing people at a greater rate than food. This is not an encouraging picture, and we know that present-day food production is of such importance that only by integrating technical, biological, socio-economic and political measures will mankind overcome the stark realism of food shortages. We also know that in the end the farmer is the one who produces the food, and that all of these integrated inputs must be aimed at assisting the farmer.

Food sources

The consensus of regional and global meetings dealing with food is that a total increase in need for cereals alone to meet food and animal demands in 1985 will be at least 30 percent more than in 1970. Some countries have very little virgin land available for cultivation and will be forced to rely heavily, indeed almost exclusively, on yield improvements. Other countries have some unused land available and will therefore be able to adopt a more flexible policy, although higher yields will continue to be a major objective. Recent experience with the "green revolution" provides startling examples of the need for an integrated approach to successful yield increases. New seed and old production practices usually result in failure. New seed requires new planting procedures, proper fertilization, improved irrigation, new pest control practices, changes in harvesting and storage procedures and facilities. A breakdown at one place along the line can result in total failure. Variation in weather is a constant source of fluctuation in crop yields, and climate sets limits upon what may be attempted, particularly in arid lands.

Approaches to food production under climatic variability

The two extremes of climatic variability for agricultural production are, at one end, a complete reliance on direct and sometimes haphazard rainfall to bring the crop to maturity, and at the other, full control of all water applications on the crop, including the removal of excess supplies, so that it receives water in carefully measured doses only when necessary. Between these two extremes are many intermediate stages characterized by partial water control. In general, the closer to full control, the more efficient is the agricultural technique, the higher the crop yields, and the greater the economy in use of water. Recent work on water inputs seeks to maximise the value of water in crop production. As an example, in water-short areas certain crops may achieve three-quarters maximum production with two-thirds maximum water input. The one-third savings in water then is used for further crop production in another area. Similar results may be realized in rain-fed agriculture where a drought or near drought may not result in complete crop failure but in a decreased unit yield.

The first steps taken to improve traditional rain-fed farming consist in the retention of rainfall in the soil by special tillage practices and by the construction of small embankments or ponds to intercept surface runoff. These are of particular use in the case of flood recession agriculture when any lengthening of the period of gradual desiccation of productive soils results in significant increases in crop yields.

The next stage consists in the more complete use of flows, either by retention and water-level control in areas protected with embankments or by conducting the water through channels or pipes to prepared production areas. These are the methods of maximum utilization of variable rainfall.

The next step is to provide facilities to maintain water supplies throughout the non-rainfall season. These processes are recognized as irrigation. As man learned to live together in arid environments along the Nile, Euphrates or Indus the variability of climate was to a great extent overcome and the minimum water input to crop production was guaranteed.

The greatest privations from drought are experienced in those countries, nearly always semiarid, in which seasonal droughts recur at more or less frequent intervals. Between these periods of stress human and animal populations build up during years of plenty and become vulnerable when the next drought occurs. The long-term relief of agricultural suffering from drought is a massive problem. Only a far-sighted and carefully planned campaign can have a significant remedial effect and many lines of attack need to be utilized and fully coordinated. These include the improvement of groundwater reserves and supplies for use in periods of drought; the provision of supplementary irrigation to cushion rainfall deficiencies; the provision of water cisterns for collecting rainfall that does occur; in some areas particularly prone to recurrent catastrophic droughts, the permanent transfer of local populations to more favored localities.

Much productive cultivatable land lies along river courses and in their flood plains, and much early irrigation development took place in these areas. These areas are liable to floods which may destroy crops and agricultural facilities on a large scale. Even when no immediate danger to crops is inflicted, agricultural production can still suffer in flood plains because of the unpredictable arrival and magnitude of the flood unless crops are suited to the inundation. In many parts of the developing world seasonal flood recession cultivations provides basic food requirements and is particularly vulnerable to irregular inundation. The defense against flood damage to agricultural production is in combinations of technical, biological and agrotechnical measures including channelization for rapid water removal or in the diversion of flood waters upstream of the main agricultural areas into natural depressions in uncultivatable zones or into artificial storages from which water later is released during subsequent dry seasons.

Irrigation contribution to food sources

The question now arises: what improvements in crop yields can be achieved from irrigated agriculture as compared with dry agriculture or rain-fed farming and what targets are acceptable as practically attainable in developing countries?

India is typical on a large scale of the conditions obtaining generally in developing countries. Table 1 shows, in the second column, crop yields obtainable under full irrigation and with all other necessary inputs applied in the correct amounts. These are yields currently recorded in agricultural experiment stations and do not show what happens on the farm. In the third column are shown average yields obtained by the farmer under average conditions and generally with very primitive irrigation facilities and other inputs, if any.

TABLE 1. *Agricultural yields—maxima and average—India*

Crops	Maximum yield (tons/ha)	Average yield (tons/ha)
Cereals		
Rice	10.0	1.6
Maize	11.0	1.1
Wheat	7.2	1.2
Other		
Potato	41.1	8.0
Tapioca	48.0	13.0
Yam	19.0	5.8

Source: FAO—*State of Food and Agriculture*, 1972.

TABLE 2. *Approximate yields at research station, farm demonstration plot, and average farm, Pakistan* (*units per acre*)

Crop	Research station	Farm demonstration plot	Average farm
Rice (paddy)	50	40	15
Cotton	35	28	8
Maize	100	80	11
Wheat	60	40	9

Source: Report to Ford Foundation on Need for Agricultural Extension Specialist in Pakistan, 1969.

A similar situation is found in Pakistan where an attempt was made to bridge the gap between research and farm by the use of demonstration plots. Table 2 shows that production from the average farm still is far below the farm demonstration plot. Water management by itself only finds response in a limited range. Irrigated agriculture at its most efficient can show enormous increases in crop yields if linked with other inputs. Even partial improvements in present techniques could result in large additions. The possible magnitude of gain may be illustrated by irrigated cereal production amounting to approximately 70 million hectares. If by more efficient farming and irrigation an average increase of only 1 ton/hectare/annum were achieved it would amount to an increased production of 70 million tons or nearly 20 percent of the estimated annual cereal output of the developing countries. This production increment would be the equivalent of about 18 million hectares of new, fully equipped irrigation projects which would otherwise require a minimum capital outlay on the order of $30 billion.

Irrigation's contribution through improvement of existing projects and by expansion is necessary to approach the food production required in the future. Probably the highest priority in the next 10 years is for productivity increase through the renovation and improvement in utilization of existing irrigation facilities. Reports presented at the 1974 World Food Conference suggest a desirable target up to 1985 would be the improvement of some 50 million hectares of existing irrigation. This would cost an estimated $25 billion at 1975 prices. The next priority may be accorded to the establishment of new irrigation schemes. In developing countries with little or no irrigated agriculture or with limited potential to expand rain-fed farming, the provision of new water development projects assumes greater importance. A deterrent is that the capital investment required for new projects is high and is rising with the increasing scarcity of cheap schemes. Projections at the 1974 World Food Conference called for the need to establish, by 1985, another 25 million hectares of irrigated land at an approximate cost of $40 billion at 1975 prices.

The long history of irrigated agriculture has not always recorded success. Some past schemes—and some very recent ones—suffered severe deterioration through salinization and waterlogging. A number were adversely affected by social and political changes or by silting (Adams 1965). Some proved excessively expensive in relation to their economic returns. In still others some of the farmers abandoned their land or never felt at home in their new setting. Any sober assessment of the opportunities to reap benefits from irrigation must take stock of the reasons for past failures as well as the resources available for further development.

Resources for production

As the medium through which moisture, micro- and macro-nutrients and other inputs to plant growth are fed to the plant roots, *soil* is distinguished by physical, chemical and biological properties. Of high importance is structure, usually denoted by layering, and by texture, usually denoted by particle sizes such as silt, sand or clay. These characteristics influence the use to which the soil may be placed, and soil survey is of necessity one of the first inventories to be obtained before deciding whether or not to go ahead with a planned irrigation development.

The FAO/UNESCO World Soil Map is the most recent attempt to survey the world's soils in a preliminary manner, using one system of classification. As additional surveys are made by cooperating countries the original data will be corrected and become more valuable for planning utilization. Preliminary estimates indicate present cultivated land ranging from 1 to 3 billion hectares with the potential ranging from 3 to 5 billion hectares, depending upon demands and allowable costs for subjugation. It has been suggested that there are about 200 million hectares presently irrigated and about 1 billion hectares which could be irrigated in the world, but this estimate ignores the many inputs, both physical and economic, which must be taken into consideration in estimating suitability. In fact, accurate data on existing and potential irrigated lands do not exist for most of the world. A world-wide inventory of existing and potential irrigation has been proposed.

Water has always been important to production of food and it is estimated that 80 percent of the water consumed by man in the world is for crops. The equitable apportionment of the total water resources is all the more necessary because the quantity of water in the world is fixed and because the runoff rarely coincides with areas of greater crop need. The volume of water on and in the earth's crust amounts to about 1.36 billion cubic kilometers. Of this, 95.5 percent is saline water in the oceans and another 2.2 percent is imprisoned in icecaps and glaciers. The effective usable water is provided by precipitation on the earth's surface and supports the rain-grown vegetation or can be abstracted from rivers or from underground. The average annual precipitation on the world's land areas amounts to about 100,000 cubic kilometers or 0.007 percent of the global water volume. About 38–40,000 cubic kilometers run off annually to the oceans and this represents the significant reserve which could be brought into use by prevention of runoff to the seas, but with environmental consequences. About one-third of the runoff is stable in the sense that it is available throughout the year whereas the remainder is chiefly flood flow. More supplies could only be obtained at considerable expense and generally in inconvenient localities by the desalination of sea water or by the tapping of underground fossil water. Although various estimates indicate that the volume needed to satisfy domestic, industrial and irrigation needs may be sufficient in total for a population at least double the present (Lvovich 1975; Falkemark 1976), the available resources are unevenly distributed among and within the continents.

Quantities used by crops vary enormously depending upon the natural factors affecting plant growth. Basically these are: precipitation, temperature, amount of sunshine, length of daylight hours and the physiology of plant being grown.

Other inputs

The introduction of improved water management in agriculture is a complex problem involving the whole of rural society and agricultural administration, and often has consequences reaching as far as national land and water laws. However, measures to accelerate benefits from water development projects for agriculture must necessarily be grouped around the farmer's field. The success of investment depends ultimately on the effectiveness with which the water is used by the farmer. It is clear that if water is to be effectively used by the farmer without long-term deterioration he must have adequate means to do so, he must have incentive to sustain his effort.

To understand how science may assist in reaching these ends requires analysis of water efficiency in irrigation, and the modifications in soil-salt-water balance and in ecosystems as a result of irrigation and drainage.

2. EFFECTIVE USE OF WATER IN IRRIGATED AGRICULTURE

Water resources

Water's importance increases with population and economic growth linked with rapid development of industry and energy and with the growing demands on water for domestic consumption. In arid and semiarid areas demands on water resources are growing with the development of irrigated agriculture.

In temperate and tropical areas with arid and semi-arid climate characterized by low mean annual rainfall, the primary role of water resources development is readily apparent. Flow discharges fluctuate considerably, their distribution is uneven, in some regions the supplies of groundwater are small, sufficing only for local requirements, and cultivation is impossible or difficult without irrigation.

It has been noted that on a global level 13 percent of the arable lands are under irrigation, using some 1,400,000 million m^3 of water per annum. Projections also show that irrigated harvested land in developing countries will increase at a much faster rate than non-irrigated arable land (Holy 1971).

In view of this situation, it is evident that new directions in water resources development and management must be pursued. This effort falls into two groups: rational use of water, and increase in the available supply.

Substantial water savings may be gained in arid and semiarid zones by applying improved irrigation and drainage systems and farming techniques. In these areas the highest demand on water is for irrigation. Currently in Egypt, it is about 75 percent of the supply. The situation in some other developing countries is similar. If the irrigation water is used with higher efficiency larger volumes may be released for future irrigation systems and industrial development. The paramount importance of such savings may be judged from the assertion that Egypt, with its supply in Lake Nasser, "may soon find itself faced with a water shortage" (p. 373).

Efficient use often implies retaining surface water in suitable areas and transporting it with minimum losses to places where it is consumed. Finding the optimum storage and transport network requires consideration of groundwater and a systems approach. This approach may be applied in multipurpose water conservancy systems of which irrigation schemes are a part. In arid and semiarid zones the principal demand placed on the multipurpose water conservancy system is guarantee of adequate water supplies for irrigation. Other pressing demands come from industry, power, and domestic water consumption, and where a system is incapable of meeting all demands they become competitive.

9

Among alternatives designed to increase the available water supply, methods of evaporation suppression, of recycling waste water, of desalination of saline water, and of possible weather modification to produce precipitation claim attention. But none of these is directly related to irrigation, where the emphasis is upon improved use.

Choice of crops

It is on irrigated land that the introduction of crop species, consisting in some cases of imported plants, has the greatest impact. With increases in the amount of available water it becomes possible to introduce crops with high water requirements, such as rice or sugar cane. The introduction of legumes, which raise soil productivity through nitrogen fixation, paves the way to establish crop rotation. This, in turn, permits addition of other new crops while crop yields are gradually increased through fertilization and improved tillage.

The choice of crop sets the range of water requirements. It may be set by tradition, by estimates of economic efficiency, and by the availability of suitable crop varieties.

Research in genetics is, for the time being, mainly oriented towards selection of high yielding varieties with a view to increasing food or revenue. However, considering arid zone characteristics, it is equally important to select varieties resistant to drought or conserving of water which have an acceptable economic value.

Efficient use of irrigation water

The efficient use of water for a given crop is normally expressed as the ratio between the quantity of irrigation water effectively used by the crop and the total quantity of water supplied from the source. To study the problems of efficient use of irrigation water, a working group of the ICID on Irrigation Efficiencies introduced the following definitions:

Overall (project) efficiency is the ratio between the quantity of water placed in the root zone (rain deficit) and the total quantity of water supplied to the irrigated areas; it represents the efficiency of the entire operation between diversion or source of flow and the root zone.

Distribution efficiency is the ratio between the quantity of water applied to the fields and the total quantity of water supplied to the irrigated area.

Farm efficiency is the ratio between the quantity of water under the farmer's control and the quantity effectively used by crops.

These concepts further enable us to study the efficiency of individual elements of the irrigation system, the losses in the respective elements of the system, and possible measures to minimize them. The need to decrease these losses results from the necessity to maximize the use of water resources; to take full advantage of the relatively high investment costs for water storage in reservoirs and for conveyance systems; and to avoid harmful environmental effects of ineffective use of water resulting in raised groundwater tables, waterlogging, salinity, alkalinity, and breeding and development of disease vectors and intermediate pests.

Technical means for reducing water losses

Reservoirs. Reservoir losses are mainly caused by seepage and evaporation. It may be uneconomic wholly to prevent seepage losses in reservoirs formed by large dams in sedimentary materials. Seepage does diminish with time.

Evaporation losses from the open surface of reservoirs are major and in some instances more tractable. In arid lands potential evaporation, affected mainly by solar radiation, is much higher than annual rainfall. Direct solar radiation supplies approximately 60 per cent of the total energy reaching the surface in the course of the day with albedo at 0.11. Some examples of evaporating losses from the ICID study (1967), "World Wide Survey of Experiments and Results on the Prevention of Evaporation Losses from Reservoirs," may be cited. In the Sudan in the vicinity of Khartoum, daily evaporation reaches 7.5 mm, and near Wadi Haifa 7.9 mm. In Egypt Lake Nasser has an annual evaporation of 2500 mm, while annual losses in Guyana reach 1597 mm, in Burma 1140–1520 mm, and in some regions of India up to 3000 mm.

In attempting to reduce these losses, and the resulting concentration of salts, subsurface storage may be effective. Where this is impracticable screening by vegetation has been used; however, this affects only a small part of large reservoirs, and may increase transpiration. Good results have been attained by monolayers forming a film on the open surface, preventing evaporation. Aliphatic monocular monolayers which do not restrict the transmission of oxygen to the water and are nontoxic are receiving appraisal (NAS 1974). In reservoirs much disturbed by wind, or with varying withdrawal of water, the need to maintain a continuous film on the water surface is acute. Hence, this method probably will be applied only in reservoirs with a slow-moving water surface.

Conveyance system. The conveyance system likewise is affected by seepage and evaporation losses and by losses caused through improper management.

The water losses by seepage from canals form, where there are high infiltration rates of soils, a considerable part of the overall losses. For instance, in the alluvial plains of Upper Pradesh and Punjab in India, the transit losses due to seepage in canals are as large as 36 percent of the supply entering a canal head (ICID 1967). Seepage losses in Algeria reach about 40 percent in the canals in sandy soils. Observations of the Kara-Kum Canal in the USSR show that in the first year the average seepage losses amounted to 43 percent of the overall discharge at the head.

These amounts vary considerably with climatic and soil conditions and with the density and length of the conveyance system. In unlevel canals they are related to the hydraulic conductivity of the soil, the depth of the groundwater table in the neighbourhood of the canals, and discharge and velocity of water. Their prevention may be achieved by lining with the type of lining selected with a view to local conditions and with regard to economy.

One of the less costly lining techniques is colmation, which is favored in porous, sandy soils. It consists of sealing the pores of porous soil by smaller spoil grains which enter the pores during water infiltration with scattered soil grains of selected size. Among chemical lining techniques there is artificial alkalinization, consisting of introducing sodium salts into the soil. When saturated by exchangeable sodium the soil swells and becomes impervious. Plastic is used to line canals with a thin, malleable, impervious membrane which requires protection against radiation. A membrane also may be made on the spot by spreading asphalt or an industrially manufactured foil. Concrete lining is watertight if joints are made properly, and is reliable. It requires checking the water table under the canal, and if pressure rises, providing the canals with vertical drains in their bottom. In some arid and semiarid areas prefabricated concrete flume canals over the surface are used, but control of joints is compulsory in order to avoid their destruction.

Finally, to reduce evaporation losses from canals it is convenient to have the water surface as small as possible, which means designing and building deeper, narrow canals and screening them by vegetation.

Distribution system. Causes of loss in the distribution system are basically the same as in the conveyance system, but due to smaller water discharges some other means of preventing these losses may be applied. Water losses by seepage, evaporation and, to a considerable degree, by operation may be prevented almost completely by introducing closed pipe conduits. In most arid and semiarid areas these are not being built, mainly due to high capital costs and mechanization requirements in building supply canals and distributors. Future supply of larger quantities of water is likely to be brought by open canals which will have to be lined.

The introduction of lined canals or of a closed conduit system for distributing water to irrigated areas is mainly a question of economy. All factors will have to be considered in a wider context—the available amount of water; costs of storage; cost of building the conveyance system; the possibility of enlarging irrigation systems in areas with restricted water resources and, hence, the possibility of providing food for a larger proportion of the population; the possible use of the saved water for other sectors of the national economy.

Social factors which cannot be expressed in economic terms also should be considered. Surface water, especially with a low velocity, provides conditions for breeding and development of disease vectors and intermediate pests. It also serves as a carrier of causative organisms of disease, such as dysentery, typhoid fever, paratyphoid and cholera. Malaria mosquitoes breed in slow-moving irrigation canals and water-logged soils. The problem is complex and its solution requires integrated social and technical solution.

Field application methods. Water losses on irrigated land include the loss of water in the soil profile, return flow, evaporation from soil and plants, and losses to the atmosphere.

Losses of water in the soil profile consist of water unused by plants and of seepage to the subsoil. The amount of physiologically unused water is higher in heavier soils. In lighter soils with low exchange capacity, it is negligible for the water balance.

Water loss by seepage into non-vegetation subsoil is affected by the soil class, soil moisture, and the evolution of the vegetation cover. If it is directly related to irrigation using suitable irrigation intensity, the loss is greatly reduced.

Water losses by return flow from the irrigated area occur when the rate of water supplied is higher than the infiltration rate in the root zone. Water then accumulates on the surface and uselessly runs off, often washing away nutrients and pesticides with which it pollutes other water resources.

Losses of irrigation water by evaporation from the surface of irrigated land depend on the method of irrigation used. In surface and sprinkler irrigation, evaporation occurs from the entire surface. Evaporation losses in the air from sprinkler irrigation depend on the temperature and relative humidity of the area, saturation deficit, wind velocity, size of the nozzle, and the working pressure. It is difficult to determine this loss, but various investigations have shown that it rarely exceeds 2 percent of the sprinkled water. The greatest losses result from wind and system design and installation.

The most frequently quoted coefficients of irrigation efficiency are 0.75 for sprinkler irrigation and 0.65 for surface irrigation. To achieve these efficiencies high technical standards and proper management systems are necessary. Sprinkler irrigation, if designed and operated properly, may be technically the most efficient. Likewise, if it can be properly graded, high efficiencies may be obtained with surface irrigation. Both systems encourage mechanization and automation. In order to eliminate many adverse environmental effects, particularly in developing countries, improved sprinkler and surface irrigation methods should be introduced.

Trickle irrigation makes possible large control over water quantity and distribution. To enable wider application, operational difficulties arising from clogging of tricklers as well as salt accumulation near the margins of the irrigated area have to be overcome. Further, water for trickle irrigation must be filtered prior to its use. Costs of this system may range from 5 to 10 times the cost of other methods.

Use of water in cropping

In dealing with irrigation practice the following ratio has proven useful:

$$r = \frac{\text{yield}}{\text{water taken from the source}}$$

This can be divided into two parts, r_1 and r_2, with $r = r_1 \times r_2$ where:

$$r_1 = \frac{\text{yield}}{\text{water retained in the root zone}} = \frac{Y}{W}$$

$$r_2 = \frac{\text{water retained in the root zone}}{\text{water taken from the source}}$$

A control problem is how r_1 may increase r.

The ratio r_1 is a function of water quality, and it may be assumed that r never increases when the quality is decreasing, insofar as reduction of the quality results from an increase in salinity.

Optimizing r_1 must be considered on a long-term basis in order to achieve a "permanent successful irrigation." The limiting constraint is to maintain soil fertility. Management resulting in an excellent ratio of yield/water consumption which would result in faster deterioration must be rejected. Otherwise, the future would be jeopardized by excessive savings of water and concomitant soil salinization or alkalinization.

Management is commonly observed at the field scale and at the project scale.

At the field scale. The word "field" means a cultivation unit, i.e., a contiguous piece of land which is devoted to one crop. On the basis of research carried out over several decades, many methods have been developed for the evaluation of water consumption by crops.

"Evapotranspiration" includes all water passing into the vapor phase either through the vegetation or directly from the soil. It should be noted that the water retained by the vegetal tissues is almost negligible in comparison with the amount dissipated in vapor.

The potential evapotranspiration (ETP) is the maximum consumption of a crop in good condition growing in a soil at such a level of humidity that water supply is not a limiting factor. This potential evapotranspiration depends mainly on the amount of solar energy reaching the field. As suggested in Chapter 1, it can be calculated either on the basis of climatic records (radiation received, temperature ...) or by the evaporation rate from water surfaces, for example in pans. Data also may be obtained from lysimeters fitted for measuring water consumption of a small plot.

With data regarding the maximum amount of water that a crop in good condition may consume and the natural supply of water to the soil, the depth of water to be applied and the time elapsing between two irrigations may then be scheduled. If water is abundant, a corrective factor greater than one, taking into account the "efficiency" of field application, may be applied. Such a method results in large water consumption. While sensible under certain socio-economic conditions for maximizing yield, it rarely yields the best value for the ratio r_2, and it may create waterlogging hazards, at least for short periods. Even for a management plan such as this some restrictions can be placed on water application without any risk of water shortage for the crop during the first growing period just after seeding, and in the preliminary phases. Sprinkling is particularly efficient at this stage when small depths of water are optimal by avoiding water waste and nutrient leaching.

Cultivation methods such as proper tillage, weed control or mulching may reduce evaporation. Likewise, the microclimate may be improved by wind breaks, either artificial or natural. These may reduce "climatic demand" of the atmosphere, utilize possible seepage from ditches and produce wood.

Starting from a regime fulfilling optimum crop requirements, it is possible to devise methods of prediction for reducing water supply without reducing yield. The target in irrigation is to maintain soil humidity at a level which optimizes the water supply to vegetation. By measuring the water retention potential of the soil, and recording either soil humidity or vegetation status, sophisticated schedules of water delivery may be prepared. This may involve applying a corrective factor less than 1 to the figures obtained for ETP.

Questions then arise: (1) when and to what extent restriction may be advisable? and (2) how to apply such a restriction? A small reduction in water supply could be made without a decrease in yield if there is another limiting factor such as tillage practice or fertilizer. For example, abundant water supply cannot replace the use of fertilizer; if fertilization is insufficient, a light restriction in water supply will not affect the yield. The same holds true for other factors, especially soil tillage.

It is well known that crop response to water shortage depends on the stage of growth. The concept of "critical period" is germane. The irrigation need may often be less than calculated according to an ETP formula; but during certain growth periods even a small shortage in supply may strongly decrease yield. In the period just before harvesting "rich regime" may result in a decrease in the quality or quantity when the plant consumes its own products. Thus, sugar-beet watering must be curtailed prior to harvest. A "modulation" of water supply therefore may be applied when accurate knowledge of crop requirements is in hand.

Under extreme climatic conditions in summer time some crops, even when amply supplied with water, return a low yield. The resulting low water efficiency may be avoided either by avoiding those crops or, in the case of perennials like alfalfa, by supplying them with an amount just sufficient for ensuring their survival until the cooler season.

The amount of water supplied may be limited either by reducing the depth applied at each irrigation, or by increasing the time elapsed between applications. When irrigating by surface methods it is difficult to apply small depths without severely reducing r_2. Assuming a choice is possible between reduction of depth and increase "period," the choice depends upon the crop and particularly its root system, a question on which more information is needed.

Salinity considerations. Problems of salt and water balance will be discussed in the following section, but a few comments are made here on salinity as it is affected by water use efficiency.

In general, an increase in soil salinity results in decreased yields only insofar as salinity is the "limiting factor." Salinity increase does not mean decrease in production as long as other factors (soil fertilization, tillage ...) are more limiting. Moreover, the impact of soil salinity on crops is not constant in all stages of growth. It may have few effects on germination, but impede emergence of the plant through clay soil.

Man may modify water salinity of supply when a tributary of the main river delivers brackish water, or when brackish groundwater use is available. In Tunisia a tributary of the Medjerdat River (named Mellegh because it is "salty") has its flow retained by a dam. Both the time of the year for discharging this water into the main river, from which irrigation water is extracted, and the extent it is blended with sweet water coming from the main river are scheduled to maintain minimum salt content. Significant results from such water management have come from research carried out in Tunisia under UNDP Projects TUNS and TUNS 29 with the assistance of UNESCO and FAO.

At the project scale. The first goal sought in planning an irrigation scheme is to have full employment of the network, i.e., to avoid peaks in field water requirements as well as idle periods during which the network is undercharged. The second is to obtain effective agronomic use of the water. The cropping pattern must be planned to achieve these two goals. The kind of crop as well as the surfaces devoted to each need to be computed in order to avoid "over demands" arising from incorrect assessment of water requirements.

Drainage

One factor of paramount importance in any irrigation scheme is proper drainage. There is temptation to overevaluate natural drainage and at the same time to underevaluate seepage at all levels (conveyance, distribution, field). Artificial drainage is costly and unattractive compared with irrigation. When working with a fixed budget, administrators are inclined to spend money for enlarging the irrigated area rather than to irrigate and drain a smaller area. Experience has shown, often quite forcefully, that it is better fully to equip a restricted area than to put water on a greater surface without sufficient provision for application efficiency and appropriate drainage.

Necessity for drainage. As has been shown, there is no irrigation method having a field efficiency of 100 per cent. Losses result from deep percolation or other factors such as field outlet design. If an irrigation scheme is poorly designed and managed, the crop is undersupplied while the groundwater is replenished, even with a small application depth.

This problem is readily apparent where surface water is applied on soils without levelling or when the levelling has been destroyed by improper tillage implements and techniques.

Drainage is essential whenever salinity is involved. In order to hold the salt concentration in the root zone at a reasonable level, leaching is compulsory. In this event the water applied must exceed the crop water requirement. Further, water must be applied over the entire cropped surface. Where levelling is uneven the higher surfaces will contain a higher proportion or salt residues.

With seasonal leaching, salinity may increase for short periods, but the crop will not necessarily suffer. In semiarid regions, where rainfall is usually concentrated in a few months, it is useful to take advantage of the wet period for leaching. Although water requirements of the crops are strongly reduced, gentle irrigation, keeping the soil at a relatively high water content, optimizes the "leaching efficiency" of rainfall. The seasonal increase in soil salinity is especially noticeable when sprinkling is applied.

Planning and design. Drainage is required in different methods according to local conditions and to the quality of groundwater. If it is not saline and consequently usable by the crops, the aim of the drainage network is to get rid of excess water in the root zone quickly, keeping the water table at a rather high level in order to benefit from capillary movement. By avoiding overdrainage, the water table can be drawn to an optimum level.

If the groundwater is brackish, water movement towards the root zone and the soil surface must be eliminated. In such cases the drawdown speed is not the first target, and instead of an optimum level, it is a critical level which must be kept in mind. A shallow drainage system would be inappropriate.

A critical choice is between open and closed drains. Open deep drainage requires wide ditches resulting in loss of land, difficulties for traffic, and maintenance problems.

Better organization and management of irrigation schemes

Experience with a wide variety of irrigation schemes in arid lands emphasizes that the organization of an irrigation scheme and its division into irrigation districts is a key to achieving higher efficiency.

A working group of the ICID in collaboration with National Committees collected data using questionnaires on the influence of individual elements of irrigation schemes on their efficiency. These elements included the size of the irrigable area where technical facilities are available, the size of the rotational unit commanded by a canal on intermittent flow, technical equipment, the network of farm ditches, the method of water distribution, the farm size, the water delivery period, the number and location of inlets, the flow rate, the field irrigation method, and water charges.

From that analysis, a few generalizations can be made as to conditions promoting overall efficiency:

avoid irrigation projects of less than 1000 ha;

divide large irrigation projects into lateral units of between 2000 and 6000 ha, depending on topography;

let each lateral unit contain a number of rational units, the size of which should vary between 70 and 300 ha, depending on topography;

operate main, lateral and sublateral canals on a schedule of continuous flow;

within a rotational unit, organize the rotation of water supply to farm inlets or group inlets independently of the distribution in adjacent units;

on large irrigation projects of more than 10,000 ha decentralize the project management so that each lateral unit has its own staff.

These suggestions, derived from 91 operating projects, may be implemented without special capital costs. They should not be taken as representative of all irrigation, but they illustrate the possible benefits of systemic examination of factors affecting efficiency.

3. MODIFICATION OF SOIL AND WATER REGIMES

In most cases, and especially in large plains, the water table is considerably raised by irrigation, and capillarity starting from the higher groundwater level causes waterlogging and accumulation of salts both in groundwater and in the soil near the surface. In these instances complete control of groundwater and soil-moisture regimes has to be achieved by supplementing irrigation schemes with effective drainage systems.

A combined program of irrigation and drainage satisfies the agricultural requirement by maintaining the soil-moisture content of the root zone near an optimum level. Drainage lowers the water table below the critical level, and thus capillarity cannot saturate the root zone. Irrigation provides water for the plants if evaporation and transpiration consume the stored water and moisture content lowers below an allowable amount.

The unsaturated root-zone is the heart of the continental branch of the hydrological cycle. Conditions prevailing in this relatively thin layer strongly influence the movement of precipitation reaching the surface of the continents by returning it to the atmosphere, causing surface runoff or permitting infiltration. The more important changes in salt and water regimes to be expected from alterations of the moisture content of the root zone may be classified as follows:

(1) *Modification of the atmospheric branch of the hydrological cycle:*
increase of actual evapotranspiration;
higher atmospheric vapor content;
change in the amount and pattern of precipitation.
(2) *Modification of surface runoff:*
increase in amount and intensity of catchment runoff resulting in higher erosion potential and greater sediment transport;
control of river discharges by reservoirs; and decrease of solids transported in streams because of reservoir retention;
resulting also in deterioration of river beds as the consequence of the smaller sediment transporting capacity.
(3) *Modification of the groundwater regime, modification of the migration of salt, and changes in hydrological processes in the unsaturated zone:*
increase in accretion and decrease in losses, as well as the rise of the water table;
development of horizontal groundwater flow from irrigated areas towards neighbouring non-irrigated lands, raising the water table of the latter and causing the development of "dry drainage" areas;
leaching of the irrigated soils, transport of salts by groundwater flow, and accumulation of salts under the "dry drainage" areas.

19

(4) *Modifications of water quality other than those occurring within the soil-moisture zone:*
increase of salt concentration due to evaporation during the storage, conveyance
and distribution of water;
other qualitative changes such as change in temperature and suspended load, pollution originating from surface runoff carrying nutrients, pesticides, herbicides into the
irrigation system, and from water of high salt content percolating into the canals;
deterioration of the quality of soils and waters caused by the effluent of drainage
systems having high salt content.

Each of these raises scientific questions about soil–salt balance which should be considered in planning, design and operation of irrigation and drainage systems if deterioration is to be prevented. A few special problems concerning salt accumulation and measures
necessary to protect the soils against harmful side-effects will be noted.

Changes in evapotranspiration and precipitation

One consequence of control of moisture content in the root zone is that actual evapotranspiration will be higher from irrigated areas than in adjoining areas. The system of
evapotranspiration is composed of two subsystems. The first is the air mass, through which
energy reaches the water, and which receives and transports the vapor. The second comprises the stored water and the reservoir containing the evaporating mass (depth and
structure of root zone, porosity, adhesion, capillarity, depth of water table, etc.). The two
subsystems meet at the evaporating surface, which forms the boundary conditions of vapor
production and vapor reception.

Irrigation does not modify the conditions of the vapor receiving system; only the surface
where evaporation of transpiration occurs may be slightly altered. The upper limit of vapor
volume which can be absorbed by the overlying air mass (potential evapotranspiration)
is not changed by the control of the moisture content. The actual flux developing between
the vapour producing and receiving subsystems depends not only on the vapor receiving
capacity of the air, but also on the water volume near the surface available for evaporation
as well as on the recharge of evaporating water from groundwater by capillary action.
These conditions are considerably enlarged by irrigation. Actual evapotranspiration
increases.

Peczely (p. 159) compares potential and actual evapotranspiration measured or calculated
in semiarid areas. The difference between the two curves represents the upper limit of
excess evapotranspiration due to irrigation. Because this parameter is important to planners
and designers in determining water demand, it is desirable to develop models and measuring methods suitable for the calculation of both potential and actual evaporation considering the conditions prevailing in the vapor producing subsystem as well. Concerning
potential evapotranspiration, the existing methods (Thornthwaite, Penman, Turc,
Christiansen) serve their purpose with modifications to consider the lower boundary
condition of the vapor receiving system. The determination of actual evapotranspiration,
depending upon the structure and instantaneous conditions of the vapor producing system,
needs further detailed investigation.

Beyond an increase in evapotranspiration, the effect of irrigation on precipitation is
still subject to conjecture. Some scientists believe that the amount and pattern of pre-

cipitation are altered. Schickedanz and Ackermann (p. 185) suggest this effect is negligible in the northern part of Texas, but its significance for other irrigated areas has yet to be established.

Sediment transport and deposition

The nutritional value of solid materials transported by streams of semiarid regions depends mainly upon their quality and quantity. It is evident that the transport of large amounts of feldspar, clay or organic matter leads to a natural fertility of sediments higher than in the case of silica or aluminium.

The Nile River illustrates this situation. The soils of its valley and delta are derived from alluvium, the richness of which constitutes the basis of Egyptian agriculture. Before the Aswan High Dam, Hamdi in 1954 reported the following analysis of alluvial deposits (Table 1). More than 72 percent of the sediment by volume was silica, aluminium and iron. It is other parts that are significant.

TABLE 1. *Chemical analysis of the clay fraction of Nile sediments for the flood of 1954*

Constituents	%
SiO_2	44.94
Al_2O_3	14.81
Fe_2O_3	13.99
CaO	3.98
MgO	1.60
K_2O	1.77
Na_2O	1.38
C org.	1.14
N org.	0.09
Carbonates	0.99

The recurrent deposit of nutrient constituents helps maintain the fertility of irrigated soils. For this reason it is preferable for sediments to reach irrigated fields rather than accumulate in reservoirs. Kovda estimates that 40 tons/ha of sediments deposited annually on the Amu Daria delta add to the soils 250 kg/ha of humus; 20 kg/ha of N; 50 kg/ha of available K_2O; and 50 kg/ha of total P_2O_5. Sediments can contribute benefits beyond fertilizer inputs. Dams may put an end to this. Nile sedimentation, which was 100 million tons/year before construction of the High Dam, is reduced to only a few tons. The silt deposited in Lake Nasser is equivalent to 13,000 tons of calcium nitrate fertilizer. Not all irrigation specialists share the same opinion on this problem. Where the irrigation canals are constructed of concrete, some specialists prefer to retain the sediments in specially designed reservoirs with a view to avoiding the difficulties resulting from canal deposition.

It is necessary to stress the important role played by soil granulometry when canals are of earth. In the case of heavy clay soils, sedimentation must be avoided, but when they are highly permeable sedimentation is beneficial. Comparative analysis of soils and sediments permits determination of the desirable level of sedimentation.

Two further aspects of the relationships of irrigation and sedimentation should be noted. The first is positive: annual deposits of sediment cause the elevation of irrigated area surfaces. The result is an amelioration of natural drainage.

The second is negative: as a result of reservoir sedimentation sediment transport is reduced below dams and, in deltas, the speed of alluvium deposition is exceeded by the speed of coastal erosion. At the same time, the carrying capacity of the river increases, and this leads to river bed and bank erosion.

Regimes of soil moisture and groundwater

The interrelation among hydrological processes occurring in the soil-moisture zone, the migration of salts, and the depth of the water table is expressed in the concept of groundwater balance as modified by irrigation. The basis for this concept is that both the positive and negative accretion of groundwater is a function of the average depth of the water table. The structure and numerical parameters of the equations describing these relationships depend on climate, surface conditions and soil characteristics.

The average amount of yearly infiltration through a unit area of land surface is determined by meteorological, surface and soil factors. One part is stored in the soil-moisture zone and evaporated directly from here. Only the remaining part of water crossing the surface reaches the water table as positive accretion. Other factors being equal, accretion decreases with depth to the average water table.

The function describing the relationship of negative accretion vs. depth is similar. If the water table is near the surface, actual evapotranspiration is equal to potential and the amount originating below the surface is provided completely from groundwater. The difference between the potential and actual evapotranspiration increases with increasing depth to the water table. Below a maximum level the actual evaporation is independent of the position of the phreatic surface, indicating that in the case of deep groundwater the latter does not influence the process of evapotranspiration. In other cases, however, evaporation can be divided into two parts: the first causes prolonged decrease of soil moisture, while the second, although it originates directly from soil moisture, is replenished from groundwater. This second part is the negative accretion of groundwater. Its value is equal to the total amount of water evaporated or transpired through the soil surface, if the water table is near the surface. It is zero if the phreatic surface is at or below the level where evapotranspiration becomes constant. Between these two positions the negative accretion can be characterized by a decreasing curve.

The groundwater balance can be expressed by equating the difference of inputs and outputs to the change in storage. Over a long period within which natural and artificial influences are not modified, the change in storage can be neglected. The time dependent values of both inputs and outputs can be approximated by their averages, neglecting seasonal fluctuations. Accepting these assumptions, a simplified form of the balance equation states that the difference between average positive and negative accretions is equal to the difference between the discharging and recharging groundwater flows. Knowing that both types of accretion are single-valued functions of the depth of the phreatic surface at a given place, their difference can be also represented as a function of the position of the water table. This relationship determines completely the general balance conditions in the groundwater zone, as is the characteristic curve of groundwater balance (see pp. 162–3).

At the equilibrium level where the curve intersects the vertical axis the positive and negative accretions are equal. The water table can develop at this level only if there is no groundwater flow, or if the inflow is equal to the outflow. Above this level is the zone of groundwater recharged by percolation. In the case of such water tables, negative accretion is greater than positive accretion and the difference is balanced by excess inflow. On the contrary, when a zone is drained by groundwater flow, the deficit is balanced by excess recharge originating from infiltration; therefore, the water table can develop only below the equilibrium level.

If the groundwater has no horizontal movement its table develops everywhere at the equilibrium level approximately parallel to the surface, if the other conditions are homogeneous. Because the terrain is generally not a horizontal plain, the groundwater would have a gradient. The flow initiated by the gradient tends to create a horizontal phreatic surface, raising the level of the latter in the valleys and lowering it under hilly areas. In reality the water tables are neither parallel to the surface nor horizontal; they follow the slope of the terrain, sinking below the equilibrium level at higher areas and rising above it in lower terrain. The horizontal flow along the slopes of the water table maintained by the gradient has to be dynamically balanced by the positive and negative accretions. The water table then develops so that recharge within relatively higher terrain equals the discharge (see p. 164).

According to this explanation the water table would be everywhere horizontal at the equilibrium level within large flat areas. However, horizontal groundwater flow can be initiated and maintained under such terrain by the materials and structures of the overlying layer.

In addition to water, very fine grained material and salts are washed out by sheet flow and accumulate in temporary ponds, marshes or in a more compact, impervious covering layer. Land reclamation and levelling may eliminate the unevenness of the surface but not these structural differences. Spots remain within arable land where water cannot infiltrate but during dry periods will evaporate water as readily as surrounding areas. The water deficit then is balanced by horizontal flow from neighboring fields. Following the concept of the characteristic curve, the equilibrium level is at a great depth or there is no such level below the impermeable spots due to weak infiltration and high capillary action. The phreatic surface is lowered below the normal equilibrium level around the spots, assuring positive accretion through excess infiltration. This amount is transported below the spots by horizontal groundwater flow to cover the deficit. Thus, differences in soil materials and structures may create the same local hydrological cycle under the plains as that maintained by the groundwater gradient under sloping terrain.

The hydrological processes explained above are closely linked to the migration of salts in the soil. Where the water table is below the equilibrium level, the resultant of the vertical movement of water in the layers overlying the water table is directed downwards. Because both the infiltrating precipitation and the evaporating water are low in salt content, the leaching of the upper layers is the characteristic process influencing the development of the soils. Among the free cations of the soluble salts, sodium is leached first, and the layer becomes calcium-dominated. If the process is continued, calcium descends to the groundwater and hydrogen enters the clay minerals and aggregates as free cation. In this fashion, first a calcareous and then an acidic leached soil develops above groundwater bodies having net positive accretion.

The cations or salts leached downwards reach the gravitational groundwater space and are carried along by the groundwater flow to deeper areas. In plains, the horizontal ground-water flow is directed towards spots having a more impervious covering layer. Reaching the areas where negative accretion is the dominating factor of the water exchange between groundwater and soil moisture, the groundwater, together with the free cations, the largest amount of which is sodium, rises toward the surface by capillarity to replenish evaporated soil moisture. This water then evaporates and leaves behind the dissolved salts.

In those areas where groundwater is drained by evaporation (Pels and Stannard, p. 171) the accumulation of salts begins as a result of hydrological processes. Since sodium is the most moveable cation, alkaline soils develop in the accumulation zones. If the gradient of the water table is the governing factor of the groundwater flow, these zones are situated generally along the margins of hilly areas.

Irrigation influences groundwater balance in two different ways. In the first, seepage from distribution systems recharges the groundwater without modifying the hydrological processes in the soil-moisture zone. It creates a horizontal flow and raises the water table, assuring that the developing dry drainage balances the recharged amount of water. The influenced zone along the canals and other recharging structures extends to limits within which dynamic equilibrium develops. The characteristic curve of groundwater balance can be applied when investigating this new equilibrium.

The second influence occurs under irrigated areas where excess water reaches the ground-water through the soil-moisture zone. The original processes governing water movement are changed, and a new characteristic curve must be constructed to take account of excess infiltration through the surface, decrease in storage capacity of the soil, and the negative accretion. These result in a higher position of the equilibrium level, assuring that leaching remains the characteristic phenomenon governing the development of the soil.

Applying this concept, under directly irrigated fields waterlogging is the main factor in soil deterioration. Salinization is less important because the irrigation water flowing through the soil maintains a downwards migration of salts. Where a rise of the water table is caused by water from distribution systems or from along the margins of irrigated areas, waterlogging is always accompanied by salt accumulation. Drainage systems designed to eliminate these harmful effects have to lower the water table below the natural equilibrium level outside the irrigated areas, while within the directly irrigated fields the requirement may be smaller, aiming only to protect against waterlogging. Examples may be cited from Fresno (California), New South Wales (Australia) and the Kon Ombo area of Egypt to show that irrigation introduced in higher terrain caused rapid salinization in lower-lying areas. In New South Wales it is expected that land likely to be lost if sufficient drainage is not provided would be slightly more than one-third of the total irrigated land. In Egypt the ratio is believed to be 1:1.

Water quality problems

In using reservoirs to increase the available water resources an increase of salt concentration is a common danger. Similar qualitative change occurs within the conveying and distributing system as a result of evaporation. Moreover, canals sometimes collect surface runoff or drain groundwater. The surface runoff may carry fertilizers, pesticides or herbicides from nearby arable lands, while groundwater percolating into the canals may contain relatively high amounts of dissolved salts apart from the chemical pollutants.

Investigations aimed at pinpointing contamination from fertilizers show that among the various nutrients only nitrates reach the groundwater and even their amount is insignificant (5–6 percent of the total amount applied on the surface). Surface runoff may carry a much higher amount of both phosphates and nitrates into river and lakes. Data collected in the USA show that the highest amount of nitrates and phosphates originate from agricultural lands. In Hungary it was found that within the catchment of a lake, only about 30 percent of the nutrients came from concentrated sources; the remaining 70 percent reach the rivers and the lakes in dispersed form, mostly from agricultural lands. These flows can be avoided by applying special farming practices (e.g., contour farming). On flat lands and within leveled irrigation schemes surface runoff hardly occurs. However, the examples draw attention to the potential hazard involved in reuse of effluent irrigation water. Careful and continuous checking is required, a mandatory procedure in irrigation management if water reuse is planned.

Among all the qualitative changes the high salt content carried by drainage water is the most serious. Hotes and Pearson report (p. 138) that along the Rio Grande (USA) the mean annual discharge decreased from 1.33 km^3 due to irrigation, while the salt concentration (in ppm) was raised from 221 to 1691. In the Yakima River Valley about 160,000 ha are irrigated. Data characterizing the quality changes there are given on pp. 140, 143.

The raise in total salts due to irrigation normally has the effect of increasing the hardness of river water (see data on p. 138).

There are instances when a separate collection system must be constructed to avoid mixing of fresh water and drained water. There are attempts made to decrease the amount of the effluent water, and while complete leaching cannot be achieved in this way, the salts can be washed down just below the root zone and they remain there within the soil profile. Such partial removal of the salts raises two serious problems: it requires a high level of farm management and continuous control. Further, there is not yet proof that limited leaching gives sufficient protection for long periods. It is important to find out whether, in the absence of a completely negative salt balance, continuously accumulating salts and their migration upwards may cause deterioration of arable lands.

Concepts of soil water–salt balance

The concepts of soil water–salt balance are outlined here on the basis of the UNESCO International Sourcebook of Irrigation and Drainage and the latest works of Kovda and Szabolcs.

The aims of water–salt balance studies (Kovda, p. 229) are:

(a) to show the sources, distribution, chemical composition and total amounts of various salts as well as the dynamics of salinization in the soil parent material, water used for irrigation, and groundwater within the irrigated area;

(b) to estimate possible changes in the water–salt balance of the area, since the beginning and at various subsequent stages of irrigation and drainage operations; and

(c) to determine the point at which a steady water–salt balance is established; the optimum salt concentrations which should be maintained in the soils and in the groundwater; the optimum and critical depth of the water table; the most effective irrigation and leaching regimes if the latter is required; the total amounts and proportions of drainage water to be removed; and the permanent elements of drainage structures.

It is also necessary to study changes in the water–salt balance in adjacent areas to determine the complete area likely to be affected by irrigation.

Water and salt balance in the root zone

The ultimate goal of water control is to regulate the moisture content of the unsaturated root zone according to the requirement of the crop. The unsaturated zone is, however, the heart of the hydrologic cycle. It is in that thin area between soil surface and the top of the saturated groundwater area that water seeps down to the groundwater table, moves back into the atmosphere through evaporation or transpiration by plants, or moves laterally into adjoining areas. As a result of these changes in the movement of the moisture in the root zone the microclimate may be altered, surface runoff may be changed, groundwater regime modified, and the quality of water in the zone and in groundwater storage may be influenced by the movement of salts.

Fundamental to the use of water by plants is the migration of salts in the soil. Where the water table is below a critical level the resultant vertical movement of capillary water in the layers overlying the water table is directed downwards. Infiltrating precipitation supports the leaching of the upper levels, and the soluble salts. Evaporation of groundwater rising by capillarity brings soluble salts back toward the soil surface.

A number of water–salt balance equations have been proposed by different authors. One of the simplest forms of the equation for inflow and outflow of salt is the following (Kovda, p. 231):

$$YC + Q_1 C_1 = Q_2 C_2 + QdrCdr,$$

where Y—amounts of irrigation water inflow during a given time limit,

C—concentration of salts in irrigation water during a given time limit,

Q_1 and Q_2—natural flow rate of the incoming and outflowing groundwater respectively,

C_1 and C_2—concentration of salts of the incoming and outflowing groundwater respectively,

Qdr—flow rate of the outflowing drainage water,

Cdr—the weighted average actual salinity of groundwater in a drained field (salt concentration of the drainage water).

A modified form of this equation is:

$$\int_0^t [QC + Q_1 C_1 + Q_2 C_2 + QdrCdr]dt = \Delta S,$$

where ΔS—change in salt content within the investigated area during a given time period /ot/,

Q—flow rate of inflowing irrigation water.

Basic concepts of soil water–salt balance have been developed to show the sources, distribution, chemical composition and total amounts of various salts as well as the dynamics of salinization in the soil and in the water used for irrigation and groundwater within the irrigated area. This makes it possible to estimate possible changes in the water–salt balance in the area from the beginning of irrigation through the various stages of continued operations, and to determine the point at which a steady water–salt balance is established. It is also practicable to estimate optimum salt concentrations, the optimal and critical depth of the water table, and the most effective irrigation and leaching regimes where the latter is required. The main factor in the regulation of the salt balance is effective drainage of saline groundwater.

Land under irrigation may deteriorate through a rise in the groundwaters so as to cause waterlogging, through salinization (whether due to high salinity of the groundwater or the dissolving of solid face salts in the root zone in the way of rising fresh groundwater), and through alkalinization. Especially difficult problems arise in the case of secondary soda salinization (alkalinization). The mechanism of oversaturation easiest to regulate is the removal of salts with drainage waters. Using the first equation, if drainage is not employed, the functioning of the irrigation system depends upon YC, Q_1C_1 and Q_2C_2. Secondary salinization of soil will not occur if $YC + Q_1C_1Q_2C_2$ due to the decompensation of groundwater, its rise, salinization and waterlogging of the soil under irrigation will occur. The introduction of artificial drainage (Qdr) and the removal of groundwater salts (Cdr)by leaching with help of a drainage network turn a positive (accumulative) salt balance into a negative (desalinization) balance. In the course of successful amelioration, the values of Cdr and Qdr must decrease considerably. The above equation is applicable in cases where initial salt reserves of the soil are removed; in opposite cases an additional parameter in the equation should be considered.

Concepts of irrigation water quality

In many countries of the world irrigation waters are characterized by increased mineralization of up to 1–5 g/l and very often by a predominance of sodium among the cations. In numerous rivers mineralization during the last decade grew from 0.2–0.3 g/l to 0.6–1.0 g/l as a result of regulation of river flow and the increasing role of evaporation; of the growing share of return water that percolates through irrigated soils; of the steady growth in volume and concentration of waste and drainage waters; and of the discharge of urban, mining and industrial waters into the river.

Future increase in drainage network construction in irrigation systems together with expansion of the area under irrigation will be accompanied by further increase in mineral and sodium content of waters in middle and lower river reaches. In the immediate future, use of mineralized water for irrigation will be more widespread due to an increasing shortage of fresh water. Irrigators will have to use drainage and artesian waters and the diluted waters of sea bays, river deltas and estuaries.

The practical and scientific experience of Tunisia, Egypt and the USSR have defined the toxic limits of chloride–sulphate concentration of soil solution in the root zone at 10–12 g/l. Under favorable drainage conditions water with a mineral content varying from 2–7 g/l can be utilized for irrigation. However, an increase of salt concentration in irrigation water will lead to a drastic increase of outflow from the irrigation system, in accordance with the water–salt balance equation. This, in turn, will elevate the required frequency of leachings, drainage rates and volumes of mineralized water to be drained from the system. Kovda's estimates of the general conditions of utilization of mineralized waters for irrigation are shown on p. 223.

If sodium reaches 60–70 percent of the cations, the process of accumulation of exchangeable sodium in the irrigated soil gradually leads to alkalinization. This does not occur if waters with mineralization of 2–5 g/l have dissolved gypsum in their composition.

Mineralized waters can be safely used for irrigation of light textured sandy or gravelly soils with high water permeability, low water-holding capacity and low absorbing power.

It is now recognized that the universal criteria for judging the quality of irrigation water may not be applicable to conditions prevailing in areas having poor natural drainage and high water table. New criteria have been worked out for the U.S.S.R. (Kovda) and India (Bhumbla, p. 283).

Waterlogging, salinization and alkalinization

Soil deterioration occurs widely due to the very low efficiency of the majority of irrigation systems which average only up to 30–50 percent. Enormous water losses result from infiltration from the canals (up to 40–45 percent of the water intake), and on fields flooded with surplus water. But the danger of waterlogging of irrigated land persists even when improved technology and better equipment are applied in the irrigation systems.

Waterlogging is a common feature of the majority of the irrigation systems around the world. The speed of the rise of the ground water may be in some cases rapid, reaching several meters a year, but usually it is measured in several centimeters a year. In any case, the process is inevitable whenever drainage is inadequate.

Most cases of secondary soil salinization are associated with rise of the water table whether due to high salinity of the groundwater or to dissolving of solid phase salts by rising fresh groundwater.

Especially serious problems arise with secondary soda salinization (alkalinization) of the irrigated soils. Many countries of Asia, Africa, Europe, and South and North America suffer from the process of alkalinization. This process is caused by (0.5–1.0–2.0 g/l) alkaline groundwaters, or by diluted soda-carrying irrigation waters (the Nile, Indus Rivers, underground waters of California, Hungary, Pakistan, South and Central Ukraine), or both (Kovda, p. 220). In extreme cases the soil absorbing complex becomes saturated with sodium up to 50–70 percent of the cation exchange capacity, and soil alkalinity rises up to pH 9–11. Soil degradation by compaction invariably results. Amelioration of such soils requires simultaneous application of deep horizontal drainage (combined with vertical drainage when the groundwater is confined), heavy doses of chemical meliorants' leachings to remove excess salt, large doses of organic manures, and permanent use of acid and physiologically acid chemical fertilizers.

Soil alkalinization may be of different origin in various natural conditions: (1) residual; (2) active, brought about by alkaline capillary fringe of groundwater; (3) secondary, after the rise of alkaline groundwater; (4) secondary, brought about by the alkaline irrigation waters; (5) secondary, appearing as a result of the cation exchange between the irrigation or rising groundwaters and the soil solid phase; (6) as a result of desalinization of the chloride-sulphate gypsum-bearing solonchaks. Each case requires study before ameliorative measures can be prescribed.

Plant nutrition and salinity response

The mechanics of oversaturation of plants by non-organic salts are not yet fully understood, but a few points are established:

There is no direct relationship between absorption intensity of water and of salts of plants;

The intensity of salt absorption is regulated by tissue permeability and their saturation level by minerals;

Plant metabolism has an effect on salt absorption; and

Salinity response leads to a change in plant metabolism.

The most important change in plant nutrition concerns the characteristics and intensity of N metabolism with ammonium and free amino-acids accumulation. The first element has an adverse effect; the latter two are involved in plant resistance to salinity. It is well established that salinity types have strong effects on plant mineral nutrition, affecting the calcium/potassium, sodium/calcium, sodium/magnesium and calcium/magnesium ratios.

When critical values of soil salinity for groups of plants are investigated, it is extremely difficult to determine universal values. This is because other factors intervene in addition to the complex salinity effects. Permissible salinity level not only depends upon the plant but upon the expected plant production. For this reason, research on the limiting factors of production needs to be integrated.

Technical measures against soil deterioration

Soil deterioration under continuous irrigation can only be prevented by a mix of ameliorative measures under proper management. The soil–water system of an irrigation field is complex and fragile. It cannot be manipulated effectively by taking any one measure alone. The best engineering will not work unless accompanied by agronomic, agrochemical and agrobiological measures.

The leading preventive and ameliorative measure is the lining of canals with impermeable screens and construction of pipelines (Kovda, p. 220). The second measure is obligatory construction of drainage where groundwaters lie deep, so that any threat of their rise would be eliminated, should it be forecast.

Third, when salty groundwaters lie near the surface and soluble salts are present in the soils, deep horizontal drainage and its operation in combination with leaching should be implemented. There would be no secondary salinization in recently constructed irrigation systems had they been provided with deep horizontal drainage, carefully levelled fields, initial leaching of soluble salts where necessary, and the proper technical management.

Finally, it is necessary to build up well-equipped irrigation systems designed to use sophisticated controls and to furnish current information for the managers. Most scientists and irrigation specialists of the world agree that excessive use should be curbed and that drainage should be proved. The only point of disagreement is on ways of so doing.

As for specific problems of soda salinization, the principal measures to combat this process include reasonable limitation of water supply, efficiency of irrigation systems, prevention of the possibility of groundwater rise or fall and removal of alkaline groundwaters by drainage.

Theoretically, a set of measures are available to combat soil deterioration under irrigation. The most important and the most effective is a complex of well-balanced irrigation and drainage networks working as a coordinated water regulating mechanism. In reality, the proper application of existing knowledge has many difficulties, such as a lack of detailed geological, soil, hydrogeological and hydrochemical survey data at the time of system design; lack of monitoring of soil and water processes during operation of the system; lack of highly qualified managers; lack of qualified specialists to help in field operations stages to help the managers; a generally low standard of irrigation knowledge on the part of many farmers; and a general conflict between the modern technology of irrigation, which is the crux of modern irrigation systems, and the traditional ancient irrigation technology.

To combat soil deterioration successfully under continuous irrigation, and make the irrigation systems profitable and effective, requires applying known preventive and ameliorative measures and this calls for overcoming the existing obstacles. Highly sophisticated knowledge and technology are available. How to utilize this knowledge is the most important question to be answered.

Estimates of extent, severity and rates of soil modification

According to El Gabaly (p. 239), salinity and waterlogging are common problems in the countries of the Near East Region. These involve inefficient water use for irrigation, lack of adequate drainage, poor water quality, and lack of proper management. The percentage of salt-affected and waterlogged soils amounts to 50 percent of the irrigated area in Iraq, 23 percent of all Pakistan, 80 percent in Punjab (Pakistan), 50 percent in the Euphrates Valley in Syria, 30 percent in Egypt, and over 15 percent in Iran.

Rise in groundwater tables reaches several meters a year, sometimes as much as 3–5 m a year. There are numerous cases throughout the world where the water table has risen within 10 years from about 25–30 m below soil surface up to 1–2 m depth. When the subsoil waters are saline, this inevitably results in secondary salinization of irrigated land.

According to FAO data, not less than 50 percent of the world's irrigated land is saline. Yields are decreasing. Much land now is completely out of cultivation. Every year several hundreds of thousands of hectares of irrigated land are abandoned as a result of salinization.

Although the growth of salinity of irrigated soils is practically universal, there are a few encouraging examples of successful prevention of deterioration and the improvement of originally saline lands. Dukhovny (1975) cites examples of irrigation systems of the Vakhsh River in Tadzhikistan, USSR; of the Chardzhou and Khorezm oases on the Amu Darya River in Turkmenia, USSR; of the Golodnaja Steepe and Fergana Valley on the Syr Darya River in Uzbekistan, USSR, where salinization processes were completely stopped and saline soils were desalinized and returned to cultivation with good results. This was achieved by deep horizontal drainage, leaching of salts in accordance with the salt balance concept, selective application of vertical pumping drainage, introduction of effective hydroisolation in the canals, and overall sound management of the water resources.

Unfortunately, the positive world experience is not widespread. As Kovda states (p. 220), the majority of the countries introducing irrigation, and especially the developing countries, continue to:

(1) Ignore the peculiarities of the natural soil situation in each region, disregarding level, mineralization and chemical composition of groundwaters, salinity of soils up to the groundwater table, and conditions of natural drainage;

(2) Go without deep drainage installations in the hope of making irrigation system construction "cheaper"; and

(3) Apply excessive water and permit substantial losses of water in the fields and unlined irrigation canals.

The total area of the salt-affected soils of the world is very large and is a good indicator of the severity of the phenomena under consideration. The table on pp. 232–4, composed on the base of FAO data from the 1:5,000,000 World Soil Map and those of Szabolcs for Europe, provides an idea of the distribution of the salt-affected soils by countries.

The general picture is one of widespread vulnerability to salinization and alkalinization. It lacks precision as to how much land in fact is abandoned and as to rates of deterioration. A step to determine those trends is recommended later after reviewing other environmental aspects of irrigation and drainage.

4. MODIFICATION OF AQUATIC ECOSYSTEMS

Irrigation is a part of a complex of practices comprising management of available water resources, controlled distribution of this water over cultivated land, and withdrawal of excessive water through drainage. Ecological consequences of this complex process include: (1) creation of new ecological systems related to water bodies (reservoirs, irrigation canals, drainage ditches, etc.), and (2) radical modification of ecological systems of the terrestrial habitat. The latter is related to practices of irrigation, ploughing, farming, etc.

Some of these ecological changes may be conceived as change from ephemeral situations to perennial situations. Water bodies in the form of pools and ponds of various size and running bodies of water in wadis (torrents) are ephemeral features that follow the incidents of cloudburst rainfall in arid lands. These provide media, though short-lived, for aquatic life. But as irrigation schemes are established, man-made lakes, irrigation canals and drainage ditches and lakes into which drains flow become perennial water bodies that provide media for types of aquatic life that are alien to the arid lands.

This chapter directs attention to major changes in aquatic ecosystems, principally man-made lakes and canals as affecting processes of energy conversion. The following chapter outlines the relationship of these changes to terrestrial ecosystems and then examines the potentials for combining irrigation with development of other resources of arid lands.

Photosynthesis

The objective of all forms of irrigation is to increase the proportion of the sun's energy which can be converted to human use through the process of photosynthesis. The proportion so converted in a grass field, for example, is about 1 percent of the total sun's energy reaching the ground; in a desert the figure will be 0, except for short periods after rain. In well-managed irrigated sugar plantations, it may reach 1.9 percent. For maize and sunflower grown under ideal agronomic conditions in Cairo figures of 10 percent have been recorded (Cooper 1975).

The process is much the same in water as on land, except that the photosynthesis is usually carried out by algae instead of higher plants. The algae are converted by consumption by fish, either direct or through organisms in the aquatic food chain, thus reducing the efficiency of primary photosynthesis by a factor of between 5 and 50 before the energy is available as human food.

This concept of energy flow through biological systems has been greatly advanced by the International Biological Programme, 1964–1974. It is applicable to natural ecosystems just as to the simpler systems of agricultural monoculture. In a warm arid climate, provided the physical character of the soil is favorable and moisture is adequate through

33

irrigation, the flow of energy, and hence the growth of plants, is normally limited by chemical factors of which nitrogen is generally the most important. Thus, the great importance of nitrogen-fixing plants emerges, mostly legumes in association with bacteria in their root nodules, but also some trees and shrubs; and in water a substantial amount of nitrogen is now known to be fixed and made available for biological production by the blue-green algae or *Cyanophyceae* (Stewart 1975; Nutman 1976).

Man-made lakes

To provide reliable water supply for irrigation schemes as well as farm and domestic supply, it is not unusual for streams and rivers to be dammed or diverted, thereby producing farm ponds and large reservoirs, sometimes with far-reaching environmental effects. In 1960 it was estimated that there were nearly 10,000 ponds in Canada and perhaps about a million in the US. In developing countries such as in the Machakos, Kenya, many small dams and ponds have been built for agricultural use.

The terrestrial ecosystem which is inundated becomes irreversibly altered. Plants die and decay, causing upsets in the balance of gases in the aquatic medium and increasing bottom deposits of organic matter. The soil texture is changed as it loses its compactness and releases particles and nutrients to the gathering waters whose quality in turn is altered chemically as well as physically. The contributions of incoming waters depends largely on the geology and the soils of the drainage basin. As the volume of water increases, especially in large basins, and the flow slows down, animal species which are not suited to life in deep static waters die out. In such circumstances the reaction of fish is most easily noticed and there are records of spectacular fish deaths during inundation of river basins. The changes in the unstable ecosystem affect aquatic life at different levels, and particularly bacteria, plankton, algae and insects which have aquatic stages in their life cycle. Such insects may flourish under the new conditions and become a nuisance. The Wadi Halfa area which adjoins Lake Nubia was once so pestered by an explosive increase of *Tanytarsus* that an asthma camp was built in the desert and the removal of the town was given serious consideration.

High sediment and silt content of inflowing waters raises the turbidity level. Erosion of the flooded areas and shores adds to suspended and dissolved matter. The wind, especially in arid sandy zones, may transport substantial amounts of soil particles to the impoundment. As a result, reservoirs may lose storage capacity. With time and settling of suspended matter, the transparency of the water gradually increases. Ecologically, this process has double significance; with the deposition of sediments, a considerable amount of nutrient matter which would have benefited downstream water usage is put temporarily or permanently out of circulation; and mineral salts like iron and manganese are lost in the sediment deposits.

Among the numerous instances where sedimentation has greatly altered the water quality in impoundments, the most publicized is the Aswan High Dam. The change in downstream water quality is alleged to have affected the irrigated land, although recent evidence shows the loss to be less than previously assumed. The sedimentation in the lake annually removes a total of 1950 tons of silt which was formerly deposited on an irrigated area of about one million hectares. It has been estimated that the loss can be replaced in part by 13,000 tons of calcium nitrate fertilizer.

On the other hand, as the turbidity of the reservoir and downstream water is reduced, sunlight penetration improves and with it increased photosynthetic activity. It is common in newly impounded waters for the influx of nutrient-rich water and the expanded medium to support extensive growth. Blooms of *Microcystis* and *Synedra* and other blue-green algae commonly develop. The taste and odor of impoundment water may be impaired by excessive growth and decay of such algae and this becomes serious when the water is required for domestic use. Blooms of the genera *Anabaena*, *Gymnodinium*, *Aphanizomenon* belonging to the blue-green algae have been recorded as causes of mammalian, avian and fish deaths.

An additional advantage of sedimentation in relatively shallow storage reservoirs is the increase in food matter for bottom feeders. In the Volta Lake the enormous amounts of organic sediment and decaying vegetable matter contributed to a major increase of fish, especially of the *Tilapia* and *Labeo* species, during the early life of the dam; great quantities were caught in the shallower Afram region. Fisheries in smaller farm ponds react similarly, especially when stocked.

The size and volume of a reservoir to a large extent determine its physical and chemical character. Where reservoirs are deep, and especially in temperate regions, stratification of the water occurs and limnochemical modifications impose limitations on the overall productivity. In such circumstances biological activities tend to be concentrated in the epilimnion which is subjected to water movements and oxygenation and which also receives adequate sunlight and warmth. The level of concentration of some dissolved chemicals and nutrients varies with depth. In general, limnological activity is progressively reduced with depth of water and in profundal hypolimnion zones light and oxygen are usually absent. Anaerobic decomposition occurs and it often produces unpleasant hydrogen sulphide odors when water is released from the depth.

Water temperatures are stratified and also tend to be low. It has been reported that the irrigation of rice with colder water drawn from great depths may decrease yield or delay germination significantly. Although stratification of temperature may not become permanently established in shallow ponds and in tropical regions, there is doubt as to how the quality of stored water varies at different levels of depth.

Some water for irrigation may be stored in underground aquifers. Even under such instances the quality of water may be significantly altered as a result of percolation of industrial, domestic and agricultural wastes.

Canals and ditches and contaminants

Perhaps it is during the conveyance and distribution phase that modifications of the character of the aquatic medium are best demonstrated. Water from a stream, dam or an underground storage reservoir is affected by evaporation as it passes through the distribution system. As noted in the preceding section the mineral content of the irrigation water may be raised. Where irrigation canals and drains also are contaminated with domestic waste, the nutrient concentration of the water may be increased and the trophic level altered.

The principal nutrients which are necessary for growth of aquatic organisms are magnesium, calcium, sodium, potassium, carbonates and bicarbonates, chlorides, nitrates, sulphates and phosphates, the specific requirements and tolerance levels differing among species of plants and animals.

The ratio of nutrients may change along a water course in an irrigation system. As water is evaporated calcium carbonate and sulphates may be precipitated, giving way to an increased concentration of sodium salts, so the irrigation water becomes more alkaline.

Pesticides, herbicides and fertilizers used for modern irrigated agriculture add a burden of chemical compounds to aquatic ecosystems. Little is known about their overall and long-term effects, but some of the established linkages may be noted. Reactions have been recorded for freshwater fauna and flora ranging from simple protozoa to complex vertebrates and from bacteria and phytoplankton to vascular plants. Most pesticides appear to have an adverse effect on non-target organisms, including man. Toxic pesticides and herbicides can cause sudden deaths. *Daphnia*, the water flea which is an essential component of many freshwater ecosystems, is known to readily succumb to a wide range of insecticides. Agricultural chemicals also destroy beneficial phytoplankton and algae. There are many examples of destruction of freshwater invertebrates. Toxaphene and rotenone reduced the populations of midges of Colorado reservoirs. Fish kills from exposure to chemicals may be acute while in the living fish body residues accumulate and may persist for a considerable time. It is generally known that pesticides have different levels of persistence. For example, the organophosphorus group is believed to break down more readily than the chlorinated hydrocarbons which may accumulate in man (Hotes and Pearson, p. 145). For some of the commoner pesticides, the concentrations which may be tolerated by man have been estimated and are in controversy. For DDT, concentration in public water supply may not exceed 0.056 mg/l. However, DDT accumulates in high quantities in fish, and gets to man who uses fish as food.

Evidence seems to suggest that contamination of water with pesticides is more the result of direct application to water than through agricultural runoff. Subsurface waters do not appear to provide direct routes into rivers and reservoirs. Generally, pesticides are believed to persist longer in water than in soils, and it is probable that most pesticides are readily absorbed onto soil particles.

Herbicides in common use have been observed to break down more readily than do insecticides. However, they can alter the ecology of aquatic ecosystems sufficiently to destroy populations of animals associated with plants.

Fertilizers have a different and more positive impact on plant growth. The discharge of wastes from fertilizers and other nutrients into rivers and reservoirs is known to cause explosive growth of algae and weeds. Nitrates and phosphates are active ingredients in fertilizer formulations, and excesses of those elements transported in surface runoff enrich fertility of irrigation waters, and together with other elements in proportions and associations which are still unclear, they create eutrophic environments. The processes which culminate in the eutrophication involve chains of assimilation and fixation of carbon and nitrogen at different trophic levels. With favorable conditions of light and warmth, organisms utilize nutrients and minerals to produce abundant plant life of algae and weeds as well as the organisms which thrive on them.

Eutrophication has been reported in many large lakes, but it occurs also in small lakes and in farm ponds which are characterized by high-level trophic productivity even though they may lack zooplankton excesses. Eutrophied waters are high in plankton and organic productivity but display low oxygen tension near dawn and after sunset. Oxygen supersaturation occurs and bubbles stream to the surface in daylight; pH for the alkaline water may be up to 9.0 and conductivity high due to concentration of dissolved salts. Under these conditions some kinds of fish cannot survive, but others thrive.

Aquatic weeds

Canals and distribution systems which are rich in organic matter and nutrients, but unsatisfactorily maintained, invariably are invaded and sometimes choked by dense growths of algae and aquatic weeds. Exceedingly large sums may be spent in the clearing of such canals.

Aquatic plants tend to be designated weeds because they have a high nuisance value in irrigation canals where they are not wanted. Irrigation systems which ensure adequate and reliable supplies of water, light and nutrients provide ideal conditions for their growth. Although canals and drains are intended to stay clear, aquatic plants will establish themselves unless prevented from doing so. They may be rooted, like the water lilies, if there is a suitable substrate, they may float (*Pistia* and *Salvinia*), or be submerged (*Ceratophyllum*). For either type, growth can be so profuse as to interfere with water flow and cause obstruction. However, under other conditions, the presence of aquatic plants may confer some benefits. Some are food for fish when alive, and protection for fish fry and of invertebrates which are essential for the ecological balance of the medium. Submerged plants contribute to aeration of waters. Such advantages do not justify their presence in man-made aquatic ecosystems where they are not wanted. They have been directly implicated in the creation of waterlogged soils. They also have been blamed for high rates of evapo-transpiration, but this has yet to be proven.

The association of aquatic weeds with the spread of certain vectors of disease in man and animals is perhaps the main cause for the general concern over their unwelcome presence in irrigation systems. Although weeds are not obligatory to the establishment of *Bulinus* and *Biomphalaria* which are intermediate hosts for schistosomiasis for example, the snails tend to flourish well under their protection and support. In the Volta Lake, off-shore growths of submerged *Ceratophyllum* harbor more *Bulinus* snails than any of the other plant communities. Reaches of streams flowing into reservoirs, shallow shores, downstream areas of dams where the volume of water generally tends to be small, and irrigation distribution systems invariably favor the establishment of schistosome-supporting snails, so establishment or aggravation of schistosomiasis is almost an invariable feature of irrigation schemes in tropical areas.

Modification of aquatic ecosystems may also increase disease-transmitting mosquitoes. The development of vegetation like *Pistia* and *Polygonum* favors the genus *Mansonia* whose larvae and pupae attach to roots, stems and leaves. Stable waters, especially when infested with weeds and polluted with organic matter, provide a suitable breeding place for *Culex*. The *Anopheles gambiae* complex prefers transient, shallow, sunlit weedless water bodies like irrigation pools, whilst *Anopheles funestus* is associated with ditches and swampy areas. As will be noted in Chapter 6, conditions created by flooding of poorly graded fields are suitable for the breeding of mosquitoes, and irrigation pools created by the use of brackish water support mosquitoes which tolerate saline water. Cultivation of densely growing crops encourages mosquitoes by providing shady resting places with suitable humidity which may increase the longevity of the adult. There are several instances where rice fields under irrigation have increased malaria vectors and especially *A. gambiae*, for example in the Kano Plain of Kenya (Hill *et al.*, p. 307).

Mosquitoes which breed on irrigation schemes also transmit arboviruses which cause human disease. Forms of encephalitis-causing viruses have been isolated from *A. funestus* and *Culex tarsalis*, both of which are favored by irrigation schemes. *C. tritaeniorhynchus* is a major rice-field mosquito in the Orient and *C. annulirostris* in eastern Australia.

5. INTEGRATION OF IRRIGATION AND OTHER RESOURCE DEVELOPMENT

The modification of aquatic ecosystems by reservoirs, canals and contaminants affects the water bodies directly but also influences the quality of adjacent lands. Some of these changes are also related to intensive management.

Sub-humid and arid ecosystems are inherently unstable and exhibit potential for dramatic changes triggered by the sudden appearance of extensive areas of irrigated crops, such as increases in insect or bird populations or weed flora. These ecosystems have limited capacity for assimilating (withstanding and responding to) inputs of water, chemicals and energy that are associated with intensive management of irrigated agroecosystems. Farming practices endeavour to extend this capacity through changing soil character by ploughing, by additional soil conditioners and fertilizers and by establishing drainage systems. The problems of salinization and other forms of fertility deterioration described in Chapter 3 are symptoms of manipulation exceeding the capacity of the ecosystem. The biological components are altered in terms of species diversity, numbers of organisms, and the stability of their inter-relationships. Aquatic organisms are most affected, namely, algae, vascular hydrophytes, aquatic invertebrates, insects with an aquatic stage in their life cycle, and fish.

In changing the natural ecology and creating new production systems on land (crop fields, orchards, wooded areas, irrigated pasture, house-gardens, etc.) and in water (fisheries and aquacultures), irrigation promotes radical changes in human population. These comprise settlement of nomadic indigenous population, re-location of farming population from other territory, and other social interactions.

Aquatic management opportunities

Biological changes may have potential benefits. Possible benefits are protein production through fish, crayfish and waterfowl, and the utilization of aquatic vegetation for water purification. Promotion of protein production in water bodies and flooded fields of irrigated systems involves new practices. It may be necessary to compromise between ideal procedures for crop production and for protein production from fish or waterfowl. The validity of such a compromise must be based on its relative economic value to the community, including control of pest organisms, as outlined in Chapter 7.

Positive interactions between fish and crop cultures may keep rice fields in a healthy condition as they consume phytoplankton, weeds and insects; and their droppings add to soil fertility. Properly selected species may have the added advantage of helping control snails and mosquito larvae. Negative interactions include conflict between requirements regarding water levels in rice fields, canals and reservoirs; inconveniencing structures of irrigation drainage networks; use of chemical pesticides proving hazardous to fish growth or marketability; control of water bodies interfering with productivity of fish species whose reproduction is dependent on natural cycles of flood seasons; and damage to cultivated plants.

It has recently been shown that swamp plants are capable of improving water quality by decreasing bacteriological load, removing excess nutrients and removing silt. The possibility of making use of this property at the inlets to irrigation systems should be actively investigated. Utilization of harvested weed plants for silage, compost or mulch may be feasible in certain situations.

Procedures for removing water weeds by chemical, mechanical or manual means are widely practiced and could be accurately costed. However, this is seldom done and incorporated into an economic evaluation of a proposed irrigation scheme at the decision-making stage. Knowledge of the ecology of the plants could lead to modification of canal design which could prevent excessive growth of nuisance plants. Possible measures include manipulation of water level and flow and training canals to dryness for periods of time to prevent the build-up of excessive plant population.

It is critically important to incorporate these and similar procedures to prevent or minimize problems at the planning stage by appropriate canal design, or design and location of human dwellings.

Biogeographical impacts

Irrigation schemes create new habitats that may promote migratory processes of organisms, especially birds and insects, and thereby break the natural barrier of arid lands. Contrarily, dams and canals may act as barriers that hinder the spread of certain aquatic species.

The vegetated areas of irrigation agriculture provide habitat for locusts and other insects, and favor a variety of plant and animal species. Populations of these species will build up and in time be ready for migration to areas that were otherwise beyond their reach. The newly formed habitat types, in turn, provide resting points for migratory birds, and may thus modify their natural routes.

Modification of terrestrial habitat

Soil fauna. Soil fauna in desert soils are sparse but consist of specially adapted species which withstand drought and take advantage of minimal gains in soil moisture from rainfall or dew formation to feed and to breed. They are mainly harvester and carpenter ants which feed on seeds and on dead decaying wood; species of sand roaches which have the unique ability of absorbing water vapor from the soil atmosphere with relative humidities as low as 28 percent; species of a host of tenebrionid beetles which feed on decaying litter; and a number of predators such as carabid beetles, ant-lions and centipedes. These animals usually have a daily rhythm of vertical migration up and down from the soil surface to deeper layers governed primarily by the changing temperature and moisture gradients which are inversed twice daily. They concentrate under desert shrubs where little could accumulate and protection against excessive heat, evaporation and predators is provided. The population densities of these animals are closely correlated with seasonal variations in the moisture and temperature regimes.

Earthworms, on the other hand, are the main constituent of soil fauna in irrigated soils, together with a host of collembola, mites, nematodes, etc. They also depend on the soil texture which determines the level of water available to the organisms.

Soil microflora. Irrigation affects physical and structural features of soil as well as soil moisture and air relationships. It thereby influences various groups of soil microflora such as fungi, actinomycetes and bacteria. The tendency is often towards reduction of number of species and increase in numbers of populations.

Soil saturation or flood is sometimes used in the control of soil-borne plant pathogens, although obviously water supply, topographical and pedological features, and land availability may be primary limiting factors. Soil saturation may directly restrict oxygen supply to the pathogen, but it is also possible that accumulation of CO_2 resulting from microbial respiration may be of considerable, or even greater, significance. The available evidence suggests that microbial activity may be directly or indirectly responsible for the destruction of certain pathogens in flooded soil: the efficiency of the antagonistic mechanism may be influenced by other factors, including oxygen availability. This is well illustrated by the investigations on the flooding of banana land to eradicate *Fusarium oxysporum f. culense* (Stover 1955). Soils were flooded with 2–5 feet of circulated water for up to 6 months. The greatest reduction of indigenous soil fungi and the greatest increase in bacterial numbers occurred in the surface mud in the first 35 days of flooding, and at least 85 percent of the indigenous fusaria were destroyed in the first 40 days. After 120 days most of the indigenous fungus flora was eradicated: however, a small fungus population was found at times on the surface of the submerged soil, and this was considered to be derived from a "migratory" flora carried by inflowing water.

Animal life. Changes in irrigation practices, as from rain-fed farming to irrigation farming, may affect populations of wild animals just as profoundly as bringing water to hitherto uncultivated lands or the creation of permanent water bodies in otherwise waterless areas. Wild animals in deserts, such as gazelles, ibexes and hyenas, depend on drinking water from time to time. When such water is provided in the middle of the desert, populations of these large animals increase rapidly. This may be regarded as a favorable change which might be utilized in conservation measures for endangered animals.

The change from basin to perennial irrigation in Upper Egypt and the disappearance of the annual Nile flood has also brought favorable and unfavorable changes. The rat is no longer the serious menace that it was, but it is replaced by another rodent, and common agricultural pests of the Delta are steadily moving in a southerly direction.

Weed flora. Crop production through irrigation is threatened among other things by unwanted growth of weeds as water and nutrients become available to the reclaimed field areas. Such weeds may invade reclaimed areas through human activities: transport of soil, crop seed impurities, but there are other sources, among which is the native flora of the area. Some of these might become a problem. So the native species of the area, flora or fauna, should be surveyed for potential pest species.

Fisheries and fish farming

The availability of water, both static and flowing, gives opportunity for fisheries, a form of land and water use which may be integrated with irrigated agriculture in favorable circumstances. Fisheries *consume* water only by losses through seepage and evaporation. Thus, water from the extensive and highly organized fish farming industry in some countries is re-used in agriculture. Moreover, in a storage reservoir or fishpond, the fish crop reduces nutrients and hence the tendency towards eutrophication.

In reservoirs in the tropics indigenous fish establish themselves naturally and large crops may be obtained. However, a weight of 20–40 kg per hectare is not often exceeded, and the same applies to canals where not affected by toxic chemicals. However, in irrigated ponds managed for fish production, with stocking of selected species, annual crops may be more than 10 times as great, say 400–1000 kg per ha, and that can be exceeded where food is supplied in addition to that produced by the water.

The combination of aquaculture and agriculture is widely practiced in the case of paddy fields where—in Madagascar, for instance—in addition to rice, a crop of fish, 20–30 kg/ha/120 days is obtained naturally, 80–200 kg/ha/120 days with the addition of fertilizers, and 200–400 kg/ha/120 days with fertilizers and some additional food (Daget, p. 295). In canals the natural fish production can be greatly increased by stocking fish in cages where they can be fed. Some engineers, however, oppose combining aquaculture with agriculture because cages and the like tend to impede flow, and require more complex design and management techniques.

There is a considerable range of fish species suitable for different conditions of aquaculture. Some are herbivores, such as grass carp and *Tilapia zillii*, and can sometimes be effective in reducing water weeds. Others, like the common carp, are omnivores or, like *Clarias*, feed on animal matter. Certain species feed on mosquito larvae, others on snails, and are sometimes used in control of malaria and schistosomiasis, although by themselves they are not fully effective as agents of biological control.

Those concerned with the management of irrigation schemes should not ignore the potential benefits of recreational fishing. Many types of leisure activities, including angling, can be important factors in the cultural development of a country.

Conservation areas and recreation

An irrigated area adjacent to arid land provides a new home for a variety of plants, some of which find their way in naturally, others introduced by design or accident. Water, food and cover attract forms of wild life, not only mammals and birds, but reptiles, amphibians, fish, insects and other invertebrates. In particular, birds are apt to be drawn from a wide area and sometimes multiply to pest proportions, e.g., *Quelea* in Africa. Wildfowl and other migrant birds which travel great distances twice a year typically follow irrigated areas, such as the Nile system, where huge numbers congregate in season, or from one oasis to another.

Thus, the establishment of conservation areas within or adjacent to irrigation schemes are important for several reasons:

(1) For scientific reasons. The variety and abundance of plant and animal life, some-

times associated with particular physical conditions, is of special importance if the area has been studied, e.g., by a center of higher education in the vicinity, so as to establish a biological "bench mark" against which to measure changes elsewhere.

(2) For teaching and training. Outdoor education whether of young or of adults is of growing importance in all countries, and this should be especially so among farming communities whose work consists of cooperating with natural processes.

(3) For recreation and cultural reasons. Particularly within reach of towns or large villages, the rise in living standards leads to demand for a variety of different types of recreational areas. This may include open spaces for group activities as well as quiet, personalized areas for walking and observing the landscape.

The International Union for Conservation of Nature (IUCN) promotes the conservation of areas as well as species on a world scale and assists in achieving this end. Although focusing mainly on natural areas, attention is also given to man-made systems. In the case of aquatic ecosystems, the "Project Aqua" has a list of areas recognized as of importance for reasons stated above, for which some form of conservation and management is highly desirable. An area of conservation need not be taken out of agricultural production, indeed it may have been created by a form of management which needs to be maintained.

Finally, the recreation demand which is always present in human communities will increase and its satisfaction needs consideration. The provision of facilities in the form of village swimming pools, where it is relatively easy to avoid vectors of disease, would be a relatively minute part of the cost of a large irrigation scheme. Like conservation areas and other health measures, they need to be considered in the planning stage.

6. PUBLIC HEALTH IMPACTS

Too often, in water resource management schemes designed to advance agricultural production in arid lands, local populations as well as resettled people are neglected in the process, and they derive little or no benefit to their health. As with other aspects, the impact of irrigation on human health has some serious deleterious effects; it also has important positive effects.

Beneficial effects on health

Provision of an abundant supply and wide distribution of water, in regions where there is little water and much of it of poor quality, has the most salient effect. Installation of an irrigation scheme often solves a problem of drinking water at least when water is pumped by tube wells. Any such improvement is of major public health importance, and, provided there is some minimal attention to the question of potable water (tightness of the tube well, adequate taps and draining traps), excellent results are obtained at minimal expense. On the contrary, when these basic precautions are neglected, a great opportunity is lost. Likewise, if properly carried out, provision can be made for waste disposal facilities. Engineering work preparatory to the setting up of an irrigation scheme can be beneficial to the local population and future inhabitants, in making the area less prone to invasion by vectors and intermediate hosts of disease.

For these aims to be achieved, close collaboration between engineers and public health workers is indispensable during the early stages of planning. In most instances, the introduction of the public health element at this stage would involve only a negligible increase in the overall cost of the project.

Where the irrigation authority deals with the construction of new villages, there is opportunity to design new houses in accordance with some minimal sanitary standards, as well as with the preferences of prospective inhabitants. This can help to keep resettlement stress within tolerable limits.

The presence of adequate quantities of water is one prerequisite for satisfactory personal and domestic hygiene, and it plays a major role as a background for the control of communicable disease. In many arid localities a sharp drop in incidence of trachoma and scabies has been noted after establishment of an adequate water supply through an irrigation scheme.

Where malnutrition is a serious problem in arid lands the development of irrigated agriculture, together with livestock raising and fishing, improves the nutrition of the population by increasing the quantity and quality of available foodstuff, particularly animal proteins.

45

Standards of living, including health standards, increase with the income of the population. There is a stronger desire for better health, hygiene, and medical care, and this "appetite for health" becomes an important aid to implementing public health programs.

However, health services in the project area, if they exist, are rarely sufficient to meet the needs of the increased population. At the same time, the area may soon become extremely vulnerable because of new epidemiological conditions induced by irrigation. Therefore, the development of regional health services is necessary, and the costs may be offset by increase in national revenues.

Deleterious effects of irrigation

Establishment of an irrigation scheme may render a region more vulnerable to diseases. Water can carry toxic chemicals and many communicable diseases, serving both as a transfer medium and as a habit for vectors and intermediate hosts. Very often the negative effects of irrigation are not related to irrigation itself, but to its misuse, and they may be avoided or alleviated by appropriate management.

Chemical pollution affecting irrigation populations may originate upstream and could endanger the health of the inhabitants. It also may occur in the irrigation scheme itself through the indiscriminate use of chemicals and pesticides. In this case it may have undesirable effects downstream as these compounds enter the food chains, accumulating in plants and fish ultimately consumed by man or domestic animals, and causing serious acute or chronic toxications.

Direct transmission of disease is the primary means for spreading cholera, typhoid, bacillery dysentery, infectious hepatitis and poliomyelitis. These diseases may be transmitted by contaminated springs or wells, soil pollution, and subsequent pollution of vegetables, as well as by direct contact. An irrigation canal network is a particularly efficient way of transporting pathogenic agents over long distance.

If latrines are not used, which is the general case in arid areas, there is a permanent fecal soil pollution by ascaris and hookworm eggs, eelworm larvae and amoeba cysts. Transmission is particularly rapid in densely populated areas typical of irrigated lands. Small marshes and ponds, muddy areas around potable water springs and wells, mostly due to bad maintenance and bad water management, are ideal sites for transmission of hookworm and eelworm, and also promote the spread of schistosomiasis and malaria.

As indicated in Chapter 4, the vectors or intermediate hosts of these diseases are either permanently aquatic or amphibious throughout their life-cycle (snails), or during their larval stage only (mosquitoes, *Simulium*), or have an ecology more and less bound to water or livestock (tse-tse flies).

The diseases under consideration are the following:

(1) *Transmitted by mosquitoes:*
   malaria
   filariasis due to: *Wuchereria bancrofti*
   　　　　　　　　　 W. malayi
   arthropod borne viruses:
   　 yellow fever
   　 dengue (and dengue hemorragic fever)
   　 equine encephalitis

(2) *Transmitted by Simulium fly:*
   onchocerciasis (river blindness)
(3) *Transmitted by tse-tse fly:*
   sleeping sickness (African trypanosomiasis)
(4) *Borne by snails:*
   schistosomiasis (urinary and intestinal)
   paragonimiasis
   distomatosis
   clonorchiasis
(5) *Borne by a freshwater crustacean:*
   dracunculosis

Disease which did not exist in an area may appear after irrigation, and the prevalence and intensity of infection of diseases that did exist, but at a reasonable level, may grow to continuous and massive infections. Severe and disabilitating cases then reduce the workers' capacity with socio-economic consequences. Malaria, schistosomiasis and onchocerciasis have caused the greatest damage in countries where irrigation has been developed with no consideration for public health effects.

Malaria. Malaria remains the most widespread disease in the world, and it is a danger in any warm and arid region with irrigation. In arid zones, mosquito larval breeding places are generally small in size, and often dry out, which interrupts transmission during the dry season. Such regions are particularly vulnerable to man-induced environmental perturbations, such as the introduction of perennial irrigation, or changing from dry farming and irrigation of winter crops to wet cultivation. This explains the high malaria incidence in the Delta region of Egypt, where rice is cultivated, and the failure to interrupt malaria transmission in perennially irrigated sections of Syria, Iraq and Iran. The very dangerous *A. gambiae* vector may adapt and invade a newly irrigated area from nearby breeding areas. The construction of the High Aswan Dam put large areas of land under the threat of an *A. gambiae* infiltration from the Sudan, and led Egypt and the Sudan to conclude an agreement to establish a *gambiae*-free buffer zone south of Lake Nasser with continuous surveillance.

All these malariogenic situations in arid countries tend to transform a seasonal and moderate transmission into a permanent hyperendemic situation. However, it is not so much the practice of irrigation that is particularly harmful, as it is poorly planned, operated and maintained irrigation schemes, with stagnant water in the canals, badly maintained drains, and permanent residual marshes and pools. What is bad for health also is bad for agricultural production.

Filariasis. Filariasis, due to *W. bancrofti*, a parasitic worm transmitted by several species of mosquitoes, is even more a consequence of poorly functioning irrigation systems and of misuse of water facilities. The larvae of the responsible mosquito species breed in water bodies rich in organic matter: marshes, poorly maintained drains, and badly built sewers. The disease affects the human lymphatic system and may result in elephantiasis, which can severely affect human work capacity.

Onchocerciasis: Another filariasis common in Africa and South America is onchocerciasis. It is transmitted by the bite of *Simulium* or black-fly, and if the bites are frequent, may result in eye disease, and finally the so-called "river blindness." As many as 30 percent of the inhabitants in some African villages are blind, and some villages and fertile lands are abandoned by the inhabitants. Treatment gives rather disappointing results, and the eye injuries, if any, are definitive. The *Simulium* larvae live in well-oxygenated disturbed waters, rapid brooks, streams and waterfalls. Irrigation may create additional breeding sites of this type. On the other hand, man-made lakes and storage reservoirs may submerge natural larval sites thereby reducing transmission.

Sleeping sickness. This may appear in a population relocated in a formerly inhabited tse-tse fly area, as near the Kariba reservoir in Zambia. Changes of floral composition, growth of trees along irrigation canals, increased air humidity, extensive introduction of cattle, may result in a tse-tse fly proliferation, as in Southern Sudan.

Schistosomiasis. Schistosomiasis in its urinary and intestinal forms, as transmitted by aquatic or amphibious snails, represents a major danger to irrigation schemes in hot arid lands. They are typically debilitating diseases, resulting in reduction of working capacity. Various drugs used in schistosomiasis treatment are expensive and have serious side-effects. If nothing is done to prevent transmission, reinfection recurs readily. In lands formerly free from schistosomiasis the diseases appear and quickly reach epidemic proportions by introduction of snails into a new irrigation system or by introduction of the disease by immigration of infected people.

In arid lands where schistosomiasis existed at a relatively modest prevalence of, say, 10–15 percent of the population, and with a correspondingly low intensity (and therefore gravity) of infection, it is common to see prevalence grow quickly to 80–90 percent. As with onchocerciasis, clinical gravity, and therefore socio-economic impact of schistosomiasis, becomes important when reinfections occur frequently and massively, which happens when prevalence and intensity of infection are high.

Invasion by schistosomiasis of irrigation schemes in arid lands is so common that there is no need to give examples. The non-invasion of schemes in a region where the disease exists, is exceptional.

Many factors converge to create favorable epidemiological conditions for the spread of schistosomiasis. These include new, widely distributed breeding places for snails; misuse of water and poor maintenance of the hydraulic system; canals with stagnant water; and non-operating drains, invaded by weeds and forming marshes. Some other factors are related to man. Intensive irrigation activities multiply human contacts with infected water. Farm work and fisheries require frequent contact with water. Concentration of people in new settlements encourages people to use the same infected contact sites for bathing, clothes washing, drinking water and other domestic purposes.

Disease control

Control of bacterial, viral and parasitic agents; control of vectors and intermediate hosts. Without entering into details of the different methods of control of the great variety of diseases connected with irrigation, a few general and important principles may be stated.

Individual or mass treatment of a population when used as the only method of control rarely leads to long-term improvement. This seems evident, but many public health problems are still dealt with by a narrow therapeutic approach.

Specific methods directed against pathogenic agent, vector, or intermediate host are different for the various diseases. The common attempts at controlling malaria or schistosomiasis use insecticides, molluscicides, mass chemotherapy, and mass eradication of snails. Environmental measures are essential as well and have the advantage of being efficient against several diseases at the same time. Thus the construction and utilization of latrines and of appropriate waste disposal protects the environment against contamination, particularly fecal pollution, and at the same time hinders the transmission of viral, bacterial and parasitic soil-borne and water-borne diseases. Provision of safe water supplies inhibits transmission of numerous diseases through drinking polluted water (mainly bacterial and viral diseases and dracunculosis) and also avoids the skin penetration of schistosomes, hookworm and eelworm infective larvae.

Hydrological engineering is an important component. Where relevant, drainage, stream channelization, lining of streams and canals, land levelling and filling to eliminate low spots, seepage control, piped or covered canals and drains, weed control, improved water management and strict discipline in the use of water promote community health. Sprinkler and drip irrigation, while extremely costly, deserve consideration in view of possible savings in public health expenses as well as saving water.

There are examples of brilliant results obtained through environmental methods in countries which did not reject such measures *a priori*, and which accepted the burden of initial costs. Thus, decisive results in the eradication of malaria in Italy and Greece were obtained only after some important drainage works were built. In southern Tunisia, where schemes were developed mainly for irrigation of date trees, because of water scarcity an important part of the irrigation network was lined with concrete, some canals were covered, and some placed underground. Preceded by these favorable environmental conditions, a vigorous anti-schistosomiasis campaign could achieve satisfactory results. Tunisia is one of the very few countries on the verge of wiping out the disease. Mollusciding had major effect in concrete-lined, weed-free basins and canals, and in many instances the destruction of snails could be achieved by periodic desiccation. In these conditions, the chances of reinfection dropped sharply.

No single method used alone is efficient against any of the listed diseases. A mixed approach must be applied. Nor can permanent results be achieved if they are not coupled with environmental methods.

Social services. New settlements created by large-scale irrigation schemes are particularly vulnerable to disease problems, especially if there is a large number of relocatees. Disruption between the previous way of life and the new one is sudden and complete. Income may be disturbed. Adaptation to the new conditions may be long and difficult, materially and psychologically. Mother and children need particular protection. To cope in part with these impacts a network of social services may be organized in the irrigated area for both curative and preventive purposes.

Disease prevention

Mass health education. Little may be expected from a program of disease prevention without establishing broad contact with the population through mass education in public health. The people, and particularly women, have to learn how to use and maintain the new wells, washing facilities and latrines; to improve the domestic management of water; and to change nutritional habits. They have to learn the danger signs of infection, circumstances and contact points, and how to avoid them. As a prerequisite, the people's needs, opinions and preferences on these matters must be determined.

Sanitary education performed by foreign staff or by nationals foreign to a region usually achieve poor results. The ordinary means of mass health education such as distribution of prospectuses (often to illiterate people), posters, radio and television messages, speeches and films, have minor observable effects if introduced from outside without involvement of the people concerned. The efforts may yield larger returns if linked to the people through their traditional and project leaders such as village sanitarians, school teachers, agricultural extension agents, traditional village chiefs, and local political and religious leaders. Specialists should be only the first link in an army of sanitary propagandists, as many and as varied as possible, chosen among the different members of the community who live in close and diverse contact with the community.

Public health surveys. As soon as an irrigation scheme is proposed and before any decision has been made, it is desirable to organize a public health survey for the project area and for regions from where people might be likely to move. This would estimate and quantify as far as possible the prevailing disease pattern, the diseases which could appear or spread, their human, social and economic effects, and the cost of the health program.

Multidisciplinary surveys. Inasmuch as health problems in irrigation schemes are entirely dependent on the design, functioning and management of the irrigation system, their solution, or better, prevention, may be found only in close cooperation between workers from several fields such as public health specialists, planners, hydrologists, engineers, agronomists, economists, geographers, sociologists and agriculture-extension agents. This cooperation should begin with the planning phase and continue through the life of the irrigation development scheme.

Epidemiological surveillance. Because of its distinctive epidemiological character, an irrigated area is a chronically vulnerable zone. There will always be need for epidemiological surveillance and for subsequent intervention, if necessary. Here again, close cooperation between the different specialists concerned is urgent.

Immunization campaigns. Immunization campaigns are particularly necessary in dense new settlements, where the contacts are close, and disease transmission easy. Young children are particularly vulnerable, and attention is required to vaccinations against tuberculosis, smallpox, diphtheria, tetanus, measles and whooping-cough.

Integration of health programs

Smooth and continuous operation of a public health component in an irrigation enterprise is an essential condition to its success. Long interruption followed by resumption of operations, owing to irregular supplies of money, may lead to disaster or to a return to the situation as it was before any action was taken. Experience has shown that an effective way of ensuring that the public health component's budget has adequate size, continuity and independence is to incorporate it in the overall project's budget.

In addition to a clearly defined and regularly renewed budget, the public health component requires administrative support and reliable staff. Where an irrigation project is administered by a special authority set up for its duration, the medical, technical and auxiliary staff on the project should come under that authority, and that authority only. Although it might seem natural to entrust health problems to the health ministry, this should be avoided whenever possible. Some day the public health problems of the project area will have to come under the routine handling of the Ministry of Health, but experience suggests that this should happen only when methods of overcoming problems have themselves become routine.

Everything that has gone before points to one cardinal concept. The public health problems that arise from an irrigation project are not solely medical problems. Their medical aspects are often not the most important, they do not govern the epidemiology of the disease or the control strategy. They are a matter of environment, associated with the far-reaching changes made by man in the environment. This underlines the necessity of integrating the health component to the project as a whole: technically, financially and administratively.

Public health specialists must not be regarded, as they are frequently, as a hindrance to the smooth operation of an irrigation development scheme and a source of complications. On the other hand, many physicians and scientists have not always been willing or able to adopt a realistic viewpoint that reconciles the health needs with the technical, economic, hydraulic and agricultural requirements of an irrigation project. It is then hardly surprising that their recommendations are not acted on.

The multidisciplinary approach to public health problems, the integration of the health component in water resource development programs, the shared search for solutions and their implementation are not merely a matter of technology, administrative structure, and finance. They embody a state of mind: the quest for common goals, language and cooperative action in project planning and operation.

7. SOCIO-ECONOMIC IMPACTS

The full impacts of irrigation and associated drainage development upon water, soil, aquatic and terrestrial ecosystems and human health extend through the social and economic fabric of local and national societies. The ways in which those societies in turn undertake improvements in irrigation and drainage have a powerful influence upon the character of the environmental effects. A nation's aims and its mode of analysis of possible means of translating those aims into irrigation schemes set the framework within which choices are made among social, economic and environmental consequences.

This chapter reviews the factors that commonly enter into a decision to undertake new irrigation and drainage works. The next reviews the difficulty of gauging and shaping the communication among the farmers, officials and scientists who are affected. In both the emphasis is upon examining how analytical methods and scientific evidence set limits and provide opportunities for establishing stable man–environment systems through arid lands irrigation.

National aims

National aims in irrigation and drainage may include (1) national economic efficiency, (2) gaining of foreign exchange through cultivation of export cash crops, (3) sedentarization of nomadic people, (4) drought damage prevention, (5) stabilization of agricultural systems, and (6) modernization of rural economy. Often these are mixed. Sometimes they follow each other in distressingly rapid sequence: the government support of irrigation projects in the Sabi Valley of Rhodesia moved from drought relief to crop production to stabilized agriculture over 20 years (Roder 1965). Depending upon national resources and aspirations, irrigation may be viewed not so much as a means of increasing net production per hectare or person or amount of water but as a means of enabling an agricultural livelihood to maintain itself in harmony with the environment.

One frequently stated aim is to use the scheme as an instrument for inducing social change among farm populations. The argument runs that people are necessarily introduced to new farm practices in new settings and thus are ready to adopt modernized modes of behaviour. In fact, this rarely happens. The difficulties of adjusting to strange environments and agricultural methods are immense. Resettlement is not a ready entry to a better life. As a result of migration or new farming practices traditional social process is disrupted. New irrigation farmers may have trouble enough surviving, let alone forging new habits and linkages. They may find the obstacles to adopting irrigation so obdurate that they never settle into the irrigation routine and leave the project lands partly uncultivated. The record of such difficulties is especially clear among semi-nomadic folk moving to the

53

Khasm El Girba in the Sudan and the Helmand project in Afghanistan. More realistic aims generally must be sought.

The decision to go ahead with an irrigation scheme depends upon (1) whether it is the most economical way of adding to food and fiber resources compared with the use of more fertilizer, pesticides, better seeds, or improved cultivation practices on land already under cultivation—dryland and irrigated, (2) whether existing projects can be improved more readily than a new project can be established, (3) whether, given a limited supply of capital, social welfare is better off with an irrigation project rather than more schools, housing, health services, roads or other forms of capital investment, and (4) whether it is the most appropriate type of land use for the land capability.

If it appears to be justified on the basis of broad budgetary considerations, the analysis proceeds to more specific matters: (1) whether it is to be simple-purpose or multipurpose; (2) the optimum size and optimum allocation of water among competing uses (even a single-purpose project may require decisions on how much water should be impounded and how much should remain in the stream); (3) the costs and benefits that cannot be measured in money and how these considerations are to be taken into account, notably (a) effects on distribution of income and wealth, (b) environmental effects, (c) socio-cultural impacts such as political responsiveness, and changes in the status of women and patterns of family life. Some of the non-monetary elements can be viewed as reinforcements of money benefits, some as offsets. To consider adequately the latter a method of computing the offsets must be devised, even if nothing more than acknowledgement that those who make the decisions should intuitively determine the tradeoffs.

Somewhere along the way the linkage between the proposed project and the national economy must be examined. Relatively large projects can have substantial economic impact through (1) source of funds (whether external or internal, if internal whether from taxes and borrowing, if from borrowing whether from savings or central bank credits), (2) changes in expenditures for consumption goods and capital goods and responsiveness of supplies, and (3) anticipated inflationary or deflationary effects. Once in operation, other multiplier effects stem from the increased demand for agricultural inputs—labor, fertilizer, seeds, pesticides, equipment and fuels, and from foods and fibers that are stored, transported or processed in adjacent communities.

Increased density of population

A new project is likely to stimulate the growth of towns or cities as well as increasing the density of rural population. Moreover, the immigrants are likely to be relatively young adults who will establish households and raise families. With enhanced food supplies, increased employment opportunities, improved amenities, reduced flood hazards, and the likelihood of improved utilities such as potable water, sanitation facilities and electric power, powerful forces for rapid population growth may come into play. These forces are likely to be augmented by the absorption of families dislocated by the project who receive new housing and promise of employment.

From the point of view of project evaluation, the distinction between those activities that are costs and those that are part of the project's output (and are, therefore, part of the project's benefits) must be maintained. For example, if the increased density of population creates new health hazards, the cost of maintaining the baseline level of health is clearly a cost of the project. If health conditions can be raised to a higher level, the improvement is a gross benefit (gross benefits can be converted to net benefits by taking account of additional expenditures made on behalf of health and subtracting these expenditures from the value of the gross benefits). Many of these values can be measured only in rough terms, but rough measures are better than none.

Loss in productivity from increased incidence of disease is just as significant a cost of the product as outlays for engineering structures by which agricultural productivity is stimulated.

Where a project is introduced to a population characterized by nomadic or extensive pastoral activity, other benefits may be taken into account such as the stimulation to economic activity that follows from provision of seed, fertilizer, insecticides, herbicides, agricultural services supplied on contract basis, equipment, fuels and labor. Much of the secondary effect grows out of increased utilization of labor that otherwise would be idle. To achieve additional economic activity the range of needs associated with creation of new cities, towns and villages must be anticipated and supplied in an orderly fashion.

In evaluating project merits the possible divergence between local and national benefits must be kept in mind. In some circumstances, a new scheme in one part of the country may adversely affect the economic welfare of competing activity elsewhere by reducing the markets and contributing to unemployment and excess capacity. Such consequences must be reflected in the benefit–cost calculations on the basis of which a decision is made.

The desirability of population policies needs no justification if the objective is an improvement in *per capita* living standards. The threat of increased population growth may come from a rise in fertility, a decline in mortality, or both. In any event, management of birth rates is a significant strategic step to be considered.

Employment, income, distribution of wealth and inflation

Whether improved productivity will be shared by the bulk of the population or concentrated among a few will depend in part on other economic circumstances that surround the financing, operation and disposition of the project's net output. Careful integration of engineering and agronomic activities with economic policies may be required if the project benefits are to accrue to those judged most urgently in need.

In the case of state-owned projects, the distribution of income can be determined by administrative action: whether in money or in kind, and if the latter, over what range of goods and services, and according to what rules of equity. Where projects are developed for privately owned land, these matters are likely mediated by market processes, and governmental action is usually limited to fiscal and monetary controls and establishment of rules or constraints over markets for labor, land, goods, fibers and agricultural inputs.

A number of devices are available to increase aggregate social welfare in the face of imperfectly operating markets and private ownership of resources of production:

(1) Limitation of size of holdings within the project.
(2) Fixing of minimum prices paid by farmers for project outputs, especially to farmers with small holdings.
(3) Fixing of maximum prices paid by farmers for agricultural inputs and consumer goods and services: housing, processed foods, other consumer goods and services.

(4) Direct supply of consumer services—potable water, sanitation facilities, schools, medical services, electricity—at suitable prices or free.

(5) Control over minimum wages and selected consumer goods prices.

(6) Taxes that recover speculative gains and apply progressively to income.

Other precautions may be necessary to assure maximum social benefits of a project. Where market forces are dominant within the economy, and incomes are very unequally distributed, landowners may find production of export is more profitable than production for the domestic market. While it is possible that the bulk of wage workers may benefit from expanded export markets, such benefits would come only if the resulting demand for labour led, ultimately, to higher real wages, a result that is neither inevitable nor even probable in the absence of appropriate governmental actions.

A new irrigation project, like any other major new capital investment, poses a threat of inflation or deflation, depending upon the way in which the project is carried out. The likelihood of deflationary pressure is relatively remote. Inflationary pressures are much more likely, and may serve to conceal the fact that little or no benefits, to say nothing of real economic losses, are experienced by broad segments of the population.

A presumed benefit of a large irrigation project is additional employment. However, unless the aggregate supply of consumer goods and services available to the workers and small farmers increases in proportion to the change in aggregate incomes, additional employment may only trigger a rise in the price level, leaving only owners of land and capital as the beneficiaries. These contingencies can be met with careful planning, but this would require a rare degree of collaborative action among cabinet-level officials. Additional consumer goods could probably be gained in the short run only through imports—an avenue likely to be opposed by those who insist upon using available foreign exchange to remedy shortages of machinery and related goods. As a consequence, foreign indebtedness is likely to increase, and the foundations are laid for inflationary pressures. As is generally the case, those who suffer most from inflation are the masses of people whose incomes press upon the minimum levels of subsistence. Only vigorous monetary, wage, fiscal and price policy can achieve an equitable distribution of the costs of capital improvement.

Social services and social arrangements

Part of a project's benefits are in the form of communally supplied social services. Those that come readily to mind are clean water for domestic use, sewerage and disposition of sewered wastes, housing, electric power, communication and transportation, recreational facilities, schools, hospitals and clinics, and nurseries. This array of services can provide amenities hitherto lacking, and stimulate changes in the tasks performed by women, in social organization related to such tasks, and to the status of women within society. With these can also come a reduction in fertility and a net decline in population growth rate.

There is no single way by which births can be reduced quickly and in conformance with socially acceptable techniques. However, stabilization of family income, reduction in need for child labor, improved education, medical information and care, elimination of onerous responsibilities that tie women to the home, provision of nurseries, and a wide range of new employment opportunities will facilitate family planning consonant with economic potential and aspirations.

Project management, farm management and the environment

Some of the decisions made on the farm are mutually supportive with and some are in conflict with decisions made at the project or national level. For many such decisions the physical environment provides the linkage. For example, when water is made available to the farm at no unit cost the farmer probably will maximize output in terms of the resource that is scarcest to him—usually land or expenditure of money for operating expenses. The effect is overuse of water, with the consequent impact of overuse on water scarcity, water-logging, salinization, alkalinization and erosion. The farmer frequently ignores the potential loss of land in computing his short-term profits and is usually totally oblivious of the costs that his actions imposed on others.

Similar conflicts arise where farmers economize on labor and capital by using fertilizers and pesticides in such volume that drainage waters are polluted or where irrigation techniques stimulate erosion and sedimentation at the cost of downstream productivity. While there is some question regarding the farmer's responsiveness to changes in water rates, it is likely that investigation into local peculiarities of pricing practices will yield useful administrative guidelines.

In some instances complementarity rather than conflict of interest is dominant. For example, improved water management may reduce the threat of malaria, and removal of weeds in canals and ditches will curb non-beneficial evapotranspiration, reduce threat of waterlogging, and eliminate favourable habitats for snails.

Water use and administration techniques pose possible conflicts between irrigation needs within a project and downstream activities and should be canvassed for in the computation of net benefits. For example, when regularization of river flow adversely affects the productivity of estuarine and offshore fisheries the losses should be assigned as a cost of the irrigation project. The beneficial scouring effect of previously unregulated flood flows are lost and should also be included as a project cost. Downstream water quality may be degraded by drainage outflow, and changes in the river's regime may reduce habitat valuable for wildlife.

Economic efficiency criteria relate to size of project, allocation of water among alternative outputs, and choice of inputs. There may, but need not be, a linkage between the costs incurred by project participants and benefits derived from the project because some costs may be subsidized as a matter of public policy. What is important is adoption of a policy by which marginal costs (i.e., incremental payments for factors of production) are designed to induce socially beneficial behavior. For example, it may be socially desirable to provide small users with a free quota of water in order to make sure that all households have sufficient for a minimum-sized garden and necessary sanitary functions or to assure proper leaching. Beyond this minimum, efficiency is served by a water user fee that conforms to the social cost of successive increments of water used. Total payments per user may fall short of total costs per user, in reflection of the subsidy, yet marginal payments per user will induce desirable behavior. This principle can be applied either by centrally planned or market-controlled systems with equal effectiveness.

Decisions regarding the optimum size of production units are equally relevant to centrally planned as well as market-controlled economies. Usually the decision rests upon the sophistication of the management system and the number of variables that can be adequately accounted for in advance. When projects are new and relatively complex, the need to learn by doing may justify a higher degree of decentralization than would be justified when available models have been successfully operated.

Within any major scheme the goals of engineers, health experts, agronomists, marketing specialists and economists concerned with successive project stages tend to be diverse, even when subject to the analytical framework outlined above. It is a responsibility of top management that these divergent expectations are reconciled to maintain orderly movement without excess delay of one phase relative to another. Ultimately it is in national policy and project management that tradeoffs are made among economic impacts that are easily reflected in money costs and those environmental impacts that must intuitively be fitted into project analysis even though money measurements are not easily made.

8. COMMUNICATION AND SOCIAL RESPONSE

Understanding of the environmental opportunities and complications arising from arid lands irrigation hinges in large measure upon communication among the people concerned —the farmer, the project manager, the national administrator, the engineer and planner, the public health administrator and the scientist. Each makes decisions that affect the total impact of a project. Often they work quite independently of each other. Finally, it is the farmer who is affected most directly by the outcome. Initially, it is the scientist who provides support for the others.

Scientific communication

Communication was the primary purpose of the COWAR Symposium, between engineers and scientists specializing in the quantity and quality of water—including physics, chemistry and biology—and those concerned with agriculture, fisheries, hygiene and disease. It was perhaps the first time that such a comprehensive exchange of views among the many disciplines involved in irrigated agriculture took place at the international level.

One clear conclusion was that to get efficient use of land and water while mitigating undesirable effects in any irrigation scheme it is essential to maintain constant communications from the outset. The need to link scientists, planners, administrators and farmers continues through the pre-planning stage of assessing feasibility, and the detailed planning stage when progressively more specialists may have to be consulted, until long after a new scheme starts to operate. Scientists cannot wholly divorce themselves from the process at any stage.

The consulting engineers planning a new scheme are normally in a dominant position. While they may be willing to consider advice from biologists, doctors or social scientists, they plan within numerous constraints, especially financial, and frequently tone down or even turn down modifications intended to improve a part of the operation or avoid a side-effect of uncertain consequences. The scientist helps identify the choices and the consequences, and his research may be shaped in part by what he knows of the needs of others.

The farmer

Basically, the person responsible for the use of water and maintenance of environmental quality is the farmer. A perfectly designed irrigation scheme in the hands of unskilled

59

farmers may yield results in terms both of economic efficiency and environmental quality that are inferior to a poorly designed scheme managed by capable people at the end of the lateral canal. Generally, irrigators must be more skilled than farmers in rainfed agriculture because their duties are more complicated and the penalties for misuse more formidable. At the field level the hazards of improper water application, soil deterioration, pests and weed growth are large.

Although the farmer may be provided with guidelines growing out of laboratory or demonstration plot experiments, he usually is obliged to make independent decisions about how and when to apply water and other inputs. He practices an art to which science can contribute only in part. Often his choice is dictated by folk experience or by the actions of agencies which govern his access to information, credit, materials, market, and other constituents of farm management. It is unrealistic to think of irrigators as dealing only with water and land. Their task is intricate and becomes acutely so when the farmer is helped with water but not with the other elements which he must manipulate. These complications enlarge as governments shift their emphasis from one social aim to another—as from drought relief to self-sufficiency or cash crop production.

Helping the farmer

Lack of adequate communication with managers accentuates water and soil problems, and the obstacles may be acute where peasant farmers are illiterate. But illiteracy in itself is not always of great consequence. Folk wisdom may be large and practical instruction in techniques of soil important. For that purpose an infrastructure of social services, including forms of juvenile and adult education, must be adapted to the local conditions. A change within the social system as it affects the farmer sometimes results in rapid improvement in the efficiency of irrigation procedure. In southern France the introduction of payments by the farmer for water actually used led to savings in water and curbed waterlogging. In areas of former basin irrigation in Egypt the change to perennial irrigation with the availability of year-round water from the High Dam encourages waste of water as former basin irrigators adapt to new systems. Reduction of waste is attempted by constructing canals *below* the level of agricultural land, thus requiring the farmer himself to lift his own water. But this may be dangerous: saline water may seep into a canal and contaminate it. Inasmuch as the farmer has many other claims on his time, it has been suggested that a possibly better incentive would come by building gravity flow canals *above* the irrigated land and charging farmers for the water drawn.

In the Near East the FAO aims at a series of demonstration studies in farmer's fields in cooperation with relevant ministries and universities. This is intended to show farmers effective ways to use water and soil in order to produce optimum yields of crops per water unit. There is no panacea for achieving better efficiency. In some areas it may be largely a matter of greater sympathy on the part of the technologist and the manager for the farmer's way of life.

Choice of farming system

Government agencies may exercise choice at the early planning stage as to whether a new irrigation scheme should be devoted primarily to economic profit, with emphasis on benefits to the investors and the state, or social benefits with emphasis on benefits to the farmers. They thereby profoundly influence the engineering works and layout, the crops to be grown, and the siting of processing factories.

The relative advantages of different plans of operation are illustrated in northern Khuzestan, Iran, where four types of agriculture compete side by side in the DEZ Irrigation Project (Ehlers, p. 85). The four types are agro-industries, traditional small-scale farming, agri-business, and farm corporations of agricultural shareholders. An agro-industry is usually government-owned and not strictly profit-oriented, for example, sugar cane plantations. Small-scale farming on traditional systems of land tenure can be strengthened with a modicum of technical improvements, including irrigation. On the other hand, agri-businesses with land concessions to companies for the development of mechanized farming can cause deep economic and social consequences for villagers, especially if their land is included in the concession. Farm corporations tend to move from individual to collective farming, with the farmers investing their own capital in the form of land, machinery and buildings. Experience suggests that only agro-industries and improved traditional farming are suited to generating social response that emphasizes social development by improving rural services, thereby reducing the rural–urban migration.

There are concrete instances where good communication helps the farmer and sharpens the national choice. In human health planning in southern Tunisian oases extensions and in the Rahad project in the Sudan, it has been shown that the necessary steps can be taken. However, the record reveals modifications in engineering designs, such as provision for changes in water level and periodic draining and drying out of canals, which could have avoided bad health situations. Infestation with weeds was not considered and was expensive to remedy at a later stage.

Research on social response

The response of human society and especially of peasant farmers to irrigation has not been studied nearly as thoroughly as has the response of soils or agricultural crops. Only in recent years has the social change provoked by irrigation schemes become a vigorous subject for study. Even so the proponents of social research are few in number compared with those in the physical, biological and medical disciplines. There are notable exceptions; the addition of a social scientist to a physical planning team has become frequent, but for the most part there continues to be insufficient examination of social responses. Irrigation farmers often have to adapt themselves to the scheme rather than the scheme being adapted to them. It is hardly surprising that the social process of adaptation in many schemes has been rough, even with the aid rendered by services the government and the scheme provide in the way of health, education and technical instruction.

Types of social response

Irrigating peoples can be divided into two categories:

(1) Those in countries so arid that land use depends almost wholly on irrigation, and the people have had a long period of adaptation before the introduction of modern technology; and

(2) Those in semi-arid areas adapted through generations to a mobile animal industry following the availability of grazing and water or to shifting cultivation of a risky nature due to unreliable rainfall.

An outstanding example in the first category is Egypt. Traditional irrigators reacted quickly to the opportunities of improved facilities, begun by Mohammed Ali who built a barrage at Cairo early in the nineteenth century. The social response was a steady, rapid increase in population from around an estimated 2.5 million in 1800. Population nearly doubled each half century to reach 18 million in 1950. Then there occurred a doubling of the *rate* of increase so that the population reached 36 million in 1975. There may be another doubling by the end of the century. If all goes well, the newly reclaimed areas and the change from basin to perennial irrigation in Upper Egypt, coupled with widespread increase in the efficiency of soil and water use, may enable Egyptian farmers to keep pace with rapid population increase before it is too late.

When an irrigation scheme is established in a semiarid area, the prospect of permanent water and prosperity inevitably attracts people from other areas with wholly different ecosystems. Those dependent upon domestic animals may wish to retain the traditional man–animal association. Many are accustomed to widely dispersed habitations, with little or no running or standing water, and have had little exposure to water-borne diseases. Disposal of excrement and waste present few problems in arid areas of dispersed human populations. The habits of the people from semiarid lands are in most ways at variance with what is required in the relatively confined, densely settled water-dominated settlements of an irrigation scheme. On the irrigation project excrement is most conveniently disposed of in water and wide-range grazing animals fit awkwardly into the land-use system. After some years, a project may not come up to expectations as in the Helmand basin, or adaptation takes place and more food, income and health service result in increased longevity and survival of children. Population begins to increase and may continue up to and sometimes beyond the limits of the new resources.

Whether the initial population of farmers in a new or extended scheme is local or comes from outside, social and psychological stresses are apt to arise. For instance, as noted in Chapter 7, large increases in the younger age groups may unleash behavior patterns which are little inhibited by the less numerous elders. In a wholly new environment the strains may be great, and create particular need for a sound infrastructure of social services. The reverse situation is when some factor, such as salinization of soils or diminution of water supply, causes the younger active farmers to go elsewhere, leaving behind an elderly population which needs geriatric services at a time when a diminishing capacity of the scheme has least ability to provide them.

Other factors may be of great importance in influencing social responses. One is a change in work pattern, resulting from almost continuous activity through the year when two to four crops are grown on the same ground under perennial irrigation. A reduction in leisure time and a change in timing of different activities through the day, month and year alter the whole pattern of life compared, say, with accompanying grazing stock. Farmers' cooperative effort in such matters as water use, mechanization, marketing and financial assistance becomes crucial to project survival.

Project manager and farmer

It is likely that the low level of efficiency in some enterprises can be attributed to the behavior of the officials who set policy or manage the scheme. Productivity might be raised sharply in some circumstances by changing the management methods or the managers. However earnest and well intentioned, they miss opportunities to help the farmer when and where needed. The general neglect of these opportunities may be charged in part to preoccupation with new capital works, the cramped vision of experts working within boundaries of their own disciplines, and the peculiar problems of water application. From observations in Ghana, Kenya and Sri Lanka, Chambers (1975) suggests that "It may be far easier to sustain an impetus for new capital works on a grand scale, than to sustain high levels of efficient and disciplined management of the water which they make available."

In contrast to knowledge about engineering features, little is known about how irrigation enterprises actually are or can be managed. There are few studies of why some projects are highly successful. A kind of no man's land exists between disciplines. Few scientists venture to ask what accounts for the ways in which public administrators, engineers, agriculturists and environmentalists go their separate ways, leaving the farmer to try to put it all together.

9. URGENT RESEARCH NEEDS

As human intervention in arid lands' ecosystems becomes more widespread and complex the need increases to understand the physical, biological and social processes which they trigger or interrupt. Any research which deepens knowledge of those processes will increase the capacity of people and nations to cope more effectively with possible environmental effects from irrigation and drainage. The significance of filling many of the gaps in information about arid land irrigation is suggested in the preceding pages. A number of plainly important gaps have been noted, as in the case of doubt as to the total evapotranspiration from water surfaces covered with floating plants. Much of the required study falls within the province of established technical and scientific disciplines. Detailed proposals for action are available in a variety of recent reports from MAB, ICOLD, ISSS and ICID. A series of reviews have been made on agricultural water research needs (Hyatt and Hopper 1970; Peterson 1972; Research Review Mission 1973). This report does not attempt to repeat those suggestions. Recommended research is pointed toward specified goals in hydrology, engineering, soils, agronomy, biology, health sciences, geography, economics, systems analysis and related fields.

The development of applied research has two main elements:

1. The adaptation of the existing technology to specific physical, cultural and political conditions which may exist in the irrigated area. This includes research on the development of simple, low-cost methods, equipment, and sources of energy. The emphasis will be on using local inputs of labour and materials.

2. The adoption of concepts and technology by the water users, local residents and others. This will require attention to the socio-economic, administrative, legal and political conditions which influence local decisions.

Nothing short of a long-term program on an international scale will bring answers to that whole array of questions in basic as well as applied research. However, one problem is of such urgency that it commands heightened attention in the near future. This is the problem of attaining integration in the research effort.

Need for integrated research

When funds are allocated for research on irrigation and associated problems the tendency is to use the customary channels of government agencies and university departments. These generally are organized along disciplinary lines which are established, convenient and effective in terms of research methodology. They nevertheless encourage work which fails to take adequate account of the linkages among sectors of acute problems inherent in irrigation and drainage.

An example is the vital question of how the incidence of schistosomiasis among irrigation farmers may be reduced. The processes of transmission are relatively well understood, as are the limitations of the available therapy. Lacking are measures which can curb continued transmission of the disease and are sufficiently cheap, simple and effective to warrant adoption on a large scale. To design such measures for a particular irrigated area requires more detailed examination of the ways in which snail distribution will be affected by canal maintenance, weed growth and molluscicides; the practicability of changing canal design and water distribution schedules; the alternative patterns of settlement and field cultivation practices in relation to water efficiency and economic return; and the circumstances in which farmers may be expected to revise their crop cultivation, domestic water use, bathing and other practices in response to new information and educational efforts. Findings from studies of snail ecology might well influence the hydraulic design of channels. Insights provided by social investigation of farmer attitudes and behavior might suggest constraints upon water delivery schedules and cultivation. The desirability of new and simplified technology in home or field might be revealed by these and other studies.

Another problem is the determination of appropriate water applications for a given soil type and for a given crop pattern in a new scheme. Beginning with investigation of soil and hydrogeological characteristics of the area and the estimated precipitation probabilities, it is desirable to find out what would be the effects upon crop yield and water–salt balance of different combinations of water application timing and volume, crops, and cultivation methods. These, in turn, are affected by circumstances of hydraulics design, the quality of available water, the likely impacts upon disease vectors, the daily behavior of farmers, and related considerations. Occasionally a project planner has sufficient information in hand to move into design without further studies, but this is the exception. It is important to fill the gaps by initiating new field investigations and by seeing to it that the results are integrated with those already in hand.

Necessary conditions for integrated research

To recommend integrated research is much easier than to carry it out, and there is no facile way of achieving it. The major difficulty often is to surmount the lack of communication among the different disciplines involved in a research program.

If the product is to be integrated it is essential to gain agreement and cooperation at several points. The central project and its aims should be studied in common by all the interested disciplines. The specific analytical methods to be used should be specified, and the types of data to be collected should be defined in advance. For example, to judge the agricultural potentialities of an arid region the several investigators should specify the data which will permit the geographer to assess the present situation and its future possibilities, the agronomist to estimate the physical and human factors affecting farm production, the sociologist to examine the social implications of methods of land cultivation, and the economist to appraise the agronomists' farm budget in relation to the national economy. The multidisciplinary team should be orchestrated under a single authority so that the several judgements can be combined. A final condition is the arrangement of periodic meetings among research workers to compare the segmented findings and to reconcile them in arriving at a shared solution.

There should be no illusions about the obstacles in the path of this type of scientific enterprise. Yet, without such attempts the pace of research on the major questions confronting irrigation and drainage development is bound to be slow.

10. PLANNING NEW OR IMPROVED IRRIGATION SCHEMES

The most glaring need in dealing with arid land irrigation is to apply the scientific knowledge already in hand. Much but certainly not all of the past efforts at irrigation and drainage in developing countries would now be yielding larger economic and social returns and contributing more to environmental stability if their planning had made full use of the technical information then available. Obstacles in the way of communication among the responsible individuals and groups have been formidable in the past and are somewhat lessened at present but are still formidable.

The institutional obstacles to achieving genuinely integrated planning of irrigation and drainage activity were outlined in Chapter 8. To a large degree the troubles are rooted in government structure and procedures. To a lesser degree, but nevertheless of major importance, the activities of scientists in both developing and developed countries can be organized and channelled so as to increase the human benefits from the projects in which they participate. Positive action is required along two lines at the international level. Available information should be more widely and rapidly disseminated among governmental and private consulting groups responsible for project planning. At the same time, a firmer policy regarding the criteria for such planning should be adopted by agencies responsible for funding new or rehabilitated irrigation enterprises.

Information flow

The flow of scientific information in the problems canvassed in this report may be expected to continue at a rather slow rate and in piece-meal fashion unless special measures are taken to speed up the process. Existing networks for sharing of technical information and experience should be strengthened by laying on them the burden of getting their findings into the hands of the planners and finance officers who decide what new schemes will be undertaken. These networks include the information activities, noted above, of the UN regional commissions, the FAO land and water services, the UNESCO/MAB Project 5, ICID meetings and publications, and sectors of the concerns of international scientific organizations such as IAHS, IGU and ISSS. The number of national agencies, inter-governmental agencies and consulting firms and their total professional personnel engaged in planning new schemes is relatively small. They should be reached by special workshops and short-term training efforts as well as by publications, and through the International Referral System of UNEP.

Criteria for public funding

The most direct and simple means of improving the planning of irrigation and drainage schemes is for the principal agencies engaged in funding such activities to insist upon careful consideration of a few criteria before going ahead. In different degree these are embraced by current project review procedures. Commonly, the funding agencies look thoroughly into the safety and feasibility of proposed engineering works, the adequacy of water supply, the prospective flow of economic costs and benefits, and the likelihood that the capital investment will be repaid at specified discount rates and time periods. Other aspects mentioned in this report often are investigated. Nevertheless, the prevalence of funded construction which falls short of taking them fully into account testifies in harsh experience to the timeliness of stressing them now.

Insofar as examination of the environmental effects is not prescribed by the procedures of the funding agencies it should be incorporated. The agencies include the International Bank for Reconstruction and Development, the Inter-American Bank, the Asian and African development banks, the United Nations Development Programme, and the various bilateral assistance organizations.

The questions which it would be essential to address to any proposed irrigation and drainage scheme, in addition to the economic questions of prospective flow of costs and benefits, are the following:

(1) Has adequate provision been made for drainage and leaching so as permanently to maintain the quality of soil and water in the root zone?
(2) Has the full range of alternative measures for achieving efficiency in water use been appraised?
(3) Has the project study examined the probable effects upon aquatic and adjoining terrestrial ecosystems of changing the hydrological and soil regime in the area?
(4) Has the study canvassed and assigned costs to the social, health and economic measures which would be required to assure that anticipated benefits from crop growth and social stability are realized?
(5) Has the well-being of the population been taken into account?

The agency providing the funds, and the scientists and engineers engaged in project planning, have a common interest in pursuing these five questions. Only by vigorous and systematic efforts can they assure that the questions are answered economically, candidly and to the extent permitted by current scientific knowledge.

11. MANAGING IRRIGATION AND DRAINAGE SYSTEMS

The rate at which irrigated land is deteriorating and going out of cultivation is sobering for all who aspire to the maintenance of permanent food sources. Unless present trends in abandonment are halted and practicable means are found to rehabilitate previously abandoned land, new investments in irrigation and drainage enterprises will barely hold the line against continued losses. More serious than abandonment in many places is the decline in mean annual yield as a result of soil and water deterioration. The key to halting much of this loss and to preventing its repetition in new projects is management of the irrigation and drainage system.

Action and research

It is apparent from the review in Chapters 2–8 of soil, water, biota and social processes that much available knowledge can be applied with benefit to current operations of irrigation and benefit schemes. Nevertheless it would be a mistake to think that there are ready answers from the domain of science to all the questions arising in seeking efficient irrigation and drainage practices in harmony with local environmental conditions. The hard fact is that at many points it is possible to arrive at solutions only through further observation, experimentation and study. Indeed, few irrigation projects do not present problems which stubbornly resist early resolution.

In these circumstances it may not be advisable to await the results of research within the project area or in a distant place. The project manager may be obliged to take one or more courses of action on a frankly trial basis. To do so may be politically difficult in situations where the impression has grown that once the construction phase of the project ends the operation of the engineering works will be a relatively simple matter. There may be reluctance to face the evidence that all is not well in an irrigated environment. Those tensions may be eased by candidly recognizing that troubles with salts or weeds or snails or farmer motivation are likely to arise and that many projects elsewhere contend with similar problems.

Annual reports

One concrete step which should be taken to improve management of irrigation and drainage enterprises is for each responsible agency to make an annual report of where individual projects stand in achieving their economic and environmental aims. It is relatively common for projects to report the amount of land served, the water supplied, and the costs and revenues from the operation of the works. Where not already stated, it would be desirable to report:

(1) Efficiency of water use: conveyance, field and farm;
(2) Area subject to waterlogging, salinization and alkalinization;
(3) Area abandoned because of soil and water quality;
(4) Population affected by selected water-borne and vectored diseases; and
(5) Births, deaths, immigration and emigration affecting the project.

Other parameters of the irrigation and drainage effort merit consideration, but these five indices would serve to focus attention upon major questions which may arise or persist in project management.

The periodic reporting of these conditions would help alert the responsible local agency to them. It also would suggest those lines along which global or regional cooperation in examining experience or promoting research should be encouraged. Appraisal of the reports would show which management questions are most troublesome and what actions are most effective. Networks for communication of the needed information—as in UN regional commissions, FAO, UNESCO/MAB, UNEP, UNDP, ICID—then could be supported in taking more vigorous measures to accelerate constructive action at the farmer and project level.

It is to be hoped that in the long run institutions responsible for education and training will have adjusted their programs so as to anticipate these and other issues. Scientific materials made available to them as part of the accelerated informational networks will assist. In the near future, however, the greatest urgency attaches to reaching those people who have it within their grasp to halt the deterioration and increase the productivity of existing irrigation and drainage. It is the farmer at the end of the irrigation ditch and the manager of a local canal or pumping system whose actions will determine whether arid environments will be degraded or enhanced by irrigation and drainage.

LITERATURE CONSULTED IN ADDITION TO PAPERS IN THIS BOOK

Abul, Ata and Azim, A. 1976. The Conversion of Basin Irrigation System into the Perennial (February), unpublished paper.

Academy of Scientific Research and Technology. 1976. *Impact of Human Activities on the Dynamics of Arid and Semi-arid Ecosystems, with Particular Attention to the Effects of Irrigation.* Egypt (February).

Ackerman, W. C., White, G. F. and Worthington, E. B., ed. 1973. Man-made lakes: their problems and environmental effects, Geophys. Monogr. No. 17, *American Geophysical Union*, pp. 847.

Adams, Robert M. 1965. *Land Behind Bagdad. The History of Settlement on the Diyala Plaines.* Chicago, Illinois: University of Chicago Press.

Anderson, R. L. and Maass, A. 1971. *A Simulation of Irrigation Systems: The Effect of Water Supply and Operating Rules on Production and Income on Irrigated Farms.* Washington, DC: US Department of Agriculture, Economic Research Service, Technical Bulletin.

Arab Republic of Egypt. 1975. *The High Dam and Its Effects,* The National Council for Production and Economic Affairs, government report.

Baker, Simon. 1976. Ancient Irrigation Systems in Southern Sri Lanka, paper presented at 72nd annual meeting of Association of American Geographers, New York City, New York (April).

Bulter, J. H. and Doehring, Donald O. 1976. Irrigation and Carbonate Terrain: Some Planning Consideration, paper presented at 72nd annual meeting of Association of American Geographers, New York City, New York (April).

Chambers, Robert. 1975. *Water Management and Paddy Production in the Dry Zone of Sri Lanka,* occasional publication No. 8, Agrarian Research and Training Institute (January).

Chambers, Robert and Moris, Jon, ed. 1973. *Mwea: An Irrigated Rice Settlement in Kenya.* Munich: Weltform Verlag, Afrika-Studien No. 83.

Cooper, J. P., ed. 1975. Photosynthesis and productivity in different environments, *IBP,* Vol. 3. Cambridge University Press, pp. 715.

Coumbaras, Alexis. 1962. L'Hydraulique du Sud Tunisien et le probleme de l'eau potable, *Extrait des Archives de l'Institut Pasteur de Tunis,* No. 3–4.

Coward, E. Walter, Jr. 1973. Institutional and social organizational factors affecting irrigation: Their application to a specific case. In *Water Management in Philippine Irrigation Systems: Research and Operations.* Los Banos, Philippines: International Rice Research Institute, pp. 207–218.

Coward, E. Walter, Jr. 1976. *Irrigation Institutions and Organizations: An International Bibliography.* New York State College of Agriculture and Life Sciences (January).

Dieleman, Pieter J. 1976. Personal correspondence, May 5.

Ehler, Eckart. 1975. Traditionelle und Moderne Formen der Landwirtschaft in Iran, *Marburger Geographische Schriften,* Heft 64.

El Hadari, A. M. 1972. Irrigated agriculture in the Sudan: New Approaches to Organization and Management. *Indian Journal of Agricultural Economics* 27, 25–37.

Frederick, Kenneth K. 1975. *Water Management and Agricultural Development.* Baltimore: Johns Hopkins Press.

Gunter, John D. 1976. Water for Western Oklahoma: Alternative Supplementation Techniques, paper presented at 72nd annual meeting of Association of American Geographers, New York City, New York (April).

Hagan, Robert M., Houston, Clyde E. and Allison, Stephen V. 1968. *Successful Irrigation.* Rome: Food and Agricultural Organization of the United Nations.

Hagan, Robert M., Houston, Clyde E. and Burgy, Robert H. More Crop per Drop: Approaches to Increasing Production for Limited Water Resources, paper presented at the International Conference on Water for Peace, no date.

Hansson, Karl-Erick. 1976. Personal correspondence, March 11.

Hellmers, H. and Bonner, J. 1959. Protosynthetic limits of forest tree yields, *Proc. Soc. Amer. For.,* pp. 32–35.

72   LITERATURE CONSULTED

Hyatt, Ellis J. and Hopper, David W. 1970. Discussion paper on Key Needs for Agricultural Water Manage-
ment Research and Training in the Developing Nations. Ottawa, Canada: International Development Research
Center.
ICOLD. 1973. *Onzième Congrès des grand barrages*, Vol. 1. Madrid, pp. 974.
Lees, Susan H. 1973. *Sociopolitical Aspects of Canal Irrigation in the Valley of Oaxaca*. Ann Arbor: University
of Michigan, Memoirs of the Museum of Anthropology, No. 6.
Lowe-McConnell, R. ed. 1966. *Man-made Lakes*. Academic Press, pp. 218.
Lu, Honathan L. 1976. Water and Rice, paper presented at 72nd annual meeting of Association of American
Geographers, New York City, New York (April).
National Academy of Sciences (NAS). 1974. *More Water for Arid Lands, Promising Technologies and Research
Opportunities*, Board of Science and Technology for International Development, Commission on International
Relations. Washington, DC: National Academy of Sciences.
Nesson, Claude. 1972. Densité des Puits et Niveaux Piezométriques dans les Palmerzies de l'Oued Righ,
Les Problèmes de Développement du Sahara Septentrional, Vol. 1. Alger (May).
Nutman, P. S. ed. 1976. *Symbiotic Nitrogen Fixation*, IBP, Vol. 7. Cambridge University Press, pp. 583.
Obeng, L., ed. 1968. *Man-made Lakes: the Accra Symposium*. Ghana University Press.
Peterson, Dean R. 1972. *Water Use and Management*. Ottawa, Canada: International Development Research
Center.
Radosevich, George E. 1974. Water user organizations for small farmers. In Huntley H. Biggs and R. L.
Tinnermeier, *Small Farm Agricultural Development Problems*. Fort Collins: Colorado State University, pp.
131–148.
Rahman, Mushtaqur. 1976. Perennial Irrigation and Hydrological Imbalance in Arid Pakistan, paper presented
at 72nd annual meeting of the Association of American Geographers, New York City, New York (April).
Research Review Mission to the Near East and North Africa. 1973. Rome, Italy: Technical Advisory Com-
mittee of the Consultative Group on International Agricultural Research.
Roder, Wolf. 1965. *The Sabi Valley Irrigation Projects*. Chicago: University of Chicago, Department of
Geography, Research Paper No. 99.
SCOPE, 1972. Man-made lakes as modified ecosystems, *ICSU*, pp. 76.
Sen, A. K. and Mann, H. S. Exploration, Utilization and Quality of Groundwater in Rajasthan. India: Central
Arid Zone Research Institute (unpublished paper).
Siddiqi, Akhtar H. 1976. Climatic Factors in the Use of Irrigation in Pakistan, paper presented at 72nd annual
meeting of Association of American Geographers, New York City, New York (April).
Stanley, M. F. and Alpers, W. P., eds. 1975. *Man-made Lakes and Human Health*.
Steward, Julian. 1955. *Irrigation Civilization: A Comparative Study*. Washington, DC: Pan-American Union,
Social Science Monograph No. 1.
Stewart, W. D. P., ed. 1975. Nitrogen fixation by free-living organisms, *IBP*, Vol. 6. Cambridge University
Press, pp. 471.
University of Reading, Department of Agricultural Economics. 1969. *The Economics of Irrigation Development:
A Symposium*. Reading, England: University of Reading, Department of Agricultural Economics, Development
Study No. 6.
Wilken, Gene C. 1976. Manual Irrigation in Meso America, paper presented at 72nd annual meeting of Associa-
tion of American Geographers. New York City, New York (April).
Wilken, Gene C. Some aspects of resource management by traditional farmers. In Huntley H. Biggs and
Ronald L. Tinnermeier, eds., *Small Farm Agricultural Development Problems*. Fort Collins, Co: Colorado
State University, no date.

SECTION II

Some Case Studies

This section is designed to give a foretaste of the various environmental factors in the total irrigation process, many of which are developed later. The case studies are drawn from Egypt, Tunisia, Syria, Persia and Mexico. Their authors include a Minister of Irrigation, an engineer, a doctor of medicine and a sociologist.

UNE APPROCHE PLURIDISCIPLINAIRE (IRRIGATION ET SANTE) DANS DEUX PROJETS INTÉGRÉS D'IRRIGATION

par ALEXIS COUMBARAS

Chargé d'Enseignement à la Faculté de Médecine de l'Université Paris VII, Paris, France

Résumé

Une coopération étroite entre les services techniques d'aménagement d'un projet d'irrigation et ceux de la santé sont nécessaires autant pour résoudre les problèmes sanitaires, que pour un meilleur fonctionnement du système d'irrigation lui-même. Cette collaboration doit jouer dès la planification du projet, mais aussi lors de sa mise en place et tout au long de son exploitation. Des exemples concrets, tirés de deux projets d'irrigation, l'un situé dans le Sud tunisien, l'autre en Syrie, montrent les succès obtenus grâce à une telle collaboration, mais aussi les échecs enregistrés, quand celle-ci vient à manquer.

Summary

Close cooperation is necessary between the technical services of an irrigation project and the public health services not only to solve the health problems that arise from irrigation, but also for good management of the irrigation system itself. This cooperation must take place not only at the planning stage of the project, but also during its implementation and throughout its operation. Some concrete examples, taken from two irrigation projects in southern Tunisia and Syria, show some success, and also some failures due to the presence or the lack of such cooperation.

Dans bien des pays chauds le développement de l'irrigation s'accompagne d'une aggravation des problèmes de santé. Dans les pays arides, où l'apport d'eau est appelé à jouer un rôle de tout premier plan pour leur développement, l'irrigation a abouti à des situations catastrophiques sur le plan sanitaire, grevant fortement les avantages escomptés pour le niveau de vie des habitants. Ces aggravations sont dues à une propagation plus facile des maladies transmises par l'eau, dont le vecteur a une biologie aquatique, comme c'est le cas de la schistosomiase (bilharziose) et du paludisme. Elles sont dues aussi aux différents problèmes d'ordre sanitaire et touchant l'habitat, créés ou aggravés par l'implantation de populations dans des agglomérations plus peuplées et plus denses qu'autrefois: problèmes tels que ceux de l'eau potable, de l'élimination des excréta et des déchets, de l'amélioration de l'habitat, de l'organisation des services médicaux de base.

Ces notions sont devenues, hélas, très banales, pour tous ceux qui ont eu à s'occuper tant soit peu d'irrigation, qu'il s'agisse de professionnels de la santé, ou non. Malheureusement les mises en garde qui se sont multipliées, plus particulièrement ces dernières années, n'ont pas encore été, dans bien des cas, suffisamment entendues et suivies d'effet, soit par ce que l'on a considéré d'emblée ces détériorations sur le plan sanitaire comme irrémédiables, soit, et c'est à mettre au passif des professionnels de la santé, parce que les remèdes proposés étaient utopiques, irréalisables sur le plan budgétaire, sans rapport avec les possibilités techniques du pays, ou inconciliables avec la pratique même de l'irrigation. De là il n'y avait qu'un pas à faire, et il a été vite franchi: "Il y a certes de graves problèmes de santé soulevés par l'irrigation", se sont dit bien des responsables de projets d'irrigation. "Mais ils sont, pour la plupart, insolubles. Les hommes de santé sont des rêveurs, quand ce ne sont pas des gêneurs, incapables de nous proposer des mesures pratiques et efficaces. Demandons leur leur avis, pour la bonne règle. Qu'ils essayent de faire ce qu'ils peuvent, mais surtout qu'ils ne s'immiscent pas dans notre travail."

Trop peu souvent une collaboration réelle et sincère a été établie entre les hommes de santé et du génie rural, pour rechercher ensemble des solutions satisfaisant toutes les parties. Ces solutions existent. On les trouve pour peu que l'on se donne la peine d'examiner les impératifs techniques et budgétaires de l'autre partie. Les intérêts de l'hydraulique de l'irrigation et ceux de la santé sont parfois opposés, il est vrai. Bien souvent cependant, ils sont complémentaires et plus souvent encore des solutions satisfaisant les deux parties peuvent être trouvées.

Dans les deux examples qui vont suivre, nous nous sommes attachés à montrer concrètement, souvent photographies à l'appui, des examples où une telle collaboration a permis de résoudre des problèmes d'irrigation, tout en ménageant au maximum les aspects de santé. Nous donnerons aussi des exemples où le manque de collaboration a eu des conséquences néfastes sur la santé, alors même que les solutions étaient simples à trouver.

I. Sud tunisien

Il s'agit d'une région semi-aride, présaharienne, formée d'une plaine sablonneuse uniforme, argileuse par endroits, avec dépôts fréquents de sel et de gypse en surface. La seule culture de quelque importance est celle du palmier—dattier, que l'on pratique en appliquant des méthodes traditionnelles out modernes.

La schistosomiase (bilharziose) urinaire est fréquente dans la région (de l'ordre de 50 pour cent chez les enfants). Le seul gîte naturel de prolifération de mollusques vecteurs de schistosomiase (*Bulinus truncatus*) sont les sources oasiennes. Il s'agit de sources artésiennes typiques. Que l'on s'imagine une cuvette d'une dizaine de mètres de diamètre, de profondeur variable, mais excédant rarement les 2 m. Au centre, un bouillonnement de sable traduit le jaillissement artésien continu. Autour, un groupement de dattiers, le plus souvent sauvages, qui signale de loin la présence d'une source. Parfois, il ne s'agit que d'une flaque d'eau de quelques cm de profondeur (photo 1), parfois encore, on est en présence d'un petit marécage à la végétation abondante. Au dernier stade enfin, ce n'est que la disposition en cercle des palmiers autour d'une cuvette, dont le fond est à peine humide, qui nous indique que jadis, il y avait là une source.

Il s'agit là d'une véritable évolution dans le temps, dont nous venons de décrire les différents stades. Une source naît, s'entoure d'une couronne de dattiers qui prolifèrent en

Photo 1: La vieille source du village de Nouil: le type même d'une collection d'eau ne servant qu'à abriter des mollusques et des larves de moustiques. Quelques sacs de sable auraient suffi pour le remblayer et pour supprimer définitivement ce gîte.

Photo 2: El Golaâ: Le captage d'une source, opération rentable, tant sur le plan agricole que sanitaire. Le gîte à bulins et à moustiques a été remplacé par une plateforme de sable et de béton.

Photo 3: El Guettar (Gouvernorat de Gafsa). Une borne-fontaine comme il y en a tant. L'eau est certes potable, mais pour y accéder il faut patauger dans un magma de boue et d'excréments d'animaux. Seuls les enfants s'y aventurent. Les adultes sont retournés au puits ancestral pollué, mais d'un abord plus propre. Résultat: création d'un nouveau gîte pour larves de moustiques et d'ankylostomes. Remède: il suffirait parfois de réparer un robinet, mais "administrativement", il paraît que c'est plus compliqué que de creuser un puits: équipement et entretien ne relèvent pas du même Ministère!

Photo 4: Conduites de distribution d'eau d'un réseau d'irrigation. Côte à côte la "bonne" et la "mauvaise" solution: conduite fermée à gauche, conduite ouverte à droite. Le bassin sert au changement de direction et de niveau de l'écoulement... et de piscine aux enfants. Véritable baignoire à attraper la schistosomiase, il suffirait de la recouvrir d'une dalle de béton pour en interdire l'accès, et pour empêcher la prolifération des mollusques, qui ne vivent pas dans l'obscurité. Noter la rapidité de courant, qui n'empêche pas la vie des mollusques.

Photo 5: Aïn Rhama (Cheikhat de Bechri): déblayage du fond d'une source; élargissement de sa cuvette, en vue de son captage. Les mottes de boue arrachées du fond sont transportées jusqu'à son bord supérieur, transmises par la chaîne humaine.

Photo 6: Aïn-Taka: Le W.C. se déverse directement dans l'eau, à 3–4 mètres seulement du point d'émergence d'une source!

présence d'eau; cette couronne la protège de l'ensablement mais en même temps constitue un obstacle contre lequel vient se briser le vent qui y abandonne sa charge de sable. C'est pourquoi, en fin de compte, la source se présente sous forme d'une butte de sable, avec au sommet sa couronne de dattiers creusée en son centre d'une cuvette au fond de laquelle jaillit l'eau.

Peu à peu, ce dispositif de protection ne suffit plus; les parois de la cuvette de plus en plus hautes s'effondrent, entraînent parfois les dattiers, la source se laisse ensabler, se colmate et finalement s'assèche complètement.

Si nous avons tant insisté sur la description de ces sources et sur leur évolution naturelle qui, sans l'intervention de l'homme, se fait inexorablement vers l'assèchement, c'est que ces sources, avec leur débit moyen de 1 à 50 l/sec, constituent le seul apport naturel d'eau utilisable à des fins agricoles et sont donc au centre des préoccupations rurales, tant à l'échelon du génie rural qu'à l'échelon individuel des cultivateurs de la région. Par ailleurs, ces sources constituent le seul gîte naturel de bulins, donc le point de départ de la schistosomiase dans la région, et la lutte contre cette maladie se rattachera de près ou de loin à l'existence de ces sources et à leur exploitation.

Depuis une trentaine d'années déjà, on a cherché à adjoindre à ces sources oasiennes naturelles des puits artésiens obtenus par enfoncement vertical d'un tube, qui permet la montée spontanée de l'eau par artésianisme. Les problèmes de santé que soulèvent ces systèmes nouveaux sont, comme on va le voir, semblables en bien des points à ceux soulevés par les systèmes anciens.

Le procédé ancestral d'irrigation à partir de ces sources oasiennes, procédé encore couramment utilisé de nos jours, consiste à pratiquer une brèche sur le bord de la cuvette, brèche que tour à tour on colmate avec de l'argile pour faire monter le niveau d'eau, et que l'on rouvre dès que le niveau requis est atteint. L'eau d'écoule alors dans une rigole ou "séguia" vers le périmètre à irriguer. Il s'agit donc d'assurer à une surface irriguée un apport d'eau important mais intermittent et de très courte durée. Si on laissait l'eau de la source en écoulement continu, cette eau, étant donné le faible débit, n'irait pas loin et s'infiltrerait rapidement dans le sable. Le périmètre irrigué en serait réduit d'autant.

Actuellement, on tend à remplacer ce système archaïque. On cherche:

(1) à augmenter le débit des sources,
(2) à diminuer les pertes d'eau par réinfiltration dans le sol, lors de l'écoulement vers le périmètre à irriguer.

Pour augmenter le débit d'une source, on procède à son captage: après avoir élargi la cuvette de la source et curé son fond (photo 5), on détermine le ou les points précis de jaillissement de l'eau, et on les capte en posant par-dessus des cylindres en béton armé appelés "buses": photo 2. L'eau recueillie dès sa sortie dans un réceptacle en béton est acheminée dans des conduites souterraines ou à ciel ouvert en béton vers les plantations à irriguer où elle est distribuée avec des pertes ainsi réduites au minimum.

Le débit est donc augmenté, devient plus facile à régler. Les pertes d'eau pendant le transport sont réduites. Mais, fait important pour nous, la cuvette-réservoir de la vieille source a été asséchée. Le gîte naturel de bulins et de larves de moustiques a disparu complément et définitivement. La fourniture d'eau potable devient possible et facile. Il suffit de brancher une pompe à main ou une conduite sur le dispositif de captage.

Encore faut-il que ce travail soit réalisé en collaboration avec les services de santé. Les résultats d'un captage, si précieux au point de vue sanitaire peuvent être remis en question:

à la suite d'un remblai insuffisant, l'eau continue à sourdre au travers de ce qui était le fond de la vieille source et reconstitue une mare stagnante. Il a suffi d'y répandre quelques centaines de paniers de sable — travail de quelques heures pour une équipe d'ouvriers agricoles-pour l'assécher et pour débarrasser à tout jamais le village attenant à cette source de l'unique foyer responsable de l'endémie schistosomienne.

Il arrive aussi, que par son évolution naturelle, selon le mécanisme que nous avons vu plus haut ou à la suite d'un forage de puits artésiens, une source oasienne se trouve asséchée, mais incomplètement. Son débit tombe à moins de 1 l/sec., et ne suffit qu'à entretenir en permanence une flaque stagnante (photo 1), sans utilisation agricole possible, foyer persistant de schistosomiase et de paludisme. Là encore, un remblayage avec du sable suffirait à réduire ces foyers, totalement et définitivement, à un prix de revient ridiculement bas.

L'aménagement d'eau potable, pour simple qu'il, doit néanmoins répondre à un minimum de salubrité. La photo 3 montre ce qu'il advient quand l'évacuation du trop-versé est défectueuse: l'apport d'eau potable est remis en cause par des réinfiltrations continues d'eaux souillées. Par ailleurs sont créés de toutes pièces, de gîtes éventuels de mollusques, de larves de moustiques, de larves d'ankylostomes et d'anguillules. Il aurait suffi de remplacer la dalle pleine qui se trouve sous le robinet par une dalle sertie d'un grillage métallique permettant l'écoulement du trop-versé dans le caniveau prévu à cet effet, ou mieux, dans un puisard.

Nous voyons donc que le "bétonnage" d'un système d'irrigation, s'il constitue un progrès pour l'irrigation, représente aussi une amélioration effective ou potentielle sur le plan de la santé. Malheureusement, et ceci n'est pas particulier au Sud tunisien, nous constatons souvent l'envahissement par les bulins des conduites ouvertes en béton du réseau d'irrigation, et encore plus, de tous les types d'ouvrages de béton qui jalonnent ces conduites. Bassins de prise, de partition et de dénivellation (photo 4) ont souvent des parois littéralement tapissées de mollusques. Or précisément, ces différents bassins servent de baignade à la population, et principalement aux enfants. Ce sont véritablement d'authentiques baignoires à attraper la schistosomiase.

Il est certain, cependant, que le "bétonnage" d'un système d'irrigation, s'il est souhaitable sur le plan agricole, l'est aussi au point de vue santé, et l'argument sanitaire peut ainsi appuyer les desiderata du génie rural. En effet, nos moyens de lutte sont bien plus facilement applicables et efficaces dans ces conditions d'aquarium: eau stagnante ou courante, assèchement à volonté rendent la vie impossible aux larves de moustiques, d'ankylostomes et d'anguillules. Les molluscicides y sont d'une efficacité presque aussi élevée qu'au laboratoire. Il existe peu ou pas de colmatage et d'envahissement par la végétation aquatique, qui gènent l'action des molluscicides. Là encore, les intérêts sanitaires et ceux de l'irrigation se rejoignent.

Dans bien des circonstances encore, l'intérêt sanitaire va de pair avec celui de l'irrigation et l'argument santé peut apporter un appui en faveur de tel ou tel dispositif technique. C'est ainsi que les conduites fermées, et si possible souterraines, doivent être, chaque fois que possible, préférées aux conduites à ciel ouvert. C'est là un moyen radical de les préserver de toute faune aquatique pouvant être préjudiciable à la santé humaine. La pratique de l'irrigation, elle aussi, y trouvera son compte. Les conduites souterraines en béton sont évidemment les plus coûteuses à la pose. Mais on s'y retrouve en économie d'eau, qui elle aussi coûte cher et est rare dans les pays arides. Il n'y a plus, en effet, de pertes par évaporation qui peuvent être énormes sous un climat chaud et sec et atteindre jusqu'à 60 pour cent de l'eau extraite. Il n'y a plus, non plus, de frais d'entretien, de curage, de réparation. Le danger de mauvaise gestion du réseau est moindre qu'en cas de situation de celui-ci entièrement en surface. Le colmatage des conduites étant éliminé, l'eau s'écoule à une vitesse accrue et constante et permet donc l'irrigation de terres plus lointaines et plus étendues. Le sol est économisé au maximum en vue des cultures, le tracé des routes est facilité. Si toutefois, malgré tous ces avantages, le coût initial d'un système de conduites souterraines paraît prohibitif, il convient de préférer des conduites en surface, mais fermées, qui coûtent à peine plus cher que les conduites à l'air libre, mais qui présentent par rapport à celles-ci, outre les avantages au point de vue sanitaire, la plupart des avantages hydrauliques et agricoles des conduites sous terre, que nous venons d'énumérer.

Certains des bassins précédemment décrits pourraient être remplacés ou transformés. Ceci étant, il est bien évident que certains de ces bassins sont indispensables tels quels au fonctionnement-même du système d'irrigation, ou ne sauraient être transformés dans le sens désiré que moyennant une augmentation du prix de revient trop importante. Dans tous ces cas, une solution simple consisterait à les recouvrir d'une dalle encastrée de béton armé, interdisant leur accès aux baigneurs. Cette mesure, maintenant le contenu du bassin à l'abri de la lumière, ferait du même coup disparaître les bulins. Le dallage des bassins aurait en même temps l'avantage d'éviter l'encrassement toujours important de ces ouvrages.

Les renseignements d'ordre géologique, hydraulique, agricole et les solutions qui ont été élaborées n'ont pu être proposées et mises en application que grâce à une collaboration étroite et une compréhension mutuelle entre les services du génie rural et de la santé. Les résultats? La schistosomiase urinaire, fréquente dans la région, aurait certainement, comme ailleurs, augmenté avec l'essor de l'irrigation. Elle est en passe d'être éradiquée, ce qui est exceptionnel, ceci grâce certainement à des conditions épidémiologiques particulièrement favorables, mais grâce aussi à un aménagement judicieux de l'irrigation, une bonne "discipline de l'eau" et une campagne de traitement des malades et de lutte contre les mollusques très efficace.

II. Syrie: région du Ghab

Le Ghab est la partie moyenne de la vallée de l'Oronte, (Nahr el Aasi) qui, dans cette partie de son cours, coule du sud au nord dans le fond d'une dépression délimitée à l'ouest par les montagnes Allaouites et à l'est par les premiers contreforts du plateau syrien, le Djebel Zawiyeh. La région est longue de 62 km et large de 22 km.

Autrefois, le Ghab était presque entièrement occupé par des marécages. La population s'occupait de pêche, de chasse, et pratiquait quelques cultures de céréales sur les hauteurs bordant la région.

Depuis une dizaine d'années, d'importants travaux de drainage ont été effectués. Le lit de l'Oronte a été canalisé. Son débit a été régularisé par une série de barrages qui permettent une distribution judicieuse de l'eau, à travers un réseau d'irrigation, à une vaste étendue de terres alluviales hautement fertiles. Le premiers résultats sur le plan agricole sont encourageants.

Sur le plan sanitaire, ils le sont moins. Toutefois, certaines améliorations pourraient y être apportées et certaines d'entre elles ont déjà reçu un début d'exécution.

La schistosomiase n'existe pas dans le Ghab, et il y a peu de chances qu'elle s'y installe. La paludisme a toujours été un fléau dans cette région marécageuse. L'irrigation et le drainage, supprimant un certain nombre de collections d'eau stagnante, auraient plutôt tendance à améliorer la situation. Malheureusement, çà et la, à la suite à un mauvais entretien, l'eau circule mal, stagne, et l'on voit se former de nouveaux marécages et de nouveaux gîtes à larves de moustiques. C'est là une situation qui n'est pas seulement préjudiciable sur le plan santé, mais aussi sur le plan agricole.

Le problème de l'eau potable pourrait être facilement résolu. La plupart des villages sont situés à la périphérie de la vallée du Ghab, aux pieds des montagnes Allaouites et du Djebel Zawiyeh qui le bordent. Or un grand nombre de sources jaillissent précisément en bordure de ces montagnes. L'eau est bactériologiquement pure et chimiquement excellente. Malheureusement elles ne sont ni captées ni protégées, mais polluées massivement dès l'émergence: photo 14 et 15. Parfois la source est aménagée d'un simulacre de protection tout à fait aléatoire. D'autres fois, il n'y en a pas du tout. Parfois, le captage est correct: l'eau arrive à l'usager bactériologiquement pure mais, laissé en écoulement libre, le point d'eau s'entoure bientôt d'un marigot pouvant servir de gîte pour les larves de moustiques, d'ankylostomes et d'anguillules.

Il est à noter que le captage de toutes ces sources eut été très facile et très peu coûteux à réaliser: de l'ordre de 200 $ USA par source. L'extraction de l'eau ne coûterait rien, l'eau s'écoulant toute seule par artésianisme.

L'élimination des excréta humains est inexistante. Quelques initiatives privées de bonne volonté, mais exécutées en dehors de toute compétence, et même du plus élémentaire bon sens, ont abouti à des situations catastrophiques. La photo 6 montre une latrine se déversant directement dans l'eau à 3–4 m à peine du point d'émergence de la source! De telles situations se passent de commentaires, sinon qu'elles n'ont rien d'exceptionnel dans les projets de développement rural, par ailleurs bien conduits, mais où les problèmes sanitaires n'ont pas reçu toute la considération qu'ils méritaient.

III. Conclusion

Par ces quelques exemples concrets pris dans deux pays aux conditions hydrologiques très différentes, nous avons voulu montrer un certain nombre de bons résultats obtenus par la collaboration entre les services techniques d'aménagement de l'irrigation et les services de santé, tant d'ailleurs pour résoudre les problèmes de santé, que pour le bon fonctionnement du système d'irrigation lui-même.

Nous avons également montré quelques échecs dûs à un manque de coopération entre ces services. Nous nous sommes placés exprès sur un plan purement technique. Nous aurons l'occasion de revenir sur la nécessité, non moins impérative, de cette collaboration sur le plan administratif, budgétaire et humain.

OPERATION OF THE IRRIGATION DISTRICTS IN MEXICO

by ABELARDO AMAYA BRONDO, LÁZARO RAMOS ESQUER and
ENRIQUE ESPINOSA DE LEÓN

Secretariat of Hydraulic Resources, Mexico

This long paper which was received late is published in summary only, with data in the form of tables and maps. The full paper can be obtained from the authors.

Summary

Mexico lies within the belt where the big deserts of the world are located. Out of the total area, 63% has all features of a desertic climate of steppes, 6% is rainy tropical lands and 31% is temperate with rains in summer. Irrigation is required in most of the country to have a fair degree of certainty in crop yields.

Mexico at the present time has an area of 4,750,000 under irrigation, and out of this total, the Secretaria de Recursos Hidraulicos manages 2,922,380 divided into 163 Irrigation Districts, and supervises the operation of 3632 Irrigation Units of medium and small acreages with an area of 800,200 ha. The total water volume that is readily available at headworks of Irrigation Districts is 46,361 million m³.

Hydraulic works for irrigation and development of agriculture began in 1926 and operation of Districts started in 1930. From that year to 1965 operation gradually evolved, but most of the Districts were lacking in water control and measuring structures, and were also without an adequate service of technical assistance. Aiming at better control a programme of Irrigation Techniques and Farm Improvement Plan (PLAMEPA) was then introduced with the main goal of raising production and yields. As a result, the efficiency of water

TABLE 1. *Irrigation districts classified according to the supply source*

Supply source	Number of districts	Area in hectares
Storage dam	56	1,260,776
Derivation	73	347,633
Deep wells pumping	16	234,071
Storage and derivation	10	220,163
Storage and pumping	6	624,777
Derivation and pumping	2	234,960
	163	2,922,380

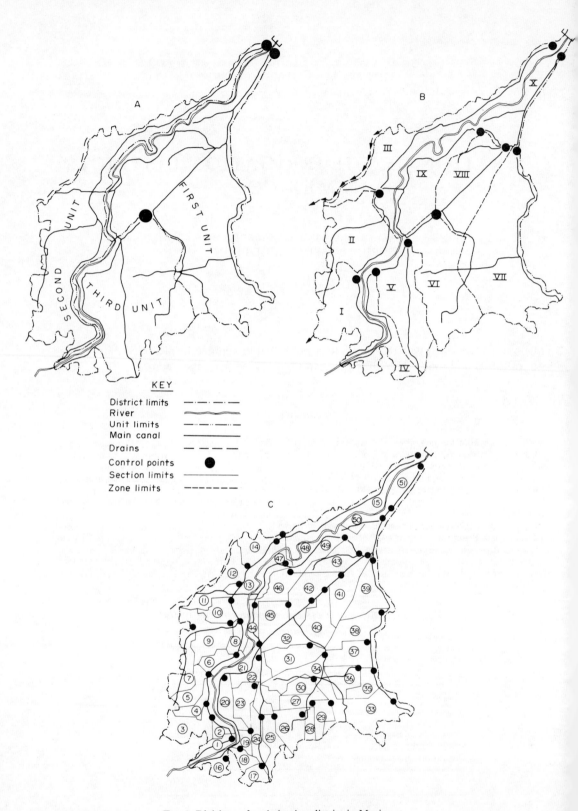

KEY

District limits	—·—·—
River	
Unit limits	—··—··—
Main canal	————
Drains	— — —
Control points	●
Section limits	————
Zone limits	- - - - -

FIG. 1. Divisions of an irrigation district in Mexico.

conveyance and of water application at the farm level has been increased and better yields in most crops have been obtained. Furthermore, the following advantages have been obtained:

(1) More public consciousness is being generated about water as a very important factor in agricultural production.
(2) Valuable statistical material has been obtained for purposes of operation and for planning new projects.
(3) Improvements have allowed a rise in incomes of farmers, aid to industrial development, enlarging home and export markets, and building up new sources of labour.

TABLE 2. *Irrigation districts classified according to the irrigable area*

Area in hectares	Number of districts
Less than 10,000	120
More than 10,000 to 30,000	23
More than 30,000 to 50,000	5
More than 50,000 to 100,000	8
More than 100,000 to 200,000	3
Bigger than 200,000	4
	163

TABLE 3. *Average yields of the main crops in the irrigation districts and in test plots (tons per hectare)*

Crops	Irrigation districts	Test plots
Wheat	4.0	4.7
Corn	2.5	4.3
Carthamus	1.6	2.5
Soya bean	2.1	2.7
Beans	1.3	1.8
Rice	4.3	5.7
Sorghum	3.7	4.5
Cotton	2.9	3.6
Sugar cane	80.3	97.2

TABLE 4. *Average conveyance efficiency obtained in the irrigation districts*

Agricultural season	Conveyance efficiency (%)
1965–1966	58.3
1966–1967	59.5
1967–1968	59.7
1968–1969	60.1
1969–1970	61.2
1970–1971	62.3
1971–1972	61.9
1972–1973	61.7
1973–1974	62.2

SOCIAL AND ECONOMIC CONSEQUENCES OF LARGE-SCALE IRRIGATION DEVELOPMENTS – THE DEZ IRRIGATION PROJECT, KHUZESTAN, IRAN

by ECKART EHLERS

Professor of Geography, 355 Marburg Geographisches Institut, Renthof 6, FRG

Summary

The Dez Irrigation Project in Khuzestan, southwestern Iran, covers an area of about 100,000 hectares of traditionally farmed agricultural land. With its completion the DIP caused a number of social and economic consequences for the population in so far as traditional agriculture was, because of its low productivity, abandoned and replaced by growth-oriented forms of agriculture:

agro-industries;
agribusiness enterprises;
farm corporations.

Thus the question arose whether to put emphasis on economic or social considerations in regard to the future development of the region. The presentation and evaluation of large-scale agriculture and their comparison with traditional farming reveals that none of these forms is suited for utmost economic *and* social benefits. The example of the DIP shows that only two forms of agriculture are suited to include both aspects: (1) the development of government-owned and not strictly profit-oriented agro-industries, as they are represented in the Haft Tappeh Sugarcane Plantation. Their advantage is to be seen in the need of a large human labour-force as well as in its substitutive effects of expensive imports of agricultural products; (2) the strengthening of the traditional agricultural sector: recent investigations prove that traditional agriculture is well suited to produce market-surplus. Preconditions for a successful development of traditional agriculture would be, however, the installation of an *effective* cooperative-system. Advantages are: diminuation of rural-urban migration, development of an independent rural economy and weakening of the cities' economic control of their rural hinterlands. Thus, only government-owned state-farms or forms of cooperative agriculture are considered to be adequate for serving both social and economic goals.

Résumé

Le périmètre d'irrigation du Dez dans le Khouzistan (Iran) intéresse une superficie d'environ 100,000 ha SAU d'agriculture traditionelle. De sa mise en valeur resultèrent de nombreux problèmes et sur le plan social et économique, étant donné que cette agriculture fut caracterisée par une rentabilité reduite. Cette agriculture était remplacé par des formes d'agriculture modernes, notamment:

agro-industries;
agribusiness;
sociétés agricoles d'actions.

L'évaluation de ces types d'agriculture moderne et leur comparaison avec l'agriculture traditionelle revèlent qu'aucune de ces formes promet des bénéfices satisfaisants sur les plans social et économique. Il va falloir trouver la solution sous forme d'un compromis. L'exemple du DIP met en relief que deux formes seulement d'agriculture sont possible pour arriver à ce but 1) la création de domaines d'Etat (agro-industries) qui ne sont pas forcement orientés vers un profit bien determiné: exemple de plantation de canne à sucre de Haft Tappeh. L'avantage de ces domaines résulte de la grande demande de main d'oeuvre ainsi que dans la substitution de l'importation de produits agricoles cause de lourdes pertes de devises; 2) le renforcement du secteur agricole traditionel. Des recherches recentes nous affirment que l'agriculture traditionelle peut très bien produire un surplus qui arrive sur les marchés. Mais pour en arriver là il est indispensable d'installer un système de co-opératives agricoles effectif. Les avantages en seraient: ralentissement de l'exode rurale; la création d'une économie rurale indépendante; l'affaiblissement du contrôle économique des villes sur leur hinterlands ruraux.

Agribusiness et sociétés agricoles d'actions sont caracterisés par l'orientation vers un profit maximal et ainsi vers la substitution de la main d'oeuvre, donc les arguments économiques dominent sur les arguments sociaux. C'est pourquoi seulement les domaines d'Etat (agro-industries) ou des cooperatives agricoles d'exploitation en commun semblent être les seules formes à réaliser des buts et sociaux et économiques.

1. Introduction

In line with the interdisciplinary character of this book, dealing with environmental problems and the effects of arid land irrigation in developing countries, I shall try to delineate a few social and economic consequences of a large irrigation project in southwestern Iran. Regionally, I am concentrating on the area of the Dez Irrigation Project (DIP), a 1959 initiated multipurpose project of energy-supply and irrigation development. The DIP is considered to be the first of several large irrigation developments in the fringe area between the Zagros Mountains and the Iranian part of the Mesopotamian Lowlands. Covering an

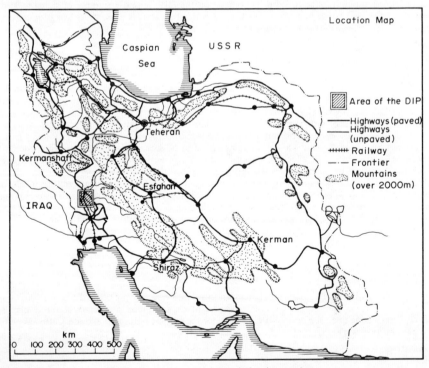

FIG. 1. Location of the Dez irrigation project.

area of approx. 100,000 hectares or 10% of the total project area, the DIP is in many respects a test and a pilot project.

Its pilot character derives from the fact that the DIP is being developed within an area of traditional farming, i.e. in a region where both irrigated farming and dry farming have been exercised for centuries. Thus the main problems of irrigation development are not to be seen in bio- or geo-ecological effects of the area in question, but in the aspects of human ecological changes and problems. The main question is: how can the extremely high input of capital and ideas be used for an optimal output of social and economic benefits within the DIP and for the whole country? It should be stressed that it is exactly this question which is at the centre of our considerations: the fact that 3 or 4 different types of agricultural holdings occur side by side under identical physical social and economic conditions as are hardly elsewhere to be found. The regional context is, thus, of only minor importance.

2. Aspects of traditional rural settlement and economy

Rural settlement and rural economy were until quite recently, i.e. until the implementation of the Iranian land reform in the 1960s, characterized by the antagonism of a few, mostly absentee landlords, who resided in Dezful, and a great number of rural sharecroppers, who lived in the villages and who were more or less dependent on the landlords.*

Tables 1 and 2 reveal certain aspects of the traditional land-rights of altogether 170 villages in the DIP. They prove that 500–600 landlords owned the whole territory of about

TABLE 1. *Traditional land-rights in the DIP, 1959*

District	Number of villages	Area (in hectares)	Number of titles	Number of proprietors
Andimeskh	3	15,823	54	?
Ben Mualla	20	15,531	60	46
Bonvar Nazer	21	18,388	182	152
Hosseinabad	29	31,056	30	13
Shahvali (partly)	4	4,462	75	?
Gheblei	29	7,537	195	140
Shamoun	18	10,557	229	35
Sharghi	46	48,211	270	153
Total	170	151,595	1,095	539

150,000 hectares and that most of these properties were concentrated in Dezful as the main urban centre in northern Khuzestan, Tehran ranking next.

Land-ownership of the villages and their territories, however, was extremely complicated by the fact that mostly not one or two landlords, but several families and their members as well as individuals, shared in the land-titles. The splitting of village lands was intensified in case of death of one of the proprietors. In this case all heirs received parts of the property of the deceased, according to the Islamic laws of inheritance. The result was, in most cases, that property-rights in a village, in certain parts of a village or

*Statistical data in Tables 1, 2 and 3 are based on *Plan Organization* (1960).

TABLE 2. *Traditional land-rights and residence of landlords in the DIP, 1959 (in hectares)*

District	Total area	Dezful	Shushtar/ Ahwaz	Tehran	Rest
Andimeshk	15,823	12,800	1,585	1,376	92
Ben Mualla	15,531	14,798	29	—	704
Bonvar Nazer	18,388	17428	960	—	—
Hosseinabad	31,056	14,990	—	16,066	—
Shahvali (partly)	4,462	2,006	2,335	—	121
Gheblei	7,537	7,513	24	—	—
Shamoun	10,557	7,934	234	1,904	485
Sharghi	48,211	30,763	462	2,516	14,470
Total	151,595	108,232	5,629	21,862	15,872

its fields were almost atomized among a great number of proprietors who only knew the *share, not the hectarage of land* which they owned within the village and its boundaries. This tendency, to possess not certain parts of the land but to possess shares of the total, was made possible by the custom to split each property or any part of it into "ideal" or "imaginary" shares or allotments. Thus, a village or a certain field could be divided as follows:

$$1 \text{ village or part of it} = 24 \text{ "peas"}$$
$$1 \text{ "pea"} = 24 \text{ "barley"}$$
$$1 \text{ "barley"} = 24 \text{ "sesame"}$$

This means that each village or any part of it could be divided into $24 \times 24 \times 24$ parts, i.e. into altogether 13,824 property-titles. It is obvious that under such circumstances regionally fixed land-allotments are hardly possible so that the "ideal" or "imaginary" land-title has become very common in northern Khuzestan.

One example may illustrate this intricate system of traditional land-ownership: the little village of Jeibar and its 140 hectares comprising fields (cf. Table 3).

TABLE 3. *Traditional land-ownership in Jeibar/DIP*

Land-owner (serial numbers)	Residence	Shares of land Peas	Barley	Sesame
1	Dezful	1	2 78/375	—
2	,,	1	10 14/25	—
3	,,	9	23 1/25	—
4	,,	1	2 78/375	—
5	,,	1	2 78/375	—
6	,,	3	10 118/125	—
7	,,	—	22 2/25	—
8	,,	1	17 59/125	—
9	,,	3	5 7/25	—
	Total	20	96	—

The great number of rural sharecroppers who work the land on the basis of more or less fixed shares were organized in working units called *bonku* (Lambton 1953). Before the land reform there were 4 *bonku* in Jeibar, each one consisting of a group of three peasants. The area which was worked by each member of the *bonku* corresponded to one *joft*, one *joft* being the amount of land which could be worked by one man and a yoke of oxen. In order to guarantee a fair distribution of land to all *bonku*, the whole arable land of Jeibar (and the other villages) was split into several parts. Each *bonku* was represented in each part, the parts being redistributed annually among the *bonku* by lottery. Within the *bonku*, a second lottery took place in order to provide each member of the working unit with his special part or share of the land. In other words: so long as the land proprietors

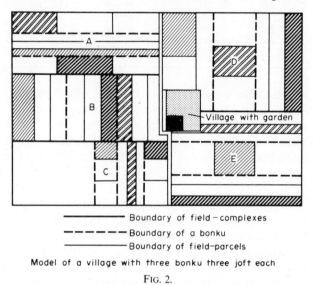

——————— Boundary of field – complexes
– – – – – – Boundary of a bonku
——————— Boundary of field–parcels

Model of a village with three bonku three joft each

FIG. 2.

could not point to any regionally fixed land-claims, the sharecroppers did not have any regionally fixed tenure-rights. Each year they received other parts of the arable land of the village, amounting, however, each time to the size of one *joft*. Thus, the sharecroppers could never develop any proprietary claims to any part of the landlords' land.

As a result of this specific land distribution system and the corresponding relationships between landlord and sharecropper we find an extremely high mobility and fluctuation of the soil within the villages. In Jeibar, for example, the arable land of which was split into 11 different parts, this would mean that, under extreme conditions, the village land of 140 hectares could be split up into

$$11 \text{ parts} \times 4 \text{ } bonku \times 12 \text{ } joft = 528 \text{ parcels of land}$$

for annual redistribution among the peasants. In reality, however, several parts of the arable lands were consolidated to larger units, so that the number of parcels redistributed decreased considerably. Nevertheless, there is a strong dependency between the size of the farm holdings and land fragmentation, as becomes obvious from Table 4.

The relationship between the large landed proprietors and the sharecropping tenants was characterized by the traditional regulation that 12.5–20% of wheat, barley and some vegetables and 50% of rice and most garden crops had to be delivered to the landlords. Under these regulations the landlords had to provide land and water, while the peasant had to contribute draught animals, seed and their own labour. Also the cleaning of the irrigation canals was part of the sharecroppers' duties.

TABLE 4. *Number and size of lots in holdings with lands in Khuzestan*

Size of holdings, (in hectares)	No. of holdings	Area of holdings	No. of lots	Average no. of lots per holding	Average area of each lot (in hectares)
0.5 to 1	21,900	11,304	38,550	1.7	0.3
1 to 5	66,460	214,035	289,764	4.3	1.66
5 to 10	47,109	341,315	319,105	6.7	1.07
10 to 20	22,660	302,828	243,709	10.7	1.2
20 to 100	7,560	221,066	100,175	13.2	2.2
Total	165,689	1,080,247	991,303	6.1	1.09

In view of these facts, the social conditions of the rural population of northern Khuzestan before the land reform must be considered antiquated and backward. Besides the unfavourable sharecropping conditions, the traditional cultivation techniques contributed to this backwardness: primitive agricultural tools (wooden nail plough!), obsolete irrigation techniques, lacking or imperfect use of fertilizer as well as no agricultural extension services seem to have been the main reasons for the deficiencies of the economic and social structure.

It is against this background that the necessity for a land reform becomes apparent. On the other hand, it is obvious that the modernization of the agricultural sector, based solely on the reorganization of land-ownership and not on the education and instruction of the rural population, would come into conflict with the construction of the DIP. Delays in the projections and hopes of the land reform legislation on one hand (Khatibi 1972) and the necessities of the expensive DIP on the other hand are probably the main reason why land reform measures were hardly implemented in the DIP area. Except for a few villages, agricultural land-use as well as the rural social structure were changed dramatically from backward and small-scale peasant farming to spectacular large-scale and market-oriented agribusiness (Ehlers 1975).

3. Dez Irrigation Project and modern forms of large-scale agriculture

When the implementation of land reform laws was in full swing in different parts of the country, the Fourth National Development Plan 1968–1972 (Tehran 1968) was published. In it we read in regard to the development of the agriculture of the country as one of the main aims:

the utilization on a large scale and with modern technology of dam-irrigated land, by encouraging the private sector, and in the absence of volunteers, knowhow or capital in the private sector, direct government initiative is needed to establish large farming units.

According to this programme, we notice from 1968 onwards also in northern Khuzestan a fast and systematic transformation of the traditionally rural landscape and social structure. Small villages and hamlets and their fields were replaced by modern agricultural enterprises, which can be divided into three different types:

(1) agro-industrial companies;
(2) agribusiness enterprises; and
(3) farm corporations or agricultural shareholders' companies.

DEZ IRRIGATION PROJECT

Areas of
traditional farming
Agribusiness
Agroindustries (Hall Tappah)
Farm corporations

Traditional Settlement Structure—Modern
Forms of Agriculture—New Towns (Shahrak)

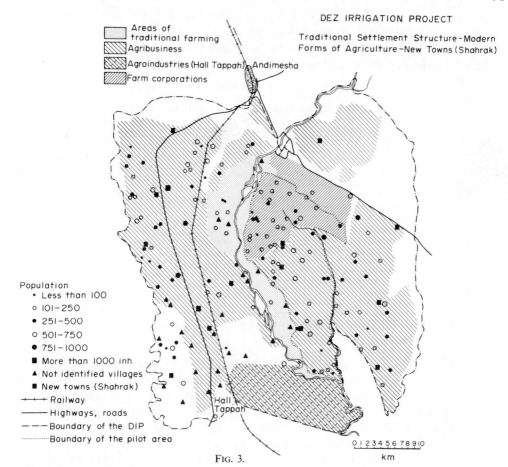

Population
• Less than 100
○ 101–250
● 251–500
○ 501–750
● 751–1000
■ More than 1000 inh.
▲ Not identified villages
■ New towns (Shahrak)
+—+—+ Railway
———— Highways, roads
—·—·— Boundary of the DIP
············ Boundary of the pilot area

0 1 2 3 4 5 6 7 8 9 10
km

FIG. 3.

The following discussion concentrates on these three modern types of agriculture in the DIP and their specific economic and social consequences.

Agro-industries and agribusiness, defined as "the sum total of all operations involved in the production operations of a large-scale farm, the storage, processing and distribution of farm commodities and items made from them" and marked off from each other by the degree of their industrial refinement, now cover an area of about 100,000 hectares. Four privately owned agribusiness companies were established recently, covering between 10,000 and 20,000 hectares each. The state-owned sugarcane plantation of Haft Tappeh, inaugurated in 1961 and extended in recent years, is the only real agro-industry in the proper sense of the word.

Agro-Industries. Haft Tappeh, considered to be the prototype of a successful agro-industrial enterprise, was developed on traditionally government-owned land (*khaliseh*) formerly belonging to Arab-speaking tribes and their sheikhs. The traditional land-use, a mixture of conventional agriculture and pastoralism, was characterized by low productivity and a sparse population-density. In other words: the "carrying-capacity" of the Haft Tappeh-area, i.e. its possibilities and potential to sustain a rural population economically and socially, was rather low.

By 1973 there were more than 1250 permanently employed labourers (agricultural workers, mechanics, drivers, etc.) and about 280 engineers, technicians, scientists and office-employees. These numbers increased with the extension of the plantation to its final size in 1975/6 by another 400–450 persons. In addition to this permanently employed labour-

TABLE 5. *Summary of production of Haft Tappeh*

Year	Productive area	Production TNC/ha	Productivity TNC/ha	Production TRS	Productivity TRS/ha
1961/2	2200	179,835	81.7	12,160.0	5.5
1962/3	2421	202,139	83.5	17,003.0	7.0
1963/4	2815	196,781	69.9	15,298.5	5.3
1964/5	3643	292,434	80.3	25,847.1	7.1
1965/6	3919	392,136	100.1	37.023.4	9.4
1966/7	3966	382,420	96.4	38,614.9	9.7
1967/8	4183	433,168	103.5	42,394.3	10.2
1968/9	4284	452,317	105.6	47,832.9	11.2
1969/70	4389	528,336	120.4	54,110.1	12.3
1970/1	4432	521,518	117.7	54,716.0	12.4
1971/2	4536	577,577	127.3	49,348.4	10.9
1972/3	6747	831,060	123.2	62,492.6	9.3

TNC = tons of net cane; TRS = tons of raw sugar.

force another 1200–1500 seasonal workers are hired for 5–7 months in connection with the sugarcane-harvest. The undoubtedly positive effect of the Haft Tappeh sugarcane project became even more apparent in 1970 with the construction of a large paper-mill. This mill, based on the use of bagasse, the refuse of the sugar-refining process, employs another 800 people. On the whole, there is no doubt, Haft Tappeh must be considered successful, although some input-output analyses point to the fact that "the project entails great constant costs so that business operation is of insignificant rentability" (Rabbani 1971).

Agribusiness. The recently established agribusiness enterprises within the area of the DIP are different from Haft Tappeh in so far as the effects of industrialization are less pronounced and in fact are not intended. So far four agribusiness enterprises have been established:

Company 1 with a total area of 20,267 ha
Company 2 with a total area of 10,536 ha
Company 3 with a total area of 15,796 ha
Company 4 with a total area of 16,680 ha

The remaining lands available, i.e. an area of 4010 ha, will be distributed among one or two of these companies.

The conditions under which these lands were assigned to the international agribusiness companies include the following aspects: lands are assigned on a 30 years' lease. While the governmental Khuzestan Water and Power Authority (KWPA) provides the main irrigation canals and part of the infrastructure (access roads, etc.), the companies are obliged to perform land-levelling, construction of the distribution canals and drainage systems as well as the performance of the internal infrastructure. Water has to be bought from the KWPA.

The annual rent comprises a fixed sum (1973: 1200–1500 rial/ha) or an amount of 2.25–2.5% of the gross agricultural production.

In view of the fact that the agribusiness enterprises were established on traditionally used agricultural lands, mechanized large-scale farming caused deep economic and social consequences for the villagers. In order to avoid too harsh a social impact the companies had to agree to employ at least one Iranian labourer per 10 hectares of agricultural land. This means, that the companies should offer jobs for at least 6500–7000 persons. On the other hand, as mechanization is one of the main advantages of large-scale farming, such a large human labour-force must be contrary to the companies' profit-oriented interests. This conflict of social and/or economic interests becomes especially perceptible in the analysis of one of the new agribusiness enterprises:

> Company 3, for example, with an area of approx. 16,000 hectares covers the territories of 26 traditional shareholders villages. In these 26 villages a total of at least 1200–1400 sharecroppers were agriculturally engaged. According to the agreement at least 1500–1600 people, i.e. the total traditional agricultural labour force of the 26 villages, should be employed in the new agribusiness. Actually, however, less than 350 people at the utmost will be needed. This means that about 75% of the traditional labour-force will be without jobs, unless new non-agricultural employment is created.

Thus, on the whole, there is a wide discrepancy between the availability of human labour and the necessities of large-scale, mechanized farming. This discrepancy will be intensified with the final stage of development of the agribusiness when land-levelling and related activities will be complete: it is anticipated that, at that time, one man will work between 50 and 100 hectares, so the necessary labour-force will shrink even more.

A second severe consequence of the installation of agribusiness is to be seen in the fact that the traditional "agricultural landscape" of the DIP will be completely reshaped. About 100 villages will have been levelled to the ground, traditional rural communities and their territories will be dissolved and their population resettled in a few large agro-towns. It is anticipated that 13 so-called *shahraks* will replace the historical settlement pattern. The *shahraks*, conceived for 800–1000 families each, are equipped with a number of installations such as schools, mosques, ambulances, bath-houses and shops.

Each family transferred to any one of these *shahraks* receives a quarter-section (approx. 40 m^2) of a house plus 750 m^2 of gardens. Animal husbandry is limited to one cow and/or two sheep or goats. Problems connected with the resettlement of several thousand traditionally thinking and agriculturally oriented families are: social distintegration of established communities; conflicts with religious thinking and belief due to the architecture of the *shahraks* (no walls); social and economic decline of the population to become agricultural labourers, sometimes only seasonal; no other activities alternative to agriculture.

On the whole, one has to agree with Rabbani's (1971) statement: "We can therefore see that, although the goal of modernizing the agricultural sector has been fulfilled, no effective remedy has been sought for the labour force released from the land!" This is especially true for the agribusiness enterprises, for which a vertical expansion towards agro-industrial activities is neither intended nor foreseeable.

Farm corporations. Farm corporations or agricultural shareholders' companies is the other

type of large-scale farming being tested in the DIP. According to Ule (1973) farm corporations are characterized by the following:

> As basic idea behind the Agricultural Shareholders' Companies reference can be made to a tendency of moving
> from individual to collective farming. It is a question of shareholders' companies, in which the farmers invest
> capital in the form of land, machinery, equipment, buildings, wells, etc. Each of them receives shares according
> to the value of the production factors made available to the companies, the aim of which is to modernize farm-
> ing and its mechanization with the simultaneous introduction of the large holdings.

In the DIP, in 1973 a total of four farm corporations was established. With a size of
approximately 2000 ha each, they cover the biggest part of the famous *9006 hectares North
of Safiabad*, i.e. an area of traditional farming in the Dez Pilot Irrigation Project (DPIP), in
which for several years traditional farming methods were combined with modern irrigation
techniques. The consolidation of the individual farm-units into four corporations resulted
in effects similar to those of agribusiness, although on a somewhat lesser scale: the
abandonment of the traditional *bonku*- and *joft*-structure; the uprooting of grown village

TABLE 6. *Farm corporations and traditional village structure, DIP*

Farm corporation	Number of villages	Total area	Number of *bonku*	Number of *joft*
I	5	2363	14	104
II	11	1894	31	174
III	8	1680	20	150
IV	7	1975	17	117
Total	33	7912	82	545

communities and their replacement by settlements in whose decision-making processes
farmers do not directly participate; the effects of unemployment through mechanization.

Comparison of the four corporations and their traditional village-structure is shown in
Table 6.

Although it is too early to make a final assessment of the economic efficiency and the
social problems involved in the establishment of farm corporations in the DIP, the results
of comparative observations in other parts of the country are not encouraging. According
to Ule (1973) three complaints are made: the farmers feel deprived of the right to cultivate
on their land; the previous property deeds are substituted by shares; these shares are
regarded as equivalent in value to production factors and the farmers have lost the right
of disposal, not *de jure* but *de facto*.

Ule's statement that there is "a change to a collectivization and socialization of the whole
of agriculture under the guise of the capitalist concept of the shareholders' company" may
be doubted, but it is certain that the farm corporation movement is alien to the sharecroppers' mentality and also to their present-day interest.

The social and economic effects of the farm corporations may be less marked than those
of agribusiness: farmers are allowed to stay within their villages; much more than in
agribusiness they can keep small flocks of sheep or goats and private agriculture may also
be possible, although only on small plots of marginal lands outside the farm corporations'
lands. On the other hand: mechanization of agriculture, with large machine parks, workshops, stables and barns at the centre of each farm corporation, point to the fact that farm
corporations, too, will replace a large amount of human labour-force by machines.

Evaluation and conclusions

This survey of traditional and modern aspects of agriculture and village life in northern Khuzestan reveals that today four different types of agriculture compete side by side: traditional small-scale farming; farm corporations; agro-industries; agribusiness.

Land reform has contributed considerably to the social awakening of the rural population and succeeded in developing a strong sense against subjugation or exploitation by landlords or any anonymous landlord-like institution. On the other hand, the government has an interest in an acceptable input-output ratio in the DIP. Thus, from the beginning, there was a conflict between economic and social goals. The whole development programme finally concentrated on the question whether to give priority to the socio-economic

TABLE 7. *Economic and social effects of different types of agriculture in the DIP*

	Agricultural productivity	Effects of industrialization	Development of human labour-force
Agro-industries	+	+	+
Agribusiness	+	(+)	−
Farm corporations	+	◯	−
Traditional farming	◯	◯	+

     +   positive development
     −   negative development
   ◯   no changes

development of the rural population or to the promotion of agricultural production of the country.

Table 7 illustrates this antagonism of social and economic consequences, which in most cases do *not* correspond. While traditional small-scale farming tends to absorb a considerable labour-force, its productivity in terms of market-oriented production is low. Agribusiness and farm corporations, on the other hand, seem to be able to fulfil the task of market delivery with agricultural products, but both tend to diminish the amount of rural population considerably. Thus, only agro-industries of the government-owned Haft Tappeh-type seem able to combine the advantages of large-scale and market-oriented industrial farming with the positive effect of binding a large human labour-force to the rural area.

Sugarcane plantations are the exception to the rule and hardly suited as a model for the agricultural development of a whole country. For this reason, the idea of farm corporations and their advantages or disadvantages has recently been discussed broadly in the German agricultural periodicals (Ehlers 1975). The opinions stated are as numerous as the statements themselves: they range from moderate support to critical warnings. Among the positive factors are positive aspects of large farming units; possibilities of broad variety of crops grown; possibilities of increased productivity per hectare; possibilities of increased income per shareholder. On the other hand, there is a catalogue of negative factors which includes lack of qualified management; lack of cooperation between management and shareholders; increase in the rural exodus; enforced enrolment in newly established corporations; fear of dependency and collectivization; alienation between farmers and their properties.

In view of these negative factors, Ule's proposal of a middle course seems worth consideration. He proposes to restrict large-scale farming and management to a few crops such as cereals and cotton which can be grown more economically on large plots, highly mechanized. Other agricultural products, however, should be grown on individual farm units with the individual disposing of his own produce.

Another possibility, so far hardly discussed seriously, could be the positive development of traditional village structures which are based on forms of cooperation and mutual assistance. The fact that in almost all cases where, within the DIP, land-reform laws were implemented, *bonku-* and *joft*-structures remained unchanged, is proof of their traditional strength. The development of cooperative farms of improved agriculture might well be achieved through collective but modernized use of the village lands by all members of the traditional *bonku*-groups. While there may be no doubt that their productivity will not be as high as the agribusiness enterprises, there should be no doubt, either, as to the potential of improved traditional agriculture. As in the past, traditional farming would not only contribute to the urban economy from the former landlord's share, but could be increased considerably by modernized techniques of land-use and marketing. The excellent results of Goodell's (1975) studies of village farming in the DIP reveal that traditional agricultural production can be market-orientated and may develop effective forms of rural productivity. Smallholdings can be extremely efficient; careful improvement of traditions, especially through the development of effective cooperatives, could serve both aims, namely increase of production and reduction of harsh social consequences.

The results of a detailed survey and a comparison between an agro-industrial enterprise and farm corporations in Fars by Bergmann and Khademadam (1975) point to a similar conclusion. These authors, too, propose a combination of different types of agriculture in order to achieve optimum effects. According to their findings the two main goals have to be: "the increase of agricultural production with the purpose of substituting import of foodstuffs; and the advancement of the rural population, i.e. the small farmers and the *khoshneshin*."

The decision for or against a growth-oriented approach in the development of rural Iran, for or against social considerations concerning the rural population, has still to be taken. There are no common rules and the decision has to vary from case to case, but a rural development strategy based not so much on a maximum of economic growth, but on an optimum of economic and social benefits, includes a great number of advantages: it lessens the psychological stress and the social implications for all villagers involved; it diminishes the push-factors for the rural–urban migration, thus also reducing the problems and effects of urban sprawl and its social consequences; it encourages the development of a genuine rural economy which could, after a certain time at least, initiate a change in the strong dependency of the rural areas on urban economies.

Such a strategy seems worth advocating in Iran, a country which has many other resources for a quick industrial development of its national economy. As soon as the process of industrialization reaches the stage where there is a strong demand for labour, a change in rural development policy could serve two purposes: first, a further intensification of the agricultural sector by large-scale mechanization; second, the economic and social integration of surplus labour from the rural areas.

References

Bergmann, H. and Khademadam, N. 1975. *The Impacts of Large-scale Farms on Development in Iran*. Publications of the Research Centre for International Agrarian Development, **4**, Saarbrücken.

Ehlers, E. 1975. Traditionelle und moderne Formen der Landwirtschaft in Iran. Ländliche Siedlung, Wirtschaft und Agrarsozialstruktur in Khuzistan seit dem späten 19. Jahrhundert. *Marburger Geographische Schriften* **64.** Selbstverlag Geogr. Institut der Universität Marburg.

Goodell, G. 1975. Agricultural production in a traditional village of Northern Khuzestan. *Marburger Geographische Schriften* **64.** Selbstverlag Geogr. Institut der Universität Marburg.

Khademadam, N. and Bergmann, H. 1973. Sozio-ökonomische Differenzierung im Gefolge der Bodenreform in Iran. *Z. Ausländische Landwirtschaft* **12,** 270–285.

Khatibi, N. 1972. Land reform in Iran and its role in rural development. *Land Reform, Land Settlement and Cooperatives* **2,** 61–68.

Lambton, A. K. S. 1953. Landlord and peasant in Persia. *A Study of Land Tenure and Land Revenue Administration.* London: Oxford University Press.

Plan Organization 1960. *Government of Iran: Dez Irrigation Project.* Report on Land- and Water-rights (with 5 appendices). Nederlandsche Heidemaatschappij, Arnhem/Holland, Jan–March.

Planck, U. 1974. Iranische Dörfer nach der Landreform. Schriften des Deutschen Orient-Instituts. Materialien und Dokumente. Opladen: Leske Verlag.

Planck, U. 1975. *Die Reintegrationsphase der iranischen Agrarreform, Erdkunde* **19,** 1–9.

Rabbani, M. 1971. A cost-benefit analysis of the Dez multi-purpose project. *Tahqiqat-e Eqtesadi VIII,* Nos. 23 u. 24 Tehran, 132–165.

Refahiyat, H. 1972. Sozialökonomische Bedeutung von agro-industriellen Kombinationsprojekten in Entwicklungsländern am Beispiel eines Zuckerrohrprojektes in Iran. *Z. Ausländische Landwirtschaft* **11,** 138–153.

Saidi, K. 1973. Landwirtschaftliche Aktiengesellschaften als Instrument der landwirtschaftlichen Entwicklung im Iran. *Z. Ausländische Landwirtschaft* **12,** 286–297.

Schowkatfard, F. D. and Fardi, M. 1972. Sozial-ökonomische Bedeutung der landwirtschaftlichen Aktiengesellschaft in Iran: Fallstudie eines Dorfes der Provinz Fars. *Z. Ausländische Landwirtschaft* **11,** 120–137.

Ule, W. 1973. Land reform in Iran and the development of agricultural shareholders' companies. Agriculture in the Near East, 97–124. Bonn–Bad Godesberg, *Verlag Neue Gesellschaft GmbH.*

THE CONVERSION OF BASIN IRRIGATION
TO PERENNIAL SYSTEMS IN EGYPT

by A. AZIM ABUL-ATA

ARE Minister of Irrigation, Cairo

Summary

One of the main objectives of constructing the High Aswan Dam was the conversion of the basin irrigation system, which was applied on about one million feddans in Upper Egypt with its one crop a year, to the perennial system with two or three crops a year. The effect is to double the yield of this significant area. The basin conversion projects have been executed side by side with the construction of the High Dam itself. This paper throws light on this project and on the changes resulting therefrom. The paper also explains how certain side-effects have resulted from the basin conversion project, how such phenomena were dealt with in some areas, and how the remaining areas are being treated.

The various positive effects of the project are demonstrated and also its consequences with regard to the national production, the social conditions of the people residing in the area, and what is yet to be done in order to realize all the fruitful objectives.

Résumé

L'un des objectifs majeurs de la construction du Barrage d'Assouan est la transformation du système d'irrigation par bassins, qui s'étend sur plus d'un million de feddans en Haute Egypte et ne produit qu'une culture par an, en un système pérenne d'irrigation entraînant la possibilité d'obtenir 2 à 3 récoltes annuelles. Cette transformation a été effectuée parallèlement à la construction du haut barrage lui-même. Cette communication met en lumière ce project et les changements qui en sont résultés. Elle indique également les effets marginaux qui se sont produits, et leur extension spatiale, ainsi que les traitements appliqués aux autres zones. Les résultats positifs du projet sont indiqués ainsi que leur conséquence sur la production nationale et les conditions sociales des populations résidant dans la zone. Les travaux encore à faire pour atteindre pleinement les objectifs sont exposés.

Introduction

Since 1953, the Government has sponsored the High Dam Construction Project, to which much attention was devoted. A special authority, including a team of specialized engineers, was entrusted with the relevant studies and the operation of experiments and researches in connection therewith. In the meantime the help of outside scientific experts and European laboratories was solicited. Consequently, the preliminary studies on the High Dam were expanded until they crystallized, and the project's features became

99

evident. This took place a long time before the Agreement to build the dam was concluded between the ARE and the USSR Governments.

Projects for the expansion of irrigation using water from the High Dam, including the project of basin conversion into the perennial system, were reckoned second in priority, pending the final and definite decision on the High Dam itself. Conversion of the basins was necessary due to the fact that the High Dam leads to decrease of the river discharges and levels downstream to an extent that makes flooding the basins impossible.

When, in 1954, the probability of constructing a dam on the Blue Nile at Roseires appeared on the scene, it was feared that this would result in a fall of the Nile levels to a degree that would not permit the irrigation of certain elevated areas in Kena basins. Thoughts were therefore directed towards bringing such areas under the perennial system through the use of the High Dam stored water in its first stage. It was proposed, as an urgent remedy, to irrigate such converted basins by pumping water from the Nile through a series of pumps to be erected on basin canal offtakes.

When agreement was concluded, late in 1958, between the ARE and the USSR Governments, it was deemed imperative to have the basin conversion projects executed at the same time as the construction of the High Dam itself. Therefore, the Higher Committee for the High Dam as well as both its Sub-committees for the Construction of the High Dam and for Irrigation, Drainage and Groundwater were formed.

Commencement of researches

Soon after the formation of the Committee for Irrigation, Drainage and Groundwater, studies and researches began in earnest for converting the basin irrigation system into the perennial. It was then scheduled to have the construction of the High Dam completed in two stages, so the research was directed towards studying the areas in which it would be difficult to practise basin irrigation after completion of the first stage and the consequent fall of the river discharges and levels. Thereby it was intended to complete the conversion of such areas while construction of the High Dam first stage was in process.

These researches were fully completed within the very short period of less than six months, and they were deeply rooted, well branched and covered all the hydrological and topographical aspects, besides the irrigation methods commonly used in basin areas. They have thrown light upon the whole question and revealed *inter alia* the following facts:

1. The area in Upper Egypt which was flooded yearly by the basin irrigation water extended to about 973,000 feddans. It included:

(a) 602,000 feddan flooded according to the usual system of basin irrigation;
(b) 257,000 feddan flooded during the flood season, but also supplied in summer with water for the propagation of cereal cultivation;
(c) 114,000 feddan flooded during the flood season and irrigated with summer water during the rest of the year.

2. Following the completion of the first stage of the High Dam, basin irrigation would not be practical in about 538,000 feddan, extending on both sides of the Nile between Nag-Hammadi and Assiut Barrages, as well as on both sides of the Ibrahimia Canal between Assiut and Dayrout. Meanwhile, the basin irrigation system would be difficult to practise

in elevated areas, unless two canals were constructed to take water from upstream of Esna Barrage and extending to Nag-Hammadi Barrage.

With such factors to be considered the Authority conducted comprehensive studies based on accurate analyses of all aspects and future prospects, including the advantages and disadvantages of all ideas. Seventeen projects, different from each other with regard to their bases and their technical advantages, were compiled. All the economic factors, including the capital cost, management and maintenance costs, and the revenue of each project, were studied.

The approved project for basin conversion

The seventeen projects were submitted to a Committee formed of prominent specialists in irrigation, whether employed by the Ministry of Irrigation or otherwise. The Committee was of the opinion that the most suitable projects were those in which the following bases applied:

1. Field irrigation should be by pumping, rather than free-flow which leads not only to extravagance in water use, but also to the rise of the groundwater table. However, it was noted that it would be possible to ensure the preservation of the groundwater table at a level suitable for the growing of plants by provision of a complete drainage network. In this connection it should be kept in mind that any drainage system is liable to inefficiency of maintenance and this hampers its functioning.

The main objective of applying the pumping field irrigation system is to reduce the amount of irrigation water. Therefore, it has been decided that the farmers should lift the irrigation water from the canals by their own means up to a limit of 0.5 m. In cases where higher lift is required the Government should provide pumps for the following reasons:

(a) In order not to overload the farmers with the difficult task of pumping water higher than 0.5 m.
(b) Installation of pumps in the upper reaches of canals to alleviate the farmers' task would need major deep canals. This would double the amount of earth to be moved and this, together with the structures, would expand the area to be expropriated and greatly increase the cost.

2. It was desirable to feed the main canals by free-flow from upstream of the major barrages, and not to pump water from the Nile, where lifting may be necessary to about seven metres. Moreover, there would be need, in such cases, for new power stations, their sites being exposed to the river's current, resulting in their exposure to erosion or precipitation.

3. It was desirable also to avoid high degrees of lift upstream of both Esna and Nag-Hammadi Barrages, which would be necessary to feed the main canals, and to avoid lowering to a great extent the upstream water levels of Esna and Nag-Hammadi Barrages, which would be necessary to save undue excavation and construction costs; to use land area on the head-side of main canals for the erection of pumping stations to the maximum in order that the command area of such stations is decreased to a minimum; and to render the distances of conveying water by main canals as short as possible.

The project finally selected from the seventeen alternatives has the following advantages:

1. It facilitates limiting the lift of water from the canals to half a metre.

2. It depends on pumps to feed the main canals from the Nile.

3. The upstream water levels proposed for Esna and Nag-Hammadi Barrages provide for the seepage level to be about one and a half metres below the cultivated land during the period of maximum requirement, i.e. four months yearly. In the rest of the requirement period, which extends to about eight months, the seepage level under the cultivated land would reach an average of about two and a half metres.

4. The scattered government and non-government pumps, already erected on the Nile downstream of the Esna Barrage, would be substituted by one pump to be installed on Assfoun Canal, thereby effecting large savings in running costs.

5. It would remain possible to utilize the water fall at the barrages for the generation of electricity.

6. No strengthening of the barrages on the Nile would be needed.

The project is divided into two stages of execution: First stage includes irrigation projects in areas where the basin system could no longer be applied after the commencement of water storage in the High Dam reservoir in 1964. This area extends between Nag-Hammadi Barrage and Dayrout City. Second stage involves irrigation projects in the remaining basin lands. Moreover, it includes installing drainage in all basin areas.

Why drainage was not installed at the same time as perennial irrigation

The Ministry of Irrigation has worked out a system for installing drainage, but it was not feasible to apply it in the projects for basin conversion, for the following reasons:

1. The area scheduled to be brought under the perennial irrigation is very big, nearly one million feddan. In view of the huge amount of excavation required for both irrigation and drainage and the deficiency of building materials and transport, and the number of engineers, the first stage of work (from 1960 to 1965) was restricted to irrigation only. Drainage works were postponed to the year 1970/1971.

2. At the beginning of basin conversion, it was important to convert half the total area, half a million feddan, by mid 1964, i.e. prior to the date fixed for the diversion of the river channel and the start of storage in the High Dam reservoir.

3. It was believed that the cessation of the flood phenomenon, due to the storage of water by the High Dam, would lead to a fall of water levels in the river from Aswan to Cairo, thus helping drainage from basin lands all the year round.

4. It was believed that the low-lift irrigation policy mentioned above would prevent the excessive use of water, thus reducing the quantity of drainage water.

5. The funds earmarked for basin conversion were not sufficient to meet the cost of irrigation and drainage projects at the same time. This was because of the huge investments which were required during the same period for the High Dam itself, industrialization and services.

The area so far brought under the perennial irrigation

The areas of land converted from basin to perennial irrigation, up till 1 December 1975, were as follows:

Serial governorate no.	Converted (feddan)	Areas still remaining (feddan)
1. Kena	282,000	—
2. Souhag	261,000	34,000*
3. Assiut	266,000	—
4. El-Menia	80,000	6,000†
5. Beni-Suef	1,000	43,000†
Totals	890,000	83,000

* This area will be converted after the expiration of the run-time of El-Khayyam Pumping Station.
† Conversion of these areas is provided for in the 1976/80 plan.

Drainage projects in the converted areas

Despite the foregoing circumstances, the construction of certain works needed for drainage was undertaken in advance as follows:

1. During the construction of major canals, eleven syphons involving big structural works were executed, in order to be ready when the drainage network is installed.

2. For certain reasons connected with the filling and draining of basins, certain major drains were constructed at the time of making the canals.

3. Public drainage projects in sugar-cane growing areas of Kena Governorate were started early in the fiscal year, 1967/68.

Soon after perennial irrigation was started in the former basins, soil deterioration began to occur in certain areas as a result of the farmers not being trained in the new system. This was in addition to difficulties occasioned by incomplete land-levelling in basin areas. Furthermore, compensation was claimed by the farmers for lack of irrigation facilities during the actual period of conversion. The rise of the groundwater table, due to the lack of any public drainage network or field drains, has brought the drainage problem into first priority among the Ministry's current projects.

Public drainage networks are now being installed in an area of 606,200 feddan, of which 417,700 were completed by 31 December 1974, in the following areas:

Governorate	Areas for public drainage networks in the converted basin lands (feddan)	Areas completed by 31 Dec. 1974 (feddan)
Kena	219,600	170,100
Souhag	100,700	97,700
Assiut	89,900	43,900
El-Menia and Beni Suef	196,000	106,000
Total	606,200	417,700

The command area of the new drains includes the lands scheduled to depend on seepage-protection drains. These were constructed to protect low lands against seepage from the elevated reclaimed lands adjacent thereto, in Esna regions, Kena Governorate, west of Tahta, Souhag Governorate and north and south of Samalut in Menia and Beni Suef Governorates. All these public drainage projects will be completed in the converted basin areas during the years of the 1976/80 plan.

Tile-drainage networks are now being constructed in Upper Egypt in an area of 171,000 feddan, out of which 35,000 lie in the converted basin areas. The 1976/80 plan includes the construction in Upper Egypt of tile-drainage projects in an area of 403,000 feddan, of which 229,000 lie in converted basin areas. This plan also includes tile-drainage projects stipulated in the IBRD agreement, and covering an additional area of 300,000 feddan of which 145,000 are in basin areas.

Effects and advantages of basin conversion projects

These can be assessed as follows:

1. The availability of summer irrigation water, at a low cost for those who use public pumping machines and free for those who employ their own private means.
2. The summer cultivation of vast areas, where cultivation used to be practised only once a year following the discharge of flood waters off the land.
3. The ripening of cotton in the field instead of being early harvested before ripening, as was the case before for fear of the crop being flooded. This increases the cotton yield considerably.
4. Improving the type of cotton as a result of its ripening before harvesting.
5. Saving cotton harvesting costs, at a rate of at least one pound per feddan, since there is no longer urgent demand for harvesting labour before the crop was flooded.
6. Considerable increase of the millet crop, as a result of being supplied with full water requirements. In the past, the plants remained unirrigated for long periods, sometimes as much as 50 days.
7. The possibility of early cultivation of clover for cattle feed, thus increasing livestock production.
8. The early cultivation of winter crops without waiting for the draining of flood water.
9. The possibility of harvesting two crops of clover instead of one, before the cultivation of cotton.
10. The possibility of cultivating certain areas three times a year, for example clover followed by summer maize followed by Nili summer millet.
11. The cultivating of gardens in the lands brought under perennial irrigation.
12. Sugar-cane cultivation. The Ministry of Industry's programme has included an expansion of sugar-cane growing areas in the Republic, to cover about 90 thousand feddan in the regions of Kouss, Esna and Dishna, and the factory scheduled to be established in Girga or Balyana.

Social benefits

In addition, basin conversion has led to the following social and urban services:

1. Employment of thousands of workers and technicians during the execution period.

2. The connection of villages and public utility services via the canal banks, which are now used as roads, thus facilitating the transport and marketing of crops.
3. As a result of such roads, public services, such as security, health and education, have increased.

The Authority has, in fact, surpassed all that was required of them in the provision of such services, to the benefit of the people in these areas, who have for long been suffering from poverty and underdevelopment. For example, the people who formerly had to build their houses above the flood level mostly did so on canal banks. The conversion projects necessitated the removal of such dwellings which, being on public property, did not qualify for compensation. However, with sympathy for those who became homeless and in accordance with socialist policy, the Ministry obtained from the Higher Committee of the High Dam a decision to aid the people by offering them money and land.

The advantages are not only restricted to the materialistic side, but they extend also to moral aspects: The Ministry of Irrigation entrusted the Ministry of Works with the building of new mosques to replace the former small ones built of hand-made bricks, and this facilitates public worship and maintains praying premises in healthy conditions.

Inside Souhag and other cities the Ministry also filled with earth the remaining parts of basin channels in order to beautify such cities and preserve public health. In cities where the removal of housing premises was imperative, the Ministry offered the land necessary for building new houses and paid compensation to the owners. The Ministry also contributed to new housing premises built to replace those in the expropriated areas, particularly in Souhag Governorate.

Conclusion

In spite of the above economic and social advantages expected from basin conversion projects, there are still some obstacles which oppose the benefits. In particular, special attention is being paid to restricting extravagance in the use of water and the absolute prohibition of free-flow irrigation. Meanwhile, public and tile-drainage projects are going forward and the funding of these will be given top priority.

DISCUSSION AND CONCLUSIONS

Compiled by E. B. WORTHINGTON

The case studies in Tunisia and Syria (p. 75) bring out the essential point that close co-operation is necessary between all technical services involved in irrigation, but specially between health services and water management. This must be in the early planning stage as well as right through the operations. The examples illustrate both the good results which can be achieved from multidisciplinary planning and collaboration between hydrologists, agriculturalists and health workers, and the bad results when such collaboration is not adequately achieved. Health hazards are created or aggravated by the increase in density of human settlements under the influence of irrigation so that problems of pure water, disposal of excreta and waste, arise of a kind which are quite unfamiliar to the natural human inhabitants of arid lands. Schistosomiasis and malaria, for example, absent from deserts, become major diseases. The interests of irrigated agriculture and of health are apt to be opposed to each other in the narrow view, but they can often be made complementary.

Mexico (p. 81) is facing a big challenge because of the very large extent of its arid lands, its lack of water resources, and a population increasing at the phenomenal rate of 3.6% yearly. To meet this challenge a major programme of irrigation techniques and of farm improvement plans has been launched with the aim of increasing production for the benefit of the water users and of the country's economy. Irrigation districts which vary greatly in size are moulded to the local orography, climatology and hydrology in a country-side where rivers in general have small catchment areas and torrential flows. There has been an impressive increase in water measurement points with the result that the efficiency of conveyance has increased significantly in the course of five years. Related to this study is another paper from the same country on the re-utilization of residual water (p. 361). The valley of Mexico, which a few centuries ago contained a very large lake of 150,000 ha, had too much water and this caused devastating floods; but successive regimes, starting in pre-hispanic times, reduced the standing water to a few relict lakes, and thereafter there has been a general water shortage. To meet this, large quantities of ground-water were extracted with consequent subsidence of the land, including even a major portion of down-town Mexico City. The effect of this succession of drastic man-made changes is analysed in terms of quantity and quality of water, its use and re-use for irrigation, recreation and other purposes.

The study of the Dez Irrigation Project in Iran (p. 85) goes to fundamentals in posing the question whether the main emphasis should be put on economic or social considerations when developing irrigated agriculture. The experience there suggests that of the various forms of agricultural development two are suited to include both aspects, namely Government-owned and not strictly profit-oriented agro-industries, and the strengthening of traditional agricultural practices with technical improvements. Many expressions of view

107

in developing countries in recent years have put emphasis in development on services to rural rather than urban areas, and endorse the advantages which are attributed in this study to the successful development of traditional agriculture.

The more specific case of conversion of basin to perennial irrigation consequent on the High Dam of Egypt likewise raises important social as well as technical problems. This subject promoted vigorous discussion at the Symposium between leading irrigation specialists in Egypt. The remainder of the discussion ranged widely but there was a recurring theme that the crux of most problems in irrigation comes down to the individual farmer, his attitudes and activities. In the mass of scientific and technical questions which have to be answered it seems that he is sometimes forgotten. The discussion is summarized below, followed by the names of those who contributed to each subject.

Environmental risks of irrigation

It is clear that the development of water resources for irrigation in developing countries will greatly increase, and with it the environmental risks such as diseases. It now seems possible to control malaria, but not yet schistosomiasis which clearly needs a trans-disciplinary approach. In the past there has been a tendency for each authority to concentrate on its particular narrow area of knowledge, and to ignore others. One specialist agency, helped by others, might initiate a truly trans-disciplinary approach in a new irrigation scheme in an arid area, the exercise to include a methodical pre-planned scheme to integrate water and agricultural development, together with the prevention of diseases and all other undesirable environmental consequences. (R. S. Odingo.)

Measures to prevent schistosomiasis are much easier in those irrigation schemes which depend on wells and underground water than on surface water conveyed by canals, but in the latter case it is certainly possible to reduce considerably the clinical gravity of these diseases by application of knowledge from several disciplines. (A. F. El Kashef & J. A. Coumbaras.)

Sanitation and education

Under pressure from population irrigation is expanding faster than sanitation, and unfortunately women continue to wash clothes in irrigation water with polluting detergents, and children to bathe in canals in spite of provision in some places of swimming pools. To remedy such situations the technician is invited to collaborate with the health worker and some examples of success in this regard have been noted. However, valves and taps are often damaged and all forms of sanitary installation are poorly maintained. Sanitary education of the people is indispensible for success, so it is essential for those responsible for education to collaborate with technicians and health workers. (I. M. Serghini, J. A. Coumbaras.)

Traditional and new systems of irrigation

The experience from Iran concerning mechanization of irrigated agricultural development, while showing the close relationship between social and technical aspects, should not scare developing countries. Rather, it indicates that technology should always be accompanied by social planning. Often, however, it is possible to propagate not a break with traditions, but a careful improvement of traditional methods by introducing modern technology. (A. M. Balba, E. Ehlers.)

Change from basin to perennial irrigation

In order to increase the efficiency of the application of water at the farm level H.E. The Minister of Irrigation in Egypt emphasizes one of the design criteria, namely to have the water level in canals 30–50 cm below field level so that the farmer must pump the water he uses. Is not this a negative attitude, adding a further problem to the many already faced by the farmer? Lifting water by animal power is economic neither to the farmer nor the country, and the many animals maintained for this purpose compete with man for food produced from limited resources of land and water. A more positive approach to prevent farmers using unnecessary excess water might be to supply water by gravity but with better operating and maintenance control. The water could be sold to individual farmers at a modest rate up to the optimum cubic metres per hectare, the price being greatly increased for any excess water taken. The Ministry of Irrigation, however, after full discussion of no less than 17 methods for changing the system, some by gravity others by lift, excluded all the gravity methods on grounds of risk of rise in water table and soil deterioration, and the health of farmers and their families.

The change from basin to perennial irrigation in Egypt is enforced by the change in regime of river level consequent on the High Dam. The introduction of modern techniques to these areas is costly, but the result is to save water for the expansion of irrigation to new areas. Delay in the installation of drains—owing to insufficient budgetary and other resources, coupled with the belief that the river itself at its new lowered level would act as a drain—has accelerated the deterioration of the areas now converted. The misuse of irrigation water, which is as bad in Upper Egypt as in Lower Egypt and in the desert oases, has accentuated the troubles. A lesson has been learned from this experience, as also from the Western Desert project where the water table has risen from 50 metres below land surface to between 1 and 1.5 metres after 2 to 5 years addition of irrigation water: whenever surface irrigation is used, field drainage is essential whatever may be the initial depth of water table. (M. M. El Gabaly, A. Arar, A. A. Kamal, A. F. El Kashef, W. N. Sefaine, M. A. Abu Zeid.)

The application of knowledge

Confucius once said, "The height of knowledge is having it, to apply it." Many countries have extensive knowledge in the natural sciences, but it is not applied rationally on a global scale, often owing to lack of proper appreciation of global problems. In Europe we consume food and fibre from irrigated agriculture in arid zones far from our national boundaries, but few of us are aware of or feel any responsibility for the problems involved in production. This meeting should develop strategies to mobilize and apply knowledge resources from the developed world to the developing world in a more rational way. (E. Danfors.)

Importance of the farmer

The crux of the irrigation problem is practising it efficiently. It does not help to blame the farmers who are doing their best to produce food from land and water which are not particularly suitable. Being well informed in our own fields, we must get down to earth, take some time from writing scientific papers, get to the farmers and show them what should be done. An irrigation extension service designed to improve water use could go a long way to increase yields in Upper Egypt, for example, by 10% to 50% without extra expense. There is hope from the FAO Regional Project on Land and Water Use in the Near East which is planning a series of demonstration studies on farmers' field in cooperation with Ministries of Agriculture and Irrigation and with Universities. (L. Obeng, N. G. Dastane, W. N. Sefaine, F. I. Massoud.)

SECTION III

Influence of Irrigation on Hydrological Processes: Quantity and Quality

CHANGES IN HYDROLOGICAL PROCESSES

by G. KOVÁCS

11 *Fo-utca* 48–50, *Budapest, Hungary*

The main objective of any human activity since the beginning of the long history of tool-making was to develop and control the environment. This objective will never change, because it is a basic demand of mankind to ensure better and more comfortable living conditions. In the past, when only a small part of the world was populated and the demands were considerably lower than now, the changes introduced caused only slight modifications to nature, change of land use and the building of habitations. The ever-growing population gradually increased the areas influenced, and isolated units coalesced to form areas where human activity dominated nature.

The relatively slow increase of man's influence was accelerated by rapid population expansion and development of the technical civilization, which started in the last century and reached a state in recent decades when its quantitative increase turned into qualitative change. Two basic causes initiated the problems arising from this new feature of development: the demand to have a truly comfortable way of living (the consumptive society) and the power and technical knowledge to execute large-scale modifications of natural conditions.

Although any direct human influence has a positive objective—to raise the living standard of a given group of human beings, the actions and interactions may have undesirable side effects which cause unexpected changes in the environment. They may initiate irreversible processes causing deterioration of the desirable balances of nature and it is for this reason that protection of the natural environment became one of the most important and urgent tasks in our time. However, the phrase *protection of the environment* is not absolutely correct. The reasonable development of the environment is often more important than its protection. Activities aiming to achieve better living conditions cannot be stopped but they can be better planned. It is necessary to survey all possible side effects, their socio-economic influences, and to determine the supplementary measures required for minimizing undesirable changes. This type of planning can be executed only after detailed investigation of the whole natural system influenced by the action in question, and such interdisciplinary study needs the close cooperation of scientists working in the different fields of natural and engineering sciences.

Irrigation in arid lands demonstrates the reasonable development and control of natural conditions as the objective, for arid lands provide mankind with the largest unexploited

113

reserve for agricultural production. This potential resource cannot be utilized, however, without water. Ancient systems in the valleys of the Nile, Indus, Tigris and Euphrates proved that relatively large populations can be supplied with sufficient food by irrigation. Some of them also demonstrated, however, that the side effects such as waterlogging or salt accumulation could deteriorate the soil. Large cultivated areas became unsuitable for any agricultural production in the Indus valley and in Mesopotamia, thereby creating worse living conditions for mankind than the natural ones.

Waterlogging and salt accumulation are only examples, for undesirable biological changes may also occur. The effectiveness of irrigation can be ensured only if the harmful effects are prevented by special design and operation of the schemes, or by supplementing them with suitable protective structures. The necessary measures, however, depend on adequate knowledge of the natural conditions and their probable modification by irrigation, and this knowledge can be achieved only by the detailed exploration of the physical, chemical and biological interrelations. The purpose of this Section is to investigate one part of this very large scientific field.

Hydrological changes due to irrigation

The soil-moisture zone between the surface and the water table is central to the terrestrial part of the hydrological cycle. Its size (depth of water table), the behaviour of the materials composing it (basic layer, soil and vegetation), the zone structure (cultivation, depth of root zone) and its condition at any time (moisture content, chemical character), are the principal parameters. The part of precipitation which is intercepted by the vegetation or stored in the soil-moisture zone is transpired or evaporated and thus returns directly to the atmosphere. The amount of water infiltrating through the soil-moisture zone and reaching the water table also depends on the above-mentioned parameters. In dry periods an upward seepage from the groundwater replenishes the water evaporated or transpired from the upper layers. This flow is similarly governed by the characteristics of the aeration zone (capillarity, unsaturated hydraulic conductivity and root suction). Thus, apart from the climatic conditions, the behaviour of this relatively thin layer is the most important factor determining both the positive and negative accretion of the groundwater. Surface runoff is composed of two components: the direct runoff originating from precipitation, and the base flow supplied from subsurface reservoirs. It is obvious that the runoff comprises the surface water from precipitation minus the amount evaporated and the infiltrated amount. The flow rate discharged by rivers from ground water reservoirs depends on their recharge. It can be stated, therefore, that surface runoff is also a function of the hydrological processes occurring in the soil-moisture zone.

The application of irrigation completely modifies the conditions prevailing above the water table. The direct purpose is to maintain the moisture content of the root zone above a given level, and this condition influences each of the three ways in which the precipitation falling on the land may continue its journey: returning to the atmosphere, runoff from the surface, or recharging the groundwater.

One of the consequences of irrigation is that water may always be easily available for evaporation or transpiration and, therefore, the actual evapotranspiration will be higher from irrigated areas than elsewhere. The increase in evapotranspiration raises the moisture content of the atmosphere and this may increase both the total amount and frequency of precipitation.

The higher soil moisture decreases the gradient of the downward flow near the surface and thus infiltration as well. Since a balance has to be maintained, this process causes an increase in surface runoff, because irrigation hardly changes the interception (and thus the direct evaporation from the surface of the plants). The result of the larger runoff may be stronger erosion and a higher suspended load in the rivers. Although the total precipitation is almost negligible in arid zones and its increase does not raise any problem, in semi-arid regions the danger of erosion and the deterioration of water quality as a consequence of larger runoff has to be considered. This topic is mentioned here as one of the side effects in order to give a complete picture of the complex interrelation between irrigation and the hydrological cycle.

Since the purpose of irrigation is to control the moisture content of the soil-moisture zone, it is obvious that this is the strongest influence on the groundwater regime. Although infiltration through the surface is smaller due to the decreased tension difference, the positive accretion is greater because the storage capacity is filled by irrigation water and, therefore, almost the total amount of the infiltrating precipitation reaches the water table. Only a few irrigation methods, such as trickle irrigation, can distribute water in the form of a continuous supply with a relatively small quantity per unit time, as it is required by the plants. Over-irrigation is therefore almost unavoidable, and its effect is to increase positive accretion. The distribution system is generally composed of earth canals which incur seepage losses between the intake works and the irrigated fields, and so the recharge of groundwater is further increased.

The other consequence of the continuous high moisture content near the surface is the decrease of negative accretion. The tension on the surface of the thicker water film is small and, therefore, the part of the total (Buckingham) potential directed upwards is also decreased in the whole profile. In most cases it becomes smaller than the gravitational potential; thus the vertical upward flow drains the gravitational groundwater only if the water table is near the surface.

The excess recharge and the considerable decrease in vertical drainage disturb the natural balance of the groundwater, so irrigation results in a continuous rise of the water table within the area of the system. Dynamic equilibrium will develop only if the water table is high enough to create new drainage paths and to maintain discharging processes to balance the sum of the excess positive accretion. Two main types of draining effects can be distinguished: high evapotranspiration from the raised level of the groundwater, and horizontal flow directed towards the neighbouring non-irrigated areas where a gradient exists between the raised and the natural water tables. The secondary effect of this horizontal groundwater flow is a rise of water table in the vicinity of the irrigation schemes, because excess horizontal recharge received by the non-irrigated areas in this way has to be balanced by higher evaporation or by an increase of horizontal flow through these areas towards nearby surface-water bodies. Both the increase in evaporation and of the horizontal flow require a local rise in the water table. The change of the position of the water table under the irrigated land and in its surroundings can be prevented by constructing artificial drainage systems to collect the amount of water originating from the excess recharge.

It is well known that a close interrelation exists between the hydrological process occurring in the soil-moisture zone and the movement, leaching and accumulation of salts dissolved in the water or bound to the soil particles. The changes initiated within the subsurface branch of the hydrological cycle disturb the natural salt balance and modify the factors that previously governed the migration of salts. The qualitative parameters are important characteristics of the hydrological cycle which cannot be separated from the quantitative data aiming at the complete description of the hydrological processes. It is necessary, therefore, to investigate the influence of irrigation on the dynamics of salt in more detail.

There are also other changes in the water quality caused by irrigation. In numerous cases water is accumulated and stored in reservoirs where storage generally reduces the sediment load and turbidity, and slightly increases the temperature. In shallow reservoirs eutrophication, and in deep ones (where stratification can develop) anaerobic processes below the thermocline may spoil the quality of water. The release of cold water from the lower strata of large reservoirs may reduce crop yields when used for irrigation. It has been suggested that the removal of plant nutrients from the water, together with the settled suspended load, is also a harmful effect of storage. The evaporation of the free water surface is high, especially in the tropical zone. When using reservoirs to increase water resources for irrigation in arid areas it is necessary to consider not only the water losses caused by the high evaporation, but the increase of salt concentration as well.

A similar qualitative change as a result of evaporation occurs within the distribution system. Moreover canals sometimes collect surface run-off which may carry pesticides or herbicides from the nearby arable lands, or they drain the groundwater which may contain a relatively high amount of dissolved salts. The biological production of aquatic weeds and algae, and especially the chemicals used to control them in canals, also influence the quality of distributed water. The qualitative changes of water within the irrigated soil layer have already been mentioned and will be discussed later in detail.

Following the path of the irrigation water from intake to recipient, the final qualitative changes that have to be mentioned are caused by the rivers taking back drainage water which generally has a high salt content. In arid zones, where the base flow of the rivers is generally small and the discharge from irrigated lands is relatively large, deterioration of the river water may hinder its re-use.

The expected influences of the numerous qualitative changes occurring as effects of irrigation are only summarized here. They are considered in detail in the paper by Hotes and Pearson (p. 127), which explains the processes and gives numerical data testifying to the importance of the environmental changes caused by the deterioration of water quality and initiated by irrigation. Excluding this topic from further discussion, and accepting that the qualitative and quantitative modifications of surface runoff are generally negligible in the arid zones, three major environmental changes caused by irrigation have to be investigated, namely modification of the atmospheric branch of the hydrological cycle; the water regime of the soil-moisture zone and groundwater; the migration of salts.

Influence of irrigation on the atmospheric branch of the hydrological cycle

One of the consequences of a permanent high moisture content of the topmost soil layer is increase of evapotranspiration. The upper potential limit of evapotranspiration is determined by the climatic and energy conditions of the air mass above, which can hold only a limited amount of vapour. There is, however, a special case when the actual evapotranspiration seems to be considerably higher than the potential as calculated from the energy balance. This occurs in relatively small irrigated areas surrounded by deserts, where incoming dry air has extremely high vapour receiving capacity. This phenomenon is called the oasis effect; it involves the energy balance of the neighbouring areas as well as the irrigated area itself.

Various terms assist in the numerical determination of the changes caused by irrigation: *evaporation* occurs on the surface of a free water body or of water films covering soil grains or inert objects; *transpiration* develops through plants; and *evapotranspiration* is the two processes combined. The transformation of water from liquid into vapour phase is governed by three main factors, namely: the energy providing the heat necessary to initiate and maintain evaporation; the mass of water which is turned from liquid into vapour; the surface where the change of phase takes place.

The entire system of evapotranspiration is composed of two subsystems: the first is the air mass through which the energy reaches the water and which receives and transports the vapour; the second is characterized by the amount of stored water and by the structure of the reservoir containing the evaporating mass (root zone and root structure, porosity, adhesion and capillarity, depth of water body, etc.). This structure determines the availability of water and the intensity of the recharging process conveying the water to the surface. The two subsystems are divided by the evaporating surface, which provides the boundary conditions.

The vapour receiving capacity of the overlying air mass is a function of the instantaneous humidity of the air, its temperature, wind velocity, radiation and other meteorological factors. It is influenced, however, by the evaporating surface as well (the specific area of surface, reflection of radiation, micro-climate at the surface, etc.). The amount of water evaporated, determined by these factors, and considering the lower boundary conditions generally calculated from the energy balance, does not represent the actual amount of evaporating water, but it gives the possible upper limit under the investigated conditions. It is called, therefore, potential evapotranspiration.

The quantity of water turned into vapour during a unit period of time is equal to the potential value only if water is always available within the vapour-producing system to the extent that recharge of water on the surface does not hinder the evaporation process. This condition occurs on free water bodies and on a completely saturated layer of soil with a high recharging capacity (shallow water table, high hydraulic conductivity). Effective or actual evapotranspiration is the flux really developing between the vapour producing and receiving subsystems. Thus Péczely's paper (p. 159) compares potential and actual evapotranspiration in semi-arid areas. The difference between the two represents the upper limit of excess evapotranspiration. For more precise numerical determination of this influence it would be necessary to develop models and methods suitable for calculating both potential and actual evapotranspiration considering the conditions prevailing in the vapour producing system.

There are many methods proposed to determine potential evapotranspiration. Some are based on the energy balance, others on climatic parameters, while in a third group the two are combined (Thornthwaite and Mather 1957, Penman 1948, 1956, Turc 1961, Christiansen 1966, Harrold and Dreibelbis 1967, Mustonen and McGuinness 1968, Antal 1968, McGuinness and Bordne 1972). Comparison of the various equations with measured data shows that no one of the methods is superior in accuracy to the others, for their applicability depends on local conditions and available data. An obvious shortcoming of the equations is that the lower boundary condition is not yet solved. There are methods valid where the condition of the evaporating surface is unambiguously fixed (e.g. Penman's values characterize a surface covered by green grass 20 cm high), but the factors necessary to calculate evapotranspiration under different conditions are not yet determined. Although it is the general opinion that variation due to surface conditions, such as type of vegetation, is no more than about 10% on average and 20% in extreme cases, further research is needed.

Knowledge concerning actual evapotranspiration is even more limited, for the flux between the two subsystems is not directly measurable. For large areas the combination of the atmospheric and the continental water balance equations provides a reliable result if data for atmospheric vapour flux are available (Palmen 1967, Peixoto 1973, Magyar 1973). Another approximation comes from using lysimeters, although this method measures only the evaporation from a very limited surface. It is necessary to investigate the representativeness of the soil column very carefully and to consider the special conditions created by the lysimeter. Preliminary results from a project initiated by IAHS concerning the evaluation of lysimeter data have been summarized (Kovács 1975).

For measurement of actual evapotranspiration in a natural environment the use of heat balance equations is proposed (conference on the evaluation of forested representative basins organized by the Technical University of Dresden in 1975), but the reliability of this method is not yet proved. Another possible way is to measure the main components separately. For measuring interception a double system of rain gauges, one above and one below the plant canopy, is generally used in representative basins (Blake 1972). It would be useful to collect a large number of measured data and attempt to establish empirical formulae for calculating interception, depending on the type and growth-phase of the plants as well as on meteorological data. The other important component is the amount of vapour originating from soil moisture, which can be determined by a combined system of sensors measuring separately both the moisture content and the tension of water at different depth of the soil profile (G. Vachaud, verbal communication). Knowing the vertical tension distribution, the gradient of the Buckingham potential can be determined. The depth where it is equal to zero divides the profile into two: in the upper part the gradient is directed upwards and the decrease of moisture content is a part of actual evapotranspiration; in the lower part the soil-moisture percolates into the groundwater space. If the sign of the gradient does not change in the profile, as in dry periods when the gradient of the whole is generally upwards, the total change of moisture content is only a part of evapotranspiration and the amount originating from the gravitational groundwater space has also to be considered. The direct measurement of moisture content distribution serves to determine change in the amount of water and within the relevant sections of the profile. Only in a few cases, for example, if there is neither surface nor subsurface runoff, can some component of evapotranspiration be determined more easily (Major 1975).

The uncertainty of measurements and the complexity of the processes are the reasons why reliable methods have not yet been developed to calculate evaporation from a unit area under given conditions or to determine the ratio of actual and potential values. Although Mustonen and McGuinness (1968) have summarized some methods giving this ratio as a function of the moisture content of the topmost soil layer, the ET/ET_0 quotient depends on a number of other biological and physical parameters, including type and development of vegetation; depth, structure and suction of roots; thickness, retention capacity and hydraulic conductivity of the soil-moisture zone; and the instantaneous condition of the system. Thus the improvement of measuring methods and the development of suitable models are the most important topics for further research concerning actual evapotranspiration.

Knowledge of the increase of evapotranspiration due to irrigation is important for determining the amount of water required and so for designing the distribution system. It has no direct influence on environmental control, but it increases the humidity of the atmosphere and may even change the pattern of precipitation. Discussions in the past have been wholly inconclusive on this point. Schickedanz and Ackermann's paper on page 185 is perhaps the first in which it is proved in an exact mathematical way that there was no very significant interrelation between the construction of large irrigation systems in a semi-arid environment and the development of the pattern of precipitation in the surroundings, although some slight influence is unquestionable. Although only preliminary investigations are yet evaluated the results of this investigation relieve the scientist and planner of the fear that irrigation could initiate large-scale modification of the climate, apart from the changes in microclimate near the surface of the irrigated land. It is advisable, however, to continue investigation and to include some other areas. It would also be useful to determine the increase of actual evaporation due to irrigation by applying the method based on atmospheric vapour balance equations. The measured data of vapour fluxes could be utilized to determine the ratio between the natural vapour transport and the excess amount introduced by human activity.

Interrelations between the hydrological processes in the soil-moisture zone: migration of salts and depth of the water table

Although the first publications dealing with the influence of the depth of water table on the water balance of the soil-moisture zone were almost 20 years ago (Kovács 1959a; 1959b) and the relationship between the hydrological processes and salt accumulation was also explained more than 10 years ago (Kovács 1960; Várallyay 1966–67), the wide application of this concept is quite recent. The second characteristic curve of the groundwater balance constructed from actual observations was by Lavrov *et al.* (1972), and the three papers in this book dealing with the hydrological changes in the soil-moisture zone follow this basic idea. Bogomolov (p. 125) and his co-workers have explained the processes governing the development of the water balance under natural and irrigated conditions in a general form. Péczely (page 159) has summarized the previously published theoretical results and their application to the influence of irrigation. The study describing the hydrological changes due to irrigation in New South Wales presented by Pels and Stannard (p. 171) shows the practical application of the same concept and testifies to its reliability.

The basis of the theory is the fact that positive and negative accretion of the groundwater is a uniform function of the average depth of the water table. Naturally the structure and the numerical parameters of the equations describing this relationship depend also on the climate (especially precipitation and radiation), the surface conditions (slope, vegetation) and the character of soil (hydraulic conductivity, capillarity) of the area.

The average amount of yearly infiltration is determined by meteorological, surface and soil parameters. One part of this amount is stored in the soil-moisture zone and evaporated directly from there. The remaining part reaches the water table as positive accretion. It is obvious that the accretion decreases with increasing depth of the average water table, because the probable volume of storage is larger in thicker layers.

The function describing the negative accretion versus depth relationship is similar. If the water table is near the surface the actual and potential evapotranspiration are equal and the amount originating below the surface is provided completely from the groundwater. It is well known that the difference between the potential and actual values of evapotranspiration increases with increasing depth of the water table, and reaches a maximum value at a given depth below which the groundwater has no further effect. Above that depth the evaporating amount can be divided into two parts: one causes the prolonged decrease of soil moisture, while the other (although it originates directly from soil moisture) is replenished from the groundwater. This second part is the negative accretion of the groundwater. Its value equals the total amount of water evaporated or transpired through the soil surface if the water table is near the surface, while it is zero if the phreatic surface is at or below the level where evapotranspiration becomes constant. Between these two positions the negative accretion can be characterized by a uniformly decreasing curve.

In any particular part of the groundwater space the water balance can be expressed by equalling the difference of inputs and output to the stored amount of water, so that a long period in which the natural and artificial influences are not modified the change in storage can be neglected. The time-dependent values of both inputs and outputs can be approximated by their averages, if the purpose is to characterize a long period and not the detailed seasonal fluctuation. Accepting this, a simplified form of the balance equation states that the difference of the average positive and negative accretions is equal to the difference of the average discharging and recharging groundwater flows. Knowing that both types of accretion are single-value functions of the depth of the phreatic surface at a given place, their difference can also be represented as a function of the position of the water table. Considering the balance equation this relationship determines completely the general balanced conditions in the groundwater space; it is called, therefore, the characteristic curve of the groundwater balance.

At the equilibrium level, where the curve intersects the vertical axis, the positive and negative accretions are equal. A water table can develop at this level only if there is no groundwater flow, or the inflow is equal to the outflow. Above this level is the zone of groundwater recharged by percolation, where negative accretion is greater than the positive, and the difference is balanced by excess inflow compared to outflow. Conversely, when the space concerned is drained by the groundwater flow, the deficit has to be balanced by excess recharge originating from infiltration and, therefore, the water table can only develop below the equilibrium level.

If the groundwater had no horizontal movement its table would develop everywhere at the equilibrium level, approximately parallel to the land surface if other conditions are uniform. The terrain, however, is often not horizontal, so the groundwater would have a gradient if its table were parallel to the surface. The flow initiated by the gradient would tend towards a horizontal phreatic surface, raising the level in the valleys and lowering it under the hills. In nature, therefore, water tables are neither parallel to the surface nor horizontal, but rather follow the slope of the terrain, sinking below the equilibrium level in higher areas and rising above it where the terrain is lower. The horizontal flow along the water table, maintained by the gradient, has to be balanced by the positive and negative accretions. The water table develops so that recharge within the relatively higher terrain should be equal to the discharge composed of negative accretion and the draining effect of river beds.

According to this explanation the water table would be horizontal at the equilibrium level everywhere within large flat areas, but horizontal groundwater flow can be initiated and maintained in such terrains by the different materials and structures of the covering layer. In large plains having no considerable slope natural runoff can hardly develop, so most of the precipitation evaporates either from the surface or, after infiltrating, from the soil-moisture zone. There are, however, local depressions, in most cases relics of old river beds filled up with sediments, which collect the surface-water from neighbouring areas.

Not only water but also very fine particles and salts washed out from elsewhere accumulate in such depressions creating temporary ponds and marshes, as well as a more compact impervious covering layer. Land reclamation and levelling may eliminate the unevenness of the surface but not the structural differences of such a layer, so there will remain patches within arable land where water cannot infiltrate but where, during dry periods, evaporation is as high as that in the surrounding land. The water deficit in such patches can be balanced by the horizontal flow from the neighbouring fields. Following the concept of the characteristic curve it can be stated that the equilibrium level is at a very great depth or that there is no such level under the impermeable patches because of the small infiltration and high capillary suction. The phreatic surface is lowered below the normal equilibrium level around the patches so that positive accretion ensures here excess infiltration which is transported below the patches by horizontal groundwater flow to cover the deficit. Thus differences of soil material and structure may create the same local hydrological process under a plain as that maintained by the ground-gradient under sloping terrain.

These hydrological processes are closely linked to the migration of salts in the soil. Where the water table is below the equilibrium level, the resultant of the vertical movement of water in the overlying layers is directed downwards. Because both the infiltrating precipitation and the evaporating water are low in salt content the leaching of the upper layers is the characteristic process influencing the soils here. Among the free cations of the soluble salts sodium is leached first, so the layer becomes calcium dominated. If the process is continued, calcium also descends to the groundwater and hydrogen enters the clay minerals and aggregates as a free cation. In this way at first calcareous, and later acidic, leached soils develop above the groundwater bodies having higher positive than negative accretion.

The cations or salts leached downwards reach the gravitational groundwater space and are carried along by the groundwater flow to lower areas. On the plains the horizontal groundwater flow is directed towards the patches having a more impervious covering layer. On reaching the areas where negative accretion is the dominating factor of the water exchange, the groundwater together with the free cations (the largest amount of which is sodium) is raised near the surface by capillarity to replenish the evaporated soil moisture. This water will then also evaporate leaving behind the dissolved salts.

It follows from this explanation that in those areas, where the groundwater is drained by evaporation (called dry drainage by Pels and Stannard), the accumulation of salts begins as the result of the hydrological processes. Since sodium is the most easily moveable cation alkaline soils develop in the accumulation zones. These zones are generally situated at the foot of hilly areas if the gradient of the water table is the factor governing groundwater flow, because the phreatic surface is there in the highest position relative to the ground surface. In the plains the areas covered by more permeable layers are the main draining points, so the highest salt accumulation is observed there.

Both infiltration and evaporation have seasonal fluctuations, so the salt migration also fluctuates. Within leached layers evaporation may become temporarily dominant, turning the movement of the salts upwards, and *vice versa* in accumulation zones leaching may occur temporarily. The strongest indicator of these salt movements is the change of sodium content in the various layers, since the most soluble cation tends always to be at the front of the migration. The lack of explained salt movements is a special feature of the very impervious drainage patches on large plains; here neither the change in the distribution of the total salt content, nor of the relative positions of the exchangeable cations, can be observed because infiltration is negligible and evaporation is the dominant process.

Irrigation influences the groundwater balance in two basically different ways:

(1) Seepage from the distribution systems recharges the groundwater without modifying the hydrological processes in the soil-moisture zone. It creates a horizontal flow and raises the water table ensuring that the developing dry drainage should be able to balance the recharged amount of water. The influenced zone along the canals and other recharging structures extends until a limit within which the dynamic equilibrium can develop, considering also the relationship between the rise of the phreatic surface and the gradient required to maintain the horizontal flow. Investigating this new equilibrium the characteristic curve of the groundwater balance determined from the observations under natural conditions can be applied, since the processes acting in and through the soil-moisture zone are not modified by the recharge.

(2) Under the irrigated areas the excess water reaches the groundwater through the soil-moisture zone, so the original processes governing water movement are changed considerably. A new characteristic curve has to be constructed taking into account the excess infiltration through the surface, the decrease of the storage capacity of the soil, and that of negative accretion. The result of these changes is a higher position of the equilibrium level; thus leaching will remain a characteristic phenomenon governing the development of the soils even in the case of a relatively high water table.

It is necessary, however, also to ensure the water balance within the irrigated area. Although the phreatic surface is raised, the accretion is still positive under the new conditions, but the new water table, being in a higher position compared to the original phreatic surface, creates a gradient which maintains horizontal flow towards the neighbouring non-irrigated areas. This groundwater flow balances the excess infiltration within the irrigated area, but raises the phreatic surface and increases both the evaporation and salt accumulation under the neighbouring unirrigated fields, if artificial drainage systems are not constructed to discharge the excess water. Pels and Stannard have pointed out that in New South Wales about 25% of the irrigated surface has to be sacrificed as dry drainage, where the high-water table and the increased accumulated salts exclude agricultural utilization of the land.

Considering that there are only few parts of the world where free areas are available without limitation and, therefore, arable land has to be protected, it must be concluded that irrigation schemes have always to be supplemented by drainage systems, especially in the arid zones. The drains collect the horizontally moving groundwater with high salt content and in this way protect the unirrigated fields against salt accumulation and the development of a high-water table. The operation of drains raises new problems: for instance the total amount of water required for irrigation is higher because of water drained from the soil to achieve the necessary leaching effect; and the high salt content of the effluent water deteriorates the quality downstream. Hotes and Pearson as well as Pels and Stannard mention that in serious cases a separate collecting system has to be constructed for drainage water, and before taking the water back into the river or reusing it the salts have to be removed, which further raises the cost of the operation. There is a general attempt, therefore, to decrease the amount of the effluent water, even if complete leaching cannot be achieved: in some circumstances the salts can be washed down just below the root zone, while they remain within the soil profile (van Schilfgaarde *et al.* 1974), though it is questionable how far such partial removal of salts can give protection to the plants.

References
(In English unless stated otherwise)

Antal, E. 1968. Irrigation Forecast on the Basis of Meteorological Data (Manuscript, thesis in Hungarian). Budapest.

Blake, G. J. 1972. Interception and Phytomorphology, Report on Projects HO/HY/6 and HO/HY/7. *Hydrological Research, Progress Report*. No. 9.

Christiansen, J. E. 1966. Estimating Pan Evaporation and Evapotranspiration from Climatic Data. *Methods for Estimating Evaporation, Irrigation and Drainage Conference, Las Vegas.*

Harrold, L. L. and Dreibelbis, F. R. 1967. Evaluation of Agricultural Hydrology by Monalith Lysimeters. *Technical Bulletin.* No. 1367. Ohio.

Kovács, G. 1959a. Determination of the Discharge of Ground-water Flow on the Basis of the Investigation of Ground-water Balance (in Hungarian). *Vizügyi Közlemények.*

Kovács, G. 1959b. Ground-water Household. *Annual Bulletin of ICID.*

Kovács, G. 1960. Relationship between Salt Accumulation and Ground-water Balance (in Hungarian). *Hidrológiai Közlöny.*

Kovács, G. 1972. Ground-water Problems Related to Agricultural Water Use. *International Post-graduate Course on Hydrological Method for Developing Water Resources Management, VITUKI.* Budapest.

Kovács, G. 1975. Application of Lysimeters to Determine Hydrological Parameters in Experimental Catchments. *IAHS Symposium on the Hydrological Characteristics of River Basins.* Tokyo.

Lavrov, A. P., Fadeyeva, M. V., Sachok, G. I., Kachovsky, A. P. and Vasilyev, V. F. 1972. Problems of Underground Water Regimes in Byelorussian Polassic. *UNESCO-IAHS Symposium on Hydrology of Marsh-ridden Areas.* Minsk.

McGuinness, J. L. and Bordne, E. F. 1972. A Comparison of Lysimeter-derived Potential Evaporation with Computed Values. *Technical Bulletin,* No. 1452. Ohio.

Magyar, P. 1973. Simulation of the Hydrologic Cycle Using Atmospheric Water Vapour Transport Data. *VITUKI.* Budapest.

Major, P. 1975. Study on the Process of Evapotranspiration from and Actual infiltration to Shallow Ground-water in Experimental Basins (in French). *IAHS Symposium on the Hydrological Characteristics of River Basins.* Tokyo.

Mustonen, S. E. and McGuinness, J. L. 1968. Estimating Evapotranspiration in a Humid Region. *Technical Bulletin,* No. 1389. Ohio.

Palmén, E. 1967. Evaluation of Atmospheric Moisture Transport for Hydrological Purposes. *Reports on WMO/IHD Projects,* No. 1.

Peixoto, J. P. 1973. Atmospheric Vapour Flux Computations for Hydrological Purposes. *Reports on WMO/IHD Projects,* No. 20.

Penman, H. L. 1948. Natural evaporation from open water, bare soil and grass.

Penman, H. L. 1956. Estimating Evapotranspiration.

van Schilfgaarde, J., Bernstein, L., Rhoades, J. D. and Rawlins, S. L. 1974. Irrigation Management for Salt Control. *Proceedings of ASCE,* No. IR.3.

Thornthwaite, C. W. and Mather, J. R. 1957. Instructions and Tables for Computing the Potential Evapotranspiration and Water Balance. *Climatology.*

Turc, L. 1961. Evaluation of the Water Demand of Irrigation, Potential Evapotranspiration (in French).

Várallyay, G. 1966–67. Salt Balance of Soils between the Danube and the Tisza (in Hungarian). *Agrokémia és Talaitan,* Parts 1 and 2, 1–2.

INFLUENCE OF IRRIGATION ON HYDROGEOLOGICAL PROCESSES IN THE AERATION ZONE

by G. V. BOGOMOLOV, A. V. LEBEDEV and Yu. G. BOGOMOLOV

Soviet Geophysical Committee, Moscow, USSR

Summary

The aeration zone, or the non-saturated zone of the soil surface, is the link between groundwaters and the atmosphere. It contains different amounts of water and soluble salts, depending on the geological structure, lithology, hydrophysical properties and temperature.

The principal hydrogeological processes in this zone comprise different types of water movement: molecular diffusion due to differences in humidity potentials, thermocapillarity due to temperature gradients, thermo-osmosis due to specific heat differences between the liquid layer and the solid body surrounding it; diffusion of the water vapour in pore spaces; infiltration of atmospheric precipitation and irrigation water by gravity. These processes are related to evaporation from the aeration zone and transpiration from vegetation.

The aeration zone is also characterized by different processes of transportation of soluble salts: transport of salts by filtration through large pore spaces of rocks which are not completely saturated; by convection diffusion; leaching of salts from the solid phase; exchange reactions.

In the aeration zone, the transfer processes of heat by conduction and convection are produced in conjunction with changes in speed of filtration.

In irrigated conditions, the regime and water balance of the saturated zone and groundwaters vary considerably, as well as the moisture pressure of interstitial waters as related to groundwaters.

Among the factors linked to irrigation, which influence the balance of water, salts and heat of the aeration zone, the following should be emphasized: the infiltration of irrigation water; the loss of water by filtration in canals; the pressure of groundwaters; runoff and drainage; transpiration of vegetation and the cultivation of soils and introductions of fertilizers.

The complex study of these factors is of primary importance in hydrogeological and soil amelioration research at all levels of irrigation work. This is conducted for the purpose of obtaining initial hydrogeological, hydrophysical and hydrodynamic parameters which, when applied, enable water regime and salt content in the aeration zone, as well as the chemical composition of the groundwaters, to be forecast.

The features of the aeration zone can be studied by the calculation of the general water balance and of the salts in the aeration zone and in the water tables in the following ways: by the lysimetric method; by the method of temperature gradient in order to measure the water which is displaced taking into account humidity and field temperature; by the hydrophysical method of evaluating movement of water in isothermic conditions and by the method of convective diffusion of soluble salts.

An important role lies in the mathematical and physical modelling of the analysed processes, because they help to solve, directly or indirectly, the problems of forecasting the water and hydrochemical regimes of the aeration zone.

The scale of the factors considered here is enormous. Under the influence of irrigation, the groundwaters rise to the surface, change in composition and often, if drainage is not effective, flood the soil and create swamps.

The basis for a solution of these problems could be found in the organization of complex research, in experimental sites, on the groundwater regimes and the aeration zone and on the heat, water and salt balances.

126   G. V. BOGOMOLOV, A. V. LEBEDEV AND YU. G. BOGOMOLOV

Résumé

La zone d'ération, ou horizon superficiel non saturé de la croûte terrestre, fait le lien entre les eaux souterraines et l'atmosphère. En fonction de la structure géologique, de la lithologie, des propriétés hydrophysiques et de la température, elle contient des quantités variables d'eau et de sels.

Les principaux processus hydrogéologiques existant dans cette zone comprennent différents types de mouvement de l'eau: diffusion moléculaire, due aux différences de potentiels d'humidité; thermocapillarité due aux gradients de température; thermo-osmotique due aux différences spécifiques de chaleur du film liquide et du corps solide qu'il entoure; diffusion de la vapeur d'eau dans l'espace poreux; infiltration par gravité des précipitations atmosphériques et de l'eau d'irrigation. Ces processus sont en relation avec l'évaporation à partir de la zone d'aération et la transpiration de la végétation.

La zone d'aération est également caractérisée par différents processus de transport des sels solubles: transport des sels par filtration au travers des larges pores durant la saturation incomplète des roches; transport des sels par diffusion de convection; lessivage des sels de la phase solide; réactions d'échange.

Dans la zone d'aération, des processus de transfert de chaleur par conduction et convection se produisent en liaison avec des changements dans la vitesse de filtration.

Dans les conditions d'irrigation, le régime et le bilan de l'eau de la zone saturée et des nappes souterraines varient considérablement, aussi bien que les tensions d'humidité des eaux interstitielles, connectées aux eaux souterraines.

Parmi les phénomènes liés à l'irrigation, qui influences les bilans d'eau, des sels et de chaleur de la zone d'aération, il faut souligner plus particulièrement: l'infiltration de l'eau d'irrigation; la perte d'eau par filtration dans les canaux; la pression des eaux souterraines; le ruissellement et le drainage; la transpiration de la végétation et la culture des sols; l'introduction d'engrais.

L'étude approfondie de ces facteurs est d'importance primordiale dans les recherches hydrogéologiques et d'amélioration des sols, ceci à tous les niveaux des travaux d'irrigation. Elle est effectuée dans le but de posséder les paramètres de base de nature hydrogéologique, hydrophysique et hydrodynamique qui, par leur prise en considération, permettront de prévoir le régime de l'eau et des sels dans la zone d'aération ainsi que la composition chimique des eaux souterraines.

Les phénomènes qui se produisent dans la zone d'aération peuvent être étudiés par le calcul du bilan général des eaux et des sels dans la zone d'aération et dans les nappes; par la méthode lysimétrique; par la méthode du gradient de température pour évaluer l'eau qui se déplace en tenant compte de l'humidité et de la température au champ; par la méthode hydrophysique d'évaluation des mouvements de l'eau en conditions isothermiques; par la méthode de la diffusion convective des sels solubles.

Un rôle important revient aux modèles mathématiques et physiques des processus analysés parce qu'ils aident à résoudre directement ou indirectement les problèmes de prévision des régimes hydriques et hydrochimiques de la zone d.'aération.

L'échelle des phénomènes considérés ici est énorme. Sous l'influence de l'irrigation, les eaux souterraines s'élèvent en surface, changent de composition et souvent, si un drainage n'est pas effectué, inondent les sols et provoquent des marécages.

Le fondement de la solution des problèmes abordés ici pourrait résider dans l'organisation de recherches approfondies, dans des sites expérimentaux, sur le régime des eaux souterraines et de la zone d'aération et sur les bilans thermique, hydrique et des sels.

EFFECTS OF IRRIGATION ON WATER QUALITY*

by FREDERICK L. HOTES

Irrigation Adviser, International Bank for Reconstruction and Development (World Bank), Washington, D.C., U.S.A.

and ERMAN A. PEARSON

Professor of Sanitary Engineering, University of California, Berkeley, California, U.S.A.

Summary

Increased attention is being given to the potential adverse effects of irrigation on water quality, such as increased salinity, turbidity, color, taste, temperature, nutrients, nematodes, bacteria and viruses, and pesticide ingredients. Eventually, if they have not already, most irrigated areas of the world will experience problems resulting from these effects, which have caused economic losses and deterioration of the environment in many places, especially in the western USA. Surface water storage in most cases does not seriously affect the water quality for irrigation, but sometimes can cause significant adverse effects for domestic and industrial users. The larger the reservoir, the greater in number and the more complex are the resulting water quality changes, but it is not possible to predict with certainty all of the influences of an impoundment, of any size, on water quality. When reasonable precautions are taken, use of underground storage for irrigation purposes generally does not pose significant water quality problems. Withdrawal of irrigation water from aquifers can, in some instances, cause intrusion of sea water and connate brines. Quality changes during conveyance of irrigation water do not justify the cost of closed conduits to prevent these changes, although a Colorado study indicates that canal linings may be economically justified where canal seepage may percolate through underground strata with high soluble salt contents. By far the greatest detrimental changes in water quality due to irrigation occur as a result of field irrigation. The resulting irrigation return flows usually show increased salinity concentrations and hardness, but the relative importance of these changes varies, as is illustrated by data from studies on the Upper Rio Grande, and the Snake, Yakima, Colorado, and San Joaquin Rivers. Changes in other quality parameters on some of these streams, including nitrogen and phosphorus content, indicate that their importance depends greatly on the circumstances in each case. A special study of the effects of pesticides in the San Joaquin Valley of California, where 38 million kg of pesticides were applied in 1970 in an area of 1.7 million ha of irrigated cropland, indicated that this usage would seem to have caused no significant adverse effects on the aquatic environment. Several hundred millions of dollars are being invested in the Colorado River Basin and San Joaquin Valley to control salinity increases due to irrigation return flows, with intense political debate and differences of opinion on the proper course of action to be taken precluding a final solution being adopted in the latter case. Several important areas of research need to be pursued to help control return flow pollution by practical, economically feasible means. The effects of irrigation on the quality of downstream and underground waters are mostly adverse, but they often are not of a serious magnitude. As irrigation development increases, the importance of these adverse effects increases also. If economic losses to water users and their nations are to be minimized, overall water quality management planning should begin now. Lack of action may increase the difficulty and cost of final solutions.

---

*The opinions expressed in this paper are those of the authors, and should not be interpreted as reflecting the attitude or opinions of their respective institutions.

Résumé

Les inconvénients que peut avoir l'irrigation pour la qualité de l'eau, tels qu'une salinité plus forte, une turbidité, une couleur, un goût, une température, une teneur plus élevée en éléments nutritifs, des nématodes, des bactéries et virus et des pesticides, sont de plus en plus reconnus et surveillés. La plupart des zones irriguées du monde connaîtront tôt ou tard, si ce n'est déjà le cas, des problèmes de cette nature, qui se soldent par des pertes économiques et détériorent souvent l'environnement, particulièrement dans l'ouest des Etats-Unis. Dans la plupart des cas, le stockage des eaux de surface n'amoindrit pas sensiblement la qualité de l'eau destinée à l'irrigation, mais il peut entraîner des effets nuisibles considérables pour les industriels et les ménages. Plus le réservoir est large, plus complexes et nombreuses sont les modifications de la qualité de l'eau, mais il n'est pas possible de prédire avec certitude tous les effets d'une rétention — de quelque ampleur qu'elle soit — sur sa qualité. Moyennant des précautions raisonnables, l'utilisation du stock des eaux souterraines à des fins d'irrigation ne pose généralement pas de problèmes particuliers en ce qui concerne la qualité de l'eau. Le prélèvement d'eau d'irrigation dans les adducteurs peut dans certains cas déclencher l'intrusion d'eau de mer et de salinité qui lui est associée. Les variations de qualité durant l'acheminement des eaux d'irrigation ne justifient pas le prix que coûterait la mise en place de canalisations fermées, bien qu'une étude au Colorado indique que le revête- ment des canalisations se justifie peut-être du point de vue économique lorsque des fuites se produisent par infiltration dans les couches souterraines à forte teneur en sol soluble. Les modifications de la qualité de l'eau dues à l'irrigation, qui sont de loin les plus nuisibles, résultent de l'irrigation des champs. Les infiltrations secondaires présentent généralement des salinités plus élevées et une dureté accrue, mais l'importance relative de ces changements varie, comme le montrent les études relatives aux cours supérieurs du Rio Grande, du Snake, du Yakima, du Colorado et du San Joaquin. Les variations d'autres paramètres de qualité de certains de ces cours d'eau, notamment la teneur en azote et en phosphore, indiquent que leur importance dépend largement des circonstances particulières à chaque cas. Une étude spéciale des effets des pesticides dans la vallée du San Joaquin en Californie, où 38 millions de kg de pesticides ont été utilisés en 1970 dans un périmètre de 1,7 million d'hectares de cultures irriguées, semble indiquer que cet usage n'a pas eu de conséquences néfastes appréciables sur l'environnement aquatique. On dépense plusieurs centaines de millions de dollars pour contrôler l'accroissement de salinité liée aux infiltrations secondaires dans le Colorado et la vallée du San Joaquin, ce qui donne lieu à de grands débats politiques et à des divergences d'opinion sur les mesures à prendre. Ceci a empêché, dans le cas du San Joaquin, l'adoption d'une solution finale. Il convient de poursuivre les recherches, dans plusieurs domaines importants, pour faciliter le contrôle de la pollution consécutive aux infiltrations secondaires, moyennant des méthodes pratiques et économiques. Les effets de l'irrigation sur la qualité des nappes phréatiques, ou des eaux situées en aval, sont la plupart du temps néfastes, mais souvent sans grande con- séquence. Avec le développement de l'irrigation, l'importance de ces effets ne cesse de s'accroître. Si l'on veut réduire au minimum les pertes économiques subies par les utilisateurs de l'eau et leurs pays, on doit amorcer sans délai la planification globale de la gestion de la qualité de l'eau. Faute d'agir, l'adoption de solutions finales peut devenir plus difficile et plus coûteuse.

INTRODUCTION

This paper is one of several being presented to an International Symposium on the Environmental Problems and Effects of Arid Lands Irrigation in Developing Countries. Consequently, the broad range of subjects which could with validity be discussed under its title has been deliberately restricted to minimize duplication with other symposium papers. However, some overlap with other presentations could not be avoided, as the subject of irrigation water quality inevitably is associated, or intimately linked, with many other aspects of the general topic of the Symposium.

Irrigation has sustained and enhanced the lives of millions of people since mankind first artificially applied water to soil and crops some 10,000 years ago. It continues to be a vital part of food and fiber production in the world and undoubtedly will be more im- portant in this connection in the future. But today increased attention is being directed to the potential negative effects of irrigation on the environment, and perhaps one of the areas of greatest concern in that regard is that of the changes in the quality of water which take place during or due to its use for irrigation. This paper attempts to summarize the present state-of-the-art and knowledge on the adverse effects of irrigation on water quality, including changes in mineral content (salinity), turbidity, color, taste, temperature, nutrients such as nitrogen and phosphate which promote aquatic growth, nematodes, bacteria and viruses, and pesticides.

While most of the examples cited in the paper are from a relatively developed country, where the pressure for more stringent environmental controls has justified greater expenditures for the collection of basic data and associated research, most irrigated areas of the world eventually will have to deal with similar problems. Indeed many of them already are so involved. Hence the sharing of information and procedures, and coordination of research efforts, will help optimize the benefits for all.

OCCURRENCE OF QUALITY CHANGES

Significant changes in the quality of irrigation water may occur during or as a result of its:

(1) storage;
(2) conveyance and distribution;
(3) application;
(4) drainage away from the areas of application;
(5) withdrawal from groundwater aquifers;

or from any combination of the foregoing. Normally the changes that take place as a result of application are of the greatest importance, but the quality changes resulting from the other processes must also be considered.

WATER QUALITY CHANGES DUE TO STORAGE

Quality changes during surface storage

Unfortunately the biology, chemistry, and physics of both shallow and deep reservoirs is complex, and our knowledge of these interrelationships so meager, that it is not possible to predict with certainty all of the influences of an impoundment on water quality (Symons et al. 1965). Such bodies of water are neither flowing streams nor lakes, and hence, while studies of streams and lakes provides some insight into the behaviour of reservoirs, more research must be conducted to improve this area of scientific knowledge (Symons et al. 1965, Silvey 1968). However, existing knowledge makes possible many important generalizations.

When stream waters are diverted into an irrigation system by means of a low diversion dam (less than 8 meters in height), the reservoir behind the dam normally is very small and the major changes in water quality to be expected normally would be reductions in the sediment load and turbidity in both the diverted water and the water continuing downstream over or through the dam. An improvement in color and a slight increase in temperature (the latter only during warm weather) may occur. Downstream turbidities may be increased during temporary sluicing of sediment from behind diversion dams. However, as a small shallow reservoir ages certain undesirable changes in the quality of the reservoir water, especially for municipal purposes, may result from increased growth of algae or other organisms, which in turn may give rise to obnoxious tastes and odors, and, occasionally, toxic blooms (Love 1961). The probability of occurrence and extent of growth of such organisms depend upon so many factors that it is almost impossible to make valid predictions in this regard.

In deeper reservoirs many of the same phenomena occur, plus others. One of the important phenomena still being studied intensely is that of water stratification, which gives rise to numerous undesirable effects on water quality. The colder, more dense, water which tends to stagnate at the bottom of the reservoir, is referred to as the "hypolimnion," while the water strata near the top which circulates is referred to as the "epilimnion." Separating these two zones is a relatively thin transition zone known as the "mesolimnion" or thermocline, in which the temperature and the dissolved oxygen content change rapidly. Horizontal movement in the hypolimnion is very slight, and vertical movements are almost nil. The oxygen content in this layer is low, making it an undesirable environment for fish. Frequently there are increases in iron and manganese near the bottom. Shallow reservoirs in warm climates may not have a true hypolimnion at any time. Others have such a zone only during the winter. Many deep reservoirs have a hypolimnion throughout the year.

Evaporation from reservoir surfaces and evapotranspiration from phreatophytes growing in the reservoir increase the mineral concentration of the water. Density currents of inflowing sediment or very cold waters can modify the prevailing currents and strata at different times of the year.

In general the chemical quality of water for domestic and industrial purposes from a reservoir will deteriorate with age (Silvey 1968). However, bacterial quality improves with storage. The quality for irrigation purposes would not change significantly over time.

The release of cold water from the lower strata of a large dam also can reduce crop yields downstream if it is used for irrigation. Rice yields were significantly decreased in some fields in northern California, USA, because releases from Shasta Dam and Reservoir had lowered the average temperature of the Sacramento River.

Entrapment of sediment in reservoirs also removes plant nutrients. Perhaps no greater debate as to the relative benefit or disadvantage of this change in the quality of the water for irrigation has ever occurred than has taken place, on an international scale, before and after the construction of the Aswan High Dam. One recent paper available to the authors on this subject (Kinawy and Shenouda 1975) referred to a study showing that the storage of water in this reservoir would not deprive the cultivated land of more than 12 percent of the inflow silt.[1] Furthermore, the plant nutrients in the silt were estimated to constitute not more than 0.04 to 0.05 percent by weight of the silt. It was concluded from the investigations that no more than the equivalent of 1950 tons of silt-borne fertilizer would be lost (for an irrigated area of more than a million hectares), which could be compensated for by 13,000 tons of calcium nitrate fertilizer worth US $830,000, produced in the fertilizer plant in Aswan using hydroelectric power generated at the dam. It was estimated also that the cost of such fertilizer would be less than one-third of the former costs of dredging and removal of silt deposited in canals by floods. The authors of this paper do not know if the figures cited are the best estimates which can be made, but they do believe that the approach used is objective and that an objective analysis is needed to resolve such issues. However, the removal of nutrients by silt entrapment apparently has had a significant adverse effect on fish production off the Nile Delta shores.

[1] The authors interpret the term "silt" to mean the silt fraction of the total sediment load.

The same paper (Kinawy and Schenouda 1975) also presents a summary description of the water characteristics of the huge reservoir behind the Aswan Dam, which has a length of 500 km, a surface area of some 6300 km^2, a maximum depth of about 130 m, and a maximum volume of 164 billion m^3. One of the interesting phenomena is that near the end of the flood season the odor of hydrogen sulphide appears in the vicinity of the dam. This is attributed to the combination of anaerobic decomposition occurring in the hypolimnion and the negative pressures occurring at the entrances to the discharge tunnels, resulting in the release of hydrogen sulfide.

Removal of sediment in a reservoir, resulting in clearer discharge waters, has been reported to be the cause in some cases of increased aquatic growth downstream, with adverse effects on salmon and steelhead spawning grounds and on river navigation (Hagan and Roberts 1975). The same reference cites the case of nitrogen supersaturation in the Columbia River in the State of Washington, USA, which has been reported as the cause of death or injury of 30 to 40 percent of the juvenile steelhead and salmon. Nitrogen gas is entrained as the water comes over the spillway and plunges into tailwater pools at the bottom. The stair-stepped series of dams on the river from Canada almost to the sea hinders release of the entrained nitrogen gas before it reaches the next dam downstream.

Seepage under dams and through earth dams can increase the salinity and hardness of downstream waters by solution of minerals in the soils and foundations through which the seepage water passes. In a study of a series of dams in the Sugar Creek Watershed in Oklahoma, USA, the total hardness of the waters below the dams, which before construction of the dams ranged from 220 to 340 mg/l, increased by amounts ranging from 80 to 720 mg/l after construction (Yost and Nancy 1975). These observed increases in mineralization downstream from the dams were traced to even more pronounced mineralization at individual dam sites. Comparisons of the hardness of reservoir water collected in 1963 from nine different reservoirs with that of seepage at the base of respective dams, showed across-the-dam increases in total hardness ranging from 150 to 660 percent. The increases were greater in areas of shale, which contained much soluble mineral material, than in areas of permeable sandstone from which most soluble material had been leached.

An analysis of 30 years of records of inflows and outflows from Elephant Butte Reservoir, a 2468 million m^3 storage basin on the Rio Grande River in New Mexico, USA, indicated that reservoirs constructed in basins with permeable sediments may have significant effects on the quantity and quality of the effluent water by the temporary bank storage of water and dissolved salts (Bliss 1965). It was estimated that from 15 to 20 per cent per year of the water in that reservoir went into bank storage during filling periods. During subsequent drawdown periods most of the bank-stored water appeared in the reservoir releases, but contained reduced amounts of the less-soluble ions, notably calcium and bicarbonate, some of which it was assumed had precipitated in both the reservoir and in the banks.

The preceding discussion illustrates some of the complexities and the continually changing nature of reservoir waters, Because of the wide range of conditions which can be encountered, only a few broad general conclusions are drawn here as to the effects of the construction of surface reservoirs. They are:

(1) Reservoir storage has both negative and positive effects on the original quality of the water in the stream. From the standpoint of use of the water for irrigation, in most instances the negative effects would not be highly significant. In the case of use for domestic and industrial purposes, some negative effects can be important, although

usually the benefits of improved bacterial quality and water availability in storage are considered to offset any negative effects.

(2) The quality of water in most new reservoirs generally first deteriorates over a period of time. Then a gradual improvement of quality may take place over a period of several years, with the reservoir water approaching the quality of the inflow waters.

(3) The larger the reservoir, the greater in number and the more complex are the resulting changes in water quality.

(4) The possible changes in water quality due to storage in surface reservoirs should be considered during the planning and design stages; and, of course, must be taken into account during operations.

Despite the fact that serious water quality problems have been associated with many surface storage reservoirs, the authors do not know of any case where a reservoir has been abandoned because of such phenomena. Usually mitigating action, which could have been avoided by appropriate design practice, can be taken at feasible costs.

Quality changes due to subsurface storage

Surface waters sometimes are stored in groundwater basins (aquifers) for extraction later by wells (boreholes) for irrigation use. Common examples of this in the western USA are: the deliberate spending of floodwaters, imported aqueduct waters and reclaimed sewage, into basins or pits for percolation to the underground aquifers, or injection through reverse wells. In Orange County in Southern California, for example, as much as 120 million cubic meters per year of water imported from the Colorado River have been recharged to the groundwater basin.

Such operations result in quality changes in both the native groundwater and the recharged water. While water levels in a groundwater basin respond to recharge operations within a few weeks, or at most a few years, changes in water quality take place much more slowly, due to the fact that normal velocities of underground flow are only a few meters per day rather than meters per second as in surface flow, and the time of travel to any point in the basin is inversely proportional to the flow velocity. The resultant quality of water at an extraction well depends upon the relative blend of recharged and native waters, and also upon changes in the quality of the recharged water as it percolates into and/or through the underground formations. Hence the quality of the extracted water cannot be predicted solely from the qualities of the two sources of water. Consideration also must be given to the physical-biochemical nature of the surface soils and substrata, i.e., precautions must be taken so that recharge does not take place through highly soluble formations. Attention should be given to keeping groundwater levels at a sufficient depth below ground surface to eliminate significant losses of the stored water by evapotranspiration by plants and from soil surfaces. The capillary action of the soil in raising water from the phreatic level to the root zone or surface level should not be neglected. When these factors are properly considered in advance, use of underground storage for irrigation purposes generally does not pose significant water quality problems. Situations can be visualized where the quality of the recharge waters would result in a blend unsuitable for domestic or industrial purposes, although suitable for irrigation.

Possibly the greatest threat to change in quality of groundwater in storage to be used for irrigation, results from industrial wastes and irrigation drainage. However, it is the source of the pollution, rather than the storage itself, which is the direct cause of such potential changes. Groundwater pollution due to irrigation return flows and drainage is discussed subsequently in this paper.

WATER QUALITY CHANGES DIRECTLY RESULTING FROM CONVEYANCE AND DISTRIBUTION

As water is conveyed from a stream, reservoir, lake, pond, or well and distributed to the fields to be irrigated, its quality may be changed because of:

(1) Evaporation from the water surfaces in open conduits, thereby slightly increasing the salinity concentrations.
(2) Solution of salts or inflow of saline groundwaters occurring in unlined canal sections.
(3) Pollution by influent surface drainage.
(4) Aquatic growths and organisms, primarily occurring in unlined canal sections but occurring also in lined sections.
(5) Operations to control aquatic weeds and pests.
(6) Canal seepage dissolving salts from underground strata through which it passes, causing deterioration of water quality in groundwater basins, streams, or other bodies of water into which the seepage flows.

It is interesting to note that all of the foregoing can be avoided or made minimal by the use of closed conduits with impervious linings. However, in most instances the amount of quality deterioration is so small, or temporary, that the very considerable expense of

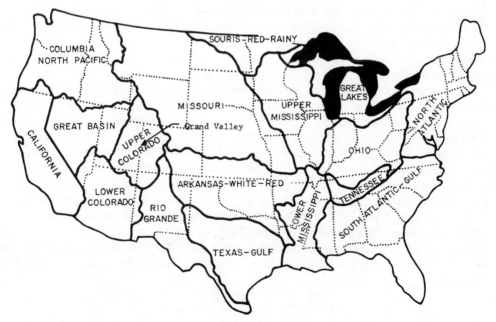

FIG. 1. Major hydrologic regions in the United States (adapted from Skogerboe and Law 1971).

covering or lining the conduits cannot be justified economically for the sole purpose of maintaining water quality. When used, closed or lined conduits normally have been justified on the basis of savings in water losses and drainage costs, although increased attention has been given recently to evaluating the benefits of canal linings to reduce saline irrigation return flows. An example of the latter is the field demonstration canal lining program carried out in a part of the Grand Valley, near Grand Junction, Colorado, in the Upper Colorado River Basin (Jenke 1974, Baer 1972, Skogerboe and Walker 1972). Figure 1 shows the location of the Basin.

The Grand Valley has a gross area of about 50,000 ha and a high desert climate with an annual rainfall of about 210 mm. Its elevation is 1500 m above sea level. About 35,000 ha are irrigated in the valley, with 2.75 m depth of water being diverted for irrigating a mixed pattern of field crops such as corn, barley, and alfalfa, plus pasture and orchards. Water use is very inefficient, and the excess water percolates downward to underlying highly mineralized formations where its salt content is greatly increased. It was estimated that this one small valley, with only 3 percent of the 1,142,000 ha of irrigated land in the entire Colorado River Basin, contributes 18 percent of the total irrigated agriculture salt load of the entire Basin!! The valley salt contribution was estimated to be from 640,000 mt to 880,000 mt/yr., or about 18 mt/ha/yr. It was further estimated that the salt load resulting from over-irrigation of the test area was 74 times as great as that resulting from canal plus lateral seepage.

About $8\frac{1}{2}$ km of canals and laterals in relatively high seepage areas were lined, out of a total length of canals and laterals of 355 km. The study estimated that this work, costing US $420,000 (1971 prices), reduced salt flows to the Colorado River by 4300 mt/yr by reducing the seepage by 1.15 million m^3/yr. The study also found that, based on an estimate that a reduction of one ton of salt at Lees Ferry would result in a saving of US $7.70 of annual damage in the Lower Basin, that this installation fully paid for itself by the reduction in Lower Basin salt load alone, neglecting the benefits of improved drainage conditions and yields in Grand Valley or elsewhere in the Upper Basin. (It should be noted that the estimate was based on an assumption which may or may not be correct; that the canal seepage water dissolved salts in a manner proportionate to all other seepage and groundwater in the area.)

Changes in water quality due to the first three categories of causes listed at the beginning of this section are normally of minor significance. Those resulting from the fourth category, aquatic growths and organisms, also generally are not important in themselves, although they could be so in poorly maintained systems, from the standpoints of color, taste, and organisms which would be undesirable for domestic uses of irrigation water. Operations to control aquatic weeds and pests (category 5) mostly involve efforts to control weeds so that canal carrying capacities are not unduly reduced, or to control organisms such as those responsible for bilharzia, onchocerciasis (river blindness), or malaria. Occasionally special pests are found which cause adverse effects and which must be controlled, such as the Asiatic clams in the California Delta–Mendota concrete-lined canal, which grew so prolifically that canal capacities were significantly reduced. Measures for control of aquatic weeds and pests involve dewatering or, more often, application of pesticides.

WATER QUALITY CHANGES DUE TO FIELD APPLICATION
OF IRRIGATION WATER

General

By far the greatest detrimental changes in water quality due to irrigation occur during or as a result of field irrigation. Of the total amount of irrigation water applied, from about 40 to 80 percent may be used for crop production. Such use is termed "consumptive use" or "evapotranspiration", since the water is discharged into the atmosphere by plant transpiration or by direct evaporation from the ground or plant surfaces. It effectively has been "consumed"; that is, placed beyond the direct control of the farmer or engineer, and removed from the runoff system of the drainage basin. The remaining 20 to 60 percent of the water either leaves the field as surface runoff or percolates below the crop root zone ("deep percolation"). Deep percolation water eventually is returned by means of ground-water flow to a stream, lake, or other body of water, usually within the same drainage basin where it was applied. In some instances deep percolation may require several years before reaching a surface water body, but in most locations where irrigation has been practiced for a period of years, its hydrographic effects normally appear within the same, or ensuing, year. Its quality effects will not appear until later, due to the long travel time of groundwater under viscous flow conditions. Deep percolation intercepted by drains appears in the receiving waters much more rapidly because of the shorter travel time by turbulent flow in the drains.

The surface runoff and deep percolation which return to the surface or the groundwater systems are termed "irrigation return flow." It has been estimated that in 1954 in the

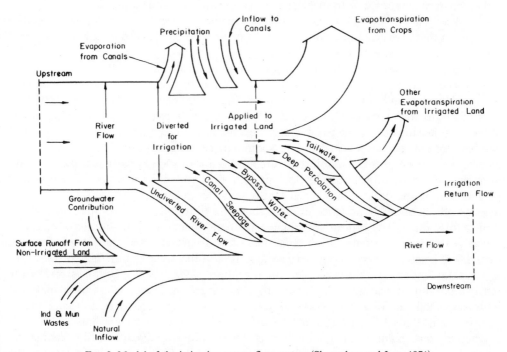

FIG. 2. Model of the irrigation return flow system (Skogerboe and Law 1971).

United States the average amount of irrigation return flow was about 40 percent of the total diversions for irrigation, with wide variations in that figure occurring because of local conditions (US Committee on Pollution 1966). Figure 2 is a model of the irrigation return flow system in a typical arid lands irrigated area.

Changes in mineral quality (salinity)

When water evaporates from the ground or plant surfaces any dissolved salts remain behind. During transpiration some salts are used by the plants, but the amount is relatively small and in most instances is neglected for the purposes of estimating salt effects or leaching requirements (US Salinity Laboratory 1954). The net result of evapotranspiration then is to increase the amount of salts in the soil in direct proportion to the salt content of the irrigation water and the depth (volume) of water applied.

The addition of more irrigation water temporarily dilutes the more concentrated soil solution, but evapotranspiration proceeds to concentrate the salinity further. At the same time any excess irrigation water applied displaces some of the soil solution downward beyond the root zone to become deep percolation water. In irrigated arid regions the salinity of the solution can be expected to be at least equal to, and normally to be greater than that of the irrigation water. Usually this concentration is 2 to 3 times greater, but often it is 5 to 10 times as much. The salinity of the soil solution should not be permitted to exceed acceptable limits for the particular crop grown. This can be accomplished by applying irrigation water in excess of that needed for consumptive use (i.e., by leaching) supplemented by available rainfall. This transfers the salinity problem caused by the irrigation from the root zone to the deep percolation and subsequent cycles of the irrigation return flow system. Normal irrigation practice includes a considerable degree of built-in leaching in that excess amounts of water must be applied to fields to assure that the root zone receives adequate moisture.

The soluble cations and anions usually of importance in irrigation and return flow water quality studies are calcium, magnesium, sodium, potassium, boron, carbonate, bicarbonate, sulfate, chloride, and nitrate. On rare occasions silicate also can be of significance. Of these ions only boron has a marked toxic effect in low concentrations ($\simeq 1$ mg/l) on plant growth, although other ions can have toxic effects, depending upon the plant, growth stage, and combinations of other chemicals present. Sodium presents an additional hazard because of its propensity to cause adverse soil structure modifications such as particle dispersion and puddling, which in turn results in low permeability and poor aeration. Calcium and magnesium, on the other hand, are generally good for soil quality, two of the principle reasons being that their presence ensures good soil tilth and decreases the relative percentage of sodium. The cations sodium, calcium, and magnesium in solution are readily exchangeable with the same cations adsorbed on soil particles, and the cation exchange capacity of a soil is a very important characteristic in the analysis of irrigated soils.

The total amount of salts present in the irrigation water or soil solution is one of the most important indicators of water quality for irrigation. Increased salinity makes it increasingly difficult for a plant to absorb water and obtain the necessary nutrients. Plants vary in their ability to resist this adverse effect, but once a particular threshold level is reached for each plant, yields gradually decrease with increased salinity. Since irrigation return flow always contains more salts than the original supply, it tends to degrade the quality of the stream system from which diverted and to which returned. In the western United States the water of many streams is used several times for irrigation, with salinity increasing during each use. Downstream users, therefore, have progressively poorer water with which to irrigate their crops. To maintain tolerable salinity limits in the soil solution, they must apply greater quantities of irrigation water. When the total amount of salts removed from a given area is equal to the total amount of salts brought into the area, the area is said to be in "salt balance." Another term used in the USA in discussing salinity phenomena is "salt loading." This is the addition of dissolved solids from natural and man-made sources.

The changes which take place in the salt content of irrigation water as it passes through the root zone are the result of a complex combination of factors. Predicting the amount and composition of the resulting return flow is not a simple exercise, although our ability to do this has improved in recent years (Christiansen 1973, van Schilfgaarde et al. 1974, Willardson et al. 1975).

The easy exchangeability of sodium, calcium, and magnesium ions already has been noted. One of the changes which takes place is that as the root zone water is consumed by evapotranspiration, the soil solution salinity increases and some of the calcium carbonate and calcium sulfate precipitates as their solubility decreases. These actions increase the relative proportion of sodium ions which are more deleterious to soil than are the calcium salts. Offsetting to some extent the preceding tendency to increase sodium percentage, is the effect of soil weathering (Christiansen 1973), which tends to decrease the sodium hazard. Some leaching of fertilizer constituents, principally nitrates, occurs during irrigation. The amount of fertilizer lost can be minimized by good coordination or irrigation and fertilization methods and timing of applications.

It has been suggested recently (van Schilfgaarde et al. 1974) that the concept of salt balance as previously explained in this paper is no longer valid as the best means of maintaining soil productivity. The referenced article suggests that the use of much smaller amounts of leaching water would keep the soil salinity at acceptable levels while at the same time greatly reducing return flow salinity. The article admits that the extreme precision in irrigation techniques required to meet the theoretical objectives described is not now available. Various discussers of the article (Willardson et al. 1975) questioned both the practicality of the proposed management method and its accuracy in fact, claiming that in many cases the overall water quality would decrease, rather than improve. It is apparent that the last word on this complex subject will not be heard for many years yet. However, most planning in the United States seems to be based on maintaining overall salt balance wherever possible.

Surface water return flow salinity

Surface runoff from irrigated lands does not show relatively as great an increase in salinity as does the subsurface return flow, the mineral composition being similar to that of the applied water, with perhaps an increase in fertilizer constituents (Jenke 1974). It is worth noting that surface flushing of salts is ineffective (US Salinity Laboratory 1954, p. 39).

138 F. L. HOTES AND E. A. PEARSON

Examples of mineral quality changes

Upper Rio Grande—New Mexico and Texas. One of the better examples of the effects of
return flow on the salinity of a major stream furnishing water to irrigated areas is that of
the Upper Rio Grande River (Wilcox 1962). Figure 3 shows the sampling station locations,
with the distance from the No. 1 station to the No. 7 station being about 725 km. Table 1
presents the summary data for the 20-year period 1934–53. Irrigated areas were as follows:

Station 1 to 2—32,000 ha
Station 4 to 5— 6,000 ha
Station 5 to 6—28,000 ha
Station 6 to 7—34,000 ha

TABLE 1. *Mineral water quality. Rio Grande River, USA, 1934–53*

					Station			
		1	2	3	4	5	6	7
A. Mean annual discharge millions of m³		1,330	1,052	975	965	918	648	251
B. Dissolved solids—ppm (weighted means)		221	449	478	515	551	787	1,691
C. Mean annual dissolved solids 1000 metric tons per year—average		294	473	468	498	507	511	424
D. Concentration of	Ca	37	61	61	65	69	85	148
dissolved constituents	Mg	7	12	13	14	15	19	37
—ppm	Na	19	65	72	79	85	145	367
	HCO_3	116	159	156	168	171	207	213
	SO_4	55	157	172	174	187	256	415
	Cl	7	39	44	54	61	124	509
E. Concentration of	Ca	57	44	42	42	41	35	28
dissolved constituents	Mg	18	15	15	14	15	13	12
—%	Na	25	41	43	44	44	52	60
	HCO_3	59	37	35	35	33	28	13
	SO_4	35	47	48	46	46	43	33
	Cl	6	16	17	19	21	29	54
F. Dissolved constituents	Ca	50	64	60	63	64	56	37
1000 metric tons per	Mg	10	13	13	14	14	13	9
year (weighted	Na	25	68	71	76	78	94	92
means)	HCO_3	76	83	75	80	77	66	26
	SO_4	74	165	168	168	172	165	107
	Cl	9	41	43	52	56	80	127

Adapted from Wilcox (1962)

The inverse relationship between discharge and salinity concentration in Table 1 indicates
a close correlation between evapotranspiration and increase in salinity. The percentages of
calcium and bicarbonate decrease downstream, indicating precipitation of calcium carbonate in the irrigated soils. The increase in sodium and chloride ions is substantial.
Favorable salt balances are indicated except for the section between Stations 6 and 7,
where an adverse salt balance of 87,000 mt per year is revealed. Unfortunately no information was reported on nitrate quantities.

A more recent article (Clarke 1972) gives data on the Rio Grande for 1963, and states

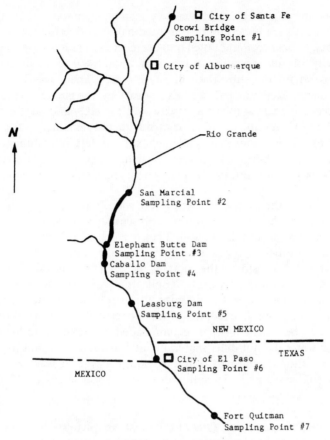

FIG. 3. Rio Grande sampling points.

that beginning in 1946 on through the 1950's the area suffered from a severe drought. Salinity concentrations.were about the same, except at Station No. 7, Fort Quitman, where the concentration was reported as 4000 ppm. Total salt loads were down greatly from the 1934–53 averages, as shown below.

Upper Rio Grande Salt Burden
(1000 metric tons)

	1963	1934–53 Average
Station 1	124	294
Station 4	282	498
Station 6	285	511
Station 7	115	424

Severe economic losses due to the drought and salinity were reported during the drought period, with one irrigation district between Station 6 and 7 practically ceasing to exist.

Snake River—Southern Idaho: The quantity and quality of irrigation return flows from a 82,000 ha tract in Southern Idaho were measured and compared to the quantity and quality of the irrigation water (Carter 1972). Moderately deep silt loam soils were irrigated primarily by furrows. These soils are underlain by fractured basalt to depths of several hundred feet. Return flows for a typical year amounted to 64 percent of the total water input to the tract. The conductivity of the irrigation water was about 460 micromhos/cm, and about 1040 micromhos/cm for the subsurface drainage water. There was a net output of all chemical components excepting PO_4 and K. For cations, the greatest relative increase was in Na; for anions the greatest relative increase was for NO_3 and least for Cl. More than 70 percent of the PO_4 entering the tract reacted with and remained in the soil. This result is in line with most observations that applied phosphorus fertilizers do not usually become part of deep percolation, but remain associated with the soil. There are reports, however, of phosphates appearing in surface drainage where considerable soil erosion is occurring during irrigation.

On the other hand the nitrate concentration was 27 times higher in the subsurface drainage water than in the irrigation water, representing a net loss of 34 kg/ha/year. This result also is typical of nitrates in the soil. They are not adsorbed by the soil, and unless used by the plants or otherwise converted by biological action, they move downward with the drainage water. A discussion of the complex nitrogen cycle in soils is beyond the scope of this paper. As mentioned previously, applications of irrigation water and nitrogen fertilizers should be coordinated to minimize nitrogen losses to return flow. It should be mentioned that in the example cited, the nitrate concentration in the drainage water was only 3.24 ppm, well below the 44 ppm recommended as a maximum for drinking water.

Surface runoff water did not differ materially in mineral quality, but had a much higher sediment concentration during part of the irrigation season. It was estimated that more than 9 mt/ha of soil were lost by irrigation erosion per season from a 1200 ha study tract in the area.

Yakima River—Washington State. One of the most complete studies reported to-date on water quality changes due to irrigation was that made in 1959–60 for an irrigated area of about 160,000 ha on the Yakima River in Washington State, USA (Sylvester and Seabloom 1963). Diversions take place at several different points along the river, with the result that some water is reused several times. The mean annual flow is about 110 m³/s. In the lower 130 km nearly the entire summer flow of 35 to 55 m³/s consists of irrigation return flow. Drainage is provided by open and subsurface drains. A major portion of the groundwater seepage reaches an open drain before it reaches the river. The average diversion represented was about 2000 mm/ha, but only about 1300 mm/ha was applied to the land. Consumptive use was about 790 mm, indicating an overall efficiency of water use of about 40 percent. Some important chemical quality changes were as follows:

	Applied water	Subsurface return flow	Surface drains near exit to river
Conductivity, micromhos/cm	83	420	283
Nitrate as NO_3	1.1	11	3.5
Total PO_4	0.32	0.86	0.83

See Table 2 (p. 143) for a complete list of water quality constituents determined.

The method of water application and the type of crops had no marked effect on return flow water quality unless there was over-irrigation or fertilizer application in excess of crop requirements. Irrigation return flow was responsible for peak concentrations of most dissolved constituents in the lower Yakima River. Nitrates and phosphates leached from the soil were not found in the lower river in comparable quantities as they were apparently assimilated by the phreatophytes and other aquatic vegetation. However, the minimum tonnage of salts discharged by the Yakima River into the Columbia River occurred during the irrigation season, while maximum quantities were discharged during periods of heavy spring and autumn runoff. An important conclusion was that "The water quality of the Yakima River has not been seriously impaired by irrigation return flow with the possible exception of high water temperature and its relation to the fishery."

Hardness. The increase in total salts due to irrigation normally has the effect of increasing the hardness of river water in the downstream direction, despite a decrease in the percentage of calcium ions and an increase in the percentage of sodium ions. The following data on five projects in the USA are typical (Eldridge 1963).

Location	Total hardness mg/1 as $CaCo_3$		Permanent hardness mg/1 as $CaCo_3$
	Above irrigation	Below irrigation	Below irrigation
Rio Grande, Texas	111	631	500
Yakima, Wash.	33	134	50
Sunnyside, Wash.	40	299	188
Arkansas River	212	890	650
Sutter Basin, Calif.	72	480	208

A hardness above 200 mg/1 as $CaCo_3$ is generally considered undesirable from the standpoint of most domestic and industrial uses.

Taste

Salinity affects taste. Most persons consider water having concentrations of chloride or sulfate of 200 mg/1 or above to taste salty, and object seriously if these concentrations exceed 500 to 600 mg/1 (Eldridge 1963). Slimes, algae and aquatic vegetation also can produce undesirable tastes. Taste changes due to irrigation are thus the result of changes in other quality characteristics.

Temperature

In the Yakima River study (Sylvester and Seabloom 1963) it was found that the temperature of the river water returned by surface routes increased an average of 4.3°C between the point of application in the fields and where it was discharged back to the river. Temperature increases of up to 11°C were reported (Eldridge 1963). However, subsurface drainage from irrigated fields had an average water temperature of 2.7°C lower than the applied water.

Turbidity and color

There is a lack of evidence that irrigation return flows significantly affect the color or turbidity of a receiving stream (Eldridge 1963). However, there is evidence that color and turbidity generally increase in surface drainage (Table 2). Both may be decreased during deep percolation. Application of excess water and subsequent erosion in particular increases color, and especially turbidity, in surface runoff. In the Yakima River, above diversions the color and turbidity were measured at 2 and 4 units respectively. During transport in the main canal over a distance of about 60 km the color increased by from 10 to 17 units, and the turbidity by from 9 to 24 units. Some surface drains at times showed color and turbidity intensities of over 200 units each. However, samples from the Yakima River below the diversions had a color measurement of only 19 units and a turbidity of only 14 units. It was also found that the color and turbidity of the River were increased more by high rates of natural runoff than they were by irrigation return flow.

Organisms and bacteriological aspects

The coliform bacteria concentration often is higher in the surface runoff from irrigated fields than in the irrigation water, especially where surface runoff comes from manured fields, pastures, feed lots, and possibly from areas with farm and community sewage disposal system. Table 2 shows average concentrations for the Yakima River study (Sylvester and Seabloom 1963).

A separate study on the Yakima Valley has shown that plant parasitic nematodes are transported in significant quantities by surface return flows (Skogerboe and Law 1971, Faulkner and Bolander 1970).

Coliform organisms are removed during percolation, mostly during travel through the first few inches or feet of the soil, by both filtration and biological action. Several studies of bacterial travel in underground formations and through the soil have clearly demonstrated this fact.

The surface water bacteriological pollution that is caused by irrigation surface return flow is not of critical significance from the standpoint of domestic water supplies since treatment by filtration and chlorination normally would be needed even in the absence of irrigation.

Nutrients and eutrophication

The presence of nitrates and phosphates are considered key indicators of potential problems in all types of water. They are nutrients which can cause or contribute to excessive growths of algae, slimes, aquatic weeds, and other organisms. Nitrates in addition can cause serious health problems in infants, if present in drinking water in concentrations greater than about 40 ppm in the water used for preparation of the infant's milk formula. It should be noted that nitrates in particular, and phosphates to a lesser degree, can be expected to be at increased concentration levels in irrigation return flow over those of the irrigation water. It is therefore in order to review briefly the significance of various concentration levels of nitrates and phosphates.

TABLE 2. *Yakima River, Washington State. Water quality constituent difference between applied and subsurface drainage water: irrigation season*

Constituent or characteristic (1)	Applied water[1] (2)	Subsurface return flow[2] (3)	Change in terms of applied water (4)	Surface drains near exit[3] (5)	Change in terms of applied water (6)
H_2O Temp °C	16.0	13.3	0.83	17.9	1.12
Dissolved Oxygen, mg/l	10.2	6.8	0.66	9.0	0.88
pH units	8.1	7.7	0.95	8.2	1.01
HCO_3 Alk. mg/l as $CaCO_3$	46	218	4.8	138	3.0
CO_3 Alk. mg/l as $CaCO_3$	1	0		2	2.0
Hardness as $CaCO_3$	46	186	4.1	121	2.6
Turbidity units	37	12	0.32	130	3.5
Color units	22	12	0.55	38	1.2
Conductivity μm^1 hos/cm	83	420	5.1	283	3.4
Chlorides mg/l	1	12	12	8	8.0
Nitrate as N mg/l	0.25	2.5	10	0.8	3.3
Total Kjeldahl N mg/l	0.27	0.32	1.2	0.25	0.92
Chem. Oxygen Demand mg/l	7	9	1.3	10	1.4
Soluble PO_4 mg/l	0.21	0.66	3.2	0.58	2.7
Total PO_4 mg/l	0.32	0.86	2.7	0.83	2.6
Sulfate (SO_4) mg/l	5.4	39	7.2	37	6.9
Calcium (Ca) mg/l	10	44	4.3	31	3.0
Magnesium (Mg) mg/l	5.0	20	3.9	12	2.3
Sodium (Na) mg/l	4.1	38	9.2	26	6.3
Potassium (K) mg/l	1.4	4.7	3.4	5.3	3.8
Coliforms per 100 ml	1,070	103	0.1	10,600	9.9

[1] Average of 7 stations.
[2] Exit from subsurface drains under irrigated fields. Average of 7 stations.
[3] Average of 5 stations. From Sylvester and Seabloom (1963).

Eutrophication and its causes are problems of major proportions in the USA. Simply defined, eutrophication is the process of enrichment with nutrients (Weiss 1969). Through usage the term "eutrophic" has come to mean highly nourished or fertile, when applied to the aqueous environment. At the opposite end of the scale, "oligotrophic" means poorly nourished. The biological life cycles in bodies of water are the result of the complex interactions of a great many factors which cannot be fully explained, especially within the scope of this paper. Both dead and live plant and animal life together support numerous food webs and food chains, which in turn support the growth of both heterotrophic and autotrophic organisms. Some of these, such as certain bacteria and blue-green algae, have the ability to fix nitrogen from the atmosphere. The presence of carbon fixed by plant life is the essential condition which begins the fertility cycle (Borchardt 1969). The decomposition of plant life produces organic acids, which in turn release phosphorus and other elements from minerals in the soil. Thus the very presence of some plant life induces the generation of additional nutrients which in turn enhances the development of more plant life and additional fertility.

Sunlight too is a very important ingredient in the biological cycle. The more sunlight that exists the more energy is available for plant growth. Hence the reduction of sediment load in a river or canal often results in increased aquatic life growth.

Frequently cited guides as to limiting concentrations for available nitrogen and phosphorus to avoid undesirable algal "blooms" based upon the classic Wisconsin lakes studies (Sawyer 1947), are 0.3 and 0.01 mg/l as N and P respectively (i.e., 1.3 and 0.03 mg/l as NO_3 and PO_4 respectively). However, it is recognized that blooms can occur at lower concentrations than those cited. For example, some algae limited by phosphorus have been shown to grow at one-half their maximum growth rate at values of phosphorus concentration as low as 5 micrograms/l (i.e., $K_s \leqslant 0.005$ mg/l). Also, it is gradually becoming recognized that nitrogen and phosphorus may not always be the growth rate limiting substance in the environment. Recently it has been recognized that nitrogen and phosphorus, while generally essential, are not solely responsible for eutrophication. For example, one study shows that massive growths of blue-green algae depend upon the availability of not less than 15 essential elements (Lange 1970). A National Technical Advisory Committee Report in the USA (US Water Quality Criteria 1968) recommended that "As a guideline, the concentration of total phosphorus should not be increased to levels exceeding 0·1 mg/l in flowing streams or 0.05 mg/l where streams enter lakes or reservoirs." The Committee noted that "Some potable surface water supplies now exceed 0.2 mg/l (P) without experiencing notable problems due to aquatic growths." Although there is no consistent relationship between total and available phosphorus, because the ratio varies with many factors, total phosphorus is usually governing as it is the reservoir that supplies the available phosphorus. The same Committee noted above did not recommend a specific range of nitrogen levels for eutrophication control, but suggested that a nitrogen-phosphorus ratio of 10 : 1, occurring in many natural waters, appeared to be a good guideline for indicating normal conditions. It was acknowledged that a wide range in this ratio occurs. A limit of 10 mg/l N (40 mg/l NO_3) was recommended for potable water supplies.

Irrigation return flows high in nitrates almost always exceed threshold limits for many organisms. Phosphorus limits also are frequently exceeded. However, it should be pointed out that once the threshold limit concentration of a nutrient for rapid growth is reached or exceeded for a particular organism, the addition of more of the same nutrient does not contribute to any increase in growth rate of the organism. Thus discharge of a eutrophic stream into a eutrophic estuary may not increase aquatic growth.

Estimate of Nutrient Contributions from Various Sources, USA

Source	Nitrogen		Phosphorus	
	10^6 kg/yr	Usual concentration mg/l	10^6 kg/yr	Usual concentration mg/l
Domestic waste	500–750	18–20	90–230	3.5–9
Industrial waste	450	0–10^4	*	*
Agricultural land	680–6,800	1–70	55–550	0.05–1.1
Rural non-agricultural	180–860	0.1–0.5	68–340	0.04–0.2
Farm animal wastes	450	*	*	*
Urban runoff	50–500	1–10	5–77	0.1–1.5
Rainfall on water surface	14–270	0.1–2.0	1.4–4	0.01–0.03

* Insufficient data.

In the Snake and Yakima River data previously mentioned, the nitrogen and phosphate concentrations in the return flows exceeded the recommended limits for eutrophication control. The available evidence in other cases confirms that irrigation provides large quantities of nutrients to downstream waters. One report states that "agricultural runoff is the greatest single contributor to nitrogen and phosphorus in water supplies (US AWWA Task Group Report 1967)." The table below, adapted from that same article, shows that there are many other important contributors to the problem, including rainfall. It should be noted, also, that irrigated areas represent only 10 percent of the crop areas in the USA.

Nitrates can be removed from waste waters, but the processes are difficult and costs are relatively high, ranging from US $0.016 to US $0.057 per m^3 (Meixner 1974). Phosphorus can be removed from waste waters easily and economically. However, the sludges produced are voluminous and difficult to dewater (Sawyer 1965).

Pesticides

Pesticides are essential to increased agricultural production in the world. There is a great lack of understanding as to their ultimate effects on the environment and on man, resulting sometimes in considerable disagreement even among experts. While irrigation is not the direct cause of their use, in most instances knowledge of their occurrence and importance in irrigation return flow and in irrigation water is important to those responsible for irrigation projects.

Definition and general characteristics. "A pesticide is a chemical used to cause the death of nonhuman organisms considered by man to be 'pests,' i.e., inimical to human interests. Rather arbitrarily the following are excluded: pathogenic microorganisms, viruses, bacteria, protozoa generally, endoparasites of man and other animals, and a host of organisms" (US Dept. Health 1969). There were in 1969 in the USA some 900 active pesticidal chemicals formulated into over 60,000 preparations. These included insecticides, fungicides, herbicides, and plant growth regulators. Total value of all pesticides in the USA increased from US $440 million in 1964 to US $12 billion in 1969, with the US production in 1969 representing from 50 to 75 percent of world pesticide manufacturing.

Major classes of pesticides may be grouped as follows (US Dept. Health 1969):

Non-persistent (1 to 12 weeks persistence time)
    Examples:
        Organophosphorus compounds
          malathion, methyl parathion, parathion
        Carbamates which contain neither chlorine nor phosphorus
          carbaryl, carbofuran, methomyl
    Note: Persistence times are periods required for a 75 percent to 100 percent loss of activity under normal environmental conditions and rates of application.

Moderately-persistent (1 to 18 months persistence time)
    Examples:
        2,4-D, atrazine, prometryne, simazine, propazine, trifluralin, fluometuron

Persistent (2 to 5 years persistence time or more)
  Examples:
    Diphenyl aliphatic chlorinated hydrocarbons
      DDT, DDD, TDE
    Chlorinated aryl hydrocarbons
      aldrin, dieldrin, endrin, chlordane, heptachlor, toxaphene

Permanent (non-degradable)
  Examples:
    mercury, arsenic, lead

Wherever insects are exposed for long periods to successive levels of the insecticide which causes less than 100 percent mortality, resistance to a particular pesticide can be

TABLE 3. *Pesticide concentrations in subsurface drain effluents in San Joaquin Valley, California—1970**

| | | | Summation of 17 stations | | | |
| | | | Reported concentration (in ppt)[1] | | | |
Pesticide	Times sampled	Times detected	Max.	Min.[2]	Avg.[3]	Avg.[4]
CHLORINATED HYDROCARBONS	60					
BHC		6	7	2	2	5
DDD		1	2	–	0	2
DDE/Dieldrin		1	7	–	0	7
DDT		19	240	2	21	35
Dacthal		3	4,780	4	302	1,608
Dieldrin		4	43	10	5	21
Heptachlor		2	28	14	3	21
Kelthane		3	45	15	6	32
Lindane		6	2,850	3	157	486
Simazine/Atrazine		11	390	5	31	67
Toxaphane		14	630	70	136	262
Complex chlorinated compounds as DDT		15	1,750	5	130	242
Unknowns as DDT		17	140	4	19	33
Summation of identified chlorinated hydrocarbon pesticides[5]		43	2,850	0[6]	129	180
ORGANIC PHOSPHORUS COMPOUNDS	60					
Parathion, Methyl		8	170	10	29	76
Thimet		1	74	–	0	74
Unknown as parathion		10	215	13	23	47

[1] ppt = parts per trillion.
[2] Detected minimum concentration.
[3] Average value includes 0 value when chlorinated hydrocarbons were not detected.
[4] Average value includes only the detected chlorinated hydrocarbons.
[5] Does not include Complex chlorinated compounds as DDT or Unknowns as DDT.
[6] Actual minimum concentration possible.
* California Dept. of Water Resources, 1970.

From Li and Fleck (1972).

developed. Some 224 species of insects and acarines in various parts of the world have developed resistance to one or more groups of insecticides. Hence new pesticides, or other methods of control must be continually under development.

Many pesticides adversely affect man, animals, birds, or plants that are not the primary target. Unfortunately the pace of change is so rapid in the pesticide field that only limited research data is available which can provide a basis for establishing precise tolerance limits to avoid such risks. However, guidelines have been developed for the use of many pesticides, which should be followed.

Allowable concentrations. Many pesticides cannot be removed by the usual water treatment processes. The allowable limits are very low for public water supplies, ranging from 0.001 mg/l for endrin to 0.042 mg/l for DDT, and 0.056 mg/l for lindane (US Water Quality Criteria 1968). These recommended limits were set with relation to human intake directly from a domestic water supply. The consequence of higher and possibly objectionable concentrations in fish to be eaten by man due to biological concentration was noted, but not reflected in the criteria recommended.

Pesticides in irrigation return flow. There are about 3.5 million ha of irrigated land in the State of California, which includes most of the cropped area. Starting in 1970 the State

TABLE 4. *Pesticide concentrations in surface drain effluents in San Joaquin Valley, California—1970**

| Pesticide | Times sampled | Times detected | Reported concentrations (in ppt)[1] | | | |
			Max.	Min.	Avg.[3]	Avg.[4]
CHLORINATED HYDROCARBONS	18					
BHC		3	10	6	5	9
DDD		3	20	3	5	9
DDT		9	450	1	62	82
Dacthal		4	4,780	13	712	1,246
Kelthane		3	75	4	24	48
Toxaphene		10	4,200	88	866	1,125
Complex chlorinated compounds as DDT		6	132,000	80	15,182	22,746
Unknown as DDT		2	320	3	65	162
Summation of identified chlorinated hydrocarbon pesticides[2]		16	7,265	5	1,415	1,592
ORGANIC PHOSPHORUS COMPOUNDS	18					
Ethion		1	225	–	17	225
Thimet		1	35	–	3	35
Methyl Parathion		3	190	10	13	72
Parathion		1	190	–	15	190
Unknown as Parathion		4	175	15	15	59

[1] ppt = parts per trillion.
[2] Does not include Complex chlorinated compounds as DDT or Unknowns as DDT.
[3] Average value includes 0 values when chlorinated hydrocarbons were not detected.
[4] Average value includes only the detected chlorinated hydrocarbons.
* California Dept. of Water Resources, 1970.

From Li and Fleck (1972).

initiated what may be the most complete record of pesticide usage of any similar area in the world, recorded on machine-readable records and compiled by electronic computer. The records indicate that some 324 different pesticides were applied in California in 1970 weighing 63 million kg on 6.2 million ha (Fielder 1970). These records were used as the basis for a detailed study of the effects of agricultural pesticides in the aquatic environment of the San Joaquin Valley (Li and Fleck 1972). The valley floor covers 3.2 million ha including about 1.7 million ha of cropland which produced in 1973 crops valued at about US $3.6 billion. In 1970 some 38 million kg of pesticides were applied in the valley, or over half of the amount used in the entire state.

Tables 3 and 4 taken from Li and Fleck (1972) show reported pesticide concentrations in subsurface and surface drain effluents in the valley in 1970. They show that the concentrations of pesticides were higher in the surface drains than in the subsurface drains. It can be noted also that many of the identified organo-phosphorus compounds found in the surface drains are not found in the subsurface drains, probably due to their biodegradation during passage through the soil.

Some of the principal conclusions of the study were:

(1) While crop irrigation does give rise to pesticide runoff in some cases, a previous report that agricultural runoff is the greatest source of pesticides reaching California waters was not confirmed for the San Joaquin Valley.
(2) Subsurface drainage appears to be a much less likely route of pesticide transport than surface runoff, and only a small proportion of pesticides applied to a field were found in tile drainage, possibly because of adsorption on unsaturated soil particles.
(3) Direct application of pesticides to water is considered to be one of the major pathways of pesticides into the aquatic environment, even though most herbicides registered for use in aquatic situations have restrictions requiring that the herbicide must dissipate at least partially before the water is again subject to normal use.
(4) The limited information available on the persistence of pesticides in water indicates that some have a longer persistence in water than in soils, and some have a shorter. Pesticides in water have a high freedom of mixing and movement. When a pesticide is added to or transported by water, most of it becomes adsorbed to fine sediments. California studies have indicated that an average pesticide concentration of 0.1 to 0.2 ppb in water may mean that bottom sediments contain 20 to 500 ppb. Generally, the persistence time is lower in water than in the sediments, and in some cases the sorbed pesticides are, for all practical purposes, removed permanently from the aquatic environment.
(5) The concentrations of oganochlorine compounds have been decreasing in the San Joaquin River, and residues found in various parts of the aquatic environment of the valley are significantly below the LC$_{50}$ values to zooplankton, benthic invertebrates, and fish.
(6) Herbicides are generally less toxic to aquatic fauna than are insecticides, and the values encountered in the aquatic environment are well below the toxic levels.
(7) The limited evidence available indicates that the agricultural use of pesticides in the San Joaquin Valley would seem to have had no significant adverse effects upon the aquatic environment.
(8) The alternatives to pesticides for pest control should be given greater attention, as over 100 pest species of agricultural importance are resistant to pesticides, pest

control costs are increasing, and environmental pollution is of worldwide concern. The nature of alternative techniques (i.e., biological control, pathogens, sterilized males, etc.) dictates that pest management will change significantly only if supported by governmental agencies. Profits can be made from pesticide sales, but are very limited if indeed existent at all, from the sale of alternative techniques.
(9) Additional data is needed before final conclusions can be drawn on the overall effects of pesticides on the environment.

WATER QUALITY CHANGES DUE TO GROUNDWATER WITHDRAWALS

In some instances the extraction of water from underground aquifers may induce undesirable intrusions of sea water or connate brines into the aquifer which, over a period of time, can cause the quality of the water in the wells in the intrusion path to deteriorate and become unsuitable for domestic, industrial, or irrigation purposes. The deterioration usually is gradual, occurring first in the wells nearest the source of the brines. If remedial action is not taken substantial portions of the aquifer may become unusable for several decades, or longer. Reduction in pumping drafts, artificial recharge of imported water by spreading grounds or injection wells, or the use of barrier injection or withdrawal wells, are methods which have been used successfully to halt, and in some cases to reverse, the intrusion.

Saline intrusion into well fields in coastal areas near saline bodies of water should always be expected to occur sooner or later. The reasons for sustained increases in the salinity of any wells should be investigated promptly, regardless of well field location, so that remedial action can be taken, if necessary, before irreparable damage is done to the aquifers.

TWO EXAMPLES OF MAJOR SALINITY CONTROL PROBLEM AREAS IN THE USA

There are several areas in the western USA where major water quality problems exist in which irrigation is one of the major, and sometimes the major, contributor to the problems. Salinity problems alone are estimated to be present in more than 25 per cent of the irrigated lands in the 17 western continental states and Hawaii. Two of these will be described briefly in this section, selected because of their special significance.

Colorado River Basin*

The Colorado River and its tributaries drain an area of approximately 583,000 km^2 in seven states and the Republic of Mexico, and thus it is an international stream (Figs. 1 and 4). The average annual virgin flow of the river at Lee Ferry, Arizona, the dividing point between the Upper and Lower Basins, has been estimated at about 18 billion m^3/year. Present consumptive use within the basin plus deliveries to Mexico approximate 14.5 billion m^3/year, with any surplus water in a given year going into storage in one

* See Skogerboe and Law 1971, Holburt 1974a, b.

of the many reservoirs on the river. The total annual salt load at Lee Ferry is 7.4 million metric tons, of which irrigated agriculture is estimated to contribute from 1.8 to 3 million metric tons 88 percent of the total salt load from irrigated agriculture in the entire basin is estimated to originate in the Upper Basin.

FIG. 4. Proposed salinity control projects in the Colorado River Basin (Skogerboe and Law 1971).

Under a 1944 treaty with the USA, Mexico has a guaranteed allotment of 1.85 billion m^3 of water per year. Between 1945 and 1961 there were no major problems resulting from the Treaty, as the salinity of the water delivered to Mexico at the Northerly International Boundary was generally within 100 mg/l of the water at Imperial Dam, the last major diversion for US users. (TDS in 1960 at Lee Ferry was about 560 mg/l and 760 mg/l at Imperial Dam). In 1961 the Wellton-Mohawk Irrigation Project in southwestern Arizona commenced operation of a system of drainage wells which discharged saline water into the Colorado below the last US diversion but above the Mexican diversion. The drainage water included a substantial proportion of highly saline groundwater that had been concentrated through re-use during the previous 50 years. Initially it had a salinity of 6000 mg/l. This resulted in a sharp increase in the salinity of the Mexican water from around 850 mg/l in 1960 to more than 1500 mg/l in 1962. At about the same time releases into Mexico were greatly reduced in anticipation of storage behind the newly constructed Glen Canyon Dam. This loss of dilution water is illustrated by the fact that for the period 1951–60 the average delivery to Mexico was 5.2 billion m³/year, while for the succeeding 10-year period to 1970, the flow averaged only 1.9 billion m³/year. Mexico raised strenuous objections.

In 1973, following protracted negotiations, Mexico and the US agreed to a series of steps which would assure Mexico of reasonably good quality water upstream from Morelos Dam, plus other measures. The water delivered upstream from Morelos would have an annual average salinity of not more than 115 mg/l ± 30 mg/l, over the annual

TABLE 5. *Estimated costs of salinity control projects. Colorado River Basin*

Project	Salt removed (thousands tons/yr)	Annual project costs[1] (thousands dollars)	Unit cost[2] (dollars/ ton/yr)
Irrigation Improvements[3]			
Grand Valley	280	3,100	5.50
Lower Gunnison River	300	3,600	5.90
Price River	80	1,000	6.30
Uncompahgre River	290	4,000	6.90
Big Sandy Creek	35	490	6.90
Roaring Fork River	45	880	9.30
Upper Colorado River	70	1,400	9.80
Henrys Fork River	35	710	9.80
Dirty Devil River	35	710	9.80
Duchesne River	250	5,700	11.40
San Rafael River	65	1,400	11.50
Ashley Creek	35	800	12.70
Subtotals	1,520	23,790	
Stream diversion			
Paradox Valley	160	700	4.30
Impoundment and evaporation			
La Verkin Springs	70	600	8.30
Desalination			
Glenwood and Dotsero Springs	340	5,000	14.90
Blue Springs	450	16,000	35.20
Totals	2,540	46,100	
Weighted Average Unit Cost			13.50

[1] Annual project costs include amortized construction, operation and maintenance costs (1971).
[2] The unit costs only include costs allocated to salinity control.
[3] Annual project costs for irrigation improvements incorporate all costs, including those allocated to the irrigation function. Costs allocated to salinity control projects were estimated to be one-half of total annual project costs.

Adapted from Skogerboe and Law (1971).

average salinity at Imperial Dam. To comply with the agreement the US has planned the following works:

(1) A major desalting plant for Wellton–Mohawk drainage waters, by December 1978.
(2) Extension of the Wellton–Mohawk drain by 85 km to the Gulf of California, by December 1976.
(3) Lining or construction of a new Coachella Canal in California, by April 1977.
(4) Reduction in Wellton-Mohawk District acreage and improved irrigation efficiency, by December 1978.
(5) Construction of a well field on the US side of the International Boundary to balance well fields recently installed by Mexico near the border.

All of the costs in money or water to achieve the new guarantee are to be borne by the US at a cost of several hundred million dollars, and the USA as well as Mexico will receive tangible benefits. The US Congress already has authorized construction of a considerable amount of these works.

Federal and State agencies also have developed plans for several other salinity control projects to reduce Colorado River salinity. Some of these are shown on Fig. 4 and listed in Table 5. The US Bureau of Reclamation estimates that an increase of one mg/l in salinity at Imperial Dam results in costs of US $240,000/year to water users in Arizona, California, and Nevada. In the absence of any measures to control salinity, the total impact of salinity increases on users in the three Lower Basin states was predicted to be about US $80 million per year by the year 2000. Dollar values of detriments to users in Mexico would be additional, but have not been estimated.

San Joaquin Valley, California[1]

The San Joaquin Valley is walled on the east, south and west by mountains. The valley floor is about 500 km long and averages 65 km in width (Fig. 5). It contains about 3.2 million ha, almost all below an elevation of 150 meters. The valley trough slopes gently from south to north and drops from an elevation of about 90 meters near Bakersfield to sea level at the Sacramento–San Joaquin Delta near Stockton. Between the Kings and San Joaquin Rivers the northward valley slope is interrupted by a low divide that prevents natural drainage of the surface waters from the southern portion into the northern part. The area north of the divide is called the San Joaquin River Basin, and that to the south is called the Tulare Lake Basin. The divide rises only to about 7.5 meters above the lowest point in the Tulare Lake Basin.

Water supplies of the Valley are almost completely under man's control, and only during major flooding does water escape the Valley without being put to use. Essentially all domestic, industrial, and municipal water supplies are obtained from groundwater. Irrigation water comes from both surface and underground sources. Most of the available water is used at least once and often is used several times. In addition to local supplies, Northern California waters are imported by the Delta–Mendota and the California Aqueducts.

In 1973 gross income from agricultural production was US $3.6 billion, almost half of

[1] See Meixner 1974.

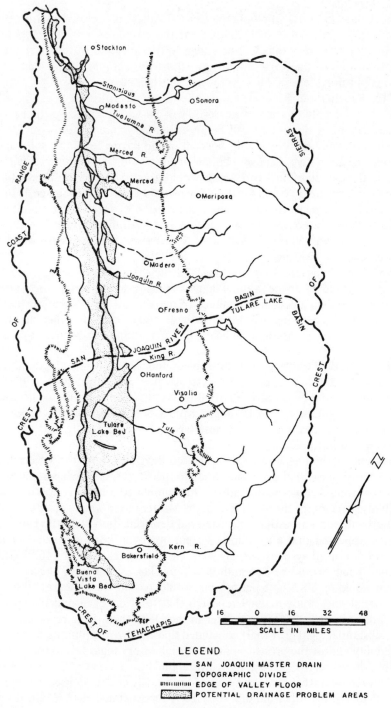

FIG. 5. San Joaquin Valley (after Skogerboe and Law 1971).

the statewide total, but this economy is seriously threatened by the growing salt management problems.

The Tulare Lake Basin (southern part of Valley) is a closed hydrologic basin, and it has a serious problem of salt balance. At the estimated 1990 development level about 1.5 million tons of new salts will be added to the soils and waters of the Basin each year. A master drain to the north could not export the amount of salt that enters the Basin, but it could provide a major step toward salt balance.

In the San Joaquin River Basin (northern part of Valley) it is estimated that by 1990 more than 1.4 million tons of new salt will be added annually to the valley floor. At the same point in time there will be an estimated annual outflow of 2.8 billion m^3/year containing 2.2 million tons of salt; hence it would have a favorable salt balance. However, without an alternative disposal channel there would be undesirable salt concentrations levels in the San Joaquin River which would greatly lower crop production.

A salt management system for the Valley would utilize tile drainage as the source of salt management. Such systems are under construction by farmers on a limited scale. During the late 1970s the composite salt content of tile drainage waters is expected to range between 6500 and 7500 mg/l. Within 30 to 50 years these concentrations would drop to between 2000 to 3000 mg/l.

Though not reusable for conventional irrigated agriculture there are other potential uses for the tile drainage waters that were considered during the State study of the problem. These included: maintenance of marshlands for recreation; cooling water for steam electric generating plants; growth of algae, fish, or other aquatic life to produce food; repressurization of oil fields; and irrigation of salt-tolerant plants for livestock feed.

Systems studied to bring the salt problem under control include: desalting, evaporation on the valley floor, transportation systems over and through the Coast Range to the Pacific Ocean, and essentially gravity flow systems from near Bakersfield to near Antioch Bridge. The total capital costs ranged from US $175 million to US $990 million (1974 costs). The lower cost plans were those having the Master Drain as an essential element, with disposal in the Delta.

Residents of the Delta and San Francisco Bay Area, however, objected to plans which would deliver untreated drainage waters to their areas, on the basis that the quality of the receiving waters would be degraded. This stimulated a large research program as to the feasibility of removing the nitrates from the Master Drain waters prior to discharge. Initially algal growth and harvesting, and bacterial denitrification were the two methods studied. Later it appeared that a symbiotic system involving both algae and bacteria would be less costly. For all systems designs were based on a capacity of 865,000 m^3/day, but the design capacity would be 2.65 million m^3/day. The estimated 1974 costs of the lowest cost system would be US $0.016 per m^3, but would require 12,000 ha of land.

A special state study in 1969 reported that total nitrogen and phosphorus concentrations are from 10 to 100 times greater in the Delta than those reported necessary to support substantial algal growth. It concluded that the growth-limiting factors then were light availability and/or the presence of toxic or inhibitory materials in the waters. It appeared feasible, therefore, to discharge agricultural tile drainage to the Bay–Delta system, without nutrient removal treatment, and it would not likely create undesirable biologic growth in the system. On the other hand, a 1971 EPA report stated that "Mixtures of San Joaquin River water and untreated agricultural drainage consistently stimulated algal growth, regardless of season," and that "nitrogen removal from agricultural drainage is definitely effective in reducing biostimulation." Obviously, professional opinions still differ.

Serious financial, social, and political problems are delaying adoption of a master plan, although considerable Federal expenditures have been made on the Master Drain. Under present state regulations, however, the drain would not be able to discharge into the Delta or the Bay.

RESEARCH NEEDS

Those working with irrigation development know well that much more information is needed, on many subjects, to enable us to better understand the nature and extent of changes in water quality which result from irrigation. Much of this must come from scientists working in specialized fields, such as in pesticides, eutrophication, soil science, and public health. However, it is known that irrigation return flows are a prime cause of salinity problems which in turn can depress downstream crop yields and economic returns. We therefore need to conduct research directed at controlling return flow pollution by practical, economically-feasible means, to disseminate the knowledge to the world, and to put that knowledge to work in irrigation and river basin projects on which we work. Furthermore, we should apply the large body of existing knowledge to both immediate and potential problems.

A brief listing of some of the more important areas of needed research follows, adopted from Skogerboe and Law (1971).

Irrigation practices

(1) Automation of surface irrigation systems.
(2) Tailwater runoff and re-use—quality degradation.
(3) Sprinkler irrigation—degradation by sediments and pollutants adsorbed on sediments.
(4) Use of high salinity waters.
(5) Drip and trickle irrigation.
(6) Fertilizer application practices and efficiencies.
(7) Increased irrigation system efficiencies involving some aspects of several of the above.

Soil-plant—Salinity relationship

(1) Salt tolerance of crops under various irrigation practices.
(2) Short-term resistance to salinity.
(3) Slow-release fertilizers.
(4) Use of surface mulches or reduced tillage.
(5) Water quality changes in and below the root zone.

Leaching

(1) Optimum leaching requirements for various types of waters, soils and crops.

Prediction of subsurface return flow

(1) Improvement of prediction techniques, including computerized models, which will describe quantity and quality of return flow.
(2) Lysimeter and field plot studies to confirm model studies and provide proper model coefficients.

Irrigation scheduling

(1) Procedures to decrease return flows and salinity, and increase production.

Pump-back systems

(1) Effects of re-circulation. Criteria for re-use.

Treatment

(1) Determine rates of denitrification in the field.
(2) Control of nitrogen in root zone.
(3) More economical nitrate reduction methods.
(4) Nutrient balance studies on typical crops under different agricultural practices and irrigation systems.

Economic evaluation

(1) Crop production and damage functions.
(2) Costs of alternative management systems.
(3) Local, regional, and national benefits and/or losses.
(4) Identification of beneficiaries.

Institutional Needs

(1) Incentives for efficient water use.
(2) Authorities and laws.

CONCLUSION

When all available evidence is considered, the conclusion is inescapable that irrigation does have significant effects on the quality of underground and downstream waters. While most of these effects are of an adverse nature, they often are not of immediate serious magnitude, depending upon the circumstances in each case. As irrigation development increases in a river basin, the importance of these adverse effects increases also. If economic losses to farmers and other water users and their nations are to be minimized, overall water quality management planning should begin now, if not already under way, followed by the execution of control measures as appropriate, but before significant damages occur. In large arid river basins the magnitude of the problem may be immense, but lack of action may only increase the difficulty and cost of a final solution.

References

Baer, T. J., Jr. 1972. Grand Valley salinity control demonstration project, Proc. Natn Conf. managing irrigated agriculture to improve water quality, Graphics Management Corporation, Washington, D.C., pp. 109–122.

Bliss, J. H. 1965. Water quality changes in Elephant Butte Reservoir. Trans. Am. Soc. Civ. Engrs, 130, 57–58.

Borchardt, J. A. 1969. Eutrophication—causes and effects. J. Am. Wat. Wks Ass. 61, 272–275.

Carter, D. L. 1972. Irrigation return flows in Southern Idaho, Proc. Natn Conf. on managing irrigated agriculture to improve water quality, Graphics Management Corporation, Washington, D.C., pp. 47–53.

Christiansen, J. E. 1973. Effect of agricultural use on water quality for downstream use for irrigation, Proc. A.S.C.E. Irrigation and Drainage Division Specialty Conference in Fort Collins, Colorado, Am. Soc. Civ. Engs, pp. 753–785.

Clark, J. W. 1972. Salinity problems in the Rio Grande basin, Proc. Natn Conf. on managing irrigated agriculture to improve water quality, Graphics Management Corporation, Washington, D.C., pp. 55–66.

Eldridge, E. F. 1963. Irrigation as a source of water pollution. Wat. Pollut. Contr. Fedn, 35, No. 5, 614–625.

Faulkner, L. R. and Bolander, W. J. 1970. Agriculturally-polluted irrigation water as a source of plant-parasitic nematode infestation. J. Nematology, 2, 368–374.

Fielder, J. W. 1970. Pesticide use report 1970, California Department of Agriculture, 107 pp.

Hagan, R. M. and Roberts, E. B. 1975. Environmental impacts of water projects, Special Session International Commission on Irrigation and Drainage in Moscow, USSR, on "Environmental control for irrigation and flood control projects," pp. 187–218.

Holburt, M. B. 1974a. International problems of the Colorado River, American Association for the Advancement of Science Meeting, San Francisco, California, February–March, 26 pp.

Holburt, M. B. 1974b. Recipe for more energy—just add water, Address to Pacific Northwest Waterways Association, November, Seattle, Washington, 28 pp.

Jenke, A. L. 1974. Evaluation of salinity created by irrigation return flows. Office of Water Program Operations, U.S. Environmental Protection Agency, 128 pp.

Kinawy, I. Z. and Shenouda, W. K. 1975. Ecological, social and economical impacts of damming the Nile at Aswan, Special Session International Commission on Irrigation and Drainage in Moscow, USSR, on "Environmental control for irrigation and flood control projects," pp. 275–293.

Lange, W. 1970. The American City Magazine, August, p. 16.

Li, M. and Fleck, R. A. 1972. The effects of agricultural pesticides in the aquatic environment, irrigated croplands, San Joaquin Valley, U.S. Environmental Protection Agency, Office of Water Programs, Pesticide Study Series 6, 268 pp.

Love, S. K. 1961. Relationship of impoundment to water quality. J. Am. Wat. Wks Ass. 53, 559–568.

Meixner, G. D. 1974. Status of San Joaquin Valley drainage problems, California Department of Water Resources, December, 66 pp.

Sawyer, C. N. 1947. Fertilization of lakes by agricultural and urban drainage. J. New Engl. Wat. Wks Ass. 61, 109.

Sawyer, C. N. 1965. Problems of phosphorus in water supplies. J. Am. Wat. Wks Ass. 57, 1431–1439.

Silvey, J. K. 1968. Effects of impoundments on water quality on the Southwest. J. Am. Wat. Wks Ass. 60, 375–379.

Skogerboe, G. V. and Law, J. P. Jr. 1971. Research needs for irrigation return flow quality control, Office of Research and Monitoring, U.S. Environmental Projection Agency, Project No. 13030. 98 pp.

Skogerboe, G. V. and Walker, W. R. 1972. Salinity control measures in the Grand Valley, Proc. Natn Conf. on managing irrigated agriculture to improve water quality, Graphics Management Corporation, Washington, D.C. 123–136.

Sylvester, R. O. and Seabloom, R. W. 1963. Quality and significance of irrigation return flow. J. Irrig. Drain. Div. Am. Soc. Civ. Engrs, 89 (No. IR3 September), 1–27.

Symons, J. M. Weibel, S. R. and Robeck, G. G. 1965. Impoundment influences on water quality, J. Am. Wat. Wks Ass. 57, 51–75.

U.S. AWWA Task Group Report 1967. Sources of nitrogen and phosphorus in water supplies. J. Am. Wat. Wks Ass. 59, 344–363.

U.S. Committee on Pollution 1966. Waste management and control, Publication 1400, NAS—National Research Council report submitted to Federal Council for Science and Technology, Washington, D.C.

U.S. Department of Health. 1969. Secretary's Commission on pesticides and their relationship to environmental health. 677 pp.

U.S. Water Quality Criteria. 1968. National Technical Advisory Committee to the Secretary of the Interior, Federal Water Pollution Control Administration, 234 pp.

U.S. Salinity Laboratory Staff. 1954. Diagnosis and improvement of saline and alkali soils, Agriculture Handbook No. 60, U.S. Department of Agriculture, 160 pp.

van Schilfgaarde, J., Bernstein, L., Rhoades, J. D. and Rawlins, S. L. 1974. Irrigation management for salt control. *J. Irrig. Drain. Div. Am. Soc. Civ. Engrs*, **100** (No. IR3, September), 321–338.

Weiss, C. M. 1969. Relation of phosphates to eutrophication. *J. Am. Wat. Wks Ass.* **61**, 387–391.

Wilcox, L. V. 1962. Salinity caused by irrigation. *J. Am. Wat. Wks Ass.* **54**, 217–222.

Williardson, L. S., Olsen, E. C. III, and Christiansen, J. E. 1975. Discussions of Reference (13). *J. Irrig. Drain. Div. Am. Soc. Civ. Engrs*, **101** (No. IR2, June), 122–128.

Yost, C. Jr. and Naney, J. W. 1975. Earth-dam seepage and related land and water problems. *J. Soil Wat. Conserv.* **30** (2), 87–91.

SOME OBSERVATIONS CONCERNING THE CHANGE IN THE HYDROLOGICAL CYCLE CAUSED BY IRRIGATION

by T. PÉCZELY

Research Institute for Water Resources Development, Budapest, Hungary

Summary

Irrigation causes the most drastic change in the short stretch of the hydrological cycle which crosses the unsaturated zone and influences evapotranspiration. On the basis of theoretical studies and observed data a characteristic curve of the groundwater balance was established to describe hydrological phenomena above the water table. Examples of this curve in semi-arid and humid regions illustrate how the unsaturated zone is modified under irrigated conditions. Seasonal fluctuation of water table and moisture content and evapotranspiration on the basis of hydrological and meteorological observations are demonstrated, and conclusions are reached concerning the summer deficit of evapotranspiration and surface evaporation under irrigated condition.

Résumé

L'irrigation cause de grands changements dans la partie terrestre du cycle de l'eau, en particulier parce qu'elle affecte la zone non saturée et influence l'évapotranspiration. Les courbes caractéristiques du régime des eaux souterraines, et qui expriment les phénomènes hydrologiques au-dessus du plan supèrieur des nappes, ont été déterminées en se foundant sue des analyses théoriques et des données d'observations. Les examples de courbes caractéristiques pour les régions semi-arides et humides illustrent bien comment la zone non saturée est modifiée par l'irrigation. Les résultats obtenus grâce aux observations hydrologiques et météorologiques sur les fluctuations saisonnières des nappes, l'humdité et l'évaporation sont présentés et des conclusions sont tirées sur le déficit d'été de l'évapotranspiration et sur l'évaporation superficielle en condition d'irrigation.

Introduction

This paper deals with two topics selected from the large field covered by the influence of irrigation on the hydrological cycle, namely forecasting of the expected change in the water table under the irrigated area and the increase of evapotranspiration as a result of irrigation. At first sight it seems that the two topics have little to do with each other. However, both depend basically on the soil moisture zone lying between the soil surface and the water table.

The importance of the soil moisture zone in the investigation of the hydrological cycle, both under natural and man-influenced conditions, accords closely with the scientific

159

objectives of the IAHS which has initiated a world-wide project, the purpose of which is to establish an inventory of the available data and methods for this subject.

Forecasting of the expected change of the water table under irrigated areas

To determine the new position of the water table under irrigation, the first step is to find a method for characterizing the natural water balance, for the changes caused by irrigation can be investigated only after knowing the natural parameters. The second step is to determine the modification of these parameters. Thereafter the new equilibrium can be characterized and this probably determines the position of the water table. For the description of the balanced condition of the groundwater and the interrelationships between the processes the characteristic curve of the groundwater balance can be used. The method was proposed by Kovács (1959a, 1959b) and some examples of its application were also published by Kovács (1960, 1966, 1971). Although there are other publications dealing with this method a summary of its main aspects is indispensable for understanding its application.

The hydrological processes occurring in the soil-moisture zone are influenced by climate (precipitation, evaporation, wind) and land surface (slope, cultivation, covering vegetation) as well as by the *composition and the instantaneous state of soil* (structure, permeability, field capacity, moisture content). If these parameters are homogeneous within the investigated area (or their local variation can be neglected, and thus they can be characterized by averages calculated from place-dependent data) the groundwater space of the area

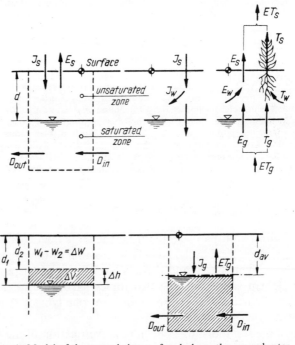

FIG. 1. Model of the water balance of and above the groundwater.

in question can be simulated by a soil column extending from the surface to the water table (Fig. 1).

The water balance of the separated column in a period from t_1 to t_2 can be described by the following equation:

$$\int_{t_1}^{t_2} I_S(t) - ET_S(t) + D_{in}(t) + D_{out}(t)\, dt = (V_2 - V_1) + (W_2 - W_1), \tag{1}$$

where I_S is infiltration through the surface;

ET_S includes both direct evaporation and transpiration originating from soil moisture and groundwater;

D_{in} and D_{out} are the discharges of groundwater flow recharging and draining the column respectively. (Note that the flow may be horizontal and/or between the shallow groundwater and deeper lying aquifers);

V_1 and V_2 are the amount of water stored below the water table at the beginning and at the end of the investigated period;

W_1 and W_2 indicate the volume of soil moisture similarly at the bordering time points of the period.

The first three parameters (I_S; ET_S and D) are time-dependent variables. Investigating a sufficiently long period, within which their fluctuation is equalized, the average calculated in time can be used instead of the changing value, and Eq. 1 can be simplified:

$$[(\bar{I}_S - \bar{E}\bar{T}_S) + (\bar{D}_{out} - \bar{D}_{in})](t_2 - t_1) = \pm \Delta V \pm \Delta W. \tag{2}$$

Further simplification can be achieved by selecting the starting and the end point of the period in a way which ensures that the closing amount of the stored water ($V_2 + W_2$) is equal to that at the beginning of the investigation ($V_1 + W_2$).

Even a further restriction could be the equality not only between the sums of the stored water but the negligible small change in the numbers as well:

$$V_2 + W_2 = V_1 + W_1;$$

and
$$V_2 = V_1; \quad W_2 = W_1. \tag{3}$$

Accepting this approximation, the investigation can be limited to the groundwater space (see the last sketch of Fig. 1), and, therefore, the average of the positive (I_g) and negative accretion (ET_g) of groundwater has to be considered instead of the infiltration and evapotranspiration through the surface. Thus the final simplified form of the balance equation,

$$I_g - ET_g = D_{in} - D_{out}, \tag{4}$$

expresses that the vertical water exchange between groundwater and soil moisture is equal to the balance of the groundwater flow.

Both the positive and negative accretions are the functions of climatic, surface and soil conditions as well as of the average depth of the water table. In a given area where the natural factors can be regarded as constant known parameters (because they are homogeneous or can be approximated by averages calculated for the area), the infiltrating recharge, the water amount raised from the groundwater to replenish the evaporated soil moisture, and also their difference can be characterized as a function of the depth of water table (Fig. 2). The latter curve is called the characteristic curve of groundwater balance.

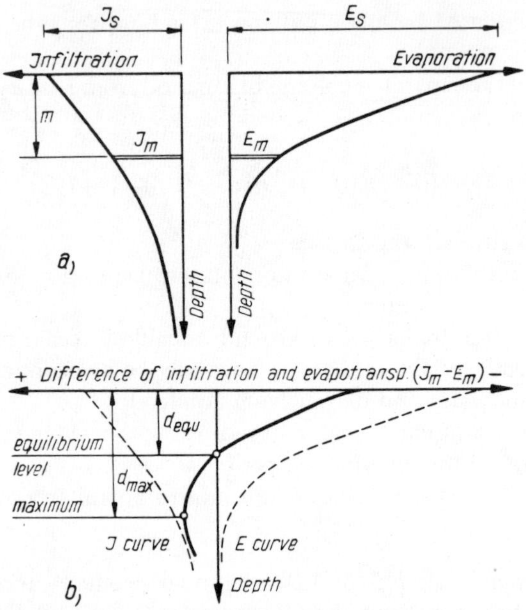

FIG. 2. Interpretation of the characteristic curve of groundwater balance.

The characteristic resultant curve has two significant points:

Its intersection with the vertical axis indicates the equilibrium level. The water table can develop at this depth only, when the result of the groundwater flow is zero (inflow is equal to outflow). Where groundwater flow recharges the area, the water table has to be above the equilibrium level to ensure the water balance, while in areas drained below the water table positive accretion has to surpass the negative one, which condition can be expected only in the case of a water table below the equilibrium level.

The maximum point of the curve divides the range of depth into two parts. Above this level the change of water exchange moves in a positive direction by increasing depth, indicating that dynamic resources become available by lowering the water table. Below this level the direction of the curve is the opposite, which means, accretion is decreased by lowering the water table.

From the point of view of the recent investigation it is important that the curve gives information about the relationship between the change of the position of the water table and accretion. Knowing the change expected as a result of irrigation or drainage, the depth of water table belonging to the new modified balance can be estimated.

Two numerically determined characteristic curves have been published: one for the humid region of Byelorussia (Lavrov et al. 1972) and one for the semi-arid area of the Great Hungarian Plain (Kovács 1959a). On the basis of observations concerning the chemical composition of groundwater at different depth of water table the probable curve in arid region can also be estimated (Fig. 3). Observations were made in Tafilalet, Morocco, by Hazan (1967).

For the application of the characteristic curve to estimate the new position of the water table under irrigation, it is necessary to note that the natural water balance is disturbed by irrigation in two different ways: (1) Seepage losses from canals, reservoirs or stretches of rivers, where the natural levels are raised by barrages, increase the discharge flowing into

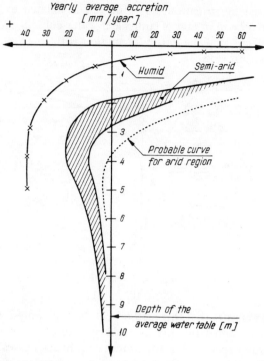

FIG. 3. Comparison of characteristic curves of groundwater balance determined for humid semi-arid and arid regions.

the investigated area below the water table; (2) the irrigation water distributed over the area reaches the groundwater space through the unsaturated zone.

If the area is influenced only by groundwater recharge arriving below the water table (non-irrigated area along canals and reservoirs, and area beside irrigated lands where the groundwater receives horizontal recharge from groundwater as a consequence of irrigation nearby) the characteristic curve can be applied in its original form. After calculating the

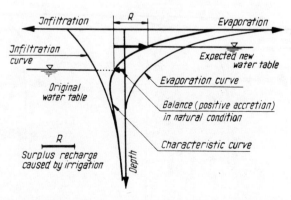

FIG. 4. Characteristic curve of groundwater balance used to forecast the expected depth of groundwater recharged artificially below the water table.

surplus recharge caused by irrigation, its value can be related to the extension of the influenced area and expressed in mm of equivalent water column. Adding this amount to the natural balance (decreasing the natural accretion with this value), the expected new water table can be forecast (Fig. 4).

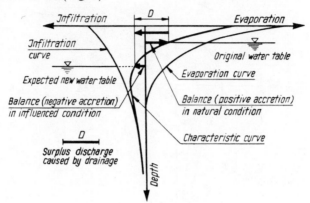

FIG. 5. Application of characteristic curve to forecast the expected position of the water table in drained areas.

It is necessary to emphasize that this method can be used only if the regime of the soil moisture zone is not modified by surplus recharge caused by artificial activity adding water to the groundwater space below the water table. In such cases the natural processes acting in the soil moisture zone (infiltration, evapotranspiration, storage) are completely modified, and, therefore, the characteristic curve is not valid.

It is worthy of mention that the influence of drainage can also be shown by the natural characteristic curve, because the drains withdraw water below the water table. Thus the estimation of the expected new water table within a drained area follows the same steps as above but with the opposite sign (Fig. 5).

As already mentioned, the natural characteristic curve cannot be applied directly to irrigated fields, because here the qualitative modification of the hydrological processes occurring in the soil moisture zone has to be considered as well. For establishing a char-

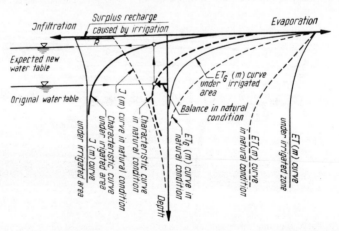

FIG. 6. Construction of a characteristic curve for the groundwater balance modified by irrigation.

acteristic curve, describing the new conditions infiltration and evapotranspiration have to be investigated separately (Fig. 6 and Kovács 1972).

The intersection of the infiltration curve with the horizontal axis (infiltration through the surface) indicates a higher value than the natural recharge. This surplus is caused by the infiltrating irrigation water. The curve runs almost vertically downwards from this starting point because the free storage capacity is almost negligible throughout the whole year in the unsaturated zone affected by irrigation. The result of this fact is that the decrease of positive accretion with increasing average depth of the table is insignificant.

The starting point of the evaporation curve is practically unchanged at the surface. In this case the total evaporation hardly depends on the depth of the water table, so this parameter may be represented by an almost vertical line in the co-ordinate system. This relationship indicates that in the case of an adequately executed irrigation the plants always have the optimum amount of water necessary for their growth. The ratio of the amount of water consumed from the groundwater related to that supplied from the unsaturated zone, however, differs considerably from the same ratio observed under natural condition. Negative accretion becomes smaller than it was originally, because the capillary suction and the tension difference ensuring the vertical groundwater movement upwards against gravity is relatively small, the top layer being almost saturated. Thus the water consumption from the gravitational groundwater is practically zero below the root zone.

From the two curves described above the characteristic curve of the groundwater balance influenced by irrigation can be easily constructed. This relationship determines the modified resultant of positive and negative accretion (infiltration and evaporation) as a function of the depth of the average water table. Thus the difference of the two curves at a given depth indicates the discharge of horizontal groundwater flow, which is balanced dynamically when the water table develops at that level.

The modified characteristic curve is an important means for investigating the salt balance of the irrigated area as well. The hydrological balance of the unsaturated zone is fundamentally changed by the modification of the processes of infiltration and evaporation respectively. As indicated by the curve, the equilibrium level, where infiltration is equal to evaporation, moves upwards. Thus there is a zone between the natural and the modified equilibrium levels, where the direction of the resultant of vertical water movement changes. In natural conditions this movement was directed upwards, indicating salt accumulation, while in the case of irrigation the unsaturated zone above a groundwater having its table in this zone becomes leached. Thus the leached area can be extended by irrigation, which is in good agreement with the practical experience that irrigation, combined with sufficient drainage, can be used for leaching salts from the soil.

As a final result of this analysis, it can be stated, that distinction has to be made between the effects of irrigation on the development of a new balanced water table, investigating whether the surplus water reaches the groundwater space below the water table or through the soil moisture zone. In the first case—similarly to the determination of the average water table in drained areas—the characteristic curve of the natural groundwater balance can be used as the basis of the estimation. Irrigation water infiltrating through the unsaturated zone changes completely the regime of soil moisture and the hydrological processes occurring in this zone. It is necessary, therefore, to determine the characteristic curve valid under the artificially influenced condition.

Increase of evapotranspiration as a result of irrigation

Various terms have been used to indicate the different types of evapotranspiration. The most important difference occurs between the potential and actual values (IAHS/HIUZ 1973, 1974), the definitions of which are as follows:

Potential evaporation and/or transpiration is the highest possible amount of water evaporated and/or transpired in a given time from the area investigated. This is determined by the vapour receiving capacity of the overlying air mass, influenced by the lower boundary conditions of the vapour receiving system. Thus the potential value of evaporation and/or transpiration of free water surface, bare soil and various kinds of vegetation cover can be distinguished.

Effective (actual, or real) evaporation and/or transpiration is the amount of water actually transformed from liquid to vapour during the investigated period within the area investigated. The upper limit of this value is the amount of vapour that can possibly be absorbed by the receiving system. The development of the highest possible evapotranspiration is hindered, however, by the water shortage in the vapour producing system. The same types of the effective parameter (free water surface, bare soil, vegetation) can be distinguished as those of the potential value.

Irrigation has very little influence on the vapour receiving system but a big effect on the vapour producing system. The water receiving capacity of the air mass overlying irrigated areas is decreased only if the area is very large and the horizontal movement of air is insignificant. In this case the higher effective evapotranspiration at the beginning of the period of irrigation raises the moisture content of the air and decreases its water receiving capacity. The arrival of dry air masses transported by wind makes this decrease of potential evaporation practically negligible.

The purpose of irrigation is to ensure within the root zone the water amount necessary for the plants during the whole growing period. If this is achieved by correct operation, water is always available for evaporation and transpiration. Thus with no water shortage in the vapour producing system maximum possible evapotranspiration can be achieved. The expected change of evapotranspiration as a result of irrigation can be calculated,

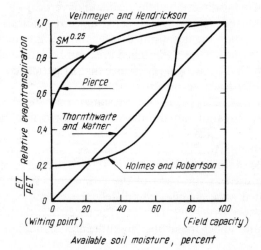

FIG. 7. Results of various studies which show the dependence of relative evapotranspiration on soil moisture.

therefore, as the difference between potential evapotranspiration within the area in question and the actual value observed or estimated under the conditions of dry farming.

There is another concept proposed by Petrasovits which divides the factors influencing evapotranspiration into three groups instead of the two mentioned above (receiving and producing), namely: energy; available water; acting surface. The energy maintaining evapotranspiration is determined by the physical characters of the vapour receiving system and is hardly influenced by irrigation. Separation of the water producing system into two parts indicates that evapotranspiration can be raised by irrigation, not only by availability of water but also by increase of acting surface consequent on more rapid growth of plants.

After explanation of the above basic ideas, the further purpose of this paper is to show some examples.

Mustonen and McGuinness (1968) have summarized previous results, supplemented by their own research, on the reduction of evapotranspiration caused by water shortage. They

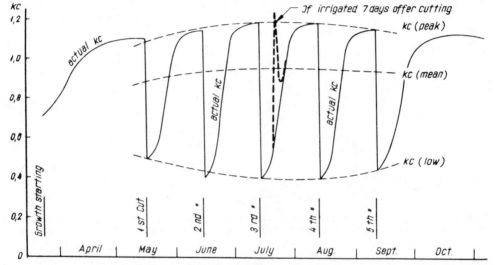

FIG. 8. k_c values for alfalfa grown in dry climate with height to moderate wind with cuttings every 4 weeks (one heavy irrigation per growth period, a week before cutting, is assumed).

supposed that actual evapotranspiration is equal to the potential value if the soil moisture is equal to or above field capacity, but it decreases with decreasing moisture content. The latter is expressed in mm of water stored in the top 1 m of soil (Fig. 7). This concept is the basis of the method proposed by FAO for irrigation practice to determine crop water requirements (Doorenbos and Pruitt 1975). A basic value has first to be determined (ET_o), which is practically equal to the potential evapotranspiration, and it has to be multiplied by a crop factor (k_c). The average of this factor is near unity and its actual value changes in time according to the development of the plants (Fig. 8).

The data observed in two groups of lysimeters at Montpellier provide numerical comparison between potential and actual evapotranspiration (Fig. 9—direct information from Y. Cormany). At both stations, A and G, the data extend to 12 years.

These graphs, constructed from the averages of the monthly values, prove the dependence of the potential value on the properties of the active evaporating surface, but this influence is not very strong. However, the surplus evapotranspiration caused by the maintenance of

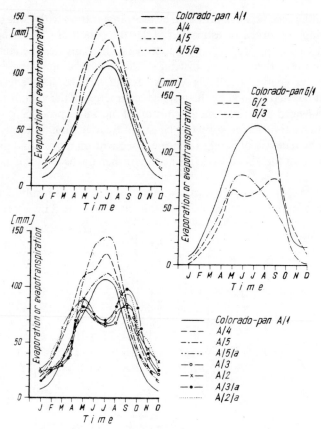

FIG. 9. Graphical representation of the monthly average values of pan evaporation, and evaporation from bare
lysimeters as well as evapotranspiration from lysimeters covered by vegetation at the experimental station
of d'Alrance in France from 1957 till 1970.

A/1 Colorado pan; A/2 bare lysimeter without gravitational groundwater (depth 50 cm); A/2/a bare lysimeter
without gravitational groundwater (depth 100 cm); A/3 lysimeter covered by vegetation without gravitational
groundwater (depth 50 cm); A/3/a lysimeter covered by vegetation without gravitational groundwater (depth
100 cm); A/4 bare lysimeter with constant water table (depth 50 cm); A/5 lysimeter covered by vegetation with
constant water table (depth 50 cm); A/5/a lysimeter covered by vegetation with constant water table (depth 100 cm).
G/1 Colorado pan; G/2 lysimeter covered by vegetation without gravitational groundwater (surface area 1 m²);
G/3 weighing lysimeter covered by vegetation with constant watertable (surface area 400 m²).

a continuous shallow water table surpasses 25% of the yearly value, and the amount of
water evaporated and transpired during the growing season is almost doubled if water is
easily available in the root zone.

The yearly average of the increase of evapotranspiration is around 250 mm in this
Mediterranean climate. In arid zones even higher influence of irrigation can be expected,
because the actual evapotranspiration under natural conditions is smaller and the potential
value is higher.

It can be proved also that the minimum of evapotranspiration, occurring in mid-summer
because of the shortage of available water, is not a result of the artificial conditions
created in the lysimeters, but is characteristic. Figure 10 shows the seasonal run of average
actual evapotranspiration in Georgia from 1958 till 1963 as determined by Magyar (1973).

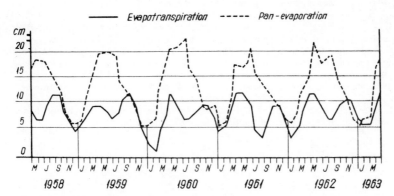

FIG. 10. Graphical representation of the seasonal run of the average actual evapotranspiration and pan evaporation in Georgia, USA, from 1958 till 1963.

The data were determined by calculating the balance of atmospheric vapour flux over a large area using meteorological observations on vertical moisture distribution, wind velocity and direction. The two yearly maxima and the summer minimum between them indicate the influence of water shortage.

Figure 10 draws attention also to changes from year to year. Similar differences occur between the annual values of potential evapotranspiration which depend on changes in meteorological processes. The difference between the averages of potential and actual values represented in Fig. 9 can be used, therefore, only as rough estimation in predicting the influence of irrigation. More accurate determination depends on the dynamic character of evapotranspiration. For this reason not only the daily meteorological conditions are needed for the calculation of instantaneous evaporation but the daily effective evaporation has to be determined as well.

Further investigations may depend on the continuous observation of moisture content in the soil, or the construction of reliable lysimeters. The better understanding and further analysis of hydrological processes occurring in the unsaturated zone is one of the most important scientific bases of irrigation.

References

Doorenbos, J. and Pruitt, W. O. 1975. Crop water requirements. FAO Irrigation and Drainage Paper No. 24. Rome.

Hazan, R. 1967. Influence of evaporation on the total salt content of groundwater at Tafilalet. Rabat. (Manuscript in French.)

IAHS/HIUZ Project. 1973. Hydrological investigation of the unsaturated zone. First circular. IAHS-VITUKI, Budapest.

IAHS/HIUZ Project. 1974. Hydrological investigation of the unsaturated zone. Second circular. International Glossary IAHS-VITUKI, Budapest.

Kovács, G. 1959a. Determination of the discharge of groundwater flow on the basis of the investigation of groundwater balance. *Vizügyi Közlemények* 3. In Hungarian.

Kovács, G. 1959b. Groundwater household. Annual Bulletin of ICID.

Kovács, G. 1960. Relationship between salt accumulation and groundwater balance. *Hidrológiai Közlöny* 2. In Hungarian.

Kovács, G. 1966. Relationship between infiltration and groundwater household. *IAHS Symposium on Water in Unsaturated Zone.* Wageningen.

Kovács, G. 1971. Salt accumulation in groundwater on soils. *IAHS Symposium on Groundwater Pollution.* Moscow.

Kovács, G. 1972. Groundwater problems related to agricultural water use. International post-graduate course on hydrological methods for developing water resources management, VITUKI, Budapest, V/2.

Lavrov, A. P., Fadeyeva, M. V., Sachok, G. I., Kakhovsky, A. P. and Vasilyev, V. F. 1972. Problems of underground water regimes in Byelorussian Polassie. *UNESCO/IAHS Symposium on Hydrology of Marsh-ridden Areas.* Minsk.

Magyar, P. 1973. Simulation of the hydrologic cycle using atmospheric water vapour transport data. VITUKI, Budapest.

Mustonen, S. and McGuinness, J. L. 1968. Estimating evapotranspiration in a humid region. Agricultural Research and Development Center. Techn. Bull. No. 1389. Coshocton, Ohio, USA.

ENVIRONMENTAL CHANGES DUE TO IRRIGATION DEVELOPMENT IN SEMI-ARID PARTS OF NEW SOUTH WALES, AUSTRALIA

by S. PELS and M. E. STANNARD

Water Conservation and Irrigation Commission, NSW, Australia

Summary

A brief description of New South Wales is given in terms of physiography and climate. Early irrigation development and the basis for this development at that time are discussed and contrasted with the present approach to land selection. An outline is given of geomorphological and soils studies of the extensive alluvial inland plains where most of Australia's irrigation is situated. Some agricultural practices peculiar to Australian irrigation, and based on the concept of the most economic utilization of water and soil resources relative to limited manpower, are discussed.

Adverse hydrologic impacts on the environment have resulted from irrigation. These include such problems as shallow water tables, surface salt accumulation, increased drainage requirements and deterioration of river flow quality, the latter being caused by both naturally occurring and artificially induced surface and sub-surface drainage inflows. The interconnected and complementary nature of deterioration of water quality in the downstream river sectors and soil salinity and waterlogging problems of irrigated lands is stressed. Hydrogeological studies and surveys have established the nature and causes of problems and indicate their present and potential extent.

Methods of investigation are discussed and an explanation of existing and potential problem areas is put forward in terms of geological and topographical features. The close relationship of geomorphology and geology on the one hand, and the distribution patterns of soils and their salt status on the other, is presented. Studies of soil stratigraphy have been used in an attempt to map areas not liable to soil salinization due to their inherent deep clay nature; the theory of this approach is described.

To overcome drainage deficiencies, four main measures are employed, based on geological and soil characteristics, and according to the intensity of land use. These are tube well drainage, tile drainage, surface drainage and "dry" drainage by evaporation. Experimental trials and the operation and efficiency of later constructed permanent installations are discussed.

The limited drainage outlet (the Murray River) of the irrigated region and the use of this river as a conveyance of both irrigation supply and drainage water have led to difficulties in river operation relative to quality control. There are four possible approaches to dealing with these difficulties. The first three involve (i) dilution of the lower river by additional releases from upstream storages, (ii) abatement of saline ground water inflow by interception at known point sources and (iii) diversion of saline surface drainage inflows. A combination of these three methods is presently practised to maintain downstream water quality. The abatement measures revolve around the principle of disposal of saline drainage waters to hydrologic sinks other than the river system and an outline is given of recent investigations into solar salt production, including technical difficulties in disposal of bitterns and the economics of the disposal of sodium chloride.

The fourth possibility involves separation of the supply and drainage functions of a river by creating a separate and artificial supply channel system adjacent to the river and above the general ground water levels of the region.

Résumé

Les Nouvelles Galles du Sud sont décrites du point de vue de la géographie physique et du climat. Les premières réalisations en matière d'irrigation, et les bases sur lesquelles elles reposaient à l'époque, sont discutées et comparées à l'approche actuelle du choix des terres. On donne un aperçu des études géomorphologiques et pédologiques des vastes plaines alluviales de l'intérieur, où se localise l'essentiel des irrigations réalisées en Australie. On discute également de certaines techniques agricoles d'irrigation propres à l'Australie et fondées sur le concept de l'utilisation la plus économique des ressources en eau et en sol en fonction des faibles disponibilités en main d'oeuvre.

L'irrigation a eu un impact hydrologique néfaste sur l'environnement. Les problèmes qui se sont posés ont trait à la faible profondeur des nappes, à l'accumulation des sels en surface, aux besoins accrus en drainage, et à la dégradation de la qualité des eaux des rivières, ce dernier aspect ayant pour origine les apports dûs au drainage superficiel et sub-superficiel, naturel ou provoqué. L'accent est mis sur le caractère complémentaire et dépendant de la dégradation de la qualité des eaux secteurs aval d'une part, et de la salure du sol et des problèmes d'engorgement des terres irriguées, d'autre part. Les études et les enquêtes hydrogéologiques ont permis d'élucider la nature et les causes des problèmes poses, et elles précisent leur extension actuelle et potentielle.

Les méthodes d'étude sont l'objet d'une discussion, et on expose les domaines d'étude actuels et potentiels en fonction des caractéristiques géologiques et topographiques. On montre l'existence d'une relation étroite entre la géomorphologie et la géologie, d'une part, et la configuration de la répartition des sols et de leur salinité, d'autre part. Il a été fait appel à des études sur la stratigraphie des sols pour tenter de cartographier les zones qui ne sont pas sujettes à la salinisation du fait de leur richesse en argile; la théorie de cette approche est décrite.

Pour pallier les déficiences de drainage, on peut utiliser quatre types principaux de mesures, fondés sur les caractéristique géologiques et pédologiques, et selon l'intensité de l'utilisation des terres. Ce sont le drainage par puits busés, le drainage par poteries, le drainage superficiel et le drainage "à sec" par évaporation. On discute des essais expérimentaux ainsi que de la mise en œuvre et de l'efficacité des installations permanentes récemment réalisées.

Les possibilités limitées de l'exutoire de la région irriguée (la rivière Murray) et l'utilisation de cette rivière à la fois comme fournisseur d'eau d'irrigation et comme collecteur des eaux de drainage, ont été à l'origine de difficultés pour contrôler la qualité des eaux. Il existe quatre méthodes possibles d'aborder ces difficultés. Les trois premières sont: (i) la dilution des eaux du cours inférieur par des apports complémentaires provenant des retenues situées en amont; (ii) la réduction des apports d'eaux des nappes salées par captage aux points de résurgence connus; (iii) la diversion des apports superficiels d'eaux de drainage salées. Actuellement, on applique une méthode consistant en une association de ces trois techniques pour maintenir la qualité des eaux en aval. Les mesures de réduction des apports des nappes salées reposent sur le principe de la dérivation des eaux de drainage salées vers des marais n'appartenant pas au système des cours d'eau, et on donne un aperçu des recherches récentes sur la production de sel grâce au soleil, sans oublier les difficultés techniques dûs au devenir des butors, ni l'aspect économique de l'utilisation du chlorure de sodium.

La quatrième possibilité est celle offerte par la séparation entre la fonction d'alimentation et la fonction de collecteur de la rivière, grâce à la création d'un système de canal d'alimentation séparé et artificiel, adjacent à la rivière, et au-dessus du niveau général des nappes de la région.

Introduction

The total area of New South Wales is about 800,000 km^2, which is four-fifths the size of the Republic of Egypt.

Out of that total area, 45% receives an average annual rainfall of less than 375 mm. Land classified as arid (less than 250 mm) comprises 20% of the State so that 25% of the State (or 200,000 km^2) is classified as semi-arid. The bulk of irrigation development has taken place in this latter area. The elevated eastern region of the State receives a higher rainfall and 15% of New South Wales has an average annual rainfall exceeding 750 mm.

The disposition of river systems is determined by the State's major physiographic feature, the Great Dividing Range. This range runs approximately parallel with the coast, some 200 km inland, and forms the line of division between the headwaters of relatively steep and short coastal rivers and those of the much longer and very low grade rivers of the Murray–Darling river system, which travel across the arid and semi-arid inland plains. This river system drains one-seventh of the Australian continent and all its tributaries combine in the south-western corner of the State to form the lower Murray River which then passes through South Australia and discharges into the Southern Ocean.

The State's programme of water conservation has as its principal aim the impounding of flow for subsequent river regulation. For this reason, construction of major storage dams has been largely concentrated on the headwaters of the inland rivers west of the Great Dividing Range.

The extent to which surface runoff is being conserved is shown by the area of catchment located above existing dams. This does not necessarily represent total control of the rivers' inflow as catchment yield may exceed the storage capacities. Over the years there has been a gradual trend to enlarge the existing storages. There are also additional future dam sites within the catchment areas. There are 9 dams on the headwaters of the inland rivers, including 7 large dimension dams. In addition there are a number of *en route* storages on the inland rivers.

On the coastal side of the Great Dividing Range, large water storages have been built to serve the main centres of population. Apart from these, storages have been constructed in two locations where the Great Dividing Range extends further inland. These are the Hunter River and the Snowy River catchments.

The Hunter River is a steep coastal river and the major storage thereon is intended for flood mitigation as well as general maintenance of regulated flow. In contrast, the storages in the Snowy River catchments have the more ambitious purpose of inland diversion of the seaward flow. These diverted flows now supplement the inland flowing Murrumbidgee and Murray Rivers. The diversion works are known as the Snowy Mountains Scheme which includes 15 major power stations to harness the 800 m fall in elevation for generating in excess of 3 million kW. The present state of water conservation on the inland flowing rivers of New South Wales is that water storages assure a supply of about 6×10^9 m^3 in most years.

The geography of the area west of the Great Dividing Range is such that maximum river regulation will never amount to total regulation. Uncontrolled additional flow is derived from the broad belt of foothills which lies between the mountain catchment areas and the plains. This is in contrast with some overseas irrigation schemes where orographic rains are restricted to the mountain ranges and an abrupt change to arid deserts results in the complete regulation of river flow.

It is well known that Australia's water resources are meagre. In comparison with major rivers of the world, the average annual flow of the Murray River is only 0.3% of that of the Amazon River. Similarly expressed figures for other rivers are 0.4% of the Zaïre River, 1.2% of the Mississippi River and 14% of the Nile. Australia is therefore in a position where special efforts by engineers and scientists are warranted to devise means for maximizing the efficient use of its water resources.

Geology and soils

The bulk of irrigation development has taken place on the northern and southern alluvial plains of New South Wales. Irrigation was developed much earlier on the southern plain and this area has been extensively studied in terms of geomorphology and soils. It extends beyond New South Wales into northern Victoria and these two areas are collectively known as the Riverine Plain of south-eastern Australia. Judging by descriptions of

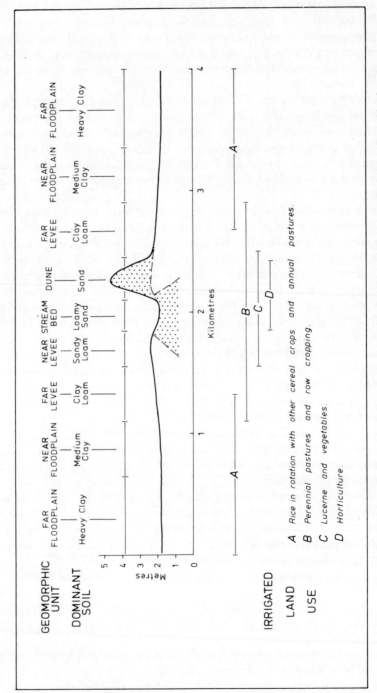

Fig. 1. Diagrammatic cross-section of prior stream illustrating inter-relationship of geomorphic units, soils and land use.

other major alluvial plains in the world, it has certain unique geomorphic features which warrant description in some detail because of their relevance to problems encountered in irrigation development.

In common with other parts of the world, early irrigation development preceded the emergence of soil science, so that soil surveys were carried out over large irrigated areas only when the hydrologic repercussions in semi-arid lands became evident in the 1930–1940 period.

Such surveys on the Riverine Plain showed that stream systems other than the present rivers were responsible for the deposition of the alluvial sediments. As a result of this work, the theory of prior streams was put forward by Butler (1950) and this has adequately explained the general distribution pattern of soils. It is now known that the Riverine Plain consists of a number of greatly sloping alluvial fans which coalesce to form a giant piedmont plain with an area, in New South Wales alone, of 65,000 km^2. The apex of each fan is situated at gaps in the foothill ranges which form the eastern boundary of the plain. Prior stream traces emerge from these gaps and radiate out across the fans, but these so-called streams are aggraded relic features which no longer convey water. The distribution of soils closely follows patterns of deposition and the stream trace normally indicates the location of linear areas of coarse sand at the surface or at shallow depth. The stream beds are flanked by levees which grade laterally from sandy to clay loams and then merge into heavy textured floodplains. Sand dunes occur as long chains adjacent to many of the stream beds on their leeward side. In the field, the stream features are hardly perceptible except for the sand dunes, but they are readily traceable on aerial photographs. An idealized cross section of a prior stream is shown in Figure 1.

Although the simple concept of soil distribution patterns being related to stream traces holds true, there are many factors which complicate the simple relationship. These include meandering of streams and superimposition of stream systems of varying age. Nevertheless the theory of prior streams has provided the key for explaining soil patterns on the plain.

Stratigraphic boring throughout the Riverine Plain has revealed that the alluvium extends to depths up to 100 m and that unconsolidated lacustrine sediments below the alluvium may continue down to depths of up to 400 m. The bulk of the alluvial plain is made up of prior stream sediments. It is an ancient land surface which is known as the "Older Alluvium." Attempts to date carbon samples from these sediments showed their age to be well beyond the C^{14} dating range.

Since the close of that phase of deposition, there has been a period of tectonics which rejuvenated the highlands of the Great Dividing Range and was responsible for a further type of stream trace over the plain surface. This period of activity created a series of wide channels which cut across the plain and broke out of the area and continued to the ocean. They subsequently aggraded back to a level somewhat below the older alluvial surface. These stream traces have been named "ancestral rivers." Their channel fill sediments are now known as the "Younger Alluvium." The sediments are the result of three periods of downcutting and backfilling and C^{14} dating showed that this period of activity goes back 30,000 years.

The present river system represents the fourth downcutting phase and the rivers generally lie on the younger alluvium which then constitutes its floodplain.

In some locations tectonic movement has diverted the present rivers from the younger alluvium which has given rise to simple channel-type river sectors cutting through the older alluvium. This has left deserted floodplains within the landscape isolated from the present rivers. These deserted floodplains are underlain by sands at shallow depth and therefore constitute further hydrologically hazardous locations where they occur in irrigated areas.

The disposition of stream traces has been mapped in greater detail by Butler *et al.* (1973). A more detailed account of the regional geology is given by Pels (1964, 1966, 1969) and by Stannard (1968).

A knowledge of sedimentary processes during the formation of the plain has led to an appreciation of their applications in the economic development of the area both in a regional and a more localized sense. Regionally it has thrown light on general suitability for irrigation insofar that it readily explains the varying soil textures and salt status of both soils and groundwaters in the eastern and western zones of the plain.

The Riverine Plain was formed under conditions of an inland basin of occluded drainage so that prior streams dissipated towards the western margin where lake relics are still common. The only relief in that area are lunettes associated with these lakes which may be partly buried by subsequent alluvial deposition. Sandhills associated with prior streams are less common in the western zone, due to the decreasing vigour and dissipation of the streams.

In the absence of surface drainage at the time, the water which accumulated in lakes and on floodplains was disposed of by evaporation which led to evaporite accumulation during the process of deposition, thus explaining the high salt status now prevailing in the western zone, both in regard to soils and sub-soils, as well as groundwater. The predominance of floodplain over streambed-levee components, the latter being only weakly developed, accounts for the large areas of heavy textured saline soils.

The eastern zone shows the reverse situation. Lighter textured soils of the streambed-levee systems are better developed and more closely spaced, so that this component predominates over the floodplain component. The soils have a lower salt status and are underlain by groundwaters of much lower salinity and sand hills are more common. The eastern zone is therefore more suitable for irrigation development and most of it has taken place there. Where irrigation was established in the western zone, salinity problems have developed within a relatively short period.

Apart from these regional considerations, there are many localized effects of geomorphic features on irrigation development and land use. In fact, the configuration of soil and sediment patterns and their topography have a direct bearing on determining the type of land use practised.

Although the landscape is of low relief and is commonly described as flat, there is a distinct micro relief. The prior streambed itself may be slightly depressed within raised levee systems which meander across the plain. Lateral changes of soil texture and topography related to prior streams have determined the distribution pattern of the vegetation under dry land conditions. The land use pattern under irrigation is similarly related.

Where horticulture is practised it occurs on the prior streambed unit or its associated sand dunes. Under irrigation agriculture, the lighter textured soils of the streambeds and near levees are used for perennial pastures or rowcropping. Annual pastures are more common on the far levee component while the flatter heavy textured floodplain component is extensively used for rice cultivation in rotation with other grain crops and annual pastures.

The stream traces on the surface represent the final phase of a period of gradually waning stream activity so that the streambed sands become laterally more extensive with depth, and shallow sand is therefore common under the near-levee formation. Irrigation channels were originally constructed along these elevated near-levees in order to command as much land as possible. High initial seepage losses were counteracted by clay lining of channels. Much of the lighter textured soil areas were preferentially developed due to the better growth response and easy access from the channel system. In these areas underlain by sand strata at shallow depth, channel losses and deep percolation under irrigation have caused high rates of accessions to the groundwater with resulting rises in water tables.

The hydrological effects of irrigation

Prior to irrigation, the groundwater surface was deep, at 30–40 m below the surface. There was a general piezometric slope towards the west with groundwater dissipating to the rivers in the far western zone. Under semi-arid dryland grazing conditions, this hydrologic system was in a state of equilibrium. Irrigation has upset this balance, the degree of imbalance being determined by the intensity and nature of irrigation.

There are three major types of irrigation development in New South Wales with a total irrigated area of about 700,000 ha.

(i) Projects known as Irrigation Areas which are Government sponsored projects where the land was originally resumed and subdivided into farms leased out for intensive irrigation. Water supply and drainage facilities are provided for 165,000 ha.
(ii) Projects known as Irrigation Districts which are also Government sponsored but only water supply systems are provided without drainage facilities. In these Districts, the land ownership was not disturbed and water was supplied originally for partial irrigation to stabilize existing production, mainly pastoral. However, parts of the Districts rapidly developed into intensively irrigated agricultural areas as a result of subdivision. Actual irrigated area in these Districts is 280,000 ha.
(iii) Private irrigation by diversion from the rivers. This development is authorized for specified areas either by individual licenses or collectively for a number of individuals by joint authorities for pumping from the rivers. 255,000 ha are so irrigated.

Despite the different natures of projects the deep water tables invariably rose following the commencement of irrigation, but there was a considerable time before shallow water tables developed. This was particularly so where dry sand beds occurred above the original groundwater levels and sub-surface drainage requirements were met by the gradual filling of these underlying sand beds. Water intake took place mainly on areas where sand beds are shallow, and the lack of surface drainage facilities in the District type of irrigation greatly accelerated the process there.

In the Irrigation Area projects, surface drainage was integrated with stock water supply schemes further west. The soil textures and salt status of the western zone made this integrated system very appropriate. Forage plants in the western zone consist of grasses and halophytic shrubs such as salt bushes (*Atriplex* and *Kochia*) which are used for dry land grazing by sheep. In this zone relatively saline groundwater was generally used for stock watering and this caused sheep to graze the less saline pasture species in preference to the halophytes. Greater utilization of natural species was therefore possible by supplying fresher surface drainage water from the Irrigation Areas for stock watering in the western zone.

Although the provision of surface drainage may slow down the rate of groundwater rise, shallow water tables will develop inevitably under irrigation. The natural counteracting factor is the environment's potential for lateral dissipation. The rate of dissipation will accelerate as lateral gradients are increased by raised groundwater levels under irrigated areas. Isolated areas irrigated by private diversions from the river therefore have the greatest potential for lateral dissipation.

Irrigation and drainage experience in Australia, in common with that in many other parts of the world, has shown that rising groundwater levels and eventual shallow water table conditions are inevitable despite such factors as lateral dissipation, surface drainage systems and low leaching requirements due to good quality irrigation waters.

It is a matter of time before problems develop and in the selection of land for irrigation it should be accepted that, if only in the long term, there will be a need for groundwater drainage to maintain the root zone favourable to crop growth. The mere removal of excess rainfall and irrigation water by shallow surface drains to prevent excessive intake will only postpone this need. Such measures will require adequate disposal systems. Where areas selected for irrigation have an inherently high salt status, the need to dispose of saline drainage water can become an acutely difficult problem, The fact that these disposal problems may only manifest themselves after many years of irrigation makes proper selection of land and environment even more important. By that time, the farming community and its secondary or supporting industries may have become well established. If the difficulties of establishing saline drainage water disposal works prove to be insuperable at that stage, a substantial proportion of the irrigated area will need to be abandoned as sacrificial "dry drainage" areas and substantial yield reductions may have to be accepted on the remaining irrigated land.

Such difficulties have developed in semi-arid New South Wales, particularly in the more saline environment of the western zone. It may be instructive to discuss the problems which are now developing under Australian conditions as similar trends must be, or will become, apparent in comparable environments elsewhere in the world.

Superimposing irrigation on a dry land hydrologic system disturbs the equilibrium and creates a dynamic hydrologic state which may be maintained for many years, but will continually move towards a new equilibrium. The eventual new equilibrium situation may be very detrimental to agricultural production.

That situation is reached when groundwater levels become sufficiently shallow to be under the influence of evaporation to the atmosphere and salt accumulation takes place in the lower topographic situations. Under these conditions the tendency is to irrigate the slightly higher land while lower areas will become areas of "dry drainage" by evaporation. On the higher land the water table might be within the critical depth for surface salt accumulation by evaporative losses from the water table through the soil to the atmosphere. Therefore, irrigation on the higher land would need to be more frequent and in greater excess of plant requirements to counteract this upward flow. This causes groundwater flow to occur towards the lower areas and develop more acute problems there.

Soil salinization

It has been found that some parts of irrigated areas in New South Wales have not been severely affected by soil salinization, despite a water table at shallow depth for many years. Attempts have been made to delineate such areas and to determine what hydrological conditions are responsible for this phenomenon.

Water drawn from the river system contains very low quantities of soluble salts (< 100 mg/l) in the upper sectors so that soil salinity problems resulting from the accumulation of salts in the applied water are not of major concern in the eastern zone. The soils and underlying strata in this environment however, contain appreciable quantities of soluble salts, being low only in the upper parts of the soil profile prior to irrigation. It is the redistribution of these salts, which can occur after shallow water tables have developed, that brings about salinity problems in this region.

In the early stages of the build-up of groundwater levels, deep percolation leaches most soluble constituents to depth, particularly the more mobile chlorides. However, because of variability in deep percolation in the various geomorphic units, some areas are leached to much greater extent than others. When high water tables have developed generally to a level in equilibrium with irrigation practice and climate, the salinity of the upper levels of the ground water will be found to vary considerably.

Where the water table is at or above the critical depth, capillary movement from the water table towards the soil surface will take place and continue under the gradient imposed by evaporation or transpiration of plants. The critical depth varies with physical characteristics of the soil, being generally less in soils of light or heavy texture and greater in soils of intermediate texture. The critical depths for a capillary flow of 1 mm per day, which is less than would occur under average winter conditions in this region, was found to range from 95 to 200 cm in typical local soils.

Capillary flow occurs at the expense of the groundwater, so that, in the absence of additions, the water table will fall eventually to below critical depth. The maintenance of the water table above critical depth can occur by way of irrigation or rainfall, but as this water contains very low quantities of soluble salts, the maintenance of capillary flow as a result will not cause any increase in soil salinization.

Alternatively, the water table can be maintained above critical depth by lateral or upward saturated flow in the general groundwater body as a result of higher aquifer pressures or gradients developed under a variation in topography or irrigation practice. Under these conditions, a high saline water table is maintained and soil salinization would occur. The rate of saturated flow under a particular gradient depends on the permeability of the soils and underlying materials. Where losses in head during transmission of the minimum evapotranspiration requirements is greater than the total head provided by topography and irrigation practice, maintenance of the water table above critical depth cannot be achieved, and thus soil salinization does not occur or only occurs to a limited extent.

Experience has shown that salinization occurs readily in depressed land underlain by shallow aquifers in which the pressure levels are high, but not in situations where a considerable thickness of superficial clay of low permeability is present.

It has been found that high ground water ridges under prior streambeds exert a hydrostatic head on laterally more extensive deeper aquifers. Lowering of groundwater levels by tube well drainage from these shallow sands is effective not only directly, by lowering water tables in the streambed environment, but also indirectly, by lowering piezometric levels in the lateral deeper sands where even small pressure reductions can be effective in reducing the saturated flow required to maintain water tables within the critical depth for salt accumulation in the overlying soils.

Reclamation measures

The position will arise where drainage requirements of the irrigated areas, formerly met by gradual saturation of the underlying sediments and lateral dissipation, will need to be met by evaporation from the low lying areas. But even this is not yet a static situation as the increasing salinity in the waters of the depressions will reduce the rate of evaporation so that comparatively large sacrificial drainage areas are required. Experiments with tube well drainage in this environment have given an indication of what percentage of the total area needs to be sacrificed to provide an equilibrium situation by evaporation drainage.

A tube well drainage flow of 2500 m^3/day safeguards about 360 ha, and if this is related to the local net annual evaporation, it would take 90 ha of evaporating surface to give an equal rate of water removal from that area. This would mean that 25% of the land would need to be sacrificed.

This would still leave the higher country with problems of salt accumulation due to uneven or infrequent irrigation, and depression of yield would result. It would also assume that continuous irrigation was practised, but this is not always the case as much of the irrigation in these regions is on annual pastures. The irrigation methods used on annual pastures may be peculiar to Australia. It really constitutes a supplementary form of irrigation, involving an extension of the winter rainfall growing season by watering subterranean clover and rye grass mixtures for germination in autumn in anticipation of the winter rainfall, and then extending the growing season again in spring by further irrigation. The pasture matures and seeds in late spring and dries off in early summer. The dry feed is grazed during the summer months.

During the non-irrigation period in summer, salt accumulation can occur in the root zone under the dried-off annual pasture to such an extent that subterranean clovers fail to germinate following initial irrigation in the following autumn. In the initial stages of the problem this can be overcome by heavy repeated waterings to leach the surface soils.

The ultimate result of soil salinization is that the productive area of farms is reduced to uneconomic size. It also detracts greatly from the environment because any natural tree growth is badly affected as this occurs mainly in the low lying areas. But even if the environmental detractions were acceptable, the economics of the situation leave only limited choice of action, namely (i) abandon the area, (ii) amalgamate affected farms into economic units, or (iii) apply drainage measures.

If drainage measures are decided on, the following options are open, providing that saline drainage waters can be disposed of:

(a) Shallow surface drains. Sacrificial areas could be reduced by shallow surface drains connecting the low lying areas, thus reducing the amount of evaporation required to maintain equilibrium between ground water inflow and evaporation. This would also reduce intake into the groundwater system by removing surface runoff after a heavy rain. It would still need salt suppressing methods of irrigation on the higher land.

(b) Deep surface drains. To reduce sacrificial areas to the minimum, the low lying areas and depression lines would require deep surface drains. In the alluvial landscape low lift pumping would be needed for satisfactory functioning, because of lack of grade. Also, heavier soils would not respond readily to this type of drainage and close spacing of drains would be required. The costs of deep drain construction, their close spacing pump requirements for their operation and the lack of control over saline water discharging from them, makes this approach inferior to tube well drainage.

(c) Tube well drainage. To avoid the necessity for sacrificial areas, tube wells could be installed into the underlying sand beds. Gradients earlier visualized as being towards sacrificial areas would then be at a lower level towards tube well screens. The method necessitates the introduction of shallow surface drains to convey pump effluent and this would further contribute towards a reduction of intake into the groundwater system. This in turn would assist the effectiveness of the tube well operation in lowering the water table to below the critical depth for surface salt accumulation, thus making irrigation practices more flexible and removing the causes of reduction in yields.

(d) Tile drainage. The use of tile drainage in New South Wales is confined to horti-culture which, in the alluvial environment, occurs on sandy soils of the prior streambeds and near-levees and on aeolian sandhills associated with prior streams. Outside the alluvial environment horticulture is commonly practised on light textured soils of aeolian origin including the dune country west of the Riverine Plain, locally known as the Mallee. Deep wind-blown soils deposited on the lower slopes of ranges surrounding the plain are also used for horticulture.

About 35,000 ha of citrus, vines, stone fruit and vegetables are irrigated in New South Wales and the tile drainage is widely practised. Some areas are safeguarded by tube well drainage where geological conditions are suitable, and pumping takes place under semi-confined conditions. Economic considerations have restricted tile drainage almost exclusively to intensive horticulture where permeable soils can be protected from water-logging and salinity problems by widely spaced tile lines. Spacing of 20 m is regarded as the minimum on economic grounds. Although the method serves the purpose of safe-guarding soils from salinization, tile drainage is also used in New South Wales to overcome or avoid waterlogging of permanent plantings during infrequently occurring wet years.

The latest large-scale introduction of tile drainage took place in the Murrumbidgee Irrigation Areas where drainage criteria of 5 mm per day and average title depth of 1.8 m were adopted. Spacings are based on hydraulic conductivity, determined by the auger hole method (Maasland and Haskew 1958), and the depth of slowly permeable layers where they occur.

The effects of irrigation on the salinity of river flow

Increased salinity of river flow is a prime example of the environmental effects of irrigation. Most of the large-scale irrigation development along the Murray River is adjacent to its middle reaches.

The water quality of the upper river section is very good (< 100 mg/l), but drainage backflow to the river in the middle sector may raise river salinity to 250 mg/l. Further increases occur in the downstream sector due to inflows both from artificial drainage of irrigated lands and natural saline ground water. The latter has been increased by raised groundwater levels under irrigated areas and by the hydrological effects of *en route* storages.

In the far downstream reaches of the river, salinity may rise as high as 1200 mg/l under conditions of low flow. This is deleterious, particularly for irrigated horticultural crops in South Australia. Furthermore, a proportion of the water supply for the City of Adelaide is drawn from the river so that the deterioration of water quality is of serious concern to South Australians.

Despite adaptations in irrigation practices to cope with more saline waters (such as conversion to undertree sprinklers in citrus orchards) there is an urgent desire to improve the quality of the water in the downstream reaches. Until now the accepted means of maintaining water quality has been by dilution from upstream sources, but the greater demands for irrigation water and the increased drainage returns to the river have made this approach increasingly difficult. The seriousness of the situation was demonstrated during a recent drought when dilution water was in very short supply and salinity in the river rose to unprecedented levels. Salt balance studies have shown that under normal regulated flow the annual salt load of the river is increased by 350,000 tonnes of salt contained in surface drainage inflows and a further 500,000 tonnes are attributed to groundwater inflow along the lower sector of the river.

Efforts to ameliorate the rivers water quality now revolve around two approaches, namely (i) increased releases of dilution flows from storages; (ii) abatement measures to reduce saline inflows.

To achieve abatement of saline inflows several methods are being applied as follows:

1. Diversion of saline surface drainage flows. This is being done at three locations where a potential inflow of 325,000 tonnes could be diverted into isolated lakes for disposal by evaporation. This aim is only partially achieved and about 25% of the above quoted amount is being diverted.

2. Interception of saline groundwater inflow. This has involved the location of point sources of groundwater inflow and the installation of tube wells together with pipe lines to convey the pump effluent to evaporation areas away from the river. Point sources of substantial groundwater inflow have been created by the construction of weirs in the Murray River to regulate flow. Whilst groundwater flowed into the river previously in a diffuse manner in long stretches of the river, the raised water levels above the weirs have suppressed this inflow and point sources of high inflow may now occur downstream of these weirs. At such locations, intercepting works are being planned following detailed environmental investigations.

3. Tube well drainage of irrigation areas with local disposal by evaporation. Large groundwater mounds have developed under some of the irrigated areas in New South Wales with water tables at depths of less than 1 m and surface salting or saline water in depressions as a common occurrence. One such area is in the Wakool Irrigation District where more than 40,000 ha are so affected. Experimental tube well drainage in this area has shown that this method is appropriate for lowering the water table, but disposal of the very saline water (up to 40 g/l) has posed a dilemma. The calculated salt content of the discharge from the 45 proposed tube wells is 300,000 tonnes per annum and to add such an amount to the surface streams is obviously unacceptable.

Feasibility studies were made into the economics of using these drainage waters for commercial salt production and it was found that it could be marketed. However, there is still the problem of disposal of bitterns which remain after the sodium chloride is harvested. Although bitterns only comprise a fraction of the original pumped drainage water, it still amounts to several hundred Ml per annum. Investigations are being carried out on disposal of the bitterns into deep aquifers.

The channel diversion approach

To by-pass all the difficulties of water quality control described so far, a different approach to the problem has been put forward. This involves the separation of the supply and drainage functions of the river by diverting the required flow into a separately constructed channel parallel to the river to supply the irrigated areas and a pipe line extension from the channel terminus to the water storage reservoirs of Adelaide in South Australia. The offtake would be in the upper sector of the river where its waters are of high quality. The channel would lie above groundwater levels along its entire length so that no deterioration by groundwater inflow could occur.

The flow down such a channel would be very much smaller than the flow at present required in the river as no dilution component would be needed. The river, being the lowest line in the landscape, acts as a natural drain and could convey all drainage waters carried by gravity from the irrigated areas. Dilution of these flows would still be practicable but scarcity of dilution water during drought periods would not be as critical as at present. Flood waters would also travel down the river in periods of excess flow.

Such a scheme was found to be technically feasible, but it has been rejected due to high construction costs, dislocation of established social patterns, and possible restrictions on future development imposed by a chosen channel capacity (Anon. 1970).

Conclusion

The problems of Australian inland river systems are by now reasonably well under-stood. However, in a developing country like Australia there is strong competition for the available funds and public opinion plays a major role in their allocation for execution of works.

Although this review is by necessity generalized, it serves to demonstrate that in Australia, as in other countries, there is an increasing awareness of the need for sound, rational development of water resources and their use. This has led to a greater apprecia-tion by the general public of the complex interdisciplinary nature of water resource manage-ment which needs continual study and adequate funding.

References

Anon. 1970. Murray Valley Salinity Investigations. River Murray Commission, Canberra, Australia.

Butler, B. E. 1950. Theory of prior streams as a causal factor of soil occurrence in the Riverine Plain of South Eastern Australia. *Aust. J. Agric. Res.* **1**, 231–252.

Butler, B. E. *et al.* 1973. A Geomorphic Map of the Riverine Plain of South Eastern Australia. ANU Press, Canberra, Australia.

Maasland, M. and Haskew, H. C. 1958. The auger hole method of measuring the hydraulic conductivity of soil and its application to tile drainage problems. *Proc. 3rd Intern. Cong. Irr. Drainage* 8.69–8.114.

Pels, S. 1964. The present and ancestral Murray River System. *Aust. Geog. Studies* **2** (2), 111–119.

Pels, S. 1966. Late Quaternary chronology of the Riverine Plain of South Eastern Australia. *J. Geol. Soc. Aust.* **13**, 27–40.

Pels, S. 1969. Radio carbon datings of ancestral river sediments on the Riverine Plain on south eastern Australia. *J. Proc. R. Soc. N.S.W.* **102**, 189–195.

Stannard, M. E. 1968. Environmental studies of the Coleambally Irrigation Area and surrounding districts. *Bull. Water Cons. Irrig. Com. N.S.W.* Land Use Series No. 2.

INFLUENCE OF IRRIGATION ON PRECIPITATION IN SEMI-ARID CLIMATES

by PAUL T. SCHICKEDANZ and WILLIAM C. ACKERMANN

Illinois State Water Survey, Box 232, Urbana, Illinois 61801, USA

Summary

A statistical-climatological study was made of whether the phenomenal growth of irrigation in the Great Plains of North America has had an appreciable effect on the climate of the region. The analysis procedures have been primarily that of Empirical Orthogonal Function (EOF) and Trend Surface analysis. These methods of pattern analyses indicate the presence of rainfall anomalies in the irrigated areas during wet months as opposed to dry and moderate months. An exception to this finding is the month of June in which the anomalies occur in all three strata of years (dry, moderate and wet). Also, the anomalies are confined to the summer months as opposed to the spring and fall months. At this early stage of the analyses, it is speculated that under very dry conditions in this climate, the addition of water vapor cannot be realized in the production of clouds and precipitation.

Résumé

Une étude statistique climatologique a été effectuée afin de déterminer si la croissance phénoménale de l'irrigation dans la région des Grandes Plaines de l'Amérique du Nord a influencé de façon importante le climat de la région. Les principes de l'analyse ont été principalement ceux de la "Fonction Orthogonale Empirique" at de la "Direction de la Surface" (Trend Surface). Les méthodes d'analyse indiquent l'existence d'anomalies pluviales dans les régions irriguées plutôt au cours des mois humides qu'au cours des mois secs ou modérément pluvieux. Le mois de juin constitue une exception à cette tendance car les anomalies surviennent en toutes périodes. Ces irrégularités se concentrent également plutôt en été qu'au printemps ou en automne. A ce stade précoce des analyses, on peut estimer que sous les conditions très sèches de ce climat, il n'est pas possible d'évaluer les effets de l'évaporation sur la formation des nuages et sur les précipitations.

Introduction

Man's inadvertent modification of weather and climate through land use changes has become the focal point of many worldwide environmental concerns. Major shifts in land use patterns, such as irrigation and cropping practices, suggest alteration in the weather and climate over substantial regions (Changnon 1973). Bryson and Baerris (1967) have suggested that over-grazing in northern India has resulted in enormous additions of dust to the atmosphere which, through sizeable alterations in radiation, have produced the extensive Rajasthan Desert. Certainly, the identification of changes in climates and their

185

causes, as well as the beneficial or detrimental aspects of such changes, is a service to mankind. If these alterations in climate are known to man, he can either alter his activities to adjust to changes, or perhaps use information concerning these changes to better manage his resources, possibly through planned weather modification efforts.

The research described in this paper is directed towards the question of whether the

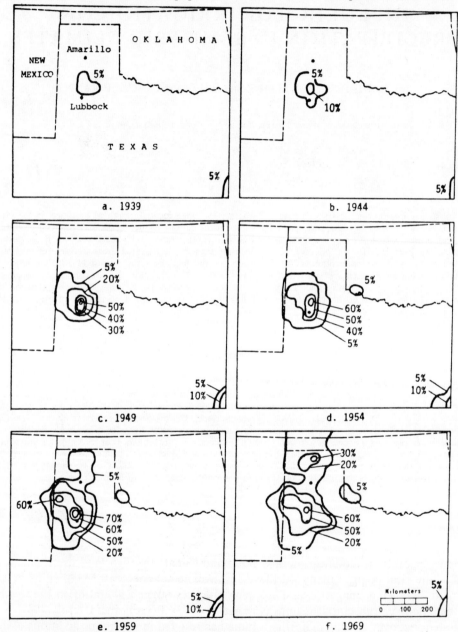

FIG. 1. The percent of total land area under irrigation for selected years during 1931–1969 in the southern study area.

phenomenal growth of irrigation in the Great Plains of North America (semi-arid climate) has had an appreciable effect on the climate of the region. The average annual rainfall of this region is 31–46 cm with 70–80 percent occurring in the warm season (April–September). The specific objective of the research is to demonstrate if rainfall alterations due to irrigation exist, and if so, to demonstrate their magnitude. Basically, the hypothesis has been that the added moisture due to sizeable evaporation in a dry climate has led to more clouds and rainfall, as indicated by Stidd (1975) and Joos (1969).

Basic data and reduction procedures

The study region covers the states of Kansas, Nebraska, Oklahoma, as well as a large portion of the state of Texas. In addition, some data from the states of Colorado and New Mexico are being used for control area purposes. The southern portion of the heavily irrigated regions and their environs are shown on Fig. 1.

Detailed data on the number of hectares irrigated in each county of the study region were found to be available from 5-year inventories (US Bureau of the Census, Census of Agriculture, 1919–1969). These data were extracted and expressed as a percent of the total land area within each county and the percent values were plotted on maps. The resulting patterns for selected years are depicted on Fig. 1. A phenomenal growth of irrigation occurred in the Lubbock–Amarillo region of the irrigated area during the 5-year period 1944–1949 with the percent of area irrigated increasing from 5–10 percent to 20–50 percent. Another expansion of the total area under irrigation occurred in the late 1950s and irrigation growth persisted throughout the 1960s, reaching northward across the state of Oklahoma. During the 1960s, the percent of land under irrigation in the central portion of the irrigated area decreased somewhat due to dwindling usable groundwater supplies and the initiation of water conservation procedures (Texas Water Development Board [1971]). However, by 1969 there was approximately 95,000 km^2 with 5 percent of the total land area irrigated, 37,000 km^2 with 20 percent irrigated, and 10,000 km^2 with 50 percent irrigated. Surface water accounted for approximately 25 percent of the total irrigation water used and groundwater accounted for 75 percent in 1969. The sprinkler method of application accounted for 19 percent of all irrigation.

Monthly precipitation data for the growing season months of April–September, 1931–1970, and for a total of 1638 reporting stations were obtained on magnetic tape from the National Weather Center, Asheville, North Carolina. These data were computer filed on magnetic disk storage so that all station amounts for a given month of a particular year were grouped together. Using this disk file, the station data were gridded on 32 km × 32 km grid. The estimation of the data at the grid points was made in the following manner. For each grid point a computer search was initiated for the nearest three stations with non-missing data. A quadratic surface was then fitted by solving the following equations simultaneously for the coefficients C_1, C_2 and C_3.

$$P_1 = \bar{P} + C_1 X_1 + C_2 Y_1 + C_3 X_1 Y_1 \tag{1}$$
$$P_2 = \bar{P} + C_1 X_2 + C_2 Y_2 + C_3 X_2 Y_2 \tag{2}$$
$$P_3 = \bar{P} + C_1 X_3 + C_2 Y_3 + C_3 X_3 Y_3 \tag{3}$$

where X_1, Y_1 are the coordinates of the 1st nearest station with a non-missing rainfall value,

X_2, Y_2 are the coordinates of the 2nd nearest station with a non-missing rainfall value,

X_3, Y_3 are the coordinates of the 3rd nearest station with a non-missing rainfall value,

P_1, P_2, P_3 are the rainfall values respectively at the 1st nearest, 2nd nearest and nearest stations,

$\bar{P}$ is the mean of the three rainfall values at the nearest three stations.

Once the coefficients are determined, the value of the grid point is estimated by substituting the values of its coordinates into Equation 1 in place of X_1, Y_1, and then solving for P_1. Under certain conditions, especially along the boundaries, the solution becomes unstable and extremely large or even negative values will be computed. Thus, whenever the computed value exceeds two standard deviations of the three-station mean or is negative, the computed value is set equal to the mean of the rainfall values at the three points.

This data reduction procedure provided a set of 750 grid point estimates for each of the six months, April–September, and for each year during the period 1931–1970. The

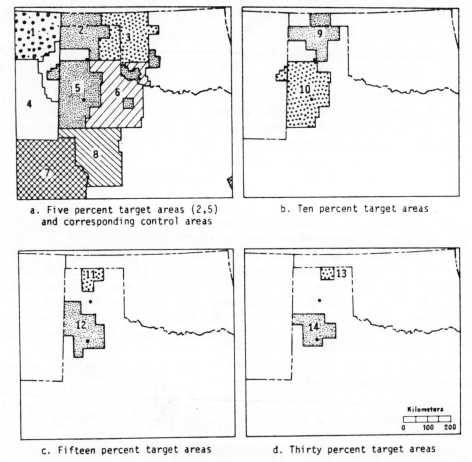

a. Five percent target areas (2,5)
   and corresponding control areas

b. Ten percent target areas

c. Fifteen percent target areas

d. Thirty percent target areas

FIG. 2. Areas with at least 5 percent of total land area under irrigation (target areas) and areas with less than 5 percent (control areas) by 1969.

gridded data were output on punch cards so that a set of cards was obtained for each of the monthly precipitation patterns (240 patterns for the southern irrigation area). These gridded data formed the data source from which all subsequent pattern analyses were performed.

Analytical techniques used in the pattern analyses

Using the irrigation inventory, areas with 5, 10, 20 and 30 percent of the total land area under irrigation by 1969 were hypothesized to be "target" areas and are shown on Fig. 2. Areas with less than 5 percent of total land area irrigated by 1969 were hypothesized to be control areas and are also shown on Fig. 2a. Although area 6 contains two small areas which are more than 5 percent irrigated, the entirety of area 6 was considered to be a control area.

The areal monthly mean rainfall for each year of the period 1931–1970 was determined for each of the areas shown on Fig. 2. The areal means from the eight areas on Fig. 2a were used to form a $m \times n$ (8 areas $\times$ 40 years) matrix X for each of the 6 months. Each matrix was then subjected to Empirical Orthogonal Function (EOF) analysis. This analysis was performed by first standardizing each of the rows (areas of the X matrix by their respective means and variances so that a $m \times n$ matrix Z of standardized variates was obtained. The correlation matrix R, which is a matrix of correlation coefficients between the rows (areas) of Z, was then determined. The eigenvalues and eigenvectors of the R matrix were calculated by the following relationship:

$$RE = DE \tag{4}$$

where E is the $m \times m$ matrix consisting of a set of orthonormal eigenvectors of R as the columns and D is a $m \times m$ diagonal matrix of the eigenvalues of R. The solution of the matrix E is the classical Characteristic Value Problem of matrix theory (Hohn 1960).

The magnitude of the eigenvalues represents the variance of the observations in the Z matrix explained by each eigenvector. The eigenvectors are then ordered so that the first diagonal element of D represents the largest eigenvalue, and the second the next largest, etc. The eigenvectors are also scaled by multiplying each orthonormal eigenvector by the square root of its eigenvalue to obtain the "principal components loading matrix":

$$A = ED^{1/2} \tag{5}$$

The matrix A is a transformation matrix which can be used to transform the matrix Z into a set of principal components:

$$F = A^{-1}Z \tag{6}$$

where F is the principal components matrix of order $m \times n$. The scaled eigenvectors of the A matrix are orthogonal (uncorrelated) functions of space and the principal components of the F matrix are orthogonal (uncorrelated) functions of time. Furthermore, the first eigenvector explains the largest amount of the combined variance of the observations in the Z matrix, the second eigenvector explains the second largest amount of variance, the third eigenvector explains the third largest amount of variance, etc. For further details the reader is referred to Massey (1965), Kutzbach (1970) and Sellers (1968).

In the present research, the first eigenvector of each month accounted for 54 to 71

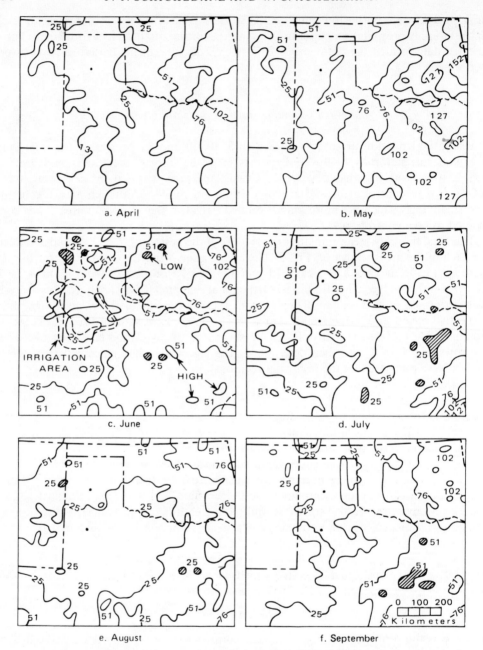

FIG. 3. The total rainfall patterns during dry years as determined by the first principal component.

percent of the total variance. Furthermore, either the largest positive or the largest negative value of the eigenvector always occurred in area 5 (see Fig. 2a). This fortuitous chain of events provided an excellent way to stratify the precipitation patterns into groups of years in which a rainfall anomaly occurred in the target areas or their immediate vicinity. This stratification was accomplished by a visual inspection of the yearly time series of the

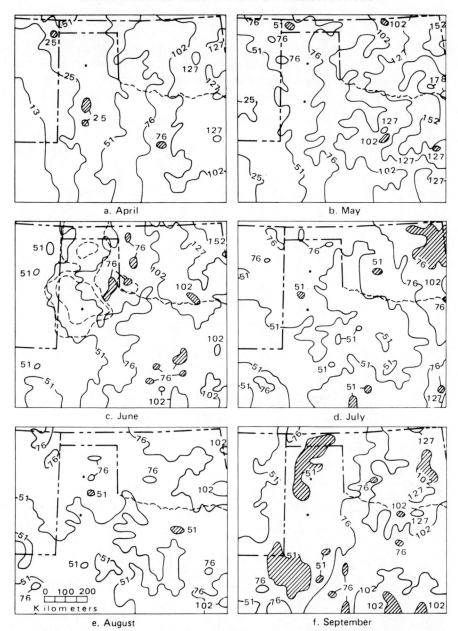

FIG. 4. The total rainfall patterns during moderate years as determined by the first principal component.

elements in the first principal component for a given month. The patterns for each year within the various strata were then summed to obtain overall patterns. The temporal stratification of the temporal space on the first principal component produced generally dry, moderate and wet years when viewed in respect to the total patterns over the entirety of the southern irrigation area.

Results of the Pattern Stratifications

The resulting patterns in the dry, moderate, and wet strata are shown on Figs. 3, 4 and 5 for all 6 months. Isolated positive rainfall anomalies occurred during dry Junes, wet Julys and wet Augusts. Although isolated anomalies did not occur in the irrigation area during the other months, there are suggestions that specific isohyets protrude into the

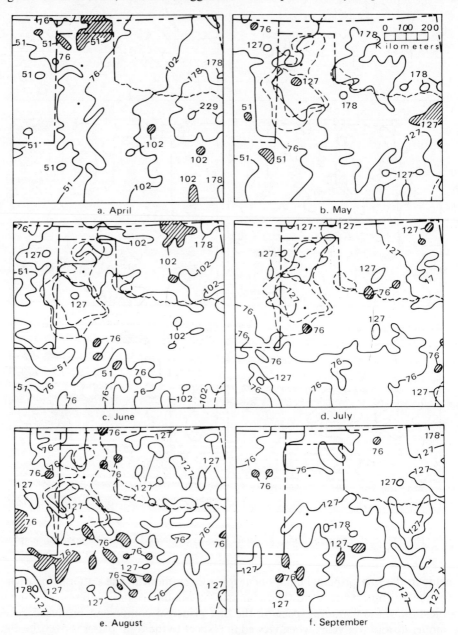

FIG. 5. The total rainfall patterns during wet years as determined by the first principal component.

irrigated areas during moderate Junes, wet Mays, wet Junes and wet Julys. With the exception of wet Mays, there is no suggestion of isolated anomalies or isohyetal protrusions in the months of April, May and September.

Further investigation was made of the month of June since there is evidence of anomalies or protrusions in all three strata. To assess the significance of the June patterns on individual years, the trend surface analysis of residuals from two-dimensional trend surfaces was performed on logarithmic transformed data (Schickedanz 1973a and b, 1974). The frequencies of when each grid point was within a 10 percent positive residual during the 10-year periods of 1931–1940, 1941–1950, 1951–1960, 1961–1970 during June are shown on Fig. 6. Clearly, there is a greater accumulation of frequencies within the irrigation area as it grew in size over time. There were also anomalies of frequencies in other areas; however, they do not have the same temporal relationship as that in the irrigation area.

The temporal relationship was also investigated by determining the average frequency within the target and control areas shown on Fig. 2a. These areas were combined to form an upwind control area (1, 4, 7), a target area (2, 5) and a downwind area (6, 7, 8). The average frequencies for the upwind, target and downwind areas are listed in Table 1.

The target area has an upward trend which extends over the four 10-year periods with the greatest increase occurring between the 1941–1950 and 1951–1960 periods. This roughly corresponds to one of the largest increases in irrigation usage which occurred in the central portions of the irrigation area in the late 1940s (see Figs. 2 and 6). There is no trend in the upwind area and the frequency appears to reflect the dry decades of the 1930s and 1950s and the wet decades of the 1940s and 1960s. In the downwind area there is an increase in frequency into the decades of the 1950s, but a decrease in the decade of the 1960s.

In addition, the Significant Rainfall Excess, as defined by Schickedanz (1973b, 1974), was determined for each of the June patterns on individual years. The accumulations of the excesses according to 10-year periods and the average accumulations for the upwind, target and control areas were determined and the results are listed in Table 1.

The target area has a generally upward trend with the greatest increase occurring

TABLE 1. *The average frequency of 10 percent residuals and the average Significant Rainfall Excess in upwind, target and downwind areas during selected 10-year June periods*

Period	Upwind	Target	Downwind
Frequency (counts per 10-year period)			
1931–1940	1.20	1.23	1.19
1941–1950	1.30	1.30	1.35
1951–1960	1.21	1.50	1.36
1961–1970	1.35	1.55	1.28
Excess (mm)			
1931–1940	24.9	28.4	27.4
1941–1950	31.5	27.7	30.0
1951–1960	52.8	32.0	62.0
1961–1970	30.2	53.3	30.7

between the 1951–1960 and the 1961–1970 periods. This roughly corresponds to the large increase in irrigation usage which occurred in the northern portion of the irrigation areas during the decade of the late 1960s. In the upwind and downwind areas, there is a generally increasing trend throughout the decade of the 1950s but a decrease during the decade of the 1970s.

Conclusions

The application of the EOF analyses has demonstrated the presence of rainfall anomalies in the irrigation area during dry Junes, wet Julys and wet Augusts. There was also an indication of protrusions of specific isohyets into the irrigation areas during moderate Junes, wet Mays, wet Junes, wet Julys and wet Augusts. Thus, with the exception of the dry and wet Junes, the presence of anomalies and protrusions is confined to the wet months. Also, with the exception of wet Mays, the presence of anomalies or protrusions appears to be confined to the summer months, as opposed to the spring and fall months.

The application of trend surface analyses to the June patterns indicated a weak temporal

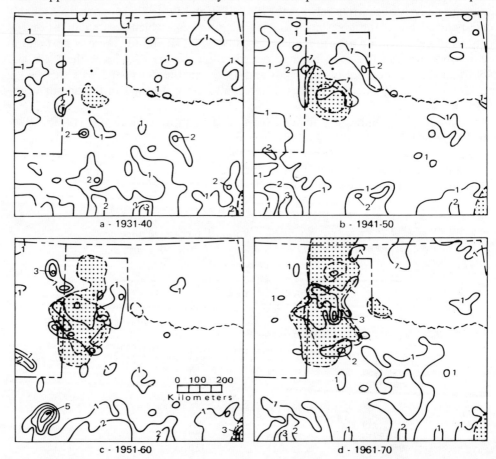

FIG. 6. The frequencies of when each grid point was included in a 10 percent significant residual.

relationship between increased frequency of significant residuals and increased use of irrigation. There was also an indication of a weak temporal relationship between increased significant rainfall excess and the growth of irrigation. The trend surface analyses will also be applied to the other months, although it is expected that any temporal relationships will be less than in June, since it was the only month in which anomalies or protrusions occurred in all three strata of years (dry, moderate and wet).

The difficult question of whether these anomalies and protrusions represent significant rainfall increases due to irrigation is in the process of being addressed. The answer to this question is being pursued through several avenues. First of all, detailed studies are being made of the second, third and fourth eigenvectors to see if they contribute additional information concerning the presence of the anomalies. The relationship between the patterns and synoptic weather conditions, as well as sounding parameters, is being pursued. Also, the number of areas used in the eigenvector analyses is being increased to include the entirety of the southern study area. Also, smaller sized areas are being used.

Furthermore, a study of several surface variables (rainfall, temperature, dewpoint temperature, cloudiness, etc.) for a limited number of stations which have long-term records (55–70 years) is being pursued. Finally, the pattern analysis of the northern study area of Kansas and Nebraska is under way. The analyses in the northern area will determine whether additional anomalies occur, and if so, whether they occur in the same months as in the southern irrigation area.

It is anticipated that these additional studies will permit a much firmer conclusion than is now possible as to whether the anomalies are, in fact, due to the growth of massive irrigation in the Great Plains.

Acknowledgments

This research was performed under the general direction of Stanley A. Changnon, Jr., Head of the Atmospheric Sciences Section of the Illinois State Water Survey. Appreciation is expressed to Marion B. Busch who performed most of the computer programming and analyses, and to Tony Barnston who performed many of the routine analyses and construction of isohyetal maps. This research was supported under Grant GI-43871, sponsored by the Weather Modification Program, RANN, National Science Foundation.

References

Bryson, R. A. and Baerris, D. A. 1967. Possibilities of major climatic modification and their implications: Northwest India, a case study. Bull. Am. Met. Soc. **48**, 136–142.

Changnon, Stanley A. 1973. Atmospheric alterations from man-made biospheric changes. Modifying the Weather, Western Geographical Series, **9**, 135–184.

Hohn, F. E. 1960. Elementary Matrix Algebra. The Macmillan Company, New York, 305 pp.

Joos, L. A. 1969. Recent rainfall patterns in the Great Plains. Paper presented at AMS Symposium on Future of the Atmosphere, Madison, Wisconsin, 8 pp.

Kutzbach, J. E. 1970. Large-scale features of monthly mean northern hemisphere anomaly maps of sea-level pressure. Mon. Weath. Rev. U.S. Dep. Agric. **6** (9), 708–716.

Massey, W. F. 1965. Principal components regression in exploratory statistical research. J. Am. Statist. Ass. **60**, 235–256.

Schickedanz, P. T. 1973a. A statistical approach to computerized rainfall patterns. Preprints, Third Conference on Probability and Statistics in the Atmospheric Sciences. Boulder, Colorado, AMS, pp. 104–109.

Schickedanz, P. T. 1973b. Climatic studies of extra-area effects from seeding. Technical Report No. 5, Illinois
    Precipitation Enhancement Program, Phase 1, US Department of Interior, Bureau of Reclamation, 53 pp.
Schickedanz, P. T. 1974. Climatic studies of extra-area effects in Midwest rainfall. Preprints, Fourth Conference
    on Weather Modification, Fort Lauderdale, Florida, AMS, pp. 505–509.
Sellers, W. D. 1968. Climatology of monthly precipitation patterns in the Western United States, 1931–1966.
    Mon. Weath. Rev. U.S. Dep. Agric. **97** (3), 173–192.
Stidd, C. K. 1975. Irrigation increases rainfall? *Science, N.Y.* **188**, 279–281.
Texas Water Development Board. 1971. Inventories of irrigation in Texas 1958, 1964 and 1969, Report No. 127,
    Austin, Texas, 229 pp.
US Bureau of the Census, Census of Agriculture, 1919–1969, Statistics for Texas, Oklahoma (and other States).
    US Government Printing Office, Washington, DC.

L'APPLICATION DE CERTAINES FORMULES EMPIRIQUES POUR LE CALCUL DE L'ÉVAPOTRANSPIRATION

par Dr. P. KARAKATSOULIS

Professeur d'Hydraulique Agricole

et G. METAXAS

Ing. Civil, Assistant à l'Ecole Nationale Supérieure Agronomique d'Athènes

Summary

Among the essentials for all irrigation projects is the determination of water needs of the crops to be grown. Many methods, both direct and indirect, have been developed. The problem is here considered for the dry region near Athens using available data on isolation and solar radiation and formulae for evaluating evapotranspiration. It is concluded that, based on actual data, good correlations exist between the empirical relations and the real needs of local crops for water. However the importance of exact climatic data is underlined in order to determine the optimum in terms of work, profit and the national economy.

Parmi les éléments les plus décisifs qui constituent la base de tout projet d'irrigation, la détermination exacte des besoins en eau des cultures possède un caractère prédominant, tant pour le dimensionnement indiqué de l'ouvrage que pour les quantités d'eau écoulées. Le problème devient plus aigu quand l'eau disponible est limitée, fait qui constitue la règle dans presque tous les pays dits arides.

Beaucoup de méthodes directes ou indirectes ont été développées dans ce but. Il est admis que les méthodes directes sont les plus valables et les plus réelles. Malheureusement, par manque de données, on emploie couramment les méthodes indirectes, qu'on doit ajuster aux conditions climatiques de la région du projet en fonction de données météorologiques quand elles existent.

Le présent rapport, pour répondre aux soucis du symposium, traite de ce problème dans la région d'Athènes, qui est une région aride et pour laquelle on dispose actuellement d'une série de mesures de l'insolation et de la radiation solaire globale.

Enfonction de ces mesures directes, on a trouvé les valeurs que doivent prendre les coefficients α et β dans les formules d'évaluation de l'évapotranspiration fondées sur l'insolation et la radiation solaire. Il est évident que ces résultats deviennent plus valables quand le nombre des séries des mesures augmente.

197

1. *Généralités:* Les méthodes de détermination des besoins en eau des cultures sont distinguées comme directes et indirectes. Bien que les méthodes indirectes (Blaney-Criddle, Penman, Thornthwaite, Turc, Lowry-Johnson, atmomètres etc.) ne soient pas si précises que les méthodes directes (Lysimètres, mesures successives de l'humidité du sol, parcelles expérimentales etc.), le manque de ces dernières données impose, dans la plupart des cas, l'utilisation des méthodes indirectes malgré leurs désavantages bien connus.

2. On sait que les méthodes de Blaney-Criddle et Thornthwaite utilisent, à la base, les mêmes données pour estimer l'évapotranspiration. Par les deux méthodes, conçues pour l'évaluation de l'évapotranspiration annuelle ou saisonnière, on obtient des valeurs surestimés pour les premières phases de développement des cultures et sous estimées pour la période de plein développement de la plante. Pour des régions arides, il semble que la méthode de Blaney-Criddle donne de meilleures valeurs de l'évapotranspiration que celle de Thornthwaite. Cependant ni l'une ni l'autre ne sont valables pour évaluer l'évapotranspiration au cours de périodes relativement courtes.

3. La méthode de Penman donne des valeurs de l'évapotranspiration plus précises. Pourtant l'application de cette méthode semble restreinte en raison de la nécessité de posséder un grand nombre de données météorologiques qui manquent souvent et ne sont pas précises. L'utilisation de la méthode dans les régions arides doit être faite avec grande réserve.

4. La méthode simplifiée de Turc, combinant en même temps les facteurs climatiques (précipitations, rayonnement solaire, température) avec les facteurs "sol et plante" semble la plus prometteuse pour les régions tropicales.

5. Nous nous sommes référés ci-dessus aux principales méthodes indirectes utilisées le plus couramment pour évaluer les besoins en eau des plantes. Nous avons voulu souligner leurs limites d'utilisation par les ingénieurs d'étude de travaux d'irrigation et d'autre part l'importance que possède l'existence des données expérimentales de l'évapotranspiration, qui permettraient l'ajustement de ces relations empiriques aux conditions particulières du terrain ainsi que le dimensionnement exact des réseaux d'irrigation.

6. Pour obtenir un ajustement plus réel il est très utile, en outre, avant de fixer les quantités d'eau nécessaires, de tenir compte de l'expérience acquise à propos des rendements obtenus par irrigation dans la région ou le projet est à l'étude. Cette expérience, quand elle existe, doit être prise sérieusement en considération. De la combinaison des éléments cités, on arrive à formuler des avis plus rationnels sur le choix du système d'irrigation le plus indiqué dans chaque cas.

7. En dépit de ce qui précède, il existe malheureusement très peu de données provenant de l'expérience directe. On explique de ce fait très facilement pourquoi, en ce qui concerne les besoins en eau différentes cultures, il existe des différences très importantes parmi nos pays méditerranéens. Il est vrai que dans tous les pays, on cherche à déterminer les besoins en eau au moyen des méthodes directes, tout en utilisant les méthodes empiriques parce que le nombre des séries des données expérimentales n'est très souvent pas suffisant. Ainsi l'ajustement de ces formules empiriques doit se faire encore, et dans la plupart des cas, à l'aide de données climatologiques de la région ou est situé le projet.

8. Dans ce but nous avons essayé, en tenant compte des données de l'Observatoire d'Athènes, d'adapter les formules de Penman (1948) et de Turc (1954) aux conditions climatologiques de la région d'Attique, région aride, afin que ces résultats possèdent une signification plus particulière pour ce symposium.

Ces formules d'évaluation de l'évapotranspiration sont:

a. *Penman:*
$$E=\frac{1}{L}\cdot\frac{Rn\frac{F'_T}{\gamma}}{1+\frac{F'_T}{\gamma}}+E\alpha\frac{1}{1+\frac{F'_T}{\gamma}}$$

Dans cette formule la quantité que nous isolons est le rayonnement net évalué par l'intermédiaire de la formule de BRUNT, soit,

$$Rn=Ig(1-\alpha)-\sigma T^4(0.56-0.08\sqrt{e})\left(0.10+0.90\frac{h}{H}\right)$$

De cette expression le terme qui nous intéresse pour notre analyse est Ig, qui est égal à:

$$Ig=Ig_A\left(0.18+0.55\frac{h}{H}\right)$$

b. *Turc:*
$$E=\frac{p+\alpha+x}{\sqrt{1+\left(\frac{p+\alpha}{1}+\frac{x}{21}\right)^2}}$$

Dans cette formule, le terme qui nous a intéressé est le terme 1 qui constitue un facteur dit heliothermique, donné par la relation,

$$1=\frac{1}{16}(t+2)\sqrt{Ig}$$

ou
$$Ig=Ig_A\left(0.18+0.62\frac{h}{H}\right)$$

Dans les formules précédentes,

Ig = radiation solaire globale que l'on peut évaluer à partir de la durée d'insolation h ou mesurer directement en cal/cm²/jour.
Ig_A = radiation solaire directe en l'absence d'atmosphère, en cal/cm²/jour.
h = durée réelle d'insolation mensuelle en heures.
H = durée maximale d'insolation mensuelle possible en heures, donnée par des tableaux en fonction de la latitude de la région considérée.

9. Dans ces formules, on constate immédiatement le rôle important des quantités Ig et Ig_A. Il est vrai que la mesure directe, systématique et précise de Ig ne se fait qu'en un relativement petit nombre de stations météorologiques. Dans les cas où ces mesures manquent, on peut surmonter la difficulté en calculant la quantité Ig par l'intermédiaire de la quantité Ig_A qui est obtenu à l'aide d'un tableau ou d'une abaque pour chaque mois en fonction de la latitude de la région considérée.

La relation qui donne la quantité Ig dans les formules susmentionnées peut être écrite sous une forme générale qui suit:

$$\left(Ig=Ig_A\,\alpha+\beta\frac{h}{H}\right)$$

ou: α et β sont des coefficients à déterminer.

TABLE 1. *Valeurs mesurées de la radiation solaire globale, Ig, par l'Observatoire d'Athènes (latitude $\phi = 37° 58.3'$).*

Année	Janvier	Fevrier	Mars	Avril	Mai	Juin	Juillet	Août	Septembre	Octobre	Novembre	Decembre
1959	187.20	278.72	355.73	463.56	554.06	604.58	573.88	525.33	409.72	299.36	188.95	140.55
1960	180.55	277.95	314.38	446.92	588.77	632.85	657.06	584.12	455.11	355.89	255.21	182.24
1961	198.23	262.41	396.09	520.12	607.75	628.46	602.05	539.67	478.98	316.55	227.64	154.67
1962	201.78	228.86	361.28	537.59	582.83	658.57	672.14	620.35	449.14	277.73	185.21	115.91
1963	168.03	254.00	336.89	481.93	487.52	645.51	640.94	590.16	466.93	273.17	232.11	159.25
1964	174.26	227.47	307.78	510.79	543.33	587.25	633.87	568.44	416.63	319.35	189.32	145.80
1965	162.14	221.08	306.56	426.19	529.83	596.13	613.48	530.07	452.32	317.27	206.31	151.83
1966	141.81	273.74	342.15	434.38	523.83	562.34	593.77	512.10	390.85	288.09	174.76	155.10
1967	179.43	225.34	336.66	424.75	528.65	553.70	553.05	490.28	389.33	251.13	176.93	143.44
1968	157.96	194.44	323.65	433.82	488.41	489.54	546.56	463.94	384.83	244.78	148.88	112.13
1969	119.63	201.51	229.51	418.59	563.38	571.97	595.17	556.67	412.92	301.05	201.54	132.06
1970	157.17	278.90	316.90	447.51	569.88	601.79	601.71	545.45	403.90	326.90	232.36	154.68
1971	142.82	197.68	351.22	436.60	516.71	571.29	549.55	509.43	377.77	282.20	188.53	157.22
1972	125.61	187.33	356.61	427.28	570.91	654.29	586.90	557.55	423.44	279.05	244.41	131.99
1973	154.09	222.45	298.88	476.96	571.78	602.98	579.85	553.46	443.88	288.39	219.46	154.05
1974	135.90	220.75	310.38	392.19	554.18	597.35	620.76	542.20	444.83	299.31	196.63	156.89
Σ	2586.61	3752.63	5242.67	7279.18	8822.43	9560.00	9620.74	8689.22	6800.58	4720.22	3268.29	2190.92
Moyenne	161.66	234.54	327.67	454.95	551.40	597.50	601.29	543.07	425.04	295.01	204.26	136.93

TABLE 2. *Valeurs mesurées de la durée réelle d'insolation mensuelle en heures, h, par l'Observatoire d'Athènes (latitude φ=37° 58.3').*

Année	Janvier	Fevrier	Mars	Avril	Mai	Juin	Juillet	Août	Septembre	Octobre	Novembre	Decembre
1959	4.48	6.16	5.72	7.78	10.54	12.11	11.98	11.63	9.03	7.18	5.93	4.19
1960	4.23	5.12	3.37	5.73	9.35	10.10	12.01	11.58	8.17	7.19	2.68	4.30
1961	4.92	5.83	7.30	9.15	10.85	11.66	12.79	12.25	10.17	6.80	6.21	4.42
1962	5.39	4.92	6.16	10.09	10.76	11.94	12.75	12.32	9.07	5.74	4.44	3.39
1963	4.16	5.19	5.80	7.87	8.15	11.74	12.32	12.14	9.94	5.51	7.26	4.25
1964	4.59	4.67	4.50	8.68	9.25	10.82	12.12	11.45	8.31	8.07	4.95	4.32
1965	3.99	4.87	5.35	6.95	9.29	10.88	12.48	10.96	9.94	8.50	6.25	4.94
1966	3.52	7.30	6.33	7.32	9.75	10.54	12.41	11.63	9.17	7.59	5.05	5.43
1967	6.03	5.51	6.85	7.51	10.23	10.85	11.94	15.52	9.41	7.10	5.98	5.01
1968	5.60	5.09	6.86	9.02	10.29	9.93	12.44	11.36	9.66	6.97	4.46	3.33
1969	3.08	5.18	4.19	8.94	11.50	10.79	12.47	12.25	8.61	7.33	6.02	3.76
1970	4.27	7.54	5.76	8.43	10.39	11.73	12.29	11.98	8.68	7.93	6.80	4.32
1971	3.07	4.05	6.29	7.82	10.11	11.57	11.51	11.84	8.39	7.74	6.00	6.24
1972	2.70	4.18	6.88	6.61	10.55	12.19	11.07	11.15	8.26	6.02	7.77	3.03
1973	3.62	4.83	4.98	8.36	10.55	11.47	11.56	11.66	9.07	6.78	7.06	5.05
1974	3.38	4.98	5.61	6.34	10.49	11.59	12.80	11.95	9.64	7.72	6.22	5.72
Σ	67.03	85.42	91.95	126.60	162.05	179.91	194.94	191.67	145.52	114.17	93.08	71.70
Moyenne	4.19	5.34	5.75	7.91	10.13	11.24	12.18	11.98	9.10	7.14	5.82	4.48

10. Ainsi, les différents auteurs donnent à ces coefficients les valeurs suivantes:

a. Penman (1948): $\alpha = 0.18$ et $\beta = 0.55$.
b. Turc (1954): $\alpha = 0.18$ et $\beta = 0.62$.
c. Glover-McCulloch (1958): $\alpha = 0.29 \cos \phi$ et $\beta = 0.52$
    ($\phi = $ la latitude en degrés).
d. Koopman (1969): pour régions tempérées $\alpha = 0.20$ et $\beta = 0.53$
    pour régions tropicales $\alpha = 0.28$ et $\beta = 0.48$
e. Baars (1973): pour la Yaugoslavie a trouvé que le coefficient α varie:
    pour l'été entre 0.21 et 0.30
    pour l'hiver „   0.14 et 0.18
    et le coefficient β varie entre 0.42 et 0.47, exceptionnellement jusqu'à 0.53. Les plus petites valeurs de β correspondent aux mois d'été.

11. Ayant à notre disposition une série de 16 années de mesure directe de la quantité Ig (désignée par Ig_m), on a essayé de voir quelles valeurs des coefficients α et β donnaient les valeurs calculées de Ig (désignée par Ig_c) les plus proches de Ig_m. Ainsi, en considérant les valeurs de α et β respectivement égales à 0.18 et 0.62 pour Turc et d'autre part égales à $0.29 \cos \phi$ et 0.52 pour Glover-McCulloch, on a trouvé que dans le premier cas $Ig_m = 0.916 \, Ig_c$ et dans le second $Ig_m = 0.936 \, Ig_c$.

12. De ce fait, on a été ensuite amené à déterminer, par la méthode des moindres carrés. les valeurs des coefficients α et β afin qu'il y ait concordance avec la valeur mesurée. En utilisant donc les tableaux de mesures 1 et 2 donnant respectivement les valeurs de la radiation solaire globale mesurée (Ig_m) et les valeurs de la durée réelle d'insolation mensuelle, on a trouvé, à l'aide des ordinateurs que:

$\alpha = 0.195$ et $\beta = 0.520$ pour toute l'année et,
$\alpha = 0.382$ et $\beta = 0.285$ pour la période des mois de mai, juin, juillet, août, septembre, période
    qui concerne particulièrement les irrigations.

En utilisant les coefficients trouvés pour calculer la radiation solaire globale Ig_c, on a obtenu des valeurs de celles-ci pratiquements égales à Ig_M soit $Ig_M = 0.9999 \, Ig_c$, fait qui vérifie l'ajustement proposé.

13. En considérant les valeurs des coefficients α et β trouvés pour toute l'année, on constate que celles-ci coïncident pratiquement avec les valeurs données par Koopman, valables pour les régions tempérées. Il est évident que si l'on disposait d'un plus grand nombre de séries de mesures, les valeurs des coefficients α et β seraient, sans aucun doute, plus représentatifs.

14. En conclusion, on pourrait citer l'intérêt évident de trouver, en se basant sur des données réelles, des corrélations entre les relations empiriques et les vrais besoins en eau de nos cultures. On a voulu d'autre part souligner l'importance que possède l'existence de données climatiques exactes, qui nous permettrons de déterminer le dimensionnement optimal de l'ouvrage en réalisation au profit de nos économies nationales.

References

Baars, C. 1973. Relation between total global radiation, relative duration of sunshine and cloudiness. Zbornik meteoroloskih i hidroloskih radira 4.

Blaney, H. F. and Criddle, W. D. 1950. Determining water requirements in irrigated areas from climatological and irrigation data. U.S.D.A., SCS–TP–96.

Brochet, P. et Gerbier, N. 1968. L'evapotranspiration monographies de la meteorologie nationale. Paris.

Chow, Ven. Te. 1964. Handbook of Applied Hydrology. McGraw-Hill. New York.

De Saint-Foulc, J. d'At. 1967. Irrigation par aspersion. Eyrolles. Paris.

Fitzgerald, P. D. and Cossens, G. G. 1966. Irrigation investigations in Otago, New Zealand. *N.Z. Jl. Agric. Res.* **9** (4).

Halkias, N. A. *et al.* 1955. Determining water needs for crops from climatic data. *Hilgardia*, **24**, 207–233.

Karakatsoulis P. 1974. Considerations sur les besoins en eau des certaines cultures dans des pays semi-arides. Newsletter G.C.I.D. Athens.

Kijne, J. W. 1974. Determining evapotranspiration. Drainage principles and applications. 3. Wageningen.

Koopman, R. W. R. 1969. De bepaling van de verdemping net behulb von nomogramen. *Cultuurtechn. Tijdschrift.* **9** (2).

Penman, H. L. 1948. National evaporation from open water, bare soil and grass. *Proc. Roy. Soc.* **193.** London.

Remenieras, G. 1972. L'Hydrologie de l'Ingĕnieur Eyrolles. Paris.

Thornthwaite, C. W. 1948. An approach towards a rational classification of climate. *Geog. Rev. N.Y.* **38** (1).

Turc, L. 1961. Evaluation des besoins en eau d'irrigation, evapotranspiration potentielle. *Annls Agron.* **12** (1). Versailles.

Wilson, E. M. 1969. Engineering Hydrology. Macmillan. London.

Climatological Bulletin 1959–1974. National Observatory of Athens Meteorological Institute.

DISCUSSION AND CONCLUSIONS

Compiled by G. KOVÁCS

At the symposium wide-ranging discussions developed around problems of the hydrological cycle, and these have been summarized by G. Kovács, Convener of the session, as follows. Participants in the discussion included Dr. Abaza, A. M. Balba, Y. Barrada, P. J. Bell, J. de Forges, M. M. El Gabaly, M. A. Ezzat, A. Fritz, F. L. Hotes, G. P. Karakatsoulis, J. Khouri, G. Kovács, F. I. Massoud, S. Mohamed, R. S. Odingo, S. Pels, H. Rofe, M. A. Zeid.

The ultimate goal of irrigation is to maintain the moisture content of the unsaturated zone (and especially that of the root zone) between two limits ensuring in this way the optimum crop-yield of the area. The same unsaturated zone is the heart of the terrestrial branch of the hydrological cycle, for conditions prevailing in this relatively thin layer determine basically the ways in which precipitation, having reached the land surface, returns directly to the atmosphere, creates surface runoff, or infiltrates. This is why the modification of the moisture content due to irrigation may considerably alter the whole hydrological cycle both quantitatively and qualitatively.

Changes in the hydrological cycle

The changes due to arid land irrigation can be classified as follows:

Modification of the atmospheric branch of the cycle:
   increase of actual evapotranspiration;
   higher atmospheric vapour content;
   change in the amount and the pattern of precipitation.

Modification of surface runoff:
   increase in the amount and intensity of catchment runoff resulting in higher erosion potential (increasing sediment transport);
   control and regulation of rivers discharges by reservoirs and decrease of the sediment load by settlement in the reservoirs;
   decrease of river flow by consuming water, and the deterioration of the river beds as the consequence of the smaller sediment transporting capacity.

Modification of the groundwater regime, migration of salts and of hydrological processes in the soil moisture:
   increase of the positive and decrease of the negative accretion, as well as rise of the water table below irrigated lands;

205

development of horizontal groundwater flow from irrigated areas towards the neighbouring non-irrigated lands, raising the water table of the latter and the development of "dry
drainage" areas; leaching of the irrigated soils, transport of salts by groundwater flow and
the acceleration of salt accumulation under the dry drainage areas.

Modification of water quality other than those occurring within the soil-moisture zone:
increase of salt concentration during storage, conveyance and distribution of water due
to evaporation;
other qualitative changes occurring during these operations, including change in temperature and suspended load; pollution caused especially by nutrients and pesticides from
surface runoff and by salts transported by water percolating into the canals;
deterioration of quality of the water downstream caused by the effluent of the drainage
systems having high salt content.

Evapotranspiration

The increase of evapotranspiration due to irrigation is limited, because the total amount
of water evaporated and transpired cannot be higher than the potential evapotranspiration
(taking into account the oasis effect). For planning and design the difference between
potential and actual evaporation has to be calculated.
The maximum possible increase of evapotranspiration provides a relatively small amount
of vapour related to the moisture transported by the air masses. This is the reason why
only insignificant changes are expected in the total amount and pattern of precipitation as
a result of irrigation. This assumption was testified by the investigations carried out in an
extensively irrigated part of Texas.

Loss of plant nutrients in reservoirs

The settling of suspended load in reservoirs causes not only decrease of the available
storage capacity, but it also removes a part of plant nutrients from the water. Recent
studies investigating this effect of the Aswan High Dam showed that no more than the
equivalent of about 2000 tons of siltborn plant nutrients would be lost annually in Lake
Nasser, and this could be offset by manufacture of up to 13,000 tons of calcium nitrate
fertilizer.

Waterlogging and drainage

Waterlogging within the irrigated area and salt accumulation under the neighbouring
fields are the results of hydrological processes modified by irrigation in the soil-moisture
zone and in the groundwater space. Examples were quoted showing that irrigation introduced on higher terrains resulted in rapid deterioration of the lower lying lands (New
South Wales and Kom Ombo area). For the protection of the arable lands against these
destructive effects the irrigation schemes have to be supplemented with drainage systems
wherever the natural drainage of the area is inadequate. Effective drainage can be achieved,

however, only if the collectors are below the water table. The combined system of irrigation and drainage can be regarded as a complete control of the groundwater and soil-moisture regime, the purpose of which is to ensure the optimum water content and flow in the root-zone. For designing this system the use of the concept of the characteristic curve of the groundwater balance was proposed, the application of which was demonstrated on the basis of data collected in Hungary, Morocco and Byelorussia.

Salinization and leaching

Among the qualitative changes those caused by high salt content of the effluent water are the most serious. Examples were quoted showing the increase of salt load along the Rio Grande, the Snake River, the Yarima River and the Colorado River. The Euphrates was mentioned as an example where the increasing salt content raises international problems. There are attempts to decrease this load by reducing the leaching requirements, but the application of limited leaching can cause the deterioration of arable land, especially if a very high level of farm management cannot be assured. It was the general opinion that both the leaching requirements and the methods to achieve leaching have to be further investigated.

Regional and international problems

It was emphasized that the hydrological changes caused by irrigation are not always local problems but they may influence large areas interconnected by the hydrological cycle. The investigations have to be extended to large regions and, especially within international river basins, the problems can be solved only by the multilateral efforts of the interested countries. It was proposed, therefore, to include the analysis of the hydrological influences of irrigation in the programme of the World Water Conference to be organized by the UN in 1977 in Buenos Aires.

Eutrophication and harmful runoff

Special attention was given to the eutrophication of waters due to fertilizers applied on cultivated land. Data quoted proved that among the various nutrients only nitrates reach the groundwater and even their amount is small, not more than 5 or 6% of the total amount applied as fertilizer. Surface runoff, however, may carry much higher amount of nutrients, both phosphates and nitrates, into rivers and lakes. Data collected in the US has shown that most of the nitrates and phosphates in rivers originates from agricultural land. Another example quoted from Hungary shows that, within the catchment of a lake, 70% of the nutrients reaches the rivers and the lake in dispersed form, mostly originating from agricultural land. Pollution from fertilizers can be avoided only by applying special farming practices. Within large plains and levelled irrigation systems natural surface runoff hardly occurs, but the possible pollution hinders the re-use of the effluent surface waters, which may contain harmful chemicals and organisms.

Further studies

On the basis of the papers presented and the discussions various scientific fields can be clearly indicated where further research is needed for the better understanding of hydrological changes caused by irrigation and the practical problems which they cause. The IAHS, as the organizer of this section, will prepare a detailed list of research needs and will include their investigation in its scientific programme. At the same time, since the fifth Chapter of IHP deals with the influence of man of hydrology in general, IAHS will ensure that the practical problems of arid land irrigation have adequate weight in the IHP.

SECTION IV

Land Use, Soil and Water

ARID LAND IRRIGATION AND SOIL FERTILITY: PROBLEMS OF SALINITY, ALKALINITY, COMPACTION

by V. A. KOVDA

Institute of Pedological Sciences, The University of Moscow, Moscow, USSR

V. A. Kovda, who had been designated Convener of this session of the symposium and had prepared the introductory keynote paper, was unforunately prevented by illness from attending. I. Szábolcs stepped into his place at a few hours' notice, presented the essence of Kovda's paper, organized the session and summarized the discussions.

Summary

Reclamation and amelioration of soils of arid and semi-arid areas by means of irrigation, drainage, chemical treatment, and washing out of salts, could free agriculture from the influence of the elements and create a highly productive agrobiogeocenosis as a stable background for double and triple yields of agricultural crops. Even today, although only 15–20% of the world's crop land is under irrigation, the production from this land amounts to 30–40% of the world agricultural output. The area of lands suitable for irrigation in the world has been estimated at 1 billion hectares, of which the area currently irrigated is of the order of 230–240 million ha, about half of them in developing countries. By the year 2000 it is expected that the area of irrigated lands will have increased to 300 million ha, but in addition it is extremely important that the operation of irrigation schemes will be improved and the unfavourable side-effects prevented or at least moderated.

Newly constructed irrigation systems do not always produce heavy agricultural crops without causing adverse effects on the soil, including erosion, salinization, alkalinization, loss of structure and waterlogging. According to estimates by FAO and UNESCO more than 50% of all irrigated lands have been damaged by these processes and as a consequence many million ha of irrigation systems have been abandoned. At present 50% of the irrigable area in Iraq, in the Euphrates valley of Syria, 80% in Pakistan and 30% in Egypt suffer to varying degrees. The reasons for these harmful processes include use of water of poor quality, rising groundwaters due to inadequate water and soil management, seepage from canals and reservoirs, uneven distribution of water, improper irrigation and drainage techniques. In many cases too much water is used, in others too little.

Some of the soil deterioration processes are very difficult to reverse and cannot be overcome by simple leaching and draining. Maximum efforts are needed to prevent these processes because the amelioration of waterlogged and secondary salinized or alkalized soils is expensive and complex. However, prevention is possible if we know about probable future changes; but to make such predictions hydrogeology, hydrology and soil science need to be invoked. The relations between soil, irrigation water and groundwater have to be worked out for each area and based on comprehensive soil and water surveys. It is necessary (i) before planning an irrigation and drainage system to have soil and hydrologic maps giving a detailed picture of the peculiarities of the natural situation and its variation from place to place; (ii) during irrigation to establish a permanent soil and hydrologic monitoring system of the changes caused; (iii) on this basis to prevent or moderate the undesirable soil processes and to reduce their causes, for example seepage or the rise of water table. To prevent alkalization, which may be even more dangerous than salinization, principle measures are a reasonable limitation of water supply, high efficiency of the system, elimination of the rise of groundwater, and efficient drainage. The paper deals with the problems of water quality, the importance of ion composition, soil conditions, climate, land use, cropping pattern, and agricultural and water management. It also summarizes such problems as seasonal salinization, the critical depth of mineralized groundwater, leaching as a means of removing salts from the soil, leaching during the growing season, and the control of the salt balance in soil water. It is concluded that:

The construction of effective drainage facilities and their proper use in areas affected or endangered by secondary salinization is the most vital and urgent problem of arid land irrigation today.
The construction of irrigation systems without drainage installations in arid countries leads to a gradual decrease of their efficiency and is a waste of labour and investment.

Résumé

La restauration et l'amélioration des sols des zones arides et semi-arides par irrigation, drainage, traitement chimique et lessivage des sels peuvent soustraire l'agriculture de l'influence de facteurs néfastes et créer une agrobiogéocénose hautement productive, capable de doubler ou de tripler les rendements des cultures. Or, bien qu'à l'heure actuelle seulement 15 à 20% des terres agricoles du monde soient soumises à l'irrigation, la production de ces terres fournit 30 à 40% des rendements agricoles mondiaux. La surface terrestre utilisable par irrigation dans le monde a été estimée à 1 billion d'hectares, dont 230 à 240 millions d'ha sont d'ores et déjà irrigués. La moitié de cette dernière surface se trouve dans les pays en voie de développement. A l'horizon des années 2000, il est prévu que la surface des terres irriguées aura augmentée de 300 millions d'ha. Il est en outre extrêmement important que la conduite des opérations en périmètres d'irrigation ait été améliorée et que les effets néfastes aient été supprimés ou tout au moins réduits.

Les nouveaux systèmes d'irrigation ne produisent pas toujours des rendements accrus sans causer des effets nuisibles aux sols, y compris des phénomènes d'érosion, de salinisation, d'alcalinisation, de destruction de la structure et de saturation en eau. Selon la FAO et l'Unesco, plus de 50% des terres irriguées ont été endommagées par ces processus et, en conséquence, bien des millions d'hectares de systèmes irrigués ont été abandonnés. A l'heure actuelle, 50% de la surface irrigable en Iraq et dans la vallée de l'Euphrate en Syrie, 80% au Pakistan et 30% en Egypte sont atteints à différents degrés. Les causes de ces processus néfastes comprennent l'utilisation d'une eau de qualité médiocre, la remontée des nappes phréatiques dûe à une utilisation inadéquate de l'eau et à un mauvais aménagement du sol, des pertes dans les canaux et dans les réservoirs, une mauvaise distribution de l'eau, des techniques impropres d'irrigation et de drainage. Dans bien des cas, beaucoup trop d'eau est utilisée et dans d'autres, beaucoup trop peu.

Quelques une des processus de détérioration du sol sont très difficiles à contrôler et ne peuvent pas être surmontés par un simple drainage ou un simple lessivage. De très grands efforts sont nécessaires pour prévenir ces processus, principalement en raison de la complexité et de la chèreté du coût de l'amélioration des sols salinisés ou alcalinisés. Cependant, cette prévention est possible à la condition que l'on sache prévoir les changements édaphologiques futurs. Mais pour faire de telles prédictions, il est nécessaire de faire appel à l'agriculture, à l'hydrologie et à la science du sol. Les relations entre le sol, l'eau d'irrigation et les nappes phréatiques ont été étudiées pour chaque région et ont été fondées sur des prospections pédologiques et hydrologiques complètes. Il est nécessaire: (1) de posséder des cartes pédologiques et hydrologiques donnant une image détaillée des caractéristiques de la situation naturelle et de ses variations spatiales avant de planifier un système d'irrigation et de drainage; (2) d'établir durant l'irrigation un système de contrôle continu des changements causés aux sols et aux eaux; (3) de prévenir ou de modérer les processus pédologiques nuisibles et de réduire leurs causes, par exemple remontée de la nappe phréatique par irrigation. Pour prévenir l'alcalinisation, qui est souvent plus dangereuse que la salinisation, les principales mesures sont une limitation des apports d'eau à un niveau raisonable, une haute efficience du système, l'élimination de la remontée de la nappe phréatique et un drainage adéquat. Cette communication concerne les problèmes de qualité des eaux, l'importance de la composition ionique, les conditions de sol, le climat, l'utilisation des terres et des eaux. Il aborde succinctement des problèmes tels la salinisation, la profondeur critique de la nappe phréatique minéralisée, le lessivage en tant que moyen d'extraire les sels de l'eau, le lessivage pendant la période de cultures et le contrôle du bilan des sels dans l'eau du sol. On peut dire en conclusion:

que la mise en place de systèmes de drainage efficaces et leur utilisation dans des zones affectées, ou risquant de l'être, par une salinisation secondaire, est un problème vital et urgent dans les terres arides irriguées à l'heure actuelle.
que la construction de systèmes d'irrigation sans prévision du drainage dans les régions arides conduit à une décroissance progressive de leur efficacité et à une perte de travail et d'investissement.

INTRODUCTION

The reclamation and amelioration of soils in arid and semi-arid areas by irrigation, drainage, use of chemicals and special processes to wash out the salts, should free agriculture from natural hazards and make a stable background for intensive cropping. The sunny climate of the arid and sub-arid zones of tropical and temperature regions, rich in light and warmth, is the most favourable one for man, and, where water supplies suffice for drinking and irrigation, also for obtaining high biological production. Thanks to the long continuity of the growing season in irrigated areas several successive crops are possible in each year—of rice, or of wheat following rice—or at least an after-harvest of forage crops. Increased or multiple crops will of course involve additional labour, energy, water and fertilizers, but not in direct proportion to yields.

Irrigation has had profound economic and specialized effects on the history of mankind. From the earliest times irrigation by artificial means has been the most effective way of coping with climatic aridity. It is significant that ancient civilizations appeared and developed at much the same time in ancient irrigation and drainage systems: in valleys and deltas of the Murgab, the Amu Darya, the Syr Darya, the Hwang, the Yangtze, the Nile, the Ganges, the Indus, the Mekong, the Tigris, the Euphrates and the Tiber. Irrigation over several thousand years in the deltas of the Nile, Murgab and Indus has allowed the creation of deep (3–7 m) artificial alluvial soils with favourable physical and biological properties, large amounts of humus, nitrogen, phosphorus and microelements. Producing two to three crops a year—sometimes even six to eight, as for example in the Pearl River delta near Canton—man has, with the aid of irrigation, created agroecosystems with surprisingly high productivity, much higher than under natural conditions.

The organization and functioning of irrigation presents an example of a complex influence on different components of the landscape aimed at the creation of highly productive agrobiogeocenoses. Dams, reservoirs, irrigation and drainage canals are built, terraces are constructed, field surfaces levelled, fertilizers applied, shelterbelt trees planted.

World irrigation systems are today servicing areas of the order of 230–240 million hectares. The main crop requiring irrigation is rice, which occupies about two-thirds of the world's irrigation area. Then follow cotton, oil-producing plants, fruit plants and grain crops.

C. E. Houston (p. 426), considering the world as a whole, has provided data showing the area already under irrigation and the large proportion (30–40%) of total agricultural output which it provides. He enlarges also on the possibilities of meeting much of the rapidly increasing demand for food by placing more land under irrigation and improving the operation and management of existing schemes. M. M. El Gabaly (p. 239) has developed these themes with reference to the Near East region. It has been estimated by H. Rush that, if the problem of water for irrigation can be solved (surface, underground and demineralized water), 3–5 billion ha of desert and semi-desert land could be transformed into productive land.

There are great resources for the development of new lands in Africa, South America and Asia, but most of these lands are situated in arid climates and suffer from droughts and desertization. Therefore it is desirable to plan systematic studies and practical work for developing irrigation of field crops, perennial fruit trees and pastures. Action is being taken in various ways and at different rates in different countries. Thus, in the USA the total area of irrigated lands has exceeded 17 million ha in the last 15 years and there are plans to increase it to 22–25 million ha by A.D. 2000–2020. In the USSR every year up to 1 million ha of land is included in newly irrigated territory; the total area of 15 million ha in the USSR includes up to 2 million ha irrigated pasture.

Irrigation of the arid regions in the Soviet Union was and is the most important and permanent line in the economic policy of the government aimed at raising the living and cultural standards of the population living in the steppes and deserts of the Soviet Republics of Asia, the Caucasus and arid regions of the south-eastern part of the country. The government has invested tens of billions of roubles in irrigation. Highly productive irrigated rice systems have been constructed and are operating successfully on previously barren lands. New centres of rice culture have been created where crops of the order of 4–4.5 and 5.0–6.0 t/ha are being grown, that is crops which exceed by 3–4 times the average world rice crops (1.6–1.8 t/ha). Rice crops of the order of 6.7–7.0 t/ha and even 9.0–10.0 t/ha are known from large areas of the USSR. With such crops rice cultivation becomes one of the most profitable branches of irrigated farming.

Wheat crops receiving careful sprinkling irrigation, good agrotechnology and fertilizers produce two to threefold compared with non-irrigated wheat. Wheat irrigation is profitable when its crops amount to 4.0–4.5 t/ha as is the case in many state and collective farms of the Ukraine, Rostov, Krasnodar and Stavropol regions. New varieties make it possible to attain crops of 7.0–8.0 t/ha under conditions of proper land cultivation and watering. However, average crops of irrigated wheat in the country are still at the 2.5 t/ha level: and much of the product is of poor quality, due to shortage of water, or over-irrigation, or soil salination.

The economic efficiency of irrigated sugar beet and legumes is particularly high. Their yield increases five to eightfold in comparison with non-irrigated plants. Irrigation of lands in the USSR southern steppes has permitted the creation of a network of extremely productive and profitable state farms producing these crops. The same can be said about the state and cooperative farms of Armehia, Kazakhstan and Uzbekistan, cultivating irrigated fruit trees and grapes that give highly profitable production.

Particularly profitable also is the cultivation of irrigated fodder—green maize, grass mixtures, alfalfa (in rotation with rice)—for irrigated fodder yields 9–10 times as much as non-irrigated. However, the physiological sensitivity of legumes, fruit trees, grape vines, maize and alfalfa to soil salinity and alkalinity and to the shortage of nutrients is even stronger than that of rice or wheat, so these crops drop their productivity considerably if grown without fertilizers and on lands suffering from salinization, alkalinity or a rise in groundwater. Cotton cultivation in the USSR is based wholly on irrigation and yields of raw cotton have risen to 3–4 t/ha and the total production amounted to 8.4 million t/yr (Alekseevski, 1975). According to data supplied by the Ministry of Water and Melioration of the USSR, the increase in production on ameliorated lands reaches 10–11% annually, which by far exceeds the 2.5% rate of population growth. The question of transferring water from northern Siberian rivers to the southern basins of the Aral and Caspian Seas with the aim of irrigating vast new territories is at present under research.

The important work in Tanzania and Kenya in creating irrigated demonstration and experimental plots directed at teaching farmers to use modern techniques of wheat and maize cultivation and to obtain heavy crops, deserves to be widely known. It is necessary to continue with the gradual organization of large-scale well-equipped cooperative and state farms. The introduction of irrigation appears to be a very important factor in raising living standards and eliminating unemployment, as irrigated farming requires substantial labour forces. It is significant that in Kenya irrigated areas increased from 2430 ha in 1959 to 5668 ha in 1969 and further growth is occurring. It has been estimated that in Kenya 160,000 ha, in Tanzania 1.6 million ha and in Uganda 200,000 ha could be irrigated.

The area of irrigated lands is growing in India, China, Pakistan, Egypt, Algeria, Tunisia, Iraq, Iran, Turkey, Mexico. Irrigation of rice, wheat and maize at the best farms in the USA, India, Pakistan, the Philippines, Mexico, Egypt and Iraq, accompanied by application of fertilizers and progressive agrotechnology, has led to grain crops of 6–8 and 9–11 t/ha (Syromyatnikova and Artemova 1973). Irrigation should allow such countries to increase two or threefold their crops of peanuts, sugar cane, sorghum, as well as fruits such as mango, papaya and banana.

Irrigation resources of the developing countries of Asia, Africa and South America have been calculated at not less than 250 million ha, whereas the area of lands suitable for irrigation in the world as a whole amount to a billion ha. The freshwater resources of the planet are also great, estimated at 37,000 km^3 of annual river discharge, from which Man uses at present only 3200 km^3 (Alekseevski 1975). But the global water resources are very irregularly distributed. Many regions do not have any freshwater reserves, and many of the large rivers are almost fully utilized (Indus, Ganges, Colorado, Nile, Amu- and Syr Darya, Dnieper, Jellow). On the other hand, latent reserves exist in the very low efficiency of water utilization (15–50%) and tremendous losses through evaporation, seepage, overwatering and wastage.

Development of irrigation is also limited by socio-economic and political factors. An 'age of land amelioration' for mankind now appears to be essential; it is already necessary to double the world production of farming and cattle breeding. Conditions of increasing aridity, lowering soil fertility, technical stagnation of farming in many developing countries, are causing the growth of crops to lag behind the growth of populations and of people's needs. Meanwhile creation of highly productive varieties of rice, wheat, sorghum and maize opens new vistas of doubling and tripling yields, but only with irrigation and fertilizer application.

Irrigation schemes do not, of course, completely free cultivation from the impact of droughts. The discharges of rivers show wide fluctuations from the average: in the case of large rivers 10–30% in some, 250–450% in others, and even up to 1840% in the case of the Hwang River. However, the creation of reservoirs and water reserves in soils and ground strata helps to overcome years of deficiency.

One may expect that by the end of this century the total area of irrigated lands will amount to 300 million ha. Of course, the type and technology of irrigation will differ depending on the level of scientific, technical and socio-economic development of a country, as well as on the type of soils and cultivated crops. The development of irrigation will improve the polluted conditions of the biosphere by generating additional volumes of oxygen, fixing carbon dioxide, and using composted wastes, by-products of industry and waste water from towns as fertilizers.

TYPES OF IRRIGATED LAND

Newly constructed irrigation systems do not always give heavy agricultural crops without causing adverse effects on soils or groundwaters. Quite often one can observe the spread of human, animal and plant diseases, and other troubles such as weeds, after the construction of reservoirs, canals and new irrigation networks in dry landscapes. These and other unexpected consequences have been stressed in a number of papers. They can be explained by the fact that many irrigation systems are designed, constructed and used without adequate preliminary studies and without understanding the complicated peculiarities of the natural environment.

Climate and irrigation

The length of the growing season varies in arid regions from 3–4 months to a whole year. The number of possible successive irrigated crops may be from 1 to 3 or 4 a year, and the number of grass mowings from 2 to 8. Potential evapotranspiration often amounts to 500–600 mm a year and sometimes reaches 3000–4000 mm a year, changing along geographic latitude and in different years. Atmospheric precipitation ranges in a similarly pronounced way: there are arid regions with a stable no-water regime of absolute deserts; there are arid regions with unstable conditions of moisture, periodic drought of indefinite degree and length. Hence the necessity for a thorough estimation of hydrothermal conditions of the region from the point of view of selection of future irrigated crops, their composition and rotation and, especially, of the quantitative needs of fields and plants for water, schedule of watering and combination of watering with atmospheric precipitation. On average, the relative humidity of non-saline soils during the growing season should not fall lower than by 65–70% of the field moisture capacity in inter-watering periods. Watering of non-saline soils should be done *only* by deficit, up to the field moisture capacity.

The type of irrigation system to be designed depends on the climate. Different systems (a) provide for plants with *no* other source of water (in arid deserts), (b) supply irrigation water *additional* to rainfall (reserve watering in autumn, watering during dry periods) to obtain optimum soil humidity. The need for irrigation water varies correspondingly from 30–50–100 mm to 500–1000–1500 mm a year. In some years, with a sufficient amount of atmospheric precipitation in steppes and savannas, irrigation is not required at all.

Overdrying and overwatering of soils are equally bad for both crops and soils. Only rice requires permanent flooding of soils and even this is now under study and being reconsidered, for rather heavy rice crops are obtained when repeated waterings maintain the relative soil humidity above 80–85%. Cultivation of a second crop after rice (e.g. barley or maize) during the same year may give a good yield without any irrigation.

Importance of natural salinization of soils and groundwaters

In arid zones soils and groundwaters are characterized by a degree of natural salinity. These questions have been studied and analysed in the works of Soviet, American, Indian, Pakistani and Egyptian scientists (UNESCO/FAO, 1968 and 1973). Toxic salts (chlorides, sulphates, carbonates) can be concentrated in the top (arable) soils or at some depth (40–100 cm or 1–2–3 m). Groundwaters of such soils are usually salinized as well and present the main reserve and sources of salts circulating in the soil profile. The concentration of salts in soils is toxic if their content reaches 0.5–1.0%. Very often the salt content amounts to 5–8–12%, groundwaters containing salts in quantities approaching sea water and greater (15–30–30 g/l). The design of irrigation systems and choice of the rate of watering under these conditions should never be limited by the normal irrigation regime suitable for non-saline soils. The design should have a special programme of desalinization and amelioration of saline or alkaline soils: the leaching of salts from the top soil before cultivation, construction of deep (2–3 m) horizontal or vertical drainage to carry leaching waters.

For leaching purposes, soil flooding is used, often combined with rice culture, with large amounts of water (1000–2000–3000 mm). This operation is sometimes repeated in the second or third year of cultivation. In so far as such soils have a tendency to restore salinity, growing season watering has to be planned to exceed field moisture capacity of soils, to drive away soil salts and saline groundwaters into drains. Leaching and draining become unnecessary if the soils of new irrigation systems are not salinized to the depth of 60–100 cm and if the groundwaters lie at the depth of 15–20 m and more. In these cases waterings should be planned *only* by deficit up to the field moisture capacity, and irrigation canals should be lined with asphalt, cement or piped. Both these measures prevent water losses through filtration and the rise of groundwater level that leads to salinization and waterlogging of soils.

It follows that the design of new irrigation systems should rest on detailed studies of the climatic conditions, and the properties of soil cover, ground and irrigation waters. Soil maps, lithology and groundwater maps (depth, chemistry, mobility) must be produced for designers as a scientific basis.

Geomorphological conditions and irrigation

The relief and geomorphology of the region play an important part in designing networks of irrigation canals and drains, in attachment of subordinate territories to the canals, in distribution of irrigation furrows for gravitational irrigation, and distribution of the water itself in all types of irrigation.

The self-silting or erosion of canals, the washing of irrigation furrows, irregularity of field moistening, washout of soils—all this results from neglect of relief and topography. But, more important, geomorphology and lithology of the territory predetermine its natural drainage capacity. According to this feature it is necessary to differentiate at least three main types of land:

(i) High piedmont plains and water check highlands, underlain below 1.5–2–3 m with permeable sands, gravels, cracked rock. These lands are characterized by a substantial free outflow of groundwaters and high natural drainage capacity. On such land irrigation very rarely causes rise of groundwaters, salinization or waterlogging.

In arid climate, however, such conditions may provoke sagging and creeping of the ground, and soil erosion by irrigation water, but irrigation may do without drainage or with only individual local drainage.

(ii) Low piedmont plains, lowlands, intermountain depressions, second and third terraces of ancient river valleys and lakes, especially those composed of homogeneous loesses, loams and clays. All these are usually rich in salt residue reserves in arid countries and are characterized by insufficient natural drainage capacity. Groundwaters which sometimes contain sodium bicarbonate or are mineralized by chlorides and sulphates, though lying at a depth of 10–20 m, can easily be drawn into movement. Irrigation water on the fields and in the canals has not enough space and time to flow away and, after accumulating, causes groundwater level to rise, first along canals and then over the whole irrigated area.

On reaching the level of 2.5–3 m and higher, groundwaters promote intensive capillary moistening and salinization of soils. To prevent this it is necessary to build irrigation systems with a closed-in (hydroinsulated-lined) network of canals and/or to plan construction of preventive deep systematic drainage.

(iii) Maritime and inland deltas, low reaches of river terraces, lowland plains and depressions without discharge composed of deposits of loams and shaly clays. These do not possess a natural drainage capacity at all. Their salty groundwaters often stagnate, are not provided with run-off, and are mainly spent by evaporation and transpiration. In an arid climate, territories of this type are subjected to the strongest natural salinization process; their soils and shallow-lying (1–3–4 m) groundwaters are strongly, though not uniformly, salinized. The land relief of such territories is convenient from the point of view of water distribution, but irrigation should necessarily start with measures directed at soil desalinization. Deep drainage must be constructed before the start of irrigation, simultaneously with the creation of the irrigation system. Without these measures irrigation systems will quickly fail. A classification of irrigated territories on the degree of natural drainage capacity has been worked out and adopted in the USSR (Kovda 1946, 1973).

The actual peculiarities of the natural drainage of any territory are much more complicated and diverse than as summarized above. They should be evaluated on the basis of thorough complex research. Unfortunately, this is rarely done and the task of designing new irrigation systems is limited to estimating the regional topography, irrigation facilities and water distribution. Disregard of the complexity of natural conditions leads to disastrous after-effects.

PROBLEM OF SECONDARY SOIL SALINIZATION

A voluminous scientific literature is devoted to analysing secondary soil salinization; it includes books and articles by the present author in which he examines the origin of the phenomena and the ways of overcoming it in irrigated soils.

Irrigation of large areas is a powerful means of technical influence of man on the environment. Through its action irrigation takes hold of the earth strata tens and hundreds of metres thick, influences tens of metres of air layer and has an impact on the water and salt balance of a considerable part of river-system basins and on deltas. Modern irrigation systems are generally built and used much as they were in ancient Egypt and Babylon. The network of canals is as a rule built without hydraulic insulation, so losses of transported water in large canals reach 40–45%. In the fields water is distributed through temporary small canals and furrows, which in almost all cases distribute enormous surplus quantities of water over the fields and vacant land. All this combines to make the majority of irrigation systems only 30–50% efficient: more than half the water coming into the system is lost. Therefore, the major and first task in preventing secondary salinization and waterlogging is technical perfection of irrigation systems themselves and a precise technology of water utilization. Improvement of technology and technical equipment (troughs, hydroinsulation, pipelines, sprinkling mechanisms, levelling, and distribution of water in accordance with the plant needs and properties of soil cover) may provide for radical improvement of water usage and may raise efficiency up to 80–85%.

But the danger of salinization of irrigated soils and waterlogging of irrigated land still persists even with improved technology and better equipment.

The point is that irrigation radically changes the natural water–salt balance (budget) of the territory (Kovda 1946, 1947, 1965, 1973). These factors are much stressed in the papers by Szábolcs, Darab, Balba, Varallyay, Hardan, Dukhovny, Bhumbla. El Gabaly has described the situation in countries of the Near and Middle East, while Pels and Stannard have shown how troubles arise even in the comparatively recent irrigation systems of Australia.

The majority of irrigated territories are not provided with natural drainage. In new irrigation systems without drainage the rise of the groundwater level sometimes reaches a speed of 3–4 m a year and subsoil salty water rises to the surface. During many centuries and even millennia only areas having a free outflow of groundwaters, as in Tashkent and Samarkand, have not undergone salinization or waterlogging. Thus increasing salinity in irrigated soils of arid lands is practically universal. In the USSR, however, overwhelming progress has been achieved in combating it since 1945 in Azerbaidzhan, Uzbekistan, Tadzhikistan and Turkmenia. The methods and achievements in the Syr Darya basin are described by Dukhovny and Litvak (p. 265); but in some regions of the South-East, Transcaucasia and Central Asia, the problem has not yet been eradicated. Scientific data pertaining to this ameliorative work are published in the author's papers (1946, 1947, 1967, 1973) and were demonstrated during the X International Congress of Soil Science in 1974 and the IX International Congress on Irrigation and Drainage in 1975, both held in the USSR.

Unfortunately this positive experience of the Soviet Union has not received publicity and wide application, one of the causes of this being the erroneous attitude of the Salinity Laboratory in Riverside, USA. For 25–30 years this laboratory has rejected indisputable conclusions of the geochemical theory of salt accumulation and the experience of land amelioration. It has underestimated the importance of the groundwater level and mineralization of the groundwater and properties of saline soils. The conception of critical level, critical regime and critical mineralization of groundwater is either rejected or ignored. Secondary salinization of soils is attributed mostly to the salts of irrigation water, which in fact are of secondary importance. The necessity of leaching of salts from salty soils independently of seasonal watering, and the necessity of desalinization of salty groundwaters, are rejected or even misunderstood. The importance of ionic composition of the soluble salts is ignored and a "cult" of electroconductivity as the way to study soil salinity has been followed.

Publications of the Riverside Laboratory by-pass all these problems, are actually cultivating the idea of a permanent domination of downwards flux and over-irrigation in order to suppress ascending capillary solution, and are supporting the anti-drainage assumptions of the advocates of "cheap" drainageless irrigation, which in fact leads to waterlogging and salinization. To this we must add the striving for introduction of rice everywhere— rice, which requires large amounts of water, and provokes irregularity in the moisture and salinity of the field surface.

Because of this misleading information the majority of countries introducing irrigation continue:

(a) to ignore the peculiarities of the natural soil situation, including level of mineralization and chemical composition of groundwaters, salinity of ground and soils, unsatisfactory natural drainage;

(b) to tend to do without deep drainage installations in the hope of making irrigation system construction "cheaper";

(c) to take in surplus water and lose substantial quantities from the fields and irrigation canals (which are generally not lined), and this leads to a rise of groundwater level.

In addition the areas of saline soils are not measured and registered in the majority of countries, so the actual position often remains unknown or is hidden.

By contrast, scientific and field experience in the Central Asian and Caucasian Republics of the USSR teaches us to accentuate the following paramount measures for combating salinity:

(a) Lining of canals with impermeable material and the construction of irrigation canals in insulated pipelines in the future.

(b) Construction of vertical mechanized drainage (tube wells) where groundwaters lie deep, so that any threat of their rise is eliminated. Note that California and Arizona in the USA have also managed to avoid the rise of groundwaters by this means.

(c) Deep horizontal drainage and its faultless operation in combination with leaching waterings, where salty groundwaters lie shallow and soluble salts are present in the soils.

There would be no salinization of irrigated soils in the recently constructed irrigation systems if they had been provided with deep horizontal drainage, carefully levelled fields, initial leaching of soluble salts and good technical management. In order to meet future needs it is now necessary to build well-equipped irrigation systems, automatically controlled with all mechanisms required and the best current information and advice for their managers.

In steppes and savannas it is not every year or every season that is arid. Under these climatic conditions irrigation is not the main source of water to plants, as is the case in deserts and semi-deserts, but a supplementary source added to atmospheric precipitation. Therefore, irrigation systems created in semi-arid climate should be highly mobile, quickly and efficiently controlled for water supply, and should irrigate only when actually necessary.

Problem of alkalization

In his publications and reports the author expressed apprehension about the potential danger of secondary alkalization (soda salinization) when irrigating soils of steppes, pampas, pushtas, savannas. This problem is of global importance and the subject of a special permanent subcommission of the ISSS. The geochemistry of sodium carbonates and bicarbonates in groundwaters as discussed in the proceedings of the Budapest and Erevan symposia (1965 and 1969).

Many countries of Asia, Africa, Europe, South and North America have soda-saline soils. These soils are formed under the influence of low mineralized (0.5–1.5–2 g/l) alkaline groundwaters, and of diluted soda-carrying irrigation waters such as the Nile, Indus, Araks rivers and underground waters in California, Hungary, Pakistan. Sodium carbonate and bicarbonate is more toxic than sodium chloride and its presence in irrigation or groundwater provokes a number of physical, chemical and mineralogical changes which adversely affect soil fertility and are in most cases extremely difficult to eliminate. The soil-absorbing complex is saturated with exchangeable sodium up to 50–70% of the cation exchange capacity, and pH reaches 9 to 11. The crumby and granular soil structure degrades, and is replaced by lumps and compactness. Peptized hydrophilic organic and mineral colloids appear. Highly dispersed minerals—smectites—with a mobile crystal lattice are formed and activated; they impart soil cementation, the development of deep cracks in dry soil and a jelly-like character in a moistened state. The productivity of soils suffering from soda salinization is meagre; their amelioration requires simultaneous application of deep horizontal drainage (combined with a vertical drainage when the groundwaters have a pressure head), big doses of chemical ameliorants (30–50 t/ha of acid or gypsum), leaching to remove surplus salts, large doses of organic fertilizers, and constant use of acid and physiologically acid fertilizers. The successful experience of California, Hungary, Armenia and Azerbaidzhan fully confirms these measures.

The formation of soda solutions is brought about by the hydrolysis of sodium alumo-silicate minerals during weathering of volcanic rocks. Desulphaction and denitrification may also provoke formation of alkali solutions, with absorption and desorption of exchange sodium in the soils as a further factor. But soda is easily neutralized by solutions of gypsum or calcium chloride and precipitates in the form of calcium carbonate. Therefore alkaline soils cannot be formed in areas containing gypsum, or where irrigation waters have weak solutions of gypsum or calcium chloride, as have parts of Iraq, Syria, Uzbekistan, Tadzhikistan, Turkmenia and the Caspian lowland depression. However, where gypsum content in soil-forming rocks, water-bearing horizons, or in rivers is low, the water usually contains weak solutions of sodium carbonates and bicarbonates which may accumulate in soils. Recent data show that alkalinization of soils may be of two different origins: residual (at depths of 70–150 cm or more), brought about by an alkaline capillary fringe of groundwaters; and secondary, after the rise of alkaline irrigation waters. Problems of amelioration are analysed in the *International Handbook on Irrigation and Drainage*, published by UNESCO/FAO (1968, 2nd ed. 1973).

Soils of alkalization are known in the irrigated soils of western Siberia, Azerbaidzhan, Armenia, at the south of the Ukraine, are frequent in the Hwang Ho delta and valleys of western China, in the Mongolian lowlands, in India, Pakistan, Iran, on the plains of the African Rift, on the river terraces of the Balkan peninsula and Slovak-Hungarian flatland, on the plains of Argentine, Arizona, California. The irrigation of steppe soils is often accompanied by the appearance of soda in the water as a result of its evaporation and exchange reactions. All this makes the task of protecting the soils of irrigation systems one of paramount importance.

It is already clear that principal measures for prevention are: reasonable limitation of water supply, high efficiency of irrigation systems, elimination of rising groundwater, lowering level and removal by drainage of groundwaters containing soda. If the main source of sodium carbonate in irrigation water is a river or underground, then a system of measures for its neutralization in the canals during transport to the fields has to be worked out. This may include additions of gypsum, nitrous, sulphuric or phosphorous acids. It is necessary to study the seasonal chemical composition of rivers from the point of view of possible "hidden soda content." This concerns primarily rivers with mountainous runoff (Indus, Nile, Kuban, Danube) that transfer the products of the hydrolysis of primary alumosilicate minerals to the lower reaches and drain deep-lying alkaline artesian waters. With alkaline irrigation waters the danger of soda appearance in soils and groundwaters after prolonged drainage and desalinization of the chloride-sulphate gypsum-bearing solonchaks should also be taken into account. When the drainage is operating with a leaching water regime the soil gypsum reserves might disappear, partly leached out and partly transformed into calcium carbonate. Then comes a period of alkali carbonates and bicarbonates in the groundwaters, and soils undergo alkalization. This process seems to have started on the long-drained areas of Mugan under the influence of the Araks alkaline water: in the process of groundwater desalinization to the level of 0.7–1.5 g/l the water acquired a sharply alkaline reaction, and to control this chemical treatment was required.

PROBLEM OF IRRIGATION WATER QUALITY

In the early history of irrigation in the USSR, the problem of water quality never arose. The Chirchik, Angren, Syr Darya, Zeravshan, Amu Darya, Kura and Volga rivers carried water with only a slight concentration of salts, 0.2–0.3 g/l, and with a favourable composition which included calcium carbonate and calcium sulphate salts. Water of this kind in the presence of even a poor natural drainage did not itself cause hazards but, on the contrary, favourably influenced the physical and chemical properties of soils, thanks to the large quantity of suspended silt and prevalence of calcium among cations—until the groundwaters started to rise. In the USA, Egypt, Algeria, Tunisia, Morocco, Arabia, Pakistan, India, the waters have more mineralization (0.5–1.5–5 g/l) and very often a predominance of sodium among cations, so the European, American, Indo-Pakistani and Egyptian schools of soil amelioration paid great attention to the chemical composition of irrigation waters.

At the present time, Soviet practice is to analyse in detail the quality and chemical composition of irrigation water, for mineralization in many rivers has increased during the last decade up to 0.6–1.0 g/l. Moreover, the ion ratio has changed towards predominance of sodium over calcium, concentration of sulphate-ion and chlor-ion increased and hydrocarbonate-ion appeared. This phenomenon is triggered by the complete regulation of river flow and increasing role of evaporation, by the return water that seeps through soils and subsoils of the irrigation systems as well as flowing in drains, and by discharge of urban, mining and industrial waters into rivers. The same process has affected the waters of other rivers such as the Danube and Nile.

Future increase of drainage in irrigation systems will certainly be accompanied by considerable growth of mineralization and sodium salts content in the waters of middle and lower reaches of rivers. In future one will more frequently have to use mineralized water for irrigation—artesian waters, diluted water of sea bays, river deltas and estuaries, and also drainage waters. The idea of using waste and drainage waters from the lower reaches of the Amu Darya for irrigating sandy pastures in Central Asia is now being realized.

Research by Soviet scientists and experience in Tunisia, Egypt and the Sahara oases testify that the physiological limit of salt concentration in the soil solution of the root zone lies near 10–12 g/l if the salt composition is chloride-sulphate, so waters with a 1.5–3 or even 5–7 g/l concentration can readily be used. However, with the rise of salt concentration in the water, the whole irrigation system must be modified to increase drastically the outflow in the water–salt budget: increase is required in frequency of leachings and drainage rates. However, the scientific problems involved have not yet been fully worked out; they are examined in the author's booklet "Soils of Arid Zone as the Irrigation Target" (1968). As a preliminary guide the following figures can be used, based on Soviet and American studies.

In the case of a non-saline soil and normal irrigation water (mineralization of 0.2–0.3 g/l with no-soda composition), natural and artificial drainage should be kept at about 10% of the water intake. For each 1 g of salts in the irrigation water it is necessary to add to the drainage runoff about 5–10% more of the water intake, as a minimum. Therefore, with the increased mineralization of irrigation water the need in drainage and leachings grows approximately in the way shown in Table 1.

TABLE 1. *Conditions of utilization of mineralized waters for irrigation*

Salt concentration (without soda) in irrigation water, g/l	Frequency of soil leachings	Removed drainage water (% of total intake)
0.5–1	Once in 1–2 years	10–15
1.0–2	1–2 times annually	20–25
2–3	Several times annually	30–35
4–5	Each watering should be a leaching	50–60

If in the ion content of dissolved salts sodium reaches 60–75% of the cations the accumulation of exchangeable sodium in the soil starts gradually to be followed by many adverse after-effects. But waters with the mineralization of 2–5 g/l have dissolved gypsum in their chemical composition, so alkalization of the soil as the result of the irrigation with mineralized waters does not take place. Mineralized waters may be used for irrigation of sandy and gravel soils with high water permeability, low water-retaining capacity and little absorptive power.

D. R. Bhumbla (p. 279) has given the most up-to-date review of the problem of water quality especially for conditions of the Indian subcontinent. His conclusion should be stressed that greater research effort is required in order to provide local criteria to suit particular water, soil, crop and climatic conditions.

USE OF SEA WATER

We must be very cautious about appeals to use sea water for irrigation. Nature herself has shown man the unfitness of sea water for the growth of normal plants on the sea coasts and in saline marshes. Diluted sea water may be used in sandy and gravel soils, if it is reduced to 5–10 g/l mineralization, especially for irrigating grasses and pastures. But even in this case we must hope for the leaching of salts from the soil during a rainy winter season or with the help of monsoon rains in summer. Waters of estuaries (5–6–7 g/l) can be used for irrigation in the deltas of large rivers, but here also it is necessary to estimate accurately the ion content and total concentration of salts during the irrigation season, mechanical composition and water properties of soils, evaporation regime and atmospheric precipitation.

The utilization of even diluted sea water with salt concentrations of 5–12 g/l requires, however, the leaching type of soil watering and frequent sufficiently deep horizontal drainage. This is necessary for regular and effective removal of salts left in the soil after each watering. The danger of sea water utilization is well illustrated by the experience of North Iran and South Iraq. There twice every 24 hours the diluted sea water flows naturally to the fields of deltas of the Karun and Shat-El-Arab rivers under the impact of tides, and large plantations of date palms grow there. Construction of dams and irrigation systems on the Karun reduced the river discharge and water dilution in the estuary; intrusion of sea water occurred and mineralization increased. Recently mass deaths of date palms started. Of course, if the matter concerned halophytes rather than normal plants, utilization of sea water is not only possible but necessary.

Water with high mineralization may be used for initial leaching of the most saline soils. Concentration of soil solutions in these soils reaches such values as 50–100 g/l. Therefore, waters with a concentration of 5–15 g/l can be used for initial leaching, but only if combined with adequate drainage. Such leachings should be followed by normal freshwater to remove residual salts completely. Positive experience exists of leaching clayish, soda-containing soils possessing an extremely low water permeability with highly mineralized water. Initially neutralization of soda (by gypsum or calcium chloride) increased the permeability due to the coagulation of colloids in the soil. But even in such a case the final leachings should be done with fresh irrigation water.

MAJOR INDICATIONS OF SOIL SALINITY, AMELIORATION AND DEVELOPMENT OF SALINE SOILS

Soils containing a mixture of easily soluble salts of chlorides and sulphates of about 0.4–0.5% in the arable (top) horizon are considered weakly saline. Germination, growth and sprouting of the majority of plants in these soils are suppressed, and the crops are reduced by 20–30%. It is desirable to use watering and drainage for reducing and then maintaining the quantity of such salts in the root zone below 0.2–0.3%. With the chloride-sulphate salt content being 1–1.5%, the majority of agricultural plants do not give normal seedlings, but perish or lose crops by 60–80%. With this or a larger content of salts, the soils can be normally irrigated and used *only after* the removal of salts by means of a preliminary capital leaching with considerable amounts of water (2000–20,000 m^3/ha), combined with deep drainage.

Seasonal salinization of soils

The content of toxic salts in soils always undergoes seasonal changes or pulsations. During a humid season salts are leached; during a dry season salts travel upwards with the capillary current and accumulate in the upper horizons, especially in the top 0–5 cm layer.

The seasonal salt regime is characterized by the coefficient of seasonal salinization (Kovda 1946), i.e. by the ratio of the sum of salts (or the content of toxic ion, e.g. chlorion) in the soil during the dry period and the sum of salts (ion) at the end of the humid period. If this ratio exceeds 1 salinization of the soil increases and has to be prevented with the help of watering combined with adequate drainage. Each watering, as well as rainfall, may cause the downward transfer of salts. After watering, during the drying of the soil, salts return in the upward movement. Experience of controlling soil salinization in the USSR has shown that maintaining the downward movement of salts in saline soils in the presence of horizontal or vertical drainage can be achieved if the groundwater level is kept at a depth of 2–3 m and the humidity of the soil after watering is kept above 70–75% of the field moisture capacity in the root zone (0–60 cm). In the case of medium loamy soils, application of water for this purpose at the relatively low rate of 500–700 m³/ha (about 30% of the field moisture capacity) completely guarantees the preservation of a sufficiently high humidity in the root zone. In the presence of deep operating drainage such moistening secures a reduction of seasonal salt accumulation and promotes increasing desalinization of soils and groundwater.

Critical depth of mineralized ground-water, installation and operation of drainage

The main source of salts in irrigated soils (if mineralized waters are not used for irrigation) is groundwater lying close to the soil surface and salts of the underlying soil. Soil-forming rocks and sub-humic horizons of the soils in arid and sub-arid regions practically always contain soluble salts (chlorides, sulphates, carbonates) which can travel comparatively easily to the top horizons in the form of capillary water solutions after the beginning of irrigation. This process occurs particularly often and with adverse effects after the rise of the groundwater level caused by irrigation.

Experience in the arid regions of Asia, Africa, Australia and America shows that, when the groundwater rises above the critical depth (1.5–2.5 m), the soils are usually strongly salinized. Only those groundwaters that contain less than 5–3 g/l of soluble sulphate-chloride do not cause natural and secondary salinization. These "freshed" groundwaters even without irrigation (and, certainly, with irrigation) favour the formation of highly fertile meadow humic hydromorphic soils. Therefore, a 3–5 g/l mineralization of the groundwaters should be viewed as a ceiling—critical for the majority of irrigated regions of chloride-sulphate salinization. Large concentrations of salts in groundwaters (10–30–70–150 g/l) play a determining part in the soil salinization of arid regions. With very intensive evaporation and transpiration capillary solutions of salty groundwaters continuously replace evaporating soil moisture, thereby introducing more and more toxic salts into the root zone and leading to plant death. Empirical generalization of the USSR experience allowed us in the first approximation to use the following formula (suggested in 1956 by the author) for estimating the probable critical depth in salty groundwater occurrence:

$$L = 170 + 8t \pm 15,$$

where L is the critical depth in cm, and
    t is the average annual temperature, °C.

The greater the annual temperature, the greater is the rate of evaporation and the mineralization of the groundwaters; the greater accordingly should be the depth at which salted groundwaters must be maintained to prevent salinization. From 4–5 m depth the groundwaters cannot provoke and maintain soil salinization under irrigated conditions. Horizontal drainage does not allow us to lower the groundwaters to such a depth, but this can be managed with the help of pumping through vertical drainage wells. The horizontal drainage, with drain depth of 3–3.5 m, can successfully freshen the groundwaters themselves during 4–5 years of operation and so create an opportunity for reducing their depth to 2.5–2.75 m. Thus, for stable development of irrigated saline soils it is necessary to employ the following:

(i) If in the newly irrigated area the groundwaters lie at a depth of 5–12 m, then in order to prevent their rise to the critical level a deep horizontal or vertical pumping drainage should be constructed and put into operation simultaneously with the irrigation network.

(ii) If mineralized groundwaters already occur close to the surface (1–2 m), it is necessary to lower their horizon with the help of a deep drainage system. Depending on the climatic, geochemical, soil and agrophysiological conditions, the horizon of salty groundwaters has to be kept at 1.75–3 m below soil surface, maintaining a level 25–50 cm lower than the critical level.

(iii) With the help of capital leachings and drainage, salinization of the rootzone of soil should be reduced from 1–2% of salt content to 0.2–0.3% maximum.

(iv) Downward currents of salt solutions in the soils should be maintained and seasonal soil salinization should be prevented or reduced as much as possible by watering.

(v) Groundwaters are to be gradually desalinized to a concentration approaching the critical one through watering and deep drainage.

Only complete desalinization of soils to a salt content of 0.2–0.3% and freshening of groundwaters to the concentration of 2–3 g/l (or lowering their level down to 4–5 m) can guarantee no reinvasion of salinization of the irrigated soils and secure maximum crops. In this situation horizontal and vertical drainage fulfil the task of soil desalinization, amelioration and prevention.

On completing the desalinization process of soils and groundwaters the effective operation of the drainage must be kept for a long time in order to remove salts newly arriving from the irrigation waters, evaporation and weathering. Unfortunately, drains and wells often fail due to silting, creeping plant growth and other processes. This immediately leads to the restoration of salinization and waterlogging of the soils. Vertical pump drainage from tube wells reduces the water pressure and level of the groundwaters and thus provides optimum condition for the downward movement and removal of salts. In cases where pumped out groundwaters are only weakly saline (1–3 g/l), they may be successfully used for irrigating adjacent areas, as shown by experience in California, Arizona, Pakistan and India. In cases where pumped-out groundwater is mineralized up to 5–7 g/l, its use for irrigation should be cautious. Mineralized drainage waters with a salt concentration of 10–15 g/l should be removed from the fields.

Leaching (removal) of salts from soils

Toxic soluble salts can be removed from soils only in the form of aqueous solutions, that is by rain, irrigation or groundwaters. In arid regions we cannot count on the radical leaching of salts from soils with the help of rain water, though in the semi-arid regions of the Mediterranean, Asia and in the humid regions of seacoasts (polders of the Netherlands) considerable salt removal is achieved during 3–5 years of drainage operation, especially if the soil is sandy. Under conditions of desert and semi-desert climate desalinization of the soils can be accomplished only with the help of special meliorative leachings coupled with adequate drainage.

(i) *Leaching waterings during the growing season.* It was mentioned above that more frequent waterings at moderate norms prevent the seasonal salinization of soils. If weakly mineralized waters (1–3 g/l) are used for irrigation, leachings have to be conducted once or twice annually or at least at the end of the growing season. With irrigation waters of concentration of 4–6 g/l it is necessary to perform a leaching every other watering, or even every watering. The watering norm capable of producing the leaching effect should exceed the volume corresponding to the field moisture capacity by 20–30%. This "surplus" of water dilutes salts left over from previous waterings and transports them to the groundwater and then drainage waters which carry them away from the system. Soil salinity does not increase in this case, for it corresponds ultimately to the composition of the product of irrigation water and evaporation. Successful experience of these types of watering is demonstrated in the USSR (Uzbekistan, Turkmenia), in the Sahara, and in the Tunisian experimental station of UNESCO/FAO.

(ii) *Meliorative leachings.* The question of leaching outside the growing season seems much more complicated. Such leachings are conducted for desalinization of highly saline soils before the beginning of their normal usage for irrigation. Without preliminary salt removal these soils do not yield good crops and their salinity grows when irrigation is applied. In the USSR leachings of this type are called "capital leachings." While preparing a development project on the irrigation of highly saline soils a special investigation of salt reserves in the soils and groundwaters has to be carried out. Then the efficiency of leachings, optimum water norms (volumes), speed of salt removal, time necessary for desalinization, and agrotechnical methods enhancing the leaching effect, such as deep loosening of soil or levelling, are studied on the soil monolith and on special plots. Separate studies are made on the risk of alkalization and accumulation of exchangeable sodium in the leached soil.

While preparing for the development of large areas of saline soils (50,000–200,000 ha) it is expedient to organize experimental pilot farms or stations for conducting field experiments on the depth and distance between drains and different variants of technique for leachings before proceeding with large-scale construction. Unfortunately, in the majority of cases the new construction of irrigation systems is carried out in haste, without preliminary experiments and studies.

In the Soviet Republics of Central Asia and Transcaucasia considerable scientific and practical experience of the meliorative leaching procedure has been gathered. This experience is reported in detail in the publications of UNESCO/FAO (Kovda 1967 and 1973).

The higher the soil's salinity, the heavier its mechanical composition, the higher the groundwater level and its salt concentration, the larger are the necessary leaching norms for salt removal. Generalizing Soviet experience the author (1957, 1967) suggested the following formula for calculating the volume of leaching water:

$$Y = n_1 . n_2 . n_3 \, 400x \pm 100,$$

where Y is the leaching water layer, mm or m^3/ha;

x is the average percentage of soluble salts in a 2 m soil thickness;

n_1 is the coefficient depending on soil mechanical composition: in sand it equals 0.5; in loams 1.0; and in clay 2.0;

n_2 is the coefficient depending on the depth of groundwater occurrence:

    at the depth of 1.5–2 m it equals 3,
    at the depth of 2–5 m it equals 1.5,
    at the depth of 7–10 m it equals 1.0;

n_3 is the coefficient depending on the groundwater mineralization,

    weak   1.0,
    medium   2.0,
    strong   3.0.

    V. R. Volobuev has given his own formula for calculating the leachings:

$$N = K_{\log}\left(\frac{Si}{So}\right)^a,$$

where N is the leaching water norm, m³/ha;

    Si is the soil salinization, % or t/ha;
    So is the permissible residual salinization, % or t/ha;
    K is the coefficient for recalculating per 1 hectare (10.000 m²);
    a is a parameter showing the salts ratio (especially of chlorides) changing from 0.9
       to 1.5.

    The norms calculated from these formulae are very close to the values obtained through direct experiments in the field.

    Fluctuations of the water volumes necessary for capital leaching are limited by the values of from 800–1000 m³/ha to 30,000–50,000 m³/ha. The leaching water must be given to the prepared field by successive applications of 2000–2500 m³/ha, each after a complete infiltration of the previous application. If the norm exceeds 14,000–15,000 m³/ha it is advisable to combine the leaching with rice culture. For this purpose the leaching has to be shifted from the non-vegetative period (autumn, winter, spring) to the growing season. Sometimes leachings have to be repeated in a second and third year. The necessity to do so is proved with the help of the analysis of soil samples taken from the field after leaching or at the end of the growing season. The leached field can be occupied by agricultural crops of high salt-resistance, and the yield of these crops (especially of rice) compensates for the expenses of amelioration. During the first years after leaching growing season waterings should be carried out in order to ensure downward direction of the currents, and this regime is necessary for several years until the groundwaters become freshened to 2–3 g/l. Success of the leachings depends on the thoroughness of drainage construction and on its efficiency and good maintenance. The leachings may often produce no or extremely little effect owing to either insufficient leaching water or the inability of the drainage to lower the level of and remove groundwater as a result of shallow depth, too wide spacing, overgrowth or silting. Without drainage the leachings of saline soils can be successful only on very small areas and if the land has natural drainage.

    The above concerns soils of a chloride-sulphate type of salinization, almost always containing gypsum, where there is no growth of alkalinity or deterioration of the physico-chemical properties after the leachings. In soils where gypsum is absent or especially in soils having free alkali carbonates, a high pH and high percentage of exchangeable sodium, chemical amendments have to precede leaching. Gypsum, sulphuric acid, iron sulphate, sulphur are introduced in large quantities. Free soda is neutralized, exchangeable sodium is replaced by calcium, alkalinity is sharply reduced and the structure and water permeability of the soil are improved. After the chemical, capital leachings combined with drainage are carried out and this completes the amelioration.

The most modern analysis and generalization of the USSR experience in the amelioration of strongly saline soils was carried out by M. M. El Mansy E. Shal (1971/1972, Moscow–Cairo). This once more confirmed that effective amelioration and productive development of strongly saline soils can be achieved by the methods described above. All the requirements for amelioration must be based on thorough studies and experiments. Appropriate formulae for the calculation of drainage, spacing, volume of water and times of performance could be applied and borrowed from the UNESCO/FAO *International Handbook on Irrigation and Drainage* (1973), and from papers by El Manci (1971/1972), Averianov and others. But the most important practical step is the establishment well in advance in areas of potential irrigation of some type of network of experimental stations or pilot farms with comprehensive programmes of field experiments and studies of the parameters and techniques of drainage, leaching, land improvement, fertilization and plant cultivation. This will give most valuable data for the elaboration of the final technology of saline soil amelioration.

Control of soil water–salt balance of irrigation systems

It is now widely recognized that a study of the water–salt balance of each irrigation system is required and that it is necessary to compute the water–salt balances separately for the initial period, the transitional ameliorative period and for the period of regular exploitation after the completion of desalinization. The aims of these studies are:

(i) To show the sources, distribution, chemical composition and total amounts of various salts as well as the dynamics of salinization in soil, groundwater and irrigation water in the area to be irrigated.
(ii) To estimate possible changes in the water–salt balance at various stages of irrigation and drainage operations. It is desirable to find out whether or not an increase in groundwater is to be expected and to determine the schedule of soil leachings to remove salts, the kinds of temporary and permanent drainage installations to be constructed, and an optimum ameliorative work schedule for the transition period.
(iii) To determine the point at which a steady water–salt balance is established, the optimum salt concentrations which should be maintained in soils and groundwaters, the optimum and critical depth of groundwater levels, the most effective irrigation and leaching regimes if required, the total amounts and proportions of removed drainage water, and the permanent elements of drainage structures. It is also necessary to study changes in the water–salt balance in adjacent areas and to determine the extent of the area likely to be affected outside the new irrigation system.

The scientific elaboration of such water–salt balances by research and design is a complicated but important and gratifying work. Up to now, few irrigation projects have taken water–salt balances into account, much less the means of their control. This is why, after several years of irrigation, unexpected and unforeseen phenomena may develop. The entire procedure of designing irrigation systems must be changed to avoid such developments.

and this requires the compilation of detailed soil and lithological maps, an analysis of the physical and chemical properties of soils and groundwaters, and the distribution of salts in various soil layers.

Predicting the water–salt balances of an irrigation system at the design stage may prove wrong. Therefore current working information should be drawn up for reference purposes. Thus water–salt balances for a whole are or part of it will enable the managers to plan and implement additional measures required for amelioration—expansion of the drainage network, additional land levelling, soil leaching, installation of pump wells for vertical drainage. At present, the necessary data are rarely made available to irrigation managers.

Drawing up current water–salt balances for annual modification of amelioration programmes requires knowledge in all the following matters: water intake and distribution data, including the chemical composition of irrigation and drainage water; the actual amount of water applied for soil leaching and removed by a network of drains; values of evaporation and transpiration; the pressure, level and chemical composition of groundwater for 3 or 4 time periods; the contours and size of saline spots and the areas of affected crops (preferably on the basis of aerial photography with ground corrections); the salt content in principal soils in spring and autumn; the discharges of drainage water, particularly the amount and composition of salts removed with this water; and reliable information on crop yields.

Each irrigation system should have a network of hydrometric and halometric stations, and a number of observation holes and plots for studying salt dynamics; it should be provided with facilities for aerial photography and should make hydrogeological, pedological, meliorative and agricultural observations, measurements and calculations in order to compare and generalize these data. All this requires skilled personnel, adequate research guidance, melioration laboratories and computing centres. In general, the proper scientific and technical management of irrigation systems and the utilization of reclaimed lands should be based on regular information including the following: every 5–10 years drawing up maps of soils and groundwaters for the entire irrigated territory on 1:10,000 or 1:25,000 scale, in comparison with actual (not average) crop yields; annual data on the progress of land levelling, soil leaching, drainage and collector operations, crop yields and losses. All these functions should be the responsibility of a special reclamation service with amelioration inspection and forecast centres.

Solutions to these problems require the application of up-to-date technological aids and techniques, such as fast and accurate soil analysis methods, machine interpretation of aerial photos, television cameras, and data recording on punched cards or magnetic tape with subsequent processing by electronic computers. These techniques would make it possible to obtain reliable information on the extent of salinization, soil moisture content, groundwater, plant growth and progress in agricultural work within 2 or 3 weeks. Without them drawing up water–salt balances, meliorative forecast and management will be deficient, as they are now, and the fight against soil salinization will become even more complicated.

In irrigation systems with operational drainage, an ample supply of water, and well-levelled fields, the component of the salt balance which is the easiest to regulate is the removal of salts with drainage water. This becomes obvious from considering the water versus salt relationship which is as follows:

$$Y . C = (Q . C_1 + Q_{dr} . C_2),$$

where Y is all kinds of irrigation water inflow;

C is the concentration of salts in irrigation water;

Q is the natural outflow of groundwaters;

C_1 is the concentration of salts in naturally drained groundwater;

C_2 is the average actual salinity of groundwater in a given field or territory;

Q_{dr} is the drainage water outflow.

If drainage is not employed, the functioning of an irrigation system depends upon both Y and C and particularly upon Q and C_1. Secondary salinization of land will not take place if $Y.C < Q.C_1$ due to adequate natural drainage. If natural drainage is poor $(Y.C > Q.C_1)$, the decompensation of groundwater, its rise, salinization and waterlogging of soils under irrigation are the result.

The introduction of artificial drainage (Q_{dr}) and the removal of groundwater salts (C_2) by leachings and a drainage network (Q_{dr}) turns a positive (accumulative) salt balance into a negative (desalinizing) balance:

$$Y.C(Q.C_1 + Q_{dr}.C_2).$$

In the course of successful amelioration, values C_1 and C_2 must decrease considerably and approach the critical groundwater salinity (C_k) which is typical of the territory under consideration. It is evident that the salt balance for an irrigated area is determined by the amount of water removed through artificial drainage and by the concentration of salts in this water. These two factors are significant for groundwater desalting, i.e. for the completion of amelioration.

In this respect the water–salt balance index (WSB) suggested by the author in 1966–1967 is a convenient and simple way of estimating amelioration efficiency. After normal completion of desalinization in the geochemical provinces of chloride-sulphate type of accumulations, the critical salinity of groundwater (C_k) must be about 3 g/l. If salinity of irrigation water is 0.3 g/l, the ratio $3:0.3=10$ which is the WSB. In this case about 10% of the total volume of water supplied to the fields must be removed by drainage in order to maintain the WSB at the level of 10 after completion of the amelioration period. If the WSB is 12–15 rather than 10, it means that salts have accumulated in groundwaters, i.e. soil leachings and drainage were ineffective. In this case the amount of drainage water outflow should be increased by 2–5% relative to the total water supply. If salt concentration in groundwater grows to 4–6 g/l, the WSB index increases up to $4/0.3=13.33$ or $6/0.3=20$, which necessitates stronger leachings and an increase of groundwater removal to 13.3–20% of the amount of water supplied to the area.

The degree of permissible evaporation of irrigation water depends on its salt concentration. If salt concentration in irrigation water is higher than 0.3 g/l, 5–8% of the total amount of salt solutions must be removed by drainage outflow per gram of salts in the waters. A general relationship for the amount of water removed by drainage in the operation of an irrigation system is as follows:

$$X = (5)8.C + \frac{C_2.10}{3},$$

where X is the percentage of the total water supply to that removed by drainage.

The example given above referred to the area of chloride-sulphate salinization. If pure sulphate salinization occurs, as in the Fergana and Bukhara areas, critical salt concentration may reach about 6 g/l. In this case, with the same mineralization of irrigation water, the WSB index will be $C_1 : C = 6 : 0.3 = 20$. The permissible degree of irrigation water evaporation is then 20 times, and coefficients in our formula must be changed accordingly. In the case of soda salinization, for instance, the rates of drainage must be much higher. Let us assume that the critical salinity of soda groundwater is about 1 g/l: with an irrigation water salinity of 0.3 g/l and a critical salinity of groundwater of 1 g/l, the WSB index would be $1 : 0.3 = 3.3$, which means that the degree of irrigation water evaporation may be permitted to increase by 3.3 times only. Thus the amount of drainage outflow must be about 33% of the total amount of water supplied to the fields. An increase in soda groundwater salinity will immediately require a substantial increase in the amount of drainage water outflow (about 33% per additional gram of salts in soda groundwater). For instance, if the average actual salt concentration of soda groundwater is about 2 g/l, then, according to our formula, 71–74% of the total amount of irrigation water containing pure soda must be removed by drainage. This is why soda salinization is so harmful and why regularly repeated chemical ameliorations are required in cases of irrigation and groundwater containing soda. The basic prerequisites for the accurate estimation of amelioration efficiency are extensive, and statistically reliable data are required together with a profound knowledge of the salt geochemistry in the area. In future research, salt balances will have to be established separately for toxic ions (SO_4, Cl, HCO_3, CO_3), but this is another aspect of the problem.

We may conclude that the construction of effective drainage facilities and their proper exploitation in areas affected or endangered by secondary salinization is the most vital and urgent problem of irrigation. Special emphasis must be placed on these measures in the development of world agriculture.

TABLE 2. *Distribution of salinity and alkalinity in countries mainly affected (areas in 1000 ha based on Soil Map of the World at 1:5 million)*

Country	Solonchaks	Saline phase	Solonetz	Alkaline phase	Total
North America					
Canada		264	6,974		7,238
USA		5,927	2,590		8,517
Mexico and Central America					
Cuba		316			316
Mexico	242	1,407			1,649
South America					
Argentina	1,905	30,568	11,818	41,321	85,612
Bolivia		5,233	716		5,949
Brazil	4,141		362		4,503
Chile	1,860	3,140		3,642	8,642
Colombia	907				907
Ecuador	387				387
Paraguay		20,008	1,894		21,902
Peru	21				21
Venezuela	1,240				1,240
Africa					
Afars and Issas Territory	59	1,682			1,741
Algeria	1,132	1,889		129	3,150
Angola	126	314	86		526

TABLE 2 (*cont.*)

Country	Solonchaks	Saline phase	Solonetz	Alkaline phase	Total
Botswana	1,131	3,878		670	5,679
Chad	2,417		3,728	2,122	8,267
Cameroon				671	671
Egypt	3,283	4,077			7,360
Ethiopia	319	10,289		425	11,033
Gambia		150			150
Ghana	200			113	318
Guinea		525			525
Kenya	3,501	909		448	4,858
Liberia		362			
Libya	905	1,552			2,457
Madagascar	37			1,287	1,324
Mali		2,770			2,770
Mauritania	150	490			640
Morocco	42	1,106			1,148
Niger			11	1,378	1,489
Nigeria	455	210		5,837	6,502
Portuguese Guinea		194			194
Rhodesia				26	26
Senegal	141	624			765
Sierra Leone		307			307
Somalia	1,043	526	3,754	279	5,602
South West Africa	562		1,751		2,313
Sudan		2,138		2,736	4,874
Tanzania		2,954		583	3,537
Tunisia	990				990
Zaire		53			53
Zambia				863	863
South Asia					
Afghanistan	2,924	177			3,101
Bangladesh		2,479		538	3,017
Burma	634				634
India	2,979	20,243		574	23,796
Iran	24,817	1,582		686	27,085
Iraq	6,679	47			6,726
Israel	28				28
Jordan	74	106			180
Kuwait	209				209
Muscat and Oman	290				290
Pakistan	1,103	9,353			10,456
Qatar	225				225
Sarawak		1,538			1,538
Saudi Arabia	6,002				6,002
Sri Lanka	180	20			200
Syria		532			532
Trucial States	1,089				1,089
North and Central Asia					
China	7,307	28,914		437	36,658
Mongolia	3,728	342			4,070
Solomon Islands		238			238
USSR	11,430	39,662	30,062	89,566	170,720
South-East Asia					
Indonesia		13,213			13,213
Khmer Republic		1,291			1,291

TABLE 2 (cont.)

Country	Solonchaks	Saline phase	Solonetz	Alkaline phase	Total
Malaysia		3,040			3,040
Thailand		1,456			1,456
Viet Nam D.R.		39			39
Viet Nam R.		944			944
Australasia					
Australia	16,567	702	38,111	301,860	357,340
Fiji		90			90
Europe					
Bulgaria					
Czechoslovakia					
France					
Hungary					
		Are not registered but frequently observed			
Italy					
Spain					
Turkey					
Yugoslavia					

References

Kovda, V. A. 1971 in English, 1946 and 1947 in Russian. *Origin and Regime of Saline Soils.* Moscow.

Kovda, V. A. (editor and co-author). 1967 and 1973. *International Source Book on Irrigation and Drainage of Arid Lands.* FAO/UNESCO, Paris.

Lunin, J., Gallatin, M. H. 1960. *Use of Brackish Water for Irrigation in Humid Regions.* Agriculture Information Bulletin No. 213, Washington.

Massoud, F. I. 1974. *Salinity and Alkalinity.* A world assessment of soil degradation. FAO, Rome.

Rapp, A. 1974. *A Review of Desertization in Africa (Water, Vegetation, Man).* Stockholm.

Richards, L. A. (editor). *Diagnosis and Improvement of Saline and Alkali Soils.* Agriculture Handbook No. 60, Dept. of Agriculture, Washington.

UNDP(SF)/UNESCO. 1970. Tunisia. Research and training on irrigation with saline water. Technical report, Paris.

In addition, for the preparation of this review, the following literature in Russian has been referred to:

Alekseevski, E. E. 1975. Melioratsiya i mezhdunarodnoe sotrudnichestvo. Gidrotekhnika i melioratsiya, No. 7. (Land reclamation and international cooperation. Hydraulic engineering and land reclamation.)

Budanov, M. F. 1956. Vliyanie orosheniya mineralizovannymi vodami na pachvi. Ukr. MIIGiM vip. 77/3, Kiev. (The impact of irrigation with mineralized waters on the land.)

Chembarisov, E. I. 1974. Izmenenie mineralizatsii vod nekotorykh rek Sredney Azii v svyazi s orosheniem. M. (The change of mineralized waters of some rivers in Central Asia in connection with irrigation.)

Dukhovnyi, V. A. 1973. Oroshenie i osvoenie Golodnovi stepi. M. "Kolos." (Irrigation and development of the Golodnaya steppe.)

Elerdashvili, S. I. 1973. Gidrogeologiya i inzhenernaya geologiya Iraka. "Nedra" M. (Hydrogeology and engineering geology of Iraq.)

Goryunov, N. S. 1973. Kak borotsya s zasoleniem orochaemikh zemel. "Kaynar". Alma-Ata. (How to combat the salinization of irrigated lands.)

Grammatikati, O. G. 1972. Nekotorye aspekty vliyaniya nauchnotekhnicheskovo progressa na proizvodstvo osnovnykh zernovȳkh kultur v stranakh Vostochnoi Afriki. CBNTI Mindvodkhkhoza SSSR, Ekspress-informatsiya, vip. 12 ser. I, M. (Some aspects of the influence of scientific-technical progress on the production of the principal grain culture in countries of East Africa.)

Ibragimov, G. A. 1973. Ispolzovanie mineralizovannykh vod na oroshenie khlopchatnika. Tashkent, "Fan" Uzb. SSR. (The utilization of mineralized waters in the irrigation of cotton.)

Ignatenok, F. V. 1974. Melioratsiya zemel. Gorki. (Reclamation of land.)

Karvalo, Kardozo. 1971. Melioratsiya zasolennykh pochv v Portugalii. V sb: Pochvy̆ sodovogo zasoleniya i ikh melioratsiya. Tr.In-ta pochvovedeniya i agrokhimii. vyp. 6. Erevan. (Land reclamation of salinized soils in Portugal. Soda salinized soils and their amelioration.)

Kovda, V. A., Rozanov, B. G. 1975. Izmeneniya pochvennogo pokrova pod vliyaniem melioratsyj. Gidrotekhnika i melioratsya. No. 7, str. 45–51. (The change of top soil under the influence of land reclamation.)

Kovda, V. A., Smoĭlova, E. M., Skudzhins, I., Tsarley, L. I. 1974. Pochvennye protsessi v aridnikh oblastyakh. 10 Mezhdunarodhyj kongress pochvovedov. Moskva. (Soil processes in arid regions. 10th International Congress of Soil Science.)

Nerpin, S. V., Michurin, B. N., Sanoyan, M. G. 1972. Zavisimost vodopotrebleniya rastenij ot fizicheskikh faktorov sredy. V sb: Issledovanie protsessov obmena energiej i veshchestvom v sisteme pochva-rastenie-vozdukh. "Nauka." (The relationship between water utilization of plants and physical factors of the media. The investigation of processes in soil–plant–air energy and matter exchanges.)

Nesterova, G. S. 1972. Vozmozhnost ispolzovaniya solenych vod dlya orosheniya selskokho-zaystvennych kultur. Obz. informatsiya, MSH SSSR, M. (The possibility of utilizing saline water for agricultural irrigation.)

Orlova, N. A. 1956. Iz opy̆ta ekspluatatsii kamenskoj orositelnoj sistemi. Ukr. NIIG iM Nauchnye trudy, vyp. 77/3, Kiev. (Experiments in exploitation in the Kamenskaya irrigation scheme.)

Sabolch, I. 1973. Sostavlenie pochvennych kart oroshaemikh territorij. (Materiali mezhdunarodnovo soveshchamiya) Budapesht. (Composition of soil maps of irrigated territories.)

Samokhvalenka, S. K. 1956. Obosnovanie norm i srokov polivov ozimoj pshenitsy dlya yuzhnoj chasti stepi Ukrainy. Ukr. NIIGiM Nauchnye trudy. vyp. 77/3, Kiev. (Establishing norms and periods of winter crop wheat in the southern part of the Ukrainian steppe.)

Sokolov, A. A., Chikolmanov, I. A. 1974. Prognoz izmeneniya stoka rek SSSR pod vliyaniem khozyajstvennoj deyatelnosti. V sb: Vliyanie mexhbassejnogo perepaspredeleniya rechnogo stoka na prirodnye usloviya Evr. territorii i sredinnogo regiona SSSR. M. (Prognosis of the influence of agricultural activity on the flow of USSR rivers in: "The influence of inter-basin redistribution of river flow on the natural resources of European territory and central regions of the USSR."

Syromyatnikova, V. A., Artemova, L. G. 1973. Melioratsiya i vodnoe khozyajstvo za rubezhom. V kn: Oroshenie v SShA. Chast I. Razmeshchenie i ispolzovanie oroshaemykh zemel. TsBNTI. Minvodkhoza SSSR. Obzornaya informatsiya No. 4. (Amelioration and irrigation economy in "Irrigation in the USA.")

Tulyakova, Z. F. 1975. Znachenie i osobennosti risoseyaniya na zasolennykh zemlyakh severnogo Kavkaza. Novocherkassk. (The importance and characteristics of rice planting on the salinized soils of northern Caucasus.)

Vaksman, E. G. 1974. Melioratsiya zasolennykh zemel v usloviyakh napornogo pitaniya gruntovykh vod (na primere yuzh Tadzhikistana i Alzhirskoj Sakhary), Dushanbe. (Amelioration of salinized soils under conditions of special groundwater pressures (as in southern Tajek and Algerian Sahara).)

The following unpublished MS have also been read and utilized in addition to the papers published in this book:

Ayers, R. S. 1975. Interpretation of quality of water for irrigation.

Balba, A. M. 1975. Predicting soil salinization, alkalization and waterlogging.

El-Mansi, El-Shal. 1971. Water–salt regime and balance of saline soils of Golodnaia steppe. Moscow.

Glantz, M. H. 1975. Nine fallacies of a natural disaster: The Case of the Sahel.

Hardan, A. 1970. Dating of soil salinity in the Mesopotamian plain.

Kassas, M. 1974. Arid and semi-arid lands.

Kovda, V. A. 1964. Problems of salinity and waterlogging of irrigated land in West Pakistan. UNESCO.

Kovda, V. A. 1964. Problems of salinity and waterlogging of irrigated land in West Pakistan (mimeograph). UNESCO, Karachi.

Massoud, F. J. 1975. Factors to be considered for prognosis: soil management and agronomic practices.

Molen van der, W. H. 1975. Factors to be considered for prognosis: natural factors.

Peters, W. B. 1975. Economic land classification for the prevention and reclamation of salt affected lands.

Rafiq, M. 1975. Use of satellite imagery for salinity appraisal in the Indus plain.

Rafiq, M. 1975. Saline, saline-alkali and waterlogged soil of the Indus plain.

Van Schiefgaarde, Jan. 1975. Water management and salinity.

Zonneveld, I. S. 1975. Survey methods for performance monitoring and prognosis ... salt-affected soils.

ON PROBABILITY OF DROUGHTS AND SECONDARY SALINIZATION OF WORLD SOILS

by V. A. KOVDA, B. G. ROZANOV and S. K. ONISHENKO

(USSR)

Note. This paper, received late, is published in extended summary. Refer to authors for the full paper which includes an extensive table on a system of complex amelioration for main soil groups of arid regions under irrigation, and a scheme of aridity, drought probability and secondary salinization of irrigated soils.

The number of countries situated in arid regions, suffering from periodic droughts or experiencing sporadic droughts, is high and continues to grow. Including deserts, semi-deserts, steppes, pampas, prairies and savannas permanently characterized by arid conditions and periodic droughts, it is probable that 70–80 countries are concerned, containing not less than 35–40% of the total world population.

The growth of population, development of human settlements and communications, growth of world trade, of industrial production, mining operations and recreation causes progressive alienation of land, degradation of soils, and movement of populations. The reclamation of sub-arid areas is more intensive during periods of moist climatic regime, and has been aided by introduction of the horse plough and then the tractor plough, mineral fertilizers, and many kinds of agricultural equipment. All this has increased the dependence of world agriculture on climatic fluctuation.

Throughout the 40–50 centuries of history and experience in world agriculture drought resistance was taken into account during plant selection and introduction, with the result that plants of moderate drought resistance predominate in world agriculture today. According to an estimation by N. I. Vavilov, nearly three-fifths of the total sown area in the world in the 1930s were cultivated with moderately drought-resistant crops: wheat, barley, corn, rye, flax, haricot bean, sunflower, sugar-beet. The most drought-resistant crops (millets, lentil, groundnut, olive, grape, almond, pistachio, fig) occupied relatively small areas, while a number of other important crops (rice, oats, soya-bean, beans, peas, mustard, hemp, tea, most vegetables and fruits) have minimal drought-resistance. This crop ratio, vulnerable from the point of view of drought influence, has changed little until the very recent introduction of new high producing varieties of wheat, rice and corn which are distinguished by their higher water and fertilizer requirements.

In order to use more effectively the dry areas, the selection of high drought-resistant plants continues to be extremely important and, in the opinion of Vavilov, the centuries

237

old agricultural civilizations of countries of Central and South Asia, Middle and Near East, the Caucasus, Mediterranean, transitional to the desert regions of Africa and Southern and Central America, retain numerous valuable stable local varieties. It is necessary that work be continued on a large scale on the collection and preservation of such plants, on their improvement and application to local and regional climates, soils, needs and traditions.

Drought resistance of plants is rarely combined with high productivity but, in order to be profitable, crops must now yield two–three–four times more than the yields of the thirties. Hence selection is now not only for drought resistance but also for high response to fertilizers and irrigation. Nevertheless, even at the highest possible rates of irrigation expansion, dry farming and dry grazing for cattle will serve mankind for a long time as basic sources of food and raw materials, and the problem of creating high productive drought-resistant varieties of plants is still far from being solved. For effective dry farming in arid and semi-arid areas the probability of droughts must be predicted and methods for conservation of moisture in the soil need to be utilized systematically, as they are in many countries.

Science so far is unable exactly to forecast the time of drought, its duration or intensity. But using existing historical, climatic, soil, botanical and hydrological data, we can compose maps of geographical distribution of droughts in the past, their frequency and severity, and this allows prediction of the areas and degree of severity of probable droughts in future. The scale of maps for such purposes should be: for districts 1 : 500,000, for large countries 1 : 1,000,000–1 : 500,000, for the continents 1 : 5,000,000, and for the whole world 1 : 10,000,000. With the aid of such maps the zones and regions where drought probability is especially high, where their frequency is less and where they occur rarely or not at all, can be computed.

Based on the above method, and other characteristics, a draft schematic map of land aridity, soil salinization and drought probability has been compiled for the continents and was demonstrated in 1975 at several international conferences. It gives fairly impressive information on the significance of desertification processes on the continents and on the need to mobilize science, technology and economies to combat this global phenomenon which is tending to destroy the biosphere.

On the basis of present data the land can be divided into nine categories in order of increasing probability of droughts and secondary salinization under irrigation:

1. Permanently moist conditions: droughts improbable;
2. Very rare drought with probability up to 5%;
3. Rare drought with probability 5–10%;
4. Relatively frequent drought with probability 10–25%;
5. Frequent drought with probability 25–30%;
6. Rather frequent drought with probability 30–50%;
7. Very frequent drought with probability 50–75%;
8. Permanent drought with probability 75–95%;
9. Absolute drought predominance with probability near 100%.

In addition to these nine groups on the maps are shown other areas where the main limiting factors for agriculture are special hazards in addition to droughts and soil salinization.

PROBLEMS AND EFFECTS OF IRRIGATION IN THE NEAR EAST REGION

by M. M. EL GABALY

FAO Regional Office for the Near East, Cairo

Summary

Irrigated areas will have to be extended to about 51% of the arable area of the Near East if they are to meet the requirements anticipated.

Salinity and waterlogging are common problems in this region. They are closely related to inefficient water use for irrigation, lack of adequate drainage and poor water quality. The problem is expected to increase as a result of expansion in irrigation which will involve the use of poor quality water and potentially saline soils. The percentage of salt-affected and waterlogged soils already amounts to 50% in the Euphrates Valley in Syria and 30% in Egypt.

Salt movement in soils is essentially the result of water movement. Design criteria should be tested under actual field conditions in pilot projects. Leaching requirements should take into account the extent of mixing between applied water and soil solution. The gross volume of water needed does not only depend on evapotranspiration and leaching requirements, but also on the application efficiency. Permissible depth of the ground water table varies with climate, soil texture, permeability, salinity, type of crop, and irrigation technique and frequency. Special attention should be paid to the management of soil and water to maintain the proper salt balance. This includes irrigation, drainage, evaporation control, erosion control and suitable tillage practices.

A field drainage network is a pre-requisite to efficient water management. Summer fallowing greatly increases the danger of resalinization. There is interaction between water management, cropping pattern, leaching requirements and groundwater depth and so a need for proper soil and water management to avoid resalinization.

Résumé

Pour atteindre les objectifs fixés, il faudra que les zones irriguées occupent une superficie de 51% environ des terres arables.

Dans les payes du Moyen-Orient, la salure et l'engorgement sont des problèmes qui se posent fréquemment. Ils sont étroitement fonction d'une mauvaise utilisation de l'eau d'irrigation, de l'absence d'un drainage convenable et de la mauvaise qualité des eaux. Il faut s'attendre à ce que le problème s'aggrave du fait de l'extension croissante de l'irrigation. Il doit s'ensuivre un recours à des eaux de mauvaise qualité et à des sols potentiellement salés. Le pourcentage des sols affectés par la salure et par l'engorgement atteint 50% des zones irrigables en Irak, 23% au Pakistan, 50% dans la vallée de l'Euphrate en Syrie et 30% en Egypte.

Le déplacement du sel dans les sols est essentiellement la conséquence du déplacement de l'eau. Il conviendrait que les normes des projets soient testées en projets pilotes dans des conditions du terrain. Les besoins nécessaires au lessivage devraient tenir compte de l'importance du mélange entre l'eau apportée et de la solution du sol.

Le volume global de l'eau d'irrigation nécessaire ne dépend pass seulement de l'évapotranspiration et du lessivage, mais aussi de l'efficience de l'application.

La profondeur de nappe admissible varie selon le climat, la texture du sol, la perméabilité, la salure, le type de culture et la technique et fréquence d'irrigation.

239

Il convient d'accorder une attention particulière à l'aménagement du sol et des eaux en vue de maintenir un équilibre convenable entre les sels. Ceci implique des pratiques telles que: irrigation, drainage, contrôle de l'évaporation, lutte contre l'érosion, emploi de pratiques culturales convenables.
– Un réseau de drainage constitue un impératif absolu, préliminaire à une utilisation efficace de l'eau.
– La jachère estivale est une pratique qui accroit fortement le danger de salinisation.
– Il existe une interaction entre l'aménagement des eaux, la distribution des cultures, les besoins du lessivage et la profondeur de la nappe. Pour éviter la resalinisation, un aménagement convenable du sol et des eaux est une nécessité.

Land and water situation

The Near East Region covers a large geographical area, covering some 20 countries from Pakistan on the East to Libya on the West. The climate, being arid and semi-arid, often limits agricultural land-use so the provision of irrigation water is basic for agricultural intensification and expansion of the cultivated land. The pattern of land use in the region (1961–63) according to the FAO Production Year Book is as follows:

	Total geographical area	Utilizable land	Forested land	Arable land	Range land	Waste land
× 1,000 ha:	1,372,388	378,103	134,024	70,964	137,115	994,285
%:	100	28	10	5	13	72

The arable land amounts to 5% of the total area and 19% of the utilizable land. The total irrigated area is about 28 million ha or 36% of the total arable area but it produces about 70% of the total crop value (UAR, 1968). The cropped rain-fed area amounts to 20 million ha or 59% of the cropped area, producing about 23% of the total gross value of agricultural production. The irrigated areas must increase to about 40 million ha or 56% of the total arable area by 1985, in order to meet the increasing demand for food within the region.

Irrigation is basic for most countries of the region; its areas can be classified into two major classes:

(a) Those depending on surface water—26,149,000 ha
(b) Those depending on groundwater—2,851,000 ha
    Although groundwater resources are less important than surface, they play an important role in some countries including Saudi Arabia, the Gulf States, Y.A.R., P.D.R.Y., Libya, Iran, Afghanistan, Pakistan, Syria.

Agricultural production

The average yield per hectare in both rain-fed and irrigated areas is very low compared with other regions or potential yield. The main crops are shown in tables opposite.

The yields of irrigated and rain-fed crops vary a great deal from country to country due to differences in level of technology, inputs used, lack of drainage, salinity, waterlogging and rainfall. The intensity of cropping of irrigated land varies from 40% to 168% and on the rain-fed from 25 to 50%. Although production has shown an increasing trend in recent years, this is not fast enough to cope with the growing demands of the increasing population.

Crops in rain-fed agriculture:

Crop	Area (× 1000 ha)	Yield (ton/ha)
Wheat	8,544	0.5
Coarse grains	5,937	0.4
Pulses	1,091	0.6
Oil seeds	1,054	
Vegetables	335	
Seed cotton	444	
Perennials	819	
Others	115	
Total	18,339	

The main crops under irrigation are:

Crop	Area (× 1000 ha)	Yield (ton/ha)
Wheat	6,603	1.24
Rice	1.937	1.62
Coarse grains	3,828	1.47
Potatoes and vegetables	762	11.60
Pulses	995	0.97
Oil seeds	509	0.74
Seed cotton	2,970	1.19

In order to fill the food gap, there is a need to focus more attention on problems of land and water use.

Soils of the arid regions

Soils of the arid and semi-arid regions possess common features due to the prevailing soil-forming processes which are dominantly the result of physical weathering with little effect of chemical decomposition. As a result of the heterogenity in parent material the soils tend to vary markedly from place to place, and because of the loose surface the soils are generally subject to both wind and water erosion. Organic matter is generally less than 0.5%. Salts tend to accumulate and show themselves very clearly following the introduction of irrigation, particularly in the absence of adequate drainage; crusts of lime and gypsum formations are often found in soils of arid regions. The most dominant clay mineral is illite with interstratified minerals. Attapulgite may dominate where the parent material is rich in lime and has been submerged under sea as in N.W. Egypt. Clay mineral degradation may take place in the presence of high soluble salts, especially Na_2CO_3. Due to the sorting action of wind, gravel and pebbles tend to accumulate on the surface forming what is known as "desert pavement" which is generally underlain by a vesicular layer of 1–3 cm thickness.

The soil profile is generally characterized by horizon stratification and heterogeneity which poses special problems when the soils are put under irrigation. The associated change in soil properties then include rapid deterioration of soil structure, decomposition of the organic matter content, decrease in permeability, salinization. The presence of soluble sodium and/or magnesium salts associated with prolonged waterlogging leads to increased exchangeable sodium and magnesium on the clay surface and hence to structural instability, lower infiltration, and the soil gradually becomes impermeable. Such unfavourable change will not take place, however, until the exchangeable sodium percentage reaches a value exceeding 15%; it is modified by the dominant type of clay mineral, exchange capacity value and soil texture. In addition, exchangeable sodium and magnesium, when exceeding a certain percentage, exert unfavourable effects on nutritional condition of the growing crop. This becomes more pronounced when these two cations are complementary in the clay.

The changes in soil structure under irrigation in the presence of soluble salts greatly influence the irrigation and drainage regime as to leaching requirements and permissible salinity levels of irrigation water, as well as soil susceptibility to erosion.

Problems of land use under irrigation

Salinity and waterlogging. Extensive areas of the presently irrigated land suffer from salinity and waterlogging, and yields of crops under such conditions are very low compared to the potential yields. To cite examples, in Pakistan, out of a total of 37 million acres of irrigated land about 27 million acres suffer from salinity, waterlogging or both to varying degrees with a pronounced reduction in the yield of the main crops. In Iraq, more than 50% of the lower Rafadain Plain suffers from salinity and waterlogging with a result that most of the area needs new reclamation measures to restore it to its original condition and to raise its productivity. In Syria about 50% of the irrigated land in the Euphrates Valley is seriously affected by salinity and waterlogging and losses to the main crops amount to 300 million dollars. In Egypt about 33% or 2 million acres suffer from salinity and poor drainage to varying degrees with a loss in crop production estimated at 30% of the potential. In Iran, over 15% of the area suffers from a combination of salinity, alkalinity and waterlogging. In all countries of the region, without exception, salinity is of prime concern.

Problems of calcereous soils. In many countries of the region, including Syria, Lebanon, Jordan, North Iraq and Egypt, soils high in $CaCO_3$ constitute an important entity of land resources. These soils are characterized by low water-holding capacity, deterioration of structure and formation of surface crust when irrigated, and specific hydrodynamic properties. The irrigation efficiency is less than 35%, and loss of water to the ground table creates problems of waterlogging and secondary salinization. An example is the newly developed land in the North Western desert of Egypt. In addition, some soils such as the newly developed soils of the Euphrates in Syria contain gypsum in varying amounts, which dissolves under irrigation, thus creating problems of failure of irrigation and drainage structures and secondary salinization.

Problems of sandy soils. Many countries in the region have large areas of sandy soils, for example Egypt, Saudi Arabia, Libya, Sudan, the Gulf States, Syria and Iraq. Such soils have very low water-holding capacity and the efficiency of water use is exceedingly low under the traditional irrigation methods. It is generally very costly to develop such lands, and there is an urgent need to find out the proper cropping pattern, system of water conveyance, distribution and use; otherwise the development of such soils on high-lying lands will create problems of waterlogging in the low-lying areas.

Soil survey and land use. The level of soil surveys and land use appraisal carried out is inadequate in many countries and must be complemented by accurate interpretation for use and management under irrigation to ensure successful development.

Technical problems related to water use

Problems of water use can be ascribed to limited supplies, poor quality, inadequate distribution networks, poor irrigation techniques and low irrigation efficiency. Small land holdings make the economic distribution technically difficult. Optimum water requirements based on cropping patterns, soil and climatic conditions are not yet established in many countries of the region. More flexibility is needed in system design to permit possible shift and change in type of farming and intensity.

The efficiency of water use is generally low, being less than 30–40% The problem is complex as it involves water management, agricultural, economic, organizational and human factors which have to be optimized with the technical aspects to obtain the maximum effectiveness of water use. Key factors in proper water use planning, including needs for agriculture, cost of irrigation and drainage, yields from irrigation, return from investments, must be optimized to ensure maximum profit. Water requirements, water application, design and operation of irrigation network are critical for effective water use and control of groundwater level. There is need to be more flexible in the operation of rotation systems on which many irrigation schemes are operating at present.

Inadequate training of farmers is one of the main causes of the misuse of water and reduced efficiency. Unless the farmer understands the value of water, it will not be possible to improve present practices.

Land and water use relationships

In most countries of the region, where there is more suitable land than water, the objective should be to obtain maximum net income per unit of water through better crop selection, early planting, water use at the stages of plant growth which provide highest yield increases, and maximum interaction of irrigation, fertility and other management practices. Where there is more suitable water than land, the objective should be to obtain maximum economic returns per unit of land through utilization of maximum quantities of water combined with sufficient quantities of fertilizers and other inputs.

Design and operation of water supply systems should provide the required amounts of water at the time needed together with measuring devices which maintain uniform flows at the farm level to cope with the water requirements of crops, leaching requirements and salt balance. Since small farms and patchy cropping pattern generally increase water losses through percolation, evaporation and misuse, planning of irrigation systems should allow for flexibility to move in the direction of consolidated holdings whenever feasible economically and socially.

Drainage and salinity control

The greatest problems facing presently irrigated land in the region are salinity and water-logging. This may be due to insufficient water supply to provide for the required leaching, the presence of high-water table caused by poor irrigation practices and the excessive seepage from canals and other water courses. To improve such situations the first step should be to provide suitable drainage which allows for the lowering of the water table below the critical level of capillary action. The leaching of accumulated salts is required at regular intervals to maintain a favourable salt balance.

Drainage systems including field drains should be planned and constructed at the same time as the irrigation system, as it is much cheaper to construct drainage systems from the beginning than to wait until after the drainage situation becomes serious. As things are, salinization threatens to put out of production considerable areas in several countries of the region within a period of years.

The groundwater table can seldom be lowered to such a depth as to prevent all capillary movement. It should be lowered to the degree that such movement becomes relatively small. To control groundwater level, improved water use and soil management practices are needed.

Shortage of water is a major factor limiting the agricultural development in the region. Many rivers are expected to be fully exploited before the end of the century, but groundwater resources are not yet fully surveyed and are less developed. The integrated approach requires the development of both underground and surface waters. Correct management of under-ground water reservoirs is essential to maintain a suitable balance between extraction and recharge.

Extent of problems related to irrigation in countries of the region and their socio-economic impacts

Afghanistan. The arable land is about 8 million hectares of which only 25% receive adequate water supply. The climate varies from sub-humid in the North to arid and semi-arid in the South-West with an average rainfall of 180 mm. The problem of salinity and waterlogging can develop in the South-West desert basins, and the irrigation project in Helmand valley is an attempt to overcome this. In the basin of the Hari Rud, groundwater development and storage are basic for irrigated agriculture. The cost of irrigation water is generally high, whereas yields per unit of water are low. Due to shortage of water, its cost is expected to rise in the future and its quality to become worse, thus resulting in increased land salinization and reduced crop yields.

Cyprus. The irrigated area amounts to 65,000 hectares and the rainfall varies between 250 mm and 1000 mm, falling from October to April. The total surface water utilized for irrigation is about 140 million m^3 and of groundwater about 160 million m^3. As a result of overpumping there is a problem of sea water intrusion. Soil salinity is confined to the alluvial and coastal plains where rainfall is 250–400 mm. The salt affected area is about 18,000 hectares, mostly in Eastern Mesaoria (10,000 ha). The saline soils are generally heavy textured with saline water table lying at 1·3–5 m during the winter season. Groundwater salinity varies from 1500 to 40,000 ppm Cl. Citrus can be irrigated with water containing

500–1000 ppm Cl on heavy and sandy soils respectively, provided that adequate leaching is carried out. Lucerne can be irrigated with water containing up to 3500 ppm Cl on soils with good drainage. Special methods of water management including irrigation methods and leaching are applied. Where traditional flood and furrow methods are used, salinization may take place even with the use of low saline water due to non-uniform leaching of salts resulting from improper levelling. Furrow irrigation is avoided whenever water is saline. Sprinkling irrigation cannot be applied for citrus where Cl exceeds 300 ppm and 500 ppm for lucerne. Rotation of crops accompanied by changes in irrigation method to the basin system are considered as salinity control measures.

Egypt. The total irrigated area is about 6·5 million acres of which about 1 million is newly reclaimed. The Nile is the main source of irrigation water with an average concentration of 200 ppm of dissolved salts. The total amount of water used for irrigation is about 54 billion m^3 at Aswan with an average water duty of 8000 m^3/acre. The climate is warm arid with average annual rainfall of 150 mm on the coast dropping quickly to 25 mm at Cairo. Salinity and waterlogging have developed as a result of change from the basin system of irrigation to the perennial system in the absence of adequate drainage. The total area affected by salinity is estimated at 2 million acres of the old land, and the rest may suffer from the presence of permanent waterlogging or perched water tables, whereas the newly reclaimed area of about one million acres has started to suffer to varying degrees from waterlogging and salinization. The overuse of irrigation in the absence of drainage caused a quick rise in the groundwater level, and the high evaporation from the shallow saline water table led to serious salinity problems. In the delta the problem is aggravated by the very poor internal drainage of the fine textured soils. In the newly reclaimed desert areas, the introduction of irrigation and the overuse of water and soils of low water-holding capacity in the absence of field drainage led to waterlogging and subsequent solution of salts of marine origin. Seepage of saline water to the low-lying irrigation canals has resulted in the salinization of the irrigation waters and hence of soils (i.e. North-West area). The use of saline drainage water on heavy textured and poorly permeable lake bed soils is the main cause of secondary salinization (NE of the Delta). In some areas with alkaline ground-water table, the introduction of irrigation produced alkalization of the soils (Wadi Tumilat, Wadi Natroun and other areas NW of the Delta).

It is estimated that waterlogging has reduced productivity of the affected areas by at least 30%, but improvement of drainage coupled with leaching of salts will restore productivity. In terms of money, the increase in agricultural production due to drainage will be more than one billion dollars. For this reason, the Government of Egypt is launching a huge programme of tile drainage which will cost at least 500 million dollars over the next ten years. The benefit cost/ratio is estimated at 2.77:1. This will be combined with a soil improvement programme including subsoiling use of gypsum and intensive use of fertilizers. Other activities in the field of land and water use include the development of new irrigation methods, water distribution, optimum water requirements and crop consolidation.

Iran. Out of a total area of 165 million ha, 16.8 million are arable, 7.3 million are saline soils, and 8.2 million are waterlogged. The problem soils are distributed mainly in the South, South-East, North and North-West. Salinization of soils and groundwaters are accelerated by the introduction of irrigation in the absence of drainage. The majority of the irrigated soils on flat topography as well as low-lying land in the arid and semi-arid zones are affected.

Salinity increases from North-West to South-West, and the coastal areas along the Gulf are salt affected. There is a need for adequate drainage in most irrigated areas.

Iraq. It is well known that Iraq's irrigated agriculture is hampered by salinity and water-logging. The Mesopotamian Plain accounts for 25% of the total arable area and has been under irrigation for more than 6000 years. The area under irrigation is about 3.6 million ha, of which more than 50% suffer from secondary salinization and waterlogging, particularly in the middle and lower Rafidain Plain. Salinity and waterlogging are attributed to misuse of irrigation water, lack of drainage, evaporation from high saline groundwater table, seep-age from canals, tidal action in coastal areas, insufficient water quantity during the summer months, low summer cropping intensity not exceeding 10%, with patchy distribution in small plots rather than consolidated holdings. The summer fallow and shallow drain depths are both inducive to seasonal salinity.

Jordan. The irrigated area is about 50,000 hectares of which 84% are supplied with surface water and 16% from groundwater. About 90% of the irrigated land lies in the Jordan Valley. The salt-affected area is about 8000 ha. The quality of irrigation water is fairly suitable (about 800 ppm) except in the desert areas where salinity may exceed 2000 ppm and the salinity of the virgin soils may exceed 20 mmohs/cm. Drainage and leaching of salts are necessary measures for the successful development of desert land. The main problem is insufficiency of water required to maintain a favourable salt balance. On the Jordan Plateau, the high salinity of the virgin soils and of groundwater makes it difficult to develop successful agriculture without reclamation and improvement measures.

Lebanon. The total irrigated area is about 65,000 ha with 60,000 ha of new land under development. Salinity and waterlogging constitute no serious problem at present but poten-tial salinity hazards may develop in the North of the Bekaa valley where groundwater resources are being exploited. Swelling clay minerals and flat topography are inducive to waterlogging, and signs of hydromorphic and pseudo-gley conditions show themselves at relatively greater depths. Along the coast sea water intrusion is a cause of salinity. The relatively high exchangeable magnesium in some soils of the Bekaa may indicate a tendency to soil deterioration.

Libya. Agriculture is concentrated in the coastal belt on Jebel Akhbar and Jebel Nafussa where both rainfall and groundwater resources are used. Some of the inland drainage basins are also cultivated. Groundwater of varying salinity, surface water diverted to the land, and water flowing from springs are the main sources for irrigation. Groundwater constitutes the main source of soil salinity. Potential salinity is expected as a result of planned expansion in irrigated agriculture being carried out at present at the expense of overuse of groundwater resources.

Pakistan. The total area is about 205 million acres of which 73 million are suitable for cultivation. The irrigated area is about 37 million acres of which 25 million are affected to varying degrees by salinity, waterlogging or both. A survey of the existing conditions indicate that about 10 million acres are poorly drained, 5 million are severely affected with salinity and 10 million acres suffer from patchy salinity. The area annually damaged by salinity and waterlogging is estimated at 100,000 acres. There is a shortage of irrigation water compared with the available land resources, and a need for the storage of unused river flow (about 29 million acre/feet), control of seepage losses and development of suitable underground water. Where underground water is of good quality, tube wells will control both waterlogging and salinity, but where water quality is poor, their continuous use leads to increased salinity and thus lowering of productivity. As a general rule crop yields are lower than expected under similar conditions elsewhere.

In the lower Indus concentrations of the groundwater may be as high as 30,000 ppm. The water quality of the tube wells has been studied and 6% of the wells have water of excellent quality, 18% poor quality and unfit for use. The remainder may salinize the soil profile to a depth of 6 feet within 12 years. Increased soil salinity is associated with a change in the cropping pattern and intensity, and as a result the area of non-tolerant crops (cotton and sugar cane), has decreased in spite of a lowering of groundwater to a depth of 18 ft. The groundwater quality is in a process of continuous deterioration with increasing use, distance from main canals, and increased depth of the wells. In areas with poor quality water, tile drainage may offer a solution to the waterlogging problem.

Saudi Arabia, Kuwait, Bahrain, Qatar and United Arab Emirates. The climate of these countries is warm and arid with rainfall not exceeding 100 mm/yr and torrential in nature. Agriculture depends mainly on the limited groundwater resources of varying salinity, and irrigation under such arid conditions generally leads to increased soil salinity. The overuse of groundwater resources results in deterioration in water quality which in turn aggravates the salinity problem, particularly where the water quantity is insufficient for adequate leaching. Where the soil texture is heavy (e.g. Qatar) salinity develops over a short period of time. There is a tendency to use sewage water in irrigation, and in some countries such water is contaminated with sea water which develops soil salinity when used. Alfalfa, being relatively salt tolerant, constitutes the main crop in most of these countries. However, because of its high water requirements the expansion in this crop depletes the limited groundwater resources and increases their salinity. When such water is used on other crops, such as fruit trees, it leads to their death. A balance between water recharge and withdrawal is basic for continuing agriculture.

Somalia. The climate ranges from arid to sub-tropical with rainfall varying between 1000 and 60 mm/yr, declining to the North. The irrigated areas are located within Juba valley, Shebelli valley and the inter-river area. Due to the flat topography and poor natural drainage these areas, especially the inter-river area, suffer from flooding which gives rise to waterlogging and potential salinity. The coastal areas may be salinized as a result of sea-water intrusion. The area of salt affected soils is about 7.5 million acres mostly distributed in the alluvial plains.

Sudan. The climate ranges from arid to tropical with rainfall varying between 1500 and 10 mm/yr. Salinity problems are evident in the Northern part of the Gezira scheme, Managil extension and the Northern province. The water table in the Gezira is not a source of secondary salinization since it lies at depths far below the critical depths and creates no waterlogging problem. Salinity is the result of the continuous accumulation of salts from irrigation water in poorly drained soils where perched water tables may rise to the surface under the influence of high evaporation. The sodic soils existing in the Northern part of the Gezira are the result of submergence under water for prolonged periods of time where anaerobic conditions may lead to alkalization. Reclamation is essential for the economic development of such soils.

Syria. The total area is about 46 million acres of which about 8 million could be cropped. The present irrigated land is about 1 million acres and will reach 2.5 million after the full development of the Euphrates project. About 50% of the present irrigated land suffers from salinity and waterlogging to varying degrees. This is the result of overuse of irrigation water in the absence of drainage, use of saline water in irrigation, poor land preparation and water management, and movements of salts when soils are put under irrigation.

Due to the aridity of the climate, with evaporation exceeding precipitation in many locations, it is estimated that 70% of the soils put under irrigation are potentially saline. The degree of salinity is related to the river flow being higher downstream and reaching its maximum where the Euphrates river flows into Iraq. In the Ghab project waterlogging still constitutes the main problem; the fast expansion in cotton cultivation and the use of well waters in the absence of drainage enhanced the spreading of both salinity and waterlogging. In the Sinn project waterlogging is caused by the heavy winter rainfall and is aggravated by the presence of an impermeable layer in the subsoil.

People's Democratic Republic of Yemen. The climate is sub-tropical with rainfall varying from 50 to 500 mm/yr. Most of the land slopes towards the Gulf of Aden, and flood water is diverted to irrigate the terraced and bunded fields. In other areas underground water of varying salinity is the main source of irrigation. Due to the aridity of the climate, salinity of groundwater and insufficient amounts of water for leaching, soil salinity develops follow-

TABLE 1. *Present and potential land use in the Near East Region area in 1000 ha*

Country	Year	Total area	Arable land	Irrigated	Potential irrigated (1985)
Afghanistan	1968	64,750	7,844	813 (1968)	3,007
Bahrain	1971	60	0.5	—	—
Cyprus	1971	960	535	102 (1967)	—
Egypt	1971	100,145	2,725	2,852 (1971)	3,156
Iran	1971	164,800	10,154	5,251 (1971)	4,894
Iraq	1970	43,492	10,000	3,675 (1963)	3,798
Jordan	1970	9,774	1,132	60 (1970)	66
Kuwait	1970	1,600	0.5	0.6 (1971)	0.7
Lebanon	1968	1,040	240	68	117
Libya	1971	175,954	2,377	120 (1971)	—
Oman	1971	21,246	16	—	—
Pakistan	1969	80,388	19,235	12,505 (1969)	24,400
Qatar	1971	2,201	0.2	—	—
Saudi Arabia	1967	214,949	765	131 (1967)	255
Somalia	1960	63,766	957	162 (1970)	0.1
Sudan	1968	250,581	7,100	711 (1967)	—
Syria	1970	18,518	5,641	450 (1970)	830
United Arab Emirates	1971	8,360	20	—	—
Yemen Arab Republic	1969	19,500	1,200	100 (1970)	—
Yemen Democratic Republic	1966	28,768	252	68	153

* Source: Indicative World Plan, FAO 1962, and Near East Statistical Directory, 1974.

ing irrigation, and this is becoming more serious with the expansion in irrigated land and over-extraction of groundwater.

Yemen Arab Republic. The climate is sub-tropical with rainfall varying between 400 and 800 mm/yr. Due to the nature of topography and the high rainfall, the problem of salinity is of minor importance, except near the sea, where sea-water intrusion may cause sporadic salinity. Terracing is practised for soil and water conservation. With the expansion in irrigation and due to the aridity of the climate, potential salinity may constitute a hazard in the future.

References

Although not specifically referred to in the text, much of the data in this paper is drawn from the following:

Arar, A. 1970. The role of drainage in agricultural development in the Ghab region. Drainage and salinity series No. 12. UNSF Ghab Project, Damascus, Syria (in preparation).
El Gabaly, M. M. 1969. Three Types of Sodic Soils in the United Arab Republic. Symp. on the Rec. of Sodic and Soda-Saline Soils, Yerevan. Agrokémia, és Talajtan. Budapest, Hungary.
E. Gabaly, M. M. 1971. Reclamation and Management of Salt Affected Soils, FAO, Salinity Seminar, Baghdad.
El Gabaly, M. M. 1975. Land and Water Use Programme in the Near East and North Africa. (FAO, Project Document.)
FAO, EPTA, No. 1932, 1964. Report on Seminar on Waterlogging in Relation to Irrigation and Salinity Problems. Lahore, Pakistan.
FAO, 1966. Indicative World Plan for Agricultural Development 1965–1985. Near East. Sub-Regional Study No. 1, Vol. 1.
FAO, 1971. Salinity Seminar, Baghdad, Methods of Amelioration of Saline and Waterlogged Soils.
UAR, Ministry of Irrigation, 1968. Irrigation and Drainage in Egypt. Survey of Egypt Series.

SOIL WATER PROBLEMS RELATED TO
SALINITY
AND ALKALINITY IN IRRIGATED LANDS

by G. VARALLYAY

Research Institute for Soil Science and Agricultural Chemistry of the Hungarian Academy of Sciences, Budapest

Introduction

The formation and accumulation of salts in soils are due to a large number of geochemical processes taking place in the upper strata of the earth's crust. As a result of weathering secondary clay minerals, oxides and other compounds are formed, including various water-soluble salts. The primary reason of the formation and occurrence of salt affected soils is the accumulation of Na^+ ions in the solid and/or liquid phases of soils[10, 11, 12, 17, 19, 20, 26, 27, 28] *

Water as reactant, solvent and transporting agent plays a decisive role in the development of salt-affected soils. Salinization and alkalinization processes are directly or indirectly related to soil water since the soluble salts move with the water and the ion composition of soil solution governs the ion composition of the soil absorption complex[6, 39].

Salinity and alkalinity are important factors of the biosphere: they limit soil fertility, hinder agricultural potential, prevent intensive agricultural development. The distribution of present salt-affected soils is closely related to climate, geological, geochemical, hydro-logical and hydrogeological conditions. As shown on the FAO-UNESCO World Map of Soils and on maps prepared in scale 1:5,000,000 within the framework of the "World Map of Salt Affected Soils"[26, 27], these soils raise serious problems in numerous developing countries in arid and semi-arid regions of Africa, Asia and South America, not only regarding the natural environment but in respect of their national economy as well[10, 13, 23, 29].

One of the most significant human soil-forming factors is irrigation. Its general aim is the maintenance of soil moisture within the range required for optimum plant growth. Irri-gation is a question of vital importance in many countries in the arid and semi-arid regions of the world, where it is an essential pre-condition of agricultural production. But it is signifi-cant in the more humid regions as well for the procurement of high yields, the production

* Owing to the large number of references in this paper they are indicated by number instead of by author and date.

of water-consumptive crops, the efficient use of high rates of fertilizers and intensive new crop varieties[8, 10].

In an ideal case of good quality water, well drained and good soils, properly planned and executed irrigation, optimum water and soil management, irrigation supplies the plants with an adequate amount of water, favourably changes the soil processes (salt regime, nutrient regime and biological processes) and increases not only the actual productivity of land but also the potential fertility of the soil[8, 10, 24]. In cases of poor quality water, poor drainage and soils, and bad management, irrigation not only directly damages the plants, limits their nutrient uptake and metabolism, but also destroys the soil structure and initiates, extends or deepens unfavourable soil processes which markedly decrease soil fertility. Among these processes secondary salinization and alkalization and waterlogging are the most widespread and harmful[2, 3, 4, 8, 9, 10, 14, 23, 25, 29]. According to the estimation of FAO and UNESCO more than 50% of all irrigated lands of the world have been damaged by these processes, and year by year many million hectares in irrigation have to be abandoned.[29]

Under arid or semi-arid conditions and in regions of poor natural drainage there is a real hazard of salt accumulation in soils or from brackish irrigation water or from saline groundwater. There is danger of alkalization as well, because the high potential evapotranspiration promotes the concentration of the soil solution which results in a harmful sodium accumulation in the liquid or solid phase[10, 23, 30]. Control of salinity and alkalinity is therefore essential to the operation of a permanently successful agriculture. It involves both reclamation of salt-affected soils and maintenance or improvement of other soils[2, 10, 13, 14, 24, 26].

Salt accumulation, salt regime

The growth-suppressing effect of salt concentration of the soil solution within the root zone is largely osmotic, caused by a decrease of the water availability to plants, but in numerous cases the accumulated salts are directly toxic for some sensitive plants. Therefore salinity limits the water and nutrient uptake by plants, retards their metabolism and causes physiological deterioration[2, 6, 8, 9, 10, 21, 23].

Whether salts accumulate or leach depends on the processes by which they move into, out of, or within the soil profile. These processes can be exactly described, quantitatively characterized and forecast by salt balances, which can be derived from water balances. They indicate the gains and losses of water or salt in a given unit over a certain time period and can be written as follows:

(incoming quantity) − (outgoing quantity) = change of storage in the soil.

In spite of their similar character water and salt balances often differ in their factors, composition and application[3, 4, 10, 17, 28, 32]. Salt balances can be calculated:

(a) for the total salt content, or for various ions (when studying specific ion effects and chemical changes in the soil solution during filtration);
(b) for the whole soil profile from the soil surface to the water table, or for various layers, horizons (when studying salt profile redistribution, hazard of resalinization, leaching efficiency) or for the root zone;

(c) for soils, mapping units or territories (having sufficiently homogeneous hydrological character);

(d) for vegetation periods, irrigation seasons, seasons, years or longer periods of time (having sufficiently homogeneous hydrological character).

The factorial salt balances reflect not only the integrated changes but also reveal the causes of the changes and quantitatively characterize the partial contribution of various factors in these changes. In this way the potential possibilities of a man-controlled salt balance regulation can be determined; a prognosis can be given for the natural salinization and alkalization processes, and the probable effect of human intervention can be forecast[10, 28, 32].

The general equation of factorial salt balances can be given as follows:

Irrespective of its source, all *irrigation water* contains dissolved salts, the kind and quantity of which depend on its origin and also on its course before use. The criteria for assessing the suitability of water depend on the specific conditions of use, including the land use, cropping pattern, salt tolerance of crops, various soil properties (especially vertical and horizontal drainage), climatic conditions, irrigation management (especially the quantity of irrigation water, frequency, method and intensity of irrigation, water distribution pattern, leaching fraction and its efficiency, etc.) and certain cultural management practices. Although there is a general tendency to use mineralized waters for irrigation, in most of the countries the existing water quality guidelines assure the prevention of a permanent harmful salt accumulation in the root zone[2, 3, 4, 10, 14, 21, 23, 24].

The *groundwater* is the main salt source first of all in low-lying, poorly or very poorly drained areas (closed basins, lowlands, low alluvial terraces and delta areas, etc.), where its horizontal flow is very low (low slope gradient, hydraulic gradient and generally very low hydraulic conductivity) and this "stagnant" character affords potential possibility of its gradual concentration. The water table is near the surface and so the capillary flow can transport relatively large quantities of water and soluble salts to the overlying horizons, to the active root zone. Under such conditions salt-affected soils may develop not only in arid regions but also under relatively humid climates[17, 25, 30, 34].

The possibility of salinization from the groundwater is determined by the dominant vertical direction of water flow and salt transport through the soil profile[2, 33, 37, 38]. The quantity of salts (Q) entering the soil profile from the groundwater is determined by the quantity (Q_s) and concentration (C_s) of the soil solution transported by capillary forces from the groundwater to the overlying horizons. C_s depends on the concentration and ion composition of the groundwater, on changes of these factors during the upward capillary flow, and on the solid phase–liquid phase interactions. Q_s is determined by the direction and velocity of the capillary flow, which is a function of the suction profile, the capillary conductivity and the depth of the water table. For the quantitative characterization of these relationships the unsaturated flow theory can be applied[6, 9, 10, 18].

The hydrophysical aspects of salinization from the groundwater have been discussed by many authors and various numerical and simulation approaches were published for the exact and quantitative description of these processes[6, 9, 10, 17, 24, 33, 37, 38].

$$\Delta S + [P + I + R + G + W + F] - [l_p + l_i + r + g + n]$$

Where:

ΔS = Salt balance

P = Quantity of salts derived from the atmosphere (air-borne salts, rainfall, wind action, etc.)

I = Quantity of salts added with the irrigation water

R = Horizontal inflow of salts transported by surface waters (runoff, flood, waterlogging)

G = Horizontal inflow of salts transported by subsurface waters (groundwaters, deep sub-surface waters, etc.)

W = Quantity of salts derived from local weathering processes

F = Quantity of salts added with fertilizers and chemical amendments

l_p = Quantity of salts leached out by atmospheric precipitation

l_i = Quantity of salts leached out by irrigation (leaching) water

r = Horizontal outflow of salts (discharge) transported by surface waters

g = Horizontal outflow of salts transported by subsurface waters (drainage)

n = Quantity of salts taken up by plants and transported from the area with the yield.

All factors can be given in tons/hectare.

The main factors of salt accumulation in irrigated soils can be summarized as follows:

(1) Water-soluble salts accumulate from saline or brackish irrigation water;
(2) The water table rises and so:
    (a) the groundwater salt content accumulates in the affected layers, the subsurface waters accumulate the soluble weathering products of large areas, from extensive watersheds into relatively small depressed lowlands;
    (b) the rising groundwater transports salts from deeper sub-surface waters, geological deposits or soil layers into the overlying horizons, to the surface layers; or
    (c) the stagnant groundwater limits the natural drainage of the area and impedes the leaching of salts derived from local weathering or irrigation water;
(3) The periodical wetting and drying of the soil promote the weathering processes and the soluble weathering processes increase salt accumulation[30, 32, 39].

If neutral sodium salt ($NaCl$, Na_2SO_4) accumulation takes place in coarse textured soils the salinization is a more or less reversible process. In such cases the salts accumulated by irrigation during the vegetation period can be washed out from the soil profile by leaching and removed from the area by horizontal drainage[2, 8, 10, 13, 14, 21, 23, 24]. This is a widely used practice in many parts of the World (Central Asia, North Africa, Near East region, USA). But this is a rather expensive procedure and has to meet the following requirements:

adequate amount of good-quality water (leaching requirement, leaching fraction);
good vertical drainage of the soil profile (light textured, permeable soils with chloride-sulphate type salinization and low Na^+ saturation);
good horizontal drainage of the area (high rate of horizontal groundwater flow);
frost-free season over the vegetative period;
drain-water reservoir with adequate storage capacity.

Under these circumstances there are two possibilities for an efficient salinity control: prevention of salt accumulation, and maintenance of salt balance in equilibrium by leaching and drainage.

Alkalization

According to the laws of the existing equilibrium between the solid and liquid phases the ion-concentration and ion-composition of the soil solution influence the exchangeable cation composition of the soil-absorption complex. The Na^+ saturation of the absorption complex (ESP value) depends on the absolute and relative concentration of Na^+ ions in the soil solution which is determined by the quantity and solubility of sodium and other salts in the soil. In the presence of sodium salts capable of alkaline hydrolysis most part of the Ca and Mg salts is precipitated, only the more soluble Na salts remain in solution and migrate within the soil profile: the Na^+ cation becomes absolutely dominant. This results in a high Na^+ saturation even at a relatively low salt concentration[11, 21, 26, 27, 28].

The high Na^+ saturation of soil colloids brings about unfavourable changes in the physical and hydrophysical properties of soils. These processes are not reversible, so they cannot be controlled and balanced by simple leaching, and in most cases the reclamation of the secondary alkaline soils is possible only by very expensive complex amelioration methods[10, 19, 20, 26]. Therefore, the most economic solution to the proper alkalinity control is prevention.

For this purpose it is necessary:

to describe exactly the salinization–alkalization and reverse processes
    to identify the main actual and potential salt sources
    to characterize quantitatively the main features of the salt regime
    to analyse the whole range of environmental factors influencing the role and importance of various salt sources and the components of the salt regime
    to determine the present and possible impact of human activity on the above mentioned factors
to elaborate a comprehensive prognosis system for the prediction of these processes;
to establish a regular monitoring system of salinity–alkalinity changes due to natural factors or human activity.

On this basis

the possibilities of establishing a satisfactory prevention of salinity–alkalinity can be revealed,
the most efficient methods can be selected according to the local natural and farming conditions,
precise technology can be elaborated for these methods.

Szabolcs, Darab and Varallyay[30, 31, 32] elaborated a comprehensive salinity–alkalinity prognosis system in Hungary. This system was successfully applied at the planning, execution and management of Tisza-irrigation systems in the Hungarian Plain and it afforded practical possibilities for the efficient prevention of soil deterioration due to secondary salinization–alkalization in the Carpathian Basin, where exists a potential hazard due to the special hydrogeological and hydrological conditions[25, 30, 32, 34, 39]. The prognosis system can be directly used or simply adapted for regions of similar physiographic conditions (closed basins, poorly drained alluvial terraces, delta areas, depressed lowlands under moderate semi-humid climate with a negative water balance during the warm and dry summer period). The general conception of the system can be applied in most of the irrigated

TABLE 1. Some characteristic data of the studied soils

No.	Horizon	Sampling depth (cm)	Loss in HCL treatment	Particle-size distribution						CaCO₃	CCC (me/100 g soil)	ESP	Organic matter	pH		Bulk density (g/cm³)		Total salt content	
				1–0.25 (mm)	0.25–0.05 (mm)	0.05–0.01 (mm)	0.01–0.005 (mm)	0.005–0.001 (mm)	<0.001 (mm)					A	B	B	AB	A	B
1	1	0–4	10.73	12.00	52.36	13.80	2.62	4.52	3.97	9.6	11.5	45.8	1.25	9.2	9.1	1.36	0.98	1.20	0.01
2	2	4–16	15.52	9.61	39.58	12.65	3.14	5.57	13.89	13.3	16.8	54.6	0.79	9.2	9.0	1.33	1.09	0.45	0.06
3	6	90–110	28.33	12.31	42.93	9.29	2.30	3.38	1.46	28.2	8.8	23.2	–	8.5	8.2	1.32	1.28	0.17	0.00
4	B₁	3–15	2.10	1.44	4.08	37.74	9.29	11.54	33.81	–	22.3	66.1	2.22	8.3	7.9	1.27	1.05	0.27	0.18
5	B₂	15–25	2.34	0.55	3.43	32.30	9.02	11.71	40.65	–	28.5	49.4	1.53	8.8	8.3	1.27	0.92	0.51	0.18
6	BC	55–66	15.02	0.01	4.15	29.31	8.44	10.55	32.52	14.3	27.4	53.4	–	9.3	8.7	1.24	0.97	0.74	0.16
7	A	0–10	1.94	0.17	30.16	32.82	4.04	6.90	23.97	–	19.7	1.9	2.30	7.3	6.7	1.10	1.22	0.03	0.02
8	C	100–125	20.84	0.00	23.98	25.88	4.51	4.20	20.59	20.2	21.4	3.5	–	8.8	8.7	1.29	1.36	0.09	0.01

Nos. 1, 2 and 3   Sodik solonchak on calcareous Danube alluvium, Hungary
4, 5 and 6   Shallow meadow solonetz on calcareous loess-like clay, Tisza Valley, Hungary
7, 8   Meadow soil on calcareous loess-like clay, Tisza Valley, Hungary

A   At the beginning of the experiment.
B   At the end of the experiment.

areas of the World threatened by salinity and alkalinity. Similar approaches were presented and discussed on the FAO/ISSS Expert Consultation on Prognosis of Salinity and Alkalinity, Rome, 1975[1, 17].

Physical deterioration

Secondary salinization–alkalization and waterlogging have significant influences on the physical properties of irrigated soils and at the same time these properties have a considerable effect on these unfavourable processes.

The porosity pattern of irrigated soils is influenced by the simple physical–mechanical effect of falling drops one- two- or three-dimensional filtration, horizontal flow and surface runoff of the irrigation water, even if it is of good quality. In such cases conductivity decreases with an increase in bulk density (decrease in bulk volume and total pore space) in accordance with the Kozeny–Carman or similar equations expressing that hydraulic conductivity is proportional to porosity. The causes of this phenomenon are structure destruction, aggregate failure, clogging of macropores by particle movement, i.e. a largely irreversible mechanical compaction[8, 35, 36].

It is well known that geometry of soil pores and pore-size distribution, rather than total pore volume, govern both saturated and unsaturated hydraulic conductivity. The mechanical compaction of irrigated soils primarily results in a considerable decrease of large pores and consequently in the formation of a surface crust or a dense surface horizon with low infiltration rate and permeability, and this creates extreme water economy in the soil, as will be discussed later.

Under the effect of the high electrolyte concentration of the soil solution a reversible flocculation of the primary particles takes place due to their low electrocinetic (zeta) potential. The floccule, forming in this way, is "stable" only in a certain concentration range and if the concentration decreases under the "threshold limit" the particles—not cemented or bound together—will be dispersed again[5, 8, 10, 22]. Consequently, the hydraulic conductivity K of highly saline soils (especially soils with chloride-sulphate type salinization) are relatively high, but during leaching K values sharply decrease with the decreasing salinity[15, 22, 35, 36, 40]. The efficiency of saline water irrigation, which has been suggested for the maintenance of favourable permeability[8, 21, 23] depends on the crop response but cannot be evaluated as soil amelioration.

The unstable character of the flocculation phenomena is clearly shown by permeability curves (changes in hydraulic conductivity as a function of time) shown in Fig. 1. Some characteristic data of the studied soils are given in Table 1. During the experiment, which was practically a distilled-water leaching, the relatively high hydraulic conductivity of salt accumulation horizons (No. 1, 5, 6) sharply decreased, parallel with a sharp decrease in the salt content of soil (Table 1).

Consequently the electrolyte concentration of the permeating soil solution fell below the "threshold" concentration. So the flocculation effect of high salinity became negligible and the physical consequences of high Na^+ saturation (Table 1) manifested themselves more expressively[35, 36]. Similar phenomena have been described by numerous authors[2, 5, 8, 9, 10, 15, 16, 22, 40]. The various mechanisms responsible for solute movement in soils (diffusion, convection or viscous flow and their combination, one and two dimensional hydrodynamic dispersion, miscible displacement, salt-sieving effect) have been exactly and accurately analysed by others[2, 5, 6, 7, 8, 9, 17, 18].

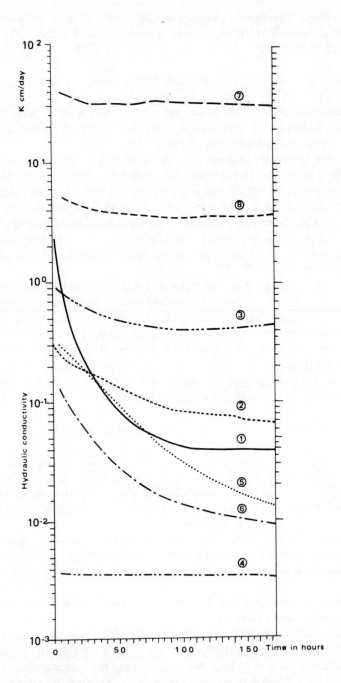

FIG. 1. Permeability curves (saturated hydraulic conductivity as a function of time) of soils with various Na+-saturation and initial salinity (salt content).

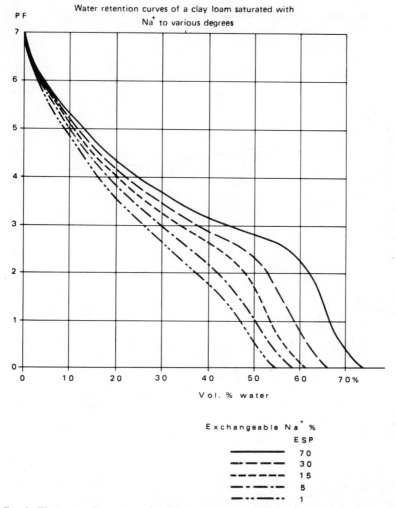

FIG. 2. Water retention curves of a clay loam saturated with Na⁺ to various degrees.

The alkalization of soils, especially of heavy-textured high-swelling clay-containing soils, causes increased hydration, swelling, dispersion of colloids, aggregate and structure destruction, clogging of macropores, i.e. significant, sometimes radical changes in the pore-size distribution: the pore size and macropore volume decrease, the micropore volume increases [40]. These changes are clearly reflected by the water retention curves. In Fig. 2, pF-curves are given for a clay loam (hydromorphic chernozem topsoil from the Hungarian Plain, Tisza Valley) saturated with Na⁺ to ESP 1, 5, 15, 30 and 70, respectively. The curves clearly indicate the positive correlation between the degree of Na⁺ saturation and the water retentivity of soil, especially in the pF 2.0–3.0 suction range. The swelling phenomenon is clearly shown as well.

As a consequence of swelling, the bulk volume and total pore volume increase and bulk density decreases with increasing Na⁺ saturation. At the same time the saturated hydraulic conductivity sharply decreases as it is shown in Fig. 3. This phenomenon is quite contrary

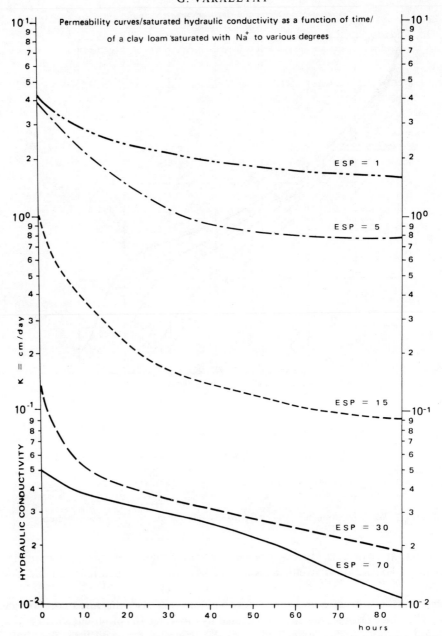

FIG. 3. Permeability curves (saturated hydraulic conductivity as a function of time) of a clay loam saturated with Na+ to various degrees.

to any expectation one may have from the Kozeny–Carman and similar equations, and can be explained only by the above mentioned pore-size redistribution[35, 36]. Similar mechanisms were described by McNeal[15, 16], Waldron and Constantin[40] and others. The size of macropores, which mainly governs hydraulic conductivity, decreases. The conductivity of "micropores" is considerably lower and the flow in fine-textured, swelling clay systems displays non-Darcian behaviour: conductivity increases with increasing hydraulic gradient, and depending on the particle and pore size, their geometry and the acting hydraulic gradient, only a certain part of the water-filled micropores can conduct water, the other part of the soil moisture shows a special "semisolid-state" character[7, 9, 18].

The extremely low permeability of the highly Na^+ saturated swelling clays can be explained only in this way, and this fact has to be taken into consideration in the evaluation and interpretation of K values for the practical purposes of drainage design, estimation of filtration losses, etc.

The conductivity curves (K values plotted against time) of Fig. 3 express sharp differences according to the ESP of soils and indicate the structural changes during filtration. Because of the extremely low vertical and horizontal saturated hydraulic conductivity of heavy-textured, swelling clay, alkali soils, their natural drainage conditions are very poor, the possibilities of their leaching are limited, and consequently their formation processes caused by sodium salts capable of alkaline hydrolysis are almost irreversible [4, 19, 20, 26, 27, 28, 35, 36].

The fertility and agricultural utility of these salt-affected soils are limited mainly by their unfavourable physical properties and extreme water economy. The main soil factors limiting the optimum (adequate and continuous) water supply of plants are summarized in Fig. 4.

In the case of a surface crust of a compact soil layer (e.g. illuvial, heavy-textured, highly Na^+ saturated solonetz B-horizon) near to the surface, not only the root penetration is retarded but the infiltration of water is also limited. The very low infiltration rate results in oversaturation, aeration problems (decreasing availability of plant nutrients, anaerobic biological processes, unfavourable reduction) and—in spite of temporary waterlogging after rainfall or irrigation—in shallow wetting zone, considerable evaporation losses and surface runoff. The water-storage capacity of this shallow wetting zone is very low and so can satisfy the water consumption of plants only for a short period. This is the main reason for the special drought-sensitivity of alkali soils even under irrigated conditions.

Crack formation in swelling clays causes another water problem in heavy-textured alkali soils. Some of the rain or irrigation water flows through these cracks directly to the groundwater. These "filtration losses" diminish water storage in the soil, decrease the rain or irrigation water efficiency and, at the same time, result in a rise of the water table which may be accompanied by unfavourable processes such as secondary salinization and alkalization. During infiltration the swelling cracks impede the uniform wetting of a thick soil layer. During dry periods deep and wide cracks are formed due to shrinkage and through these cracks the soil dries out deeply so that evaporation losses increase.

In these soils only a part of the soil moisture is available for plants, for at least three reasons:

High water retention, high wilting percentage (WP) and low available moisture content (AMC);

High osmotic potential (Y_s) due to the high concentration of soil solution in comparison to the electrolyte concentration in the plant root tissues;

Extremely low capillary conductivity ($k = 10^{-5}$ to 10^{-6} cm/day). As a consequence of the water extraction by plants a thin film-like depletion zone is formed around the plant roots. If the capillary conductivity is low, this depletion zone will extend to a significant extent within short periods. The flux from the wet soil to the plant roots is extremely slow through this extracted layer. As a result the water supply to plants, especially those

SOIL FACTORS LIMITING THE WATER SUPPLY OF PLANTS

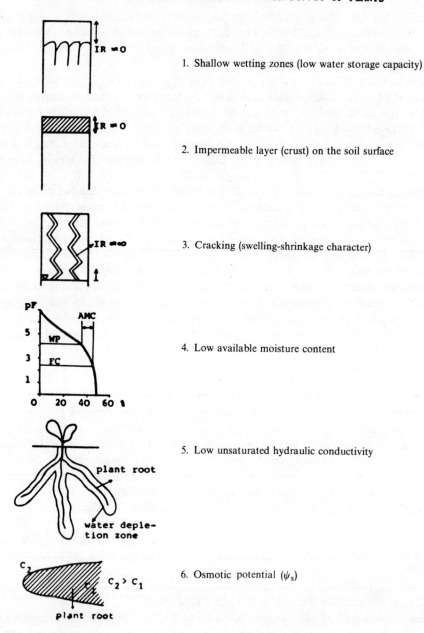

1. Shallow wetting zones (low water storage capacity)

2. Impermeable layer (crust) on the soil surface

3. Cracking (swelling-shrinkage character)

4. Low available moisture content

5. Low unsaturated hydraulic conductivity

6. Osmotic potential (ψ_s)

Fig. 4. Soil factors limiting the water supply of plants.

with widely spaced and sparse root systems, is limited, and they show water deficiency symptoms even if the soil as a whole has considerable moisture content[10, 39].

It can be concluded that in these soils the prevention of salinization and alkalization has great importance and sometimes this is the only way for permanent successful agricultural utilization.

References

1. Balba, M. A. 1975. Predicting soil salinization, alkalization and waterlogging. FAO/ISSS Subcommission on Salt Affected Soils Expert Consultation on Prognosis of Salinity and Alkalinity, Rome.
2. Bresler, E. 1972. Control of soil salinity. *In:* Hillel, D., *Optimising the soil physical environment toward greater crop yields*, pp. 101–132. Academic Press, New York–London.
3. Darab, K. 1962. Application of the principles of soil genetics to irrigation in the Great Hungarian Lowland (H, r, e, g)*. OMMI Genetikus Talajtérképek Sorozat No. 4. Budapest.
4. Darab, K. and Ferencz, K. 1969. Soil mapping and control of irrigated areas. (H, e, r). OMMI Genetikus Talajtérképek Sorozat No. 10. Budapest.
5. Felhendler, R., Shainberg, I. and Krenkel, H. 1974. Dispersion and the hydraulic conductivity of soils in mixed solutions. *Trans. 10th Int. Congr. Soil Sci.*, Moscow, **I**, pp. 103–111.
6. Gardner, W. R. 1960. Soil–water relations in arid and semi-arid conditions. *In: Plant–water relationships in arid and semi-arid conditions*, pp. 37–61. Arid Zone Res. XV. UNESCO, Paris.
7. Globusz, A. M. 1969. Experimental hydrophysics of soils. (R). Gidrometizdat, Leningrade.
8. Hagan, R. M., Haise, H. R. and Edminster, T. W. 1967. Irrigation of agricultural lands. *Am. Soc. Agron.*, Ser. Agronomy No. 11. Madison.
9. Hillel, D. 1971. Soil and water. Physical principles and processes. Academic Press, New York–London.
10. International Source-Book. 1973. Irrigation, drainage and salinity. FAO/UNESCO Hutchinson, Paris.
11. Kovda, V. A. 1947. Origin and regime of saline soils I–II. (R. e. g). Izd. Akad. Nuak SSSR, Moskva.
12. Kovda, V. A. 1954. Geochemistry of deserts in the USSR (R). Izd. Akad. Nauk SSSR, Moskva.
13. Kovda, V. A. 1961. Principles of the theory and practice of reclamation and utilization of saline soils in the arid zones, *Arid Zone Res.*, **14**, 201–213.
14. Kovda, V. A. and Minashine, N. G. 1967. Irrigation and drainage of salt affected soils. (R). Izd. Nauka, Moskva.
15. McNeal, B. L. and Coleman, N. T. 1966. Effect of solution composition on soil hydraulic conductivity. *Proc. Soil Sci. Soc. Am.*, **30**, 307–312.
16. McNeal, B. L. *et al.* 1968. Factors influencing hydraulic conductivity of soils in the presence of mixed-salt solutions. *Proc. Soil Sci. Soc. Am.*, **32**, 187–190.
17. Molen, W. H. van der, 1975. Factors to be considered for prognosis. FAO/ISSS Subcommission on Salt Affected Soils Expert Consultation on Prognosis of Salinity and Alkalinity, Rome.
18. Nielsen, D. R. *et al.* 1972. Soil water. *Am. Soc. Agron., Soil Sci. Soc. Am.*, Madison.
19. Proceedings of the Symposium on Sodic Soils. Budapest 1964. *Agrokémia és Talajtan*, **14**, Suppl. 1–480. 1965.
20. Proceedings of the Symposium on the Reclamation of Sodic and Soda-Saline Soils. Yereven, 1969. *Agrokémia és Talajtan*, **18**, Suppl. 1–392. 1969.
21. Richards, L. A. 1964. Diagnosis and improvement of saline and alkali soils. US Dep. Agr. Handbook No. 60. Riverside.
22. Quirk, J. P. and Schofield, R. K. 1955. The effect of electrolyte concentration on soil permeability. *J. Soil. Sci.*, **6**, 163–178.
23. Salinity problems in the arid zones. 1961. Proceedings of the Teheran Symposium. *Arid Zone Res.*, **14**. UNESCO, Paris.
24. Schilfgaarde, J. van. 1975. Water management and salinity. FAO/ISSS Subcommission on Salt Affected Soils Expert Consultation on Prognosis of Salinity and Alkalinity, Rome.
25. Szabolcs, I. 1961. Effect of water regulations and irrigation on soil formation processes in the region beyond the river Tisza (H). Akadémiai Kiadó, Budapest.
26. Szabolcs, I. 1971. European solonetz soils and their reclamation. Akadémiai Kiadó. Budapest.
27. Szabolcs, I. 1974. Salt affected soils in Europe. Martinus Nijhoff, The Hague—Research Institute for Soil Sci. Agric. Chem. Hung. Acad. Sci., Budapest.
28. Szabolcs, I. 1974. Sodium balance in alkaline soils. *Trans. 10th Int. Congr. Soil Sci.*, Moscow, **10**, 49–56.
29. Szabolcs, I. 1975. Present and potential salt affected soils. FAO/ISSS Submission on Salt Affected Soils Expert Consultation on Prognosis of Salinity and Alkalinity, Rome.

*Denotes languages: In Hungarian (H) with summaries in Russian (r), English (e), German (g).

30. Szabolcs, I. Darab, K. and Varallyay, G. 1969. Methods of predicting salinization and alkalinization processes due to irrigation on the Hungarian Plain. *Agrokémia és Talajtan*, **18**, Suppl. 351–376.
31. Szabolcs, I. Darab, K. and Varallyay, G. 1973. Methods of predicting salinization and alkalinization processes due to irrigation. 9th Eur. Reg. Conf. ICID Q2, R:2.1/5. 1–15. Budapest.
32. Szabolcs, I., Varallyay, G. and Darab, K. 1975. Soil and hydrologic surveys for the prognosis and monitoring of salinity and alkalinity. FAO/ISSS Subcommission on Salt Affected Soils Expert Consultation on Prognosis of Salinity and Alkalinity. Rome.
33. Talsma, T. 1963. The control of saline groundwater. Meded. LandbHoogesch. Wageningen.
34. Varallyay, G. 1968. Salt accumulation processes in the Hungarian Danube Valley. *Trans. 9th Int. Congr. Soil Sci.*, Adelaide, **1**, 371–380.
35. Varallyay, G. 1972. Hydraulic conductivity of salt affected soils in the Hungarian Plain (H, e, r, f). *Agrokémia és Talajtan*, **21**, 57–88.
36. Varallyay, G. 1974. Hydraulic conductivity studies on Hungarian salt affected soils. *Trans. 10th Int. Congr. Soil Sci.*, Moscow, **1**, 112–120.
37. Varallyay, G. 1974. Unsaturated flow studies in layered soil profiles. (H, e, r, f). *Agrokémia és Talajtan*, **23**, 261–296.
38. Varallyay, G. 1974. Hydrophysical aspects of salinization processes from the groundwater. *Agrokémia és Talajtan*, **23**, Suppl. 29–44.
39. Varallyay, G. and Szabolcs, I. 1974. Special water problems in salt affected soils. *Agrochimica*, **18**, 277–287.
40. Waldron, L. J. and Constantin, G. K. 1970. Soil hydraulic conductivity and bulk volume changes during cyclic calcium-sodium exchange. *Soil Sci.*, **110**, 81–85.

EFFECT OF IRRIGATION ON SYR DARYA WATER REGIME AND WATER QUALITY

by V. DUKHOVNY

Director of the Middle Asian Research Institute of Irrigation, USSR

and L. LITVAK

Chief Engineer of the Middle Asian Designing Survey and Research Institute for Irrigation and Drainage Construction, USSR

Summary

The Syr Darya River is one of the major irrigation sources in Middle Asia. Its regulation permits the use for irrigation of a volume of water which exceeds the value of many years' mean river flow. The area under irrigation in the basin grows constantly and has now reached 2.6 million ha, within four Middle Asian Union republics. Nearly 50% of the country's seed cotton and one-fourth of its rice are grown here. Irrigation development in the Syr Darya basin has drastically changed the natural conditions within this area in the following ways:

1. Intensification of irrigation, particularly during the last thirty years, has resulted in disturbing the natural drainage conditions and thus in a rise of groundwater level. In turn, underflow returning to the river network has increased as well as surface inflow from drains.
2. In 1950–1970 intensive drainage was started within the basin territory and this increased the return water; but improvements involving a transition to subsurface horizontal and vertical drainage will somewhat hinder this effect. The proportion of return water is reduced also by improved efficiency which has been accomplished recently. Therefore, while there will be a general increase in the volume of return water due to the expansion of irrigated lands, a decrease is expected in the ratio of return water to head water diversion.
3. A step-type arrangement of the irrigation systems proceeding from the upper part of the basin to the river mouth permits multiple use of return water.
4. Development of irrigation, and particularly of drainage, has changed the chemical composition of the river water, and at some sites mineralization has increased two to three times during the last 60 years. The increase of water mineralization is further due to the creation of large reservoirs, and is particularly pronounced in low-flow years. To control this tendency experiments are being carried out to intensify the intersystem use of saline drainage water and to prevent its return to the river.
5. Large-scale water withdrawal for irrigation has drastically reduced the Syr Darya flow reaching the Aral Sea. Taking into account similar conditions existing on the Amu Darya, this causes the progressive lowering of the water level in the Aral Sea.

Resumé

Le Syr Darya constitue l'une des sources principales d'irrigation de l'Asie Centrale. La régularisation de son écoulement permet maintenant l'utilisation d'un volume d'eau dépassant la valeur de l'écoulement annuel moyen de la rivière. Les zones irriguées du bassin sont en constante extension, et elles atteignent actuellement 2.6 millions d'hectares répartis sur le territoire de quatre Républiques d'Asie Centrale. Près de 50% du coton-graine et un quart du riz produits par le pays proviennent de ces zones. Les progrès de l'irrigation dans le bassin du Syr-Darya ont brutalement modifié les conditions naturelles de la région.

1. L'intensification de l'irrigation, en particulier au cours des 30 dernières années, s'est traduite par une perturbation des conditions du drainage naturel, et par conséquent par une remontée du niveau des nappes. Il s'en est suivi une intensification des apports d'eau sous forme d'écoulement souterrain vers le réseau hydrographique, ainsi qu'une augmentation de la quantité d'eau drainée par les chenaux des rivières.
2. En 1950–1970, on a mis en chantier un drainage intensif de la superficie du bassin contribuant à ameliorer la récupération des eaux. Mais l'amélioration du drainage entraine le passage à un drainage subsuperficiel horizontal et vertical qui retarde dans une certaine mesure la vitesse d'extension des surfaces irriguées. La proportion des eaux de récupération est reduite par l'accroissement récent de l'efficacité des installations. Par conséquent, malgré l'accroissement généralisé du volume des eaux qui retournent au bassin du fait de l'extension croissante des terres irriguées, on s'attend à une diminution du pourcentage de retour de ces eaux du fait de leur détournement en amont pour l'irrigation.
3. La disposition des systèmes d'irrigation en marches d'escalier depuis la partie supérieure du bassin jusqu'à l'embouchure permet un usage multiple des eaux de récupération dans le bassin.
4. Le développement de l'irrigation, et plus particulièrement du drainage, sur les terres irrigables, a provoqué une transformation chimique de l'eau du fleuve. En certains endroits, la minéralisation de l'eau s'est accrue de 2 à 3 fois en l'intervalle de 60 ans. Cet accroissement de la minéralisation de l'eau est imputable en partie à la création de vastes retenues. Ce phénomène est particulièrement sensible lors des années à faible écoulement. Des expériences ont été entreprises pour contrôler ce phénomène, dans le but d'intensifier l'utilisation des eaux de drainage salées, pour éviter que ces eaux ne retournent dans le Syr Darya, et pour faire ainsi décroître sa salinité.
5. Le pompage de l'eau sur une grande échelle à des fins d'irrigation a réduit de façon brutale le volume de l'écoulement qui parvient à la Mer d'Aral. Si l'on tient compte de l'existence de conditions comparables sur l'Amou Darya, le résultat général est un abaissement progressif du niveau de l'eau dans la Mer d'Aral.

Environmental factors

The Syr Darya river is the most important water source for irrigation purposes in the USSR's Middle Asia. Flowing through the territories of four republics (Uzbekistan, Tadjikistan, Kazakhstan and Kirghizia) it supplies water to a huge area where agricultural crops—cotton, rice, melons, grapes and others—are grown. Half of the USSR's raw cotton and a quarter of its rice are produced from the Syr Darya basin. Moreover, the largest industrial complexes of the USSR's Middle Asia, Chirchik, Angren, Tashkent and Fergana plants, which are located in this basin, have enormous water demands. At least half of the total population of Middle Asia lives in this basin; and its annual growth is too high, ranging from 3.6% to 4.2%.

The largest part of the basin has a high mean annual temperature of $+12°C$ to $15°C$ with mean temperature in July of $+26°C$ to $31°C$; mean temperature in January is $+4°C$ to $6°C$. The maximum recorded temperature in the shade was $+48°C$; the minimum temperature $-37°C$. In the lower reach of the Syr Darya river mean annual temperature is $+4°C$ to $6°C$. Wind velocity is up to $40\,m/sec$ and causes high evaporation—up to $1500\,mm/year$. In the summer night-time humidity is 30 to 50%; day-time 15 to 30%. Rainfall fluctuates from $575\,mm$ in the upper collecting area to $100–150\,mm$ in the lower part of the river basin.

According to calculations made by the Sredazgiprovodschlopok Institute the total land resource of the Syr Darya river basin is 44.4 million ha, of which 13.3 million ha (net) are suitable for irrigation (see Table 1). These lands range in altitude from 800 m to 55 m above sea level. The climatic and natural conditions of the lands located in the Fergana Valley, middle course and lower reach of the Syr Darya river and in Chakir, are the most suitable areas for irrigation. Cotton can be grown in most of the basin territory; rice in the lower reaches only.

TABLE 1. *Land resources of the Syr Darya river basin (1000 ha)*

District	Gross area	Unsuitable area	Area suitable for irrigation		Existing irrigated area	Free area
			gross	net		
1. Upper reach of the Narim river	5,228	4,952	276	138	120	18
2. Fergana Valley	8,954	7,092	1,862	1,538	1,235	303
3. Middle course of the Syr Darya river	3,404	1,845	1,559	1,309	576	733
4. Chakir	2,564	1,558	1,006	722	359	363
5. Lower reach of the Syr Darya river	24,242	12,437	11,805	9,554	408	9,146
Total	44,392	27,884	16,508	13,261	2,698	10,563

TABLE 2. *Characteristics of the main rivers of the Syr Darya river basin*

River	Length (km)	Area (km^2)	Average river flow rate (m^3/sec)	Average specific discharge (l/sec/km^2)	Runoff distribution			%
					III–VI	VII–IX		X–II
1. Narim	534	58370	430	7.38	44.9	35.9		19.2
2. Kara Darya	177	24040	270	9.17	51.9	29.8		18.2
3. Soshch	94	3270	43	13.1	23.4	60.5		14.1
4. Chirchik	174	11940	270	20.1	57.4	32.1		15.5
5. Angren	236	4010	43	10.7	75.4	12.9		11.7
6. Arys	339	7170	65	2.07	53	6.5		40.5

TABLE 3. *Characteristics of the Syr Darya river runoff* (m^3/sec)

Characteristics	Runoff through the periods		
	XI–IV	V–X	Total
1. Year with average water probability	588	1613	1101
2. Dry year (1916–1917)	401	957	679
3. Wet year (1968–1969)	801	2842	1821
4. Year with 75% water probability (1944)	537	1304	920
5. Year with 25% water probability (1949)	594	1847	1220

The development of irrigation is limited by water resources which include surface runoff, underground waters and return waters. The surface runoff of the Syr Darya river is estimated as 37.2 cubic km: 52.9% of this is from Narim river, 15.4% from the Kara Darya river and 31.7% from side tributaries of the Syr Darya. The characteristics of the main affluents are given in Table 2, and Table 3 shows the characteristics of the years having various water probabilities. The average specific discharge of the whole basin is 8.1 l/sec/km^2 which is slightly less than the Amu Darya river which originates in glaciers. Syr Darya river has a characteristic alternation of 4–5 year periods of dry and wet years. A situation may occur when two extra dry years run in succession.

Along the total length of the Syr Darya three zones can be distinguished: (1) a runoff collecting, highland area down to the confluence of the Kara Darya river and the Narim river; this zone generates 68% of the Syr Darya runoff, (2) from the confluence down to Chardary site; in this zone water is received from collectors and side tributaries which include the Chirchik and Angren rivers, (3) from Chardary site to the delta; runoff losses of 1.5 km³/year occur in this zone.

History of Irrigation

From time immemorial irrigation systems have existed in the Syr Darya river basin. By the beginning of the twentieth century the total area of irrigated lands was more than 1 million ha of which the Fergana Valley contributed 750,000 ha. Cohakir district and the middle course of the Syr Darya had 250,000 ha. As a result of the First World War and Civil War the irrigated lands decreased to 700,000 ha by 1922, but they had been redeveloped by 1928. In this period water was provided mainly from the tributaries without impoundments. With the establishment of Soviet power the irrigation systems have been reconstructed and equipped with new installations. In 1933–1936 construction of the first dams were begun on the Kara Darya, Chirchik and Soshch rivers, and intensive development took place in the Fergana Valley and the Golodnaya (Hungary) Steppe in the period of 1937–1941. During this time the largest irrigation channels were constructed: Big Fergana channel, Kirov channel, Tashkent channel and others. By the beginning of the Great Patriotic War (World War II) the Syr Darya basin had 1.3 million ha of irrigated lands from which 900,000 ha were in the Fergana Valley. During the War (1941–1945) construction of the irrigation systems was continued, especially in the Golodnaya Steppe and Tashkent zone. The first dam with a hydropower station (Farshchad water complex) was built in 1943–1948. This complex provided a gravity water supply to the largest part of the Golodnaya Steppe and Dalverskaya Steppe. The Kairakum reservoir was built in 1950–1956; the Chardan reservoir in 1958–1963.

TABLE 4. *Comparison of water resources available and water demands* (km³) *in the Syr Darya river basin*

Categories of water resources and water demands	Years			
	1970	1971	1972	1973
1. Surface water resources	40.8	35.9	34.5	38.9
2. Return waters	13.3	13.7	13.6	14.7
Total water resources	54.1	49.6	48.1	53.6
3. Ground water which can be extracted without harmful effects	7.5	7.5	7.5	7.5
Total water resources available	61.6	57.1	55.6	61.1
4. Irrigation draft	38.1	39.3	39.0	42.4
5. Water intake for industrial purposes	2.6	2.8	3.0	3.1
6. Water intake for municipal purposes	1.1	1.2	1.3	1.4
7. Fish demands	3.5	3.0	3.5	3.5
Total water	45.3	46.3	46.8	50.4

These reservoirs allowed irrigation in the areas of the lower and middle course of the Syr Darya river where irrigation has been especially carried out over the last 20 years: approximately 280,000 ha of irrigated lands have been developed in the Golodnaya Steppe, 260,000 ha in the lower reach of the Syr Darya, 130,000 ha in the Fergana Valley. The irrigated area rose to 2.7 million ha by 1975 and this drastic increase demanded a supply of water which exceeded the surface resources available in some years. Table 4 shows that over the last four years the water demands have been met only by usage of underground and return waters in addition to the regulation river runoff.

TABLE 5. *Reservoirs in the Syr Darya basin*

Reservoir	Sources (rivers)	Capacity (km^3)		Character of stream flow control
		Total	Available	
1. Toktogul reservoir (under construction)	Narim	19.5	14	multi-year control
2. Andijan reservoir (under construction)	Karadarya	1.8	1.6	multi-year control
3. Kairakum reservoir	Syr Darya	4.0	2.6	seasonal control
4. Chardar reservoir	Syr Darya	5.7	4.7	seasonal control
5. Charvarksk reservoir	Chirchik	2.0	1.6	seasonal control
6. Reservoirs in Fergana Valley	tributaries	1.6	1.4	seasonal control
7. Reservoirs in Chakir area	tributaries	0.7	0.6	seasonal control
8. Reservoirs in Artur area	tributaries	1.7	1.4	seasonal control
Total		37.0	27.9	

At present seasonal flow control is being carried out by means of a cascade of large reservoirs constructed on the Syr Darya river and its tributaries. After completion of construction of the Toktogul reservoir and Andijan reservoir multi-year flow control will also be provided (see Table 5). In the exceptionally dry year of 1974 the water demands were met by using the dead storage of the reservoirs (3.8 billion m^3) and by application of drainage waters for irrigation purposes (1.8 billion m^3).

Water withdrawal and return

Return waters provide an essential resource in the Syr Darya basin; in 1965 they were 32% of water intake volumes; 36% in 1973. In order to forecast future trends in the use of return water it is necessary to understand how they are generated. An increase of natural river flow can be divided into two components: natural and artificial. The natural increase is generated by inflows entering the hydrographic network from the catchment area, as streams or groundwater discharge. The artificial increase is from return waters generated by artificial draining of the catchment area, which is carried as groundwater or by installed collectors and drains, directing water back to the river. It is worth noting that the downthrow waters entering drains from the irrigation systems through escape canals or directly from the fields form an integrated part of the return waters. A lot of scientists are trying to separate these from other return waters, but from qualitative and quantitative points of view this is impossible to do; in fact, the drainage waters and downthrow waters which enter the collectors are mixed and discharged as one body into the hydrographic network. As regards their further usage it doesn't matter how these return water resources have been generated, either from the drains or direct from the fields. Generation of the return waters (together with the downthrow waters) can be expressed as follows:

$$Wb = \overline{\bigwedge} + \sqcap\!\!\sqcup + C; \quad (1)$$

where: Wb = volume of return waters;

$\overline{\bigwedge}$ = inflow into hydrographic network which is considered as natural drain;

$\sqcap\!\!\sqcup$ = drainage waters discharged by the drainage units;

C = downthrow waters entering the water receiving collectors.

The first component of equation (1) depends upon permeability of the soils, hydrogeological structure, availability of the natural water and position of the water table. Since all the natural factors are stable over a time, the dynamics of this part of the return waters will be mainly affected only by the levels of ground and underground waters in the drainage area drained by rivers and, therefore, the dynamics of the return waters depends upon the man-made processes to the least extent.

The second component of the equation (1) directly depends upon the availability of drainage systems in the basin, the type of system, and the area which it serves. The third component is determined by the irrigation equipment, and water application technique as well as proper management of the systems.

All the above-mentioned can be expressed by the following equation:

$$Wb = f_1 \Sigma h_i ipb \ F + f_2 qgp \ Fn + P_3 \ W \ (1-\eta_{mn})(1-\eta_{opi}) \quad (2)$$

where: h_i = water table in area Fi;

qgp = drainage specific discharge;

F = area drained by river;

n = factor of availability of drainage systems in the area;

W = water supply for irrigation purposes;

η_{mn} = efficiency of irrigation technique;

η_{opi} = management efficiency of irrigation system.

Taking into account the fact that the drainage waters can be expressed as part of water supply, and at the same time the water table can be expressed as a function of annual water probability, equation (2) can be transformed into the following:

$$Wb = a_1 p + W[a_2 n + a_3 (1-\eta_{mn})(1-\eta_{opi})] \quad (3)$$

From this one can find the percent ratio of the return waters and irrigation draft which is as follows:

$$\frac{Wb}{W} = \frac{a_1 p}{W} + a_2 n + a_3 (1-\eta_{mn})(1-\eta_{opi}) \quad (4)$$

where: p = annual water probability;

a_1, a_2, a_3 = appropriate factors of proportionality.

Analysing equation (4) one can deduce that the return waters are increasing with the increase of irrigation draft and availability of drainage systems in the area, but the return waters are decreasing with an improvement in the state of drainage and irrigation and their operating conditions.

In this connection the forecasts show that there will be no increase of return waters in the middle and lower reaches of the Syr Darya river, because the designed density of drainage systems has been already achieved. Drainage activity is now directed to constructing vertical and horizontal closed drainage systems. The ratio of the return water to supply water will decrease by lowering the drainage discharges and raising the efficiency of the systems. As a result, the 1990-year outlook on return water resources is now estimated at the level of 32% in the Syr Darya river basin.

In order to analyse the process of generation of the return waters Table 6 and Fig. 1 give the dynamics of growth in its volumes over the period 1936–1973 in the Fergana Valley. The table shows that the irrigation draft and availability of drainage are the main factors determining the return waters. Before extending the drainage systems in this valley the return waters were 9–16% of the irrigation draft and they increased to 48% in 1973. It is interesting to note that "accounted downthrows" increased with development of the drainage network in the period 1950–1960 and they achieved a maximum proportion of 43.8% in the period 1966–1970; after that they decreased to 39.7% owing to introduction of new vertical and horizontal closed drainage systems. The proportion of accounted downthrows to total return waters fluctuated from 3.3 to 13.6%. It is clear that this share of the return waters decreased in the wet years and increased in the dry years, and this can be explained by the lower water level of the river in the dry years and thus more inflow from the surrounding area.

TABLE 6. *Analysis of interconnection between return waters and various factors in the Fergana Valley Basin*

Years	Inflow into valley (km³)	Irriga- tion draft (km³)	Area (1,000 ha)	Return waters (km³)	Account- ed down- throws (km³)	Differ- ence between return waters and down- throws (km³)	Percent- age of return waters	Avail- ability of drainage systems (linear m/ha)	Effici- ency of drain- age system	Percent- age of account- ed down- throws	Percent- age of non- account- ed down- throws
1. 1936–1940	22.29	10.8	822	0.72	0.10	0.62	7.0	5.6	0.45	1.0	6
2. 1941–1945	26.7	11.3	860	1.89	1.40	0.49	16.7	7.8	0.45	12.3	4.4
3. 1945–1950	25.4	11.6	820	3.3	1.72	1.58	28.6	12.7	0.45	14.8	13.6
4. 1951–1955	31.1	12.0	877	3.96	2.89	0.93	33.0	14.8	0.47	24.0	9
5. 1956–1960	29.4	15.1	952	6.08	4.44	1.64	40.02	16.2	0.49	29.4	10.8
6. 1961–1965	23.3	16.4	983	7.02	5.31	1.71	42.8	18.8	0.52	32.3	10.5
7. 1966–1970	30.9	17.8	1022	8.40	7.80	0.6	47.1	22.8	0.56	43.8	3.3
8. 1971–1973	25.5	18.5	1187	9.0	7.36	1.64	48.6	24.9	0.58	39.7	8.8

A reduction of water inflow connected with lowering the water table in the river valley is similarly explained. Such a high ratio of the return waters occurs only in the Fergana Valley due to the specific hydrogeological, soil and topographic conditions there. In other parts of the basin—Chakir, Golodnaya Steppe, lower reach of the river and Dalverzskaya Steppe—the ratio of the return waters is much less. The Golodnaya Steppe represents a special interest: in recent years the share of return waters is 39% and has a tendency to decrease in spite of extended drainage systems in this area. At the present time in the total of 530,000 ha of irrigated lands 208,000 ha have horizontal closed drainage systems and

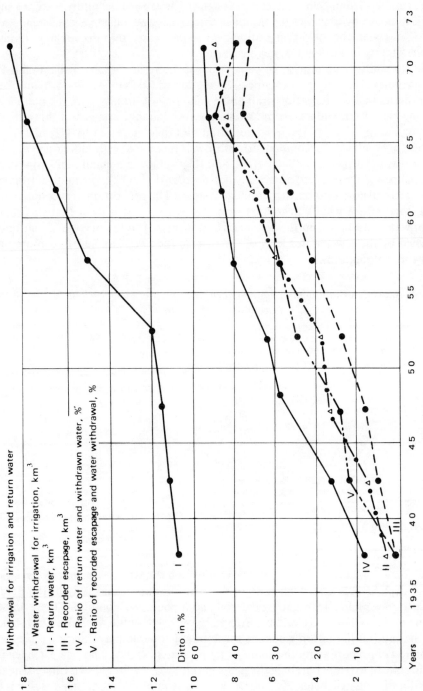

FIG. 1. Withdrawal for irrigation and return water.

126,000 ha have vertical drainage systems. A substantial proportion of water used for irrigation is drawn from wells.

In the cotton-producing zone of the Syr Darya river up to the Chardarin storage reservoir
the return waters are at present estimated as 33–36%, and by 1990 will be reduced to 28–
30% due to construction of vertical and horizontal closed drainage and to increasing the
efficiency of the drainage systems up to 0.75–0.8. In the rice-producing zone, from Chardary
site to Kazalinsk, the return waters are only 22–25% because the river itself and the narrow
strip of irrigated lands adjacent to it charge the ground waters of surrounding non-irrigated
lands.

The irrigated lands which descend in altitude from the Fergana Valley to the lower reach
of the river provide a multi-purpose, stepped usage of return waters. A substantial part
of the return waters, however, having a salt content greater than 2 g/l, is used only in dry
years, and necessitates an increased number of leaching operations.

The intensive development of irrigation provides both the return waters and a considerable change of runoff along the course of the river. Although from 1910 up to now the
runoff from the collecting highland area has not changed, at a site located in the lower
reach of the river, near Kazalinsk city, the runoff has decreased by 34% in 1961–70 and by
47% in 1971–1974 as compared with the period of 1910–1938.

The storage reservoirs (existing and under construction) are of great importance for runoff
redistribution within the seasonal and multi-year periods. At the same time they consume
runoff to meet water losses caused by increased infiltration through the bottom of reservoirs,
evaporation losses from the water surface and water volumes needed for initial filling of
reservoirs. At present the annual evaporation and infiltration losses of all the reservoirs
are $1.5 \, \text{km}^3$ and by 1990 these losses will be 1.9–$2.0 \, \text{km}^3$. In the period 1975–1980, 3.6–
$3.8 \, \text{km}^3$ of water will be required for filling new reservoirs and in 1981–1987, 3.1–$3.3 \, \text{km}^3$
of water will be required for this purpose. Thereafter water will be discharged only to meet
necessary water demands and evaporation and infiltration losses. In the wet year of 1969
the reservoirs were used to prevent the hazardous flooding in the lower reach of the Syr
Darya river. Twenty km^3 of runoff were discharged into Arnasausk lowland through the
emergency by-pass canal of the Chardarinsk storage reservoir.

Factors of irrigation

It has been shown that the total water consumption for irrigation purposes is determined
by many factors. This can be expressed by the following equation:

$$Op = \left[\frac{(U + Tp) - Oc - Mip.bof}{Kng_{mn}} + Nnp \right] Kng$$

where: Op = irrigation rate
$U + Tp$ = total evapotranspiration;
Oc = rainfall;
$Mip.bof$ = addition from ground waters;
Nnp = washing discharges;
Kng_{mn} = efficiency of irrigation technique;
Kng_{me} = efficiency of system.

The first two terms of this equation are determined by the climatic conditions of the area (temperature, rainfall, deficit of saturation, frequency and strength of winds) as well as by the crop rotation and its yield productivity.

Science and practice have determined that the semi-automorphic regime is the optimum meliorative regime for the greater part of Middle Asia. This requires minimum water discharges for washing operations; it maintains a water table under which, on the one hand, a downthrow of surplus waters below the root zone is not allowed while irrigating and, on the other hand, upward movement of salts is prevented.

The savings of irrigation water due to control systems in the drainage network now average 2000 to 2500 m^3/ha. The efficiency of the total system is now on average 0.62–0.63 in the Syr Darya river basin and 0.78–0.80 in the Golodnaya Steppe. This has been achieved by systems having leakproof coatings, flumes, pipelines, etc. Efficiency of technique has been considerably raised also by using the sprinkler irrigation method and flexible pipelines.

As a result of such measures it is now suggested that the gross irrigation rate be reduced from 15,300 m^3/ha to 11,000 m^3/ha in the Syr Darya river basin and this will allow extension of the irrigation system in the future. It is worth noting that the increase of irrigation is accompanied by a reduction of evaporation losses caused by natural vegetation: in the period 1928–1932 useless evaporation losses were 4.7 km^3, and useful losses were 9.0 km^3; in the period of 1963–1967 the corresponding volumes were 1.6 and 17 km^3, respectively; in 1971–1974 0.9 and 21 km^3. There is a tendency also for the total water consumption by vegetation to decrease in relation to the total river runoff, and this can be explained by the irrigation draft, generation of return waters, and construction of storage reservoirs.

During the 60-year period of intensive development of the Syr Darya river basin mineralization of the river water has drastically changed along the river course. In 1912 total mineralization of the river water at two gauging stations located in the upper and lower reaches, Kal and Jazalinsk stations, was 0.3 to 0.37 g/l respectively, and at the present time in the upper site (confluence of Narim and Karadarya rivers) mineralization is 0.64 g/l and in the lower site 1.6 g/l. The river water has on average a small amount of sulphates and chlorides: sulphate content 0.242 g/l and chlorides 0.038 g/l. Proceeding downstream the total solids increase up to 0.9 g/l on average at the Kzil–Kishlak gauging station. From Kairakum to Begovat the runoff from a great number of collectors increases mineralization, on average from 0.32 g/l to 0.95 g/l. The content of sulphates and chlorides also increases to 0.39 g/l, and up to 0.108 g/l, respectively. Below the Begovat station total solids increase to 1.2 g/l at Nadezda.

Mineralization of the river is sharply affected by the return waters and the construction of storage reservoirs, which may raise the salt content by 3% to 7%. Mineralization drastically increases in dry years, for instance in 1974–1975, when mineralization was 1.4–1.6 g/l in the middle course of the river and more than 2 g/l in the lower reach. Measures now proposed to reduce the mineralization of return waters include intensified usage of drainage water inside the system by mixing the return water with irrigation water.

Another measure is to adopt the optimum meliorative regime for the area, because large quantities of salts reach the drains from washing waters, even where the groundwater is of low salt content (2–3 g/l). On testing plots located in Khorezm 40–56 t/ha of salts have been washed out and brought into the active zone, and on the Fergana testing plot 20–32 t/ha of salts. After changing to the semi-automorphic regime the salt exchange decreased to 11–25 t/ha and 8/15 t/ha respectively on these testing plots. Thus development of the drainage systems in order to apply the optimum meliorative regime lowers the salt content of the return waters.

The factors mentioned above have resulted in a considerable reduction of the Syr Darya river's inflow entering the Aral sea. It has caused decrease of the size of the Aral sea and increase of its salt content. In the period 1910–1938 a mean multi-year inflow from the Syr Darya river to the Aral sea was $15 \, km^3/year$; in the period 1961–1970 it was reduced to $9.78 \, km^3/year$ and in 1971–1973 to $8.01 \, km^3/year$. This increasing deficit has raised a question of transferring $15–25 \, km^3$ of the Siberian rivers' runoff into the Syr Darya river basin to provide for future development of agriculture, a work which is planned to be done by 1990–1995. The transfer of this volume of fresh water into the Syr Darya river basin coupled with further development of irrigation is estimated to increase the net volume of the return waters by $5.8 \, km^3/year$. If all this return water is directed to the river, it will considerably retard the lowering of the Aral sea level.

OBSERVATIONS SUR DEUX SOLS ARGILEUX (VERTISOLS) IRRIGUES DEPUIS 25 ANS DANS LA PLAINE DU GHARB (MAROC)

by P. ANTOINE

Professeur Assistant à l'Université du Minnesota, détaché à l'Institut Agronomique et Vétérinaire Hassan II, Rabat

et A. DELACOURT

Ingénieur Agronome, Enseignant à l'Institut Agronomique et Vétérinaire Hassan II, Rabat

Note. The full paper, in French, which includes detailed descriptions of soil profiles and their interpretation, was included in the pre-printed papers for the Alexandria symposium. For copies refer to the authors.

Résumé

Les propriétés morphologiques actuelles et l'évolution pédogénétique de deux vertisols irrigués situés dans la plaine du Gharb (Maroc) semblent avoir été surtout influencées durant les vingt dernières années par les alluvionnements consécutifs aux inondations de 1963 et 1970, et par les pratiques culturales (spécialement la riziculture ou pratique de la jachère).

Sous l'action des phénomènes de vertisolisation (autobrassage) une réincorporation nette des sédiments récents dans la masse du sol ancien s'observe actuellement. Ce phénomène est parallèle à une tirsification (mélanisation) et à une évolution notable de la stabilité structurale des dépôts récents. La pratique de la riziculture paraît induire un début de formation "d'horizon placique" discontinu. Ce phénomène semble assez rapide mais réversible.

Enfin, la pratique de la jachère sur ce type de sol contribue à une augmentation assez élevée des taux de sodium échangeable. L'entraînement de l'excès d'ions Na$^+$ au moyen des eaux d'irrigation dont la qualité est satisfaisante devrait permettre d'éliminer cette contrainte culturale potentielle.

Summary

Recent deposition of sediments (severe floodings in 1963 and 1970), vertisol-forming processes, and agricultural practices (mainly rice cultivation and fallow) have had a major influence, during the past twenty years, on the properties of two irrigated soils (Typic Chromoxererts) located in the Gharb plain of Morocco. Argillopedoturbation has resulted in the reincorporation by the soil matrix of many recent sediments. This phenomenon parallels an increase in the degree of melanisation and in the stability of the structure. Early formation of fragments of a placic horizon typical of rice-cultivated soils can be observed. That formation seems to be rapid, but reversible.

The practice of fallow results in an increase of exchangeable sodium ions in those soils. The use of available irrigation waters (non alkaline) before the commencement of a new cycle of cultivation practices may solve the problem.

CHEMICAL COMPOSITION OF IRRIGATION WATER AND ITS EFFECT ON CROP GROWTH AND SOIL PROPERTIES

by D. R. BHUMBLA

Indian Council of Agricultural Research—New Delhi, India

Summary

Out of a total cultivated area of about 141 million hectares in India 31 million hectares are irrigated through canals, tanks and wells. The irrigation by wells using groundwater has increased by nearly 6 million hectares during the last 20 years. The canals' and tanks' waters have generally very low salinity and sodium hazards but the chemical composition of waters varies a great deal. In the alluvial regions with rainfall more than 60 cm the groundwaters have low salinity hazard but in some cases may have high sodium hazard because of the high residual sodium carbonate (more than 5 me/l). Most of the waters in the arid areas have high salinity hazard but usually have low sodium hazard. Some of the waters being used for irrigation may have electrical conductivity of more than 10,000 micromhos. The relative proportion of sodium, magnesium and chloride increases with increasing salt content. Generally magnesium is more than calcium. Boron content may vary from less than 1 ppm to as much as 5 ppm. Highly saline water may also contain high amount of nitrate. In sandy areas waters of 6000 to 8000 micromhos conductivity can be used for tolerant crops like wheat and mustard (*Brassica campestris*), but even in light-textured soils, sensitive crops suffer. In clay and clay loam soils, water with salt content of more than 2000 micromhos proves hazardous.

Résumé

Aux Indes, sur une surface totale cultivée de 141 millions d'hectares, 31 millions sont irrigués. L'irrigation se fait à l'aide de canaux, de réservoirs et de puits. Ces derniers ont provoqué, au cours des 20 dernières années, un accroissement de 6 millions d'ha de la surface irriguée en utilisant les eaux souterraines. L'eau des canaux et des réservoirs présente généralement une salinité faible et contient peu de sodium, mais sa composition chimique varie largement. Dans les régions alluvionnaires recevant annuellement moins de 60 cm de pluie, le danger de salinisation des eaux souterraines est petit, mais en certains cas, elles peuvent contenir beaucoup de sodium en raison d'une forte teneur en carbonate de sodium (plus de 5 me/l.). La plupart des eaux des régions arides risquent de présenter une salinité élevée mais généralement ne risquent pas d'être riches en sodium. Des eaux utilisées pour l'irrigation peuvent avoir une conductivité électrique de plus de 10.000 micromhos. La proportion relative en sodium, magnesium et chlorures croît avec la teneur en sels. Généralement le magnesium l'emporte sur le calcium. La teneur en bore peut aller de moins d'1 ppm à plus de 5 ppm. Les eaux hautement salines peuvent également contenir une grande quantité de nitrates. Dans les régions sableuses, des eaux d'une conductivité de 6.000 à 8.000 micromhos peuvent être utilisées pour les cultures tolérantes comme le blé ou la moutarde (*Brassica campestris*). Même dans des sols à texture légère, les cultures sensibles souffrent. Dans les sols argileux et limons-argileux, des eaux salines d'une conductivité de plus de 2000 micromhos surviennent rarement.

Introduction

Out of a total cultivated area of about 141 million hectares in India 31 million ha are irrigated, canals, tanks and wells being the main sources of water. The total annual recharge of groundwater was estimated by Raghaw Reo in 1962 as 42.57 million hectare metres. The total annual canal seepage is about 5.46 million ha m, assuming an average seepage of 115 ha m per running km of canal and 38.2 ha m per km of the distributory. After deducting the losses due to the sub-soil flow the net availability has been estimated to be 27.0 million ha m.

TABLE 1. *Area irrigated by different sources in India (million hectares)*

Year	Canals	Wells	Tanks	Others	Total
1950–1951	8.29 (39.8)	5.98 (28.7)	3.61 (17.3)	2.97 (14.2)	20.85
1955–1956	9.38 (41.3)	6.74 (29.6)	4.42 (19.4)	2.21 (9.7)	22.76
1960–1961	10.37 (42.1)	7.29 (29.6)	4.56 (18.5)	2.44 (9.8)	24.66
1965–1966	10.95 (41.7)	8.65 (32.8)	4.27 (16.2)	2.47 (9.4)	26.34
1970–1971	12.52 (40.0)	11.85 (37.8)	4.54 (14.5)	2.40	31.29

Figures in parenthesis denote percentage of the total.

TABLE 2. *Net area irrigated by sources, 1970–1971—Statewise (thousand hectares)*

State/union Territory	Canals	Tanks	Wells	Others	Total
Andhra Pradesh	1,579	1,112	509	113	3,313
Assam	362	—	—	210	572
Bihar	814	169	551	626	2,160
Gujarat	207	30	962	10	1,209
Haryana	951	1	575	5	1,532
Himachal Pradesh	(a)	(a)	1	90	91
Jammu & Kashmir	272	(a)	1	6	279
Karnataka	421	365	259	92	1,137
Kerala	211	73	5	142	431
Madhya Pradesh	710	130	562	78	1,480
Maharashtra	312	225	821	69	1,427
Manipur	—	—	—	65	65
Maghalaya	—	—	—	37	37
Nagaland	—	—	—	12	12
Orissa	263	583	45	258	1,149
Punjab	1,292	—	1,591	5	2,888
Rajasthan	756	270	1,083	23	2,132
Tamilnadu	884	898	774	36	2,592
Tripura	—	—	—	22	22
Uttar Pradesh	2,498	371	4,034	287	7,190
West Bengal	960	303	17	209	1,489
Union Territories	25	7	44	9	85
All India	12,517	4,537	11,834	2,404	31,292

From *Indian Agriculture in brief*, 13th ed., 1974.

The irrigated area increased by more than 12 million ha during the period 1950–1951 to 1970–1971, and nearly 50% of this increase was due to wells (Table 1). Well irrigation has certain obvious advantages: the period for completion is much shorter than for the big irrigation projects and control is with the farmer who has not to depend upon water from a canal. Secondary salinization as a result of rise in water table is much less common than in canal irrigated areas. Well irrigation has been practised for centuries in India using bullock labour for water lifting, but bullocks are gradually being replaced by small diesel or electric pumps. About two-thirds of the well irrigated area is in the alluvial flat plains of the Uttar Pradesh, Punjab, Haryana, Bihar and parts of Rajasthan (Table 2).

Exploitation of groundwater is mainly in the alluvial region; in other areas it is not possible because of either the poor quality of water (high salinity or sodium hazard) or poor rate of recharge. The water quality problem is well illustrated in the State of Haryana. In the eastern part of this State groundwater is of satisfactory chemical composition and the water table is going down because of the over exploitation by tube wells; but in the canal irrigated areas in the Western part, where the groundwater is saline, the water table is rising because of the seepage from canals and in many areas land has gone out of cultivation because of secondary salinization.

Chemical composition of irrigation waters

Considerable work on the chemical composition of the waters used for irrigation has been carried out during the last twenty years. Rivers of the north of India fed from snow of the Himalayan mountains have low salt content—usually less than 500 micromhos (Paliwal, 1972); calcium and magnesium are the predominant cations, with calcium usually more than magnesium. Though waters contain HCO_3, SAR is very low. In South India the salt content of some rivers varies a great deal not only from season to season but also at different points along the rivers. In the Western part of the country the salt content of some seasonal rivers in the arid areas of Gujarat and Rajasthan may be high, particularly in periods of low flow. Bapat and Shukla (1971) have reported salt content in some of the rivers as very high, but such rivers are not used for irrigation, and the salt content of canal waters is generally low (Table 3). The water of most of the tanks used for irrigation has low salinity and sodium but some have high bicarbonate content, the source of which is usually the sodium felspar in the rocks or soils of the catchment area.

Generally the groundwaters in areas with rainfall of 70 cm or more in alluvial regions have low salt content; the chloride and sulphate concentration decreases with an increase in rainfall (Table 4). They may contain relatively higher proportions of carbonate and bicarbonate and present problems of sodium hazard (Sharma et al. 1975). Detailed investigations on the source of carbonate and bicarbonate have not been carried out. In the rainfall zone of 40–60 cm groundwater may show considerable difference in chemical composition at different depths, particularly if the aquifers are not connected. In some areas as in parts of Haryena the salt content at depths down to 80 cm is low but the waters in deeper layers are saline.

In arid and semi-arid areas the chemical composition of groundwaters have been given in a number of reviews (Bhumbla 1969, Paliwal 1972). The electrical conductivities of the waters in these areas may vary from 1000 to 16,000 micromhos. Waters of low salt concentration are usually in the beds of rivers of earlier times. Introduction of canal irrigation

TABLE 3. Quality of waters of some canals in North India

Name of canal	No. of samples	pH	EC (micro-mhos/cm)	SAR	Na$^+$	Ca^{++}	Mg^{++}	Cl	SO$_4^-$	CO$_3^-$	HCO$_3^-$
						me/l					
1. Lower Ganges canal, Etah	9	8.2	365	1.45	1.54	1.78	0.44	0.24	1.35	0.13	2.00
2. Lower Ganges canal, Farrukhabad	14	7.9	289	0.40	0.41	1.31	0.82	0.15	0.32	0.10	1.81
3. Lower Ganges canal, Kanpur	23	8.5	444	2.52	2.35	1.15	0.60	0.26	0.32	0.34	2.74
4. Lower Ganges canal, Unnao	18	8.0	438	1.20	1.30	2.20	1.20	0.80	2.20	0.30	2.00
5. Yamuna canal, Mathura	1	7.4	280	1.02	0.85	1.40	0.70	0.50	—	—	2.50
6. Rajasthan canal	4	7.8	200	1.02	0.68	1.45	0.59	0.47	—	—	1.67

TABLE 4. Mean composition and quality of ground waters of Punjab

District	Normal rainfall (cm)	No. of samples tested	EC (mmhos/cm)	Carbonate (me/l)	Bicarbonate (me/l)	CL (me/l)	Sulphur	Ca+Mg (me/l)	Na (me/l)	SAR	RSC	Boron (ppm)	NO$_3$ (ppm)
Ferosepur	40.5	1,702	2,062	0.35	8.52	5.78	5.77	5.80	14.82	8.70	3.37	—	—
Shatinda	52.2	1,151	2,668	0.77	10.15	10.72	5.04	8.20	18.48	9.14	2.72	—	—
Kapurthala	54.2	45	899	—	5.59	2.19	1.21	3.65	5.31	2.40	1.91	—	—
Sangrur	58.8	400	1,521	0.80	11.21	15.75	−10.59	3.50	11.67	8.10	8.00	0.53	27.68
Amritsar	63.5	533	1,014	0.43	8.47	1.73	−0.49	3.76	6.38	4.60	5.14	—	—
Patiala	70.0	76	1,006	0.20	6.59	2.19	1.08	5.01	5.05	3.10	1.78	—	—
Jullundur	70.0	143	768	0.30	6.63	2.47	−1.72	4.60	3.08	1.71	2.33	—	—
Ludhiana	70.6	642	810	0.45	6.89	1.41	−0.66	4.92	3.18	2.00	2.43	—	—
Rupar	100.0	31	770	—	6.72	1.06	−0.08	5.01	2.69	1.70	1.71	—	—
Hoshiarpur	104.7	91	856	—	6.59	1.90	0.07	3.67	4.89	5.26	2.92	—	—
Gurdaspur	112.7	66	794	—	5.49	2.54	−0.09	4.43	3.51	2.35	1.06	—	—

TABLE 5. *Average composition of irrigation waters of various salt content* (me/l)

Electrical conductivity (mm/cm)	Ca^+	Mg^{++}	Na^+	K^+	CO_3^- HCO_3^-	Cl^-	SO_4^-	NO_3^- (ppm)
<2	2.8	4.8	7.1	0.3	6.1	5.2	3.8	1.4
2–4	4.0	8.5	17.8	0.3	9.4	17.4	8.5	3.3
4–6	7.5	14.3	27.8	0.6	9.4	28.3	11.5	3.1
6–8	11.2	22.9	37.6	1.0	8.6	42.7	18.0	2.6
8–10	14.0	27.5	51.3	1.1	8.6	62.8	19.6	5.1
10–12	15.8	35.9	61.7	2.4	12.4	78.2	22.5	4.4
12–14	18.3	40.1	74.0	2.5	6.1	102.0	17.9	4.1
7–14	23.8	50.6	95.1	2.0	8.0	133.8	28.0	8.5

Source: Abrol *et al.* (1972).

in these areas may result in improvement of groundwater quality as reported by Sharma *et al.* (1975). In other areas, where the soils contain salt layers at shallow depths, the lateral movement of saline water in the soil has rendered some of the wells unsuitable for irrigation.

In the groundwaters of arid and semi-arid areas sodium constitutes 50–60% of the total cations, and magnesium is generally more than calcium. The relative concentration of magnesium increases with increasing salinity. Amongst anions the concentration of carbonate and bicarbonate remains more or less constant, sulphate increases slightly with increase in salinity, and chloride concentration increases sharply (Table 5). Some of the highly saline waters in these areas contain high nitrate content and may add 20–30 kg N/per ha per irrigation. The boron content may vary from 1 to 5 parts per million.

Criteria for judging the quality of irrigation water

The earliest classification was that of Asghar *et al.* (1936) who considered all waters with salt content of less than 10 me/l salt as suitable for irrigation and all waters with more than 20 me/l as unsuitable. Water between these limits was considered suitable or unsuitable depending upon the relative amounts of sodium and calcium. They proposed a salt index which, however, proved inadequate as most of the waters which had been used for centuries without any harmful effects were classified as unsuitable.

In the late fifties the criteria as given in USDA handbook 60 for judging the quality of irrigation water were adopted in many laboratories, but it was soon realized that these criteria were not applicable under conditions prevailing in most parts of India. Kanwar (1961) proposed a modification to the USDA classification by suggesting that C_4 and C_5 conductivity classes be used for waters with EC × 106 of 2250–5000 and 500–20,000 respectively. He also suggested that, while judging the quality of irrigation water, the texture of the soil and the salt tolerance of the crop should also be taken into consideration. Inclusion of all waters with EC × 106 of 5000 and above in one class presented problems. In sandy areas waters of EC × 106 of 5000 could be used for irrigating tolerant crops, but in the same areas most waters containing salts above 10,000 micromhos created serious problems. A few that were used contained either very high amounts of nitrate or relatively high proportions of calcium. Moreover, out of the sixteen classes originally proposed by USDA six (C1S2, C1S3, C1S4, C2S4, C4S1, C4S2) are seldom found in nature.

Recently the workers in this field in India got together and, on the basis of available data on the effect of quality of irrigation water on crop growth and soil properties, they proposed the following classification (Bhumbla and Abrol 1972).

Nature of soil	Crops to be grown	Upper permissible limit of EC of water for safe use for irrigation
Deep black soils and alluvial soils having clay content greater than 50%.	Semi-tolerant	1500
Soils that are fairly to moderately well drained.	Tolerant	2000
Heavy-textured soils having a clay content of 20–30%. Soils that are	Semi-tolerant	2000
well drained internally and have a good surface drainage system.	Tolerant	4000
Medium-textured soils having a clay content of 10–20%. Soils that are	Semi-tolerant	4000
very well drained internally and have a good surface drainage system.	Tolerant	6000
Light-textured soils having a clay content less than 10%. Soils that have	Semi-tolerant	6000
excellent internal and surface drainage.	Tolerant	8000

In connection with the above classification it should be noted:

1. It is presumed that the groundwater table at no time of the year is within 1.5 m from the surface. If the water table should rise to the root zone the limits need to be reduced to half the values stated.
2. If the soils have impeded internal drainage, either on account of hard pans, unusually high amounts of clay or other morphologic reasons, the limit of water quality should likewise be reduced to half.
3. If the water has more than 70% soluble sodium gypsum should be added to soil occasionally.
4. If supplemental canal irrigation is available, waters of high electrical conductivity could be used profitably.

It must be pointed out also that there is need for greater research effort so that the criteria evolved could be modified to suit particular soils, crops and climatic conditions.

Effect of irrigation water on crop growth and soil properties

Information about the effect of irrigation water on soil properties and crop growth in different climatic conditions is meagre. Bhumbla (1969) reported that in sandy soils with rainfall of about 60 cm appreciable reduction in yield of wheat grain and maize fodder occurred only when the EC of the water was more than 11,000 micromhos.

Detailed investigations under the Coordinated Project on the Use of Saline Waters in Agriculture sponsored by the India Council of Agricultural Research have been started in the different agro-climatic regions of the country on different crops, and some resulting data are given in Table 6. The soil of the experimental station at Jobner is sandy to loamy sand, at Agra sandy loam to loam, and at Indore and Siruguppa clay loam to clay. The mean annual rainfall at Jobner is about 60 cm and at Agra, Siruguppa and Indore 70 cm, 65 cm and 95 cm respectively.

In the light-textured soils the yield of crops like wheat is affected very little even when irrigation water has electrical conductivity of 4000–6000 micromhos. The accumulation

TABLE 6. *Effect of saline water on the grain yield of different crops*

E.C (mmhos/cm)	Jobner	Agra		Indore		Siruguppa	
				Grain yield (C/ha)			
	Wheat	Wheat	Mustard	Wheat	Maize	Wheat	Sorghum
0	—	39.0	27.2	39.1	—	13.3	—
2	36.9	42.2	26.1	35.2	51.3	11.2	27.0
3	—	—	—	—	—	—	—
4	31.7	40.5	24.6	34.4	44.2	12.2	21.2
6	—	38.4	24.1	31.4	39.8	—	—
8	33.0	36.2	17.0	28.8	37.3	3.7	13.9
10	—	—	—	—	—	—	—
12	30.0	27.7	—	—	—	—	—
16	28.1	12.2	—	20.7	31.8	0.4	4.4
C.D at 5%	2.7	3.7	—	2.8	4.6	3.8	10.9

of salts is rather low because the salts accumulated during winter are leached during the subsequent rainy season. The water table in absence of canal irrigation may be as low as 20 m. In heavier textured soils, even if the irrigation water has electrical conductivity as low as 2000 micromhos, sensitive crops suffer because of the poor internal drainage. In arid areas it is a common practice to grow crops like wheat and barley in one season with highly saline water and leave the land fallow for 2–3 seasons to leach the accumulated salts. As mentioned above, many of these highly saline waters contain nitrates which may add 20–30 kg N/ha with each irrigation. The exact role of high amount of nitrate in salinity tolerance by plants is not clearly understood (Paliwal 1972).

It has also been reported that boron concentration of more than 2 ppm may affect adversely the growth of pea and barley, and more than 1 ppm of mustard and wheat (Paliwal 1972). But apart from the absolute concentration, the relative concentration of Ca and B in the plant tissue is important: growth is affected adversely if the Ca:B ratio is less than 200 (Deo and Ruhal 1971).

The effect of irrigation water on soil properties is dependent not only on chemical composition of the water but also on permeability of the soil and on the climatic conditions. Singh and Kanwar (1963) reported that lack of internal drainage resulted in salinity build

TABLE 7. *Effect of irrigation water on salt and boron content*

Location	EC $\times 10^6$		Boron (ppm)	
	Irrigation water	Saturation extract EC $\times 10^6$ (0–15 cm)	Irrigation water	Saturation extract
Bhoorekunhan	462	720	0.30	1.60
Bhikhiwind	970	2,180	0.20	1.82
Sursingh	1,080	2,400	0.08	2.75
Khemkaran	1,206	800	0.30	1.42
Bhikhiwind	1,295	1,550	0.20	1.95
Sangwen	1,300	1,820	0.30	1.90
Sursingh	1,600	3,000	0.31	2.35
Bhikhiwind	1,729	1,150	0.50	1.50
Khemkaran	1,859	3,600	0.46	1.90
Average	1,278	1,913	0.29	1.91

up even with the use of otherwise good quality water (EC 1500 micromhos). Their data indicate that, whereas the salt content of the saturation extract was about one and a half times that of the irrigation water, the boron concentration was about seven times (Table 7).

Singh and Bhumbla (1968) reported the results of a study in which they had analysed soil samples collected from the fields irrigated with saline waters for a period of 5–20 years in Hissar district of Haryana. The mean annual rainfall of the district is about 45 cm most of which is received during three months, July to September. The temperatures range from near freezing in winter to as high as 45°C in June. They reported that accumulation of salt depended upon the clay content of the soil: in the light-textured soil (10% clay) the salt content of the saturation extract was less than half of the irrigation water but in soils with more than 20% clay it was more than 1.5 times. The effect of texture and the salt accumulation in the surface layers is shown below:

Clay content	Value of r	Y
10%	0.64	$0.43x+0.47$
10–20%	0.61	$0.77x+2.19$
20%	0.86	$1.52x+0.75$

where x is the electrical conductivity of the irrigation water and Y the electrical conductivity of the saturation extract.

The effect of SAR on the soil ESP under field conditions is dependent on the associated anion. If the predominant anion is chloride there is very little increase in ESP if the soil contains calcium carbonate, as is generally the case in most of the arid and semi-arid areas. If the associated anion is carbonate and bicarbonate the ESP increases rapidly and the soil becomes alkaline. It has generally been the experience that sodium hazard is high if the residual sodium carbonate (RSC) of the irrigation water is more than 5 me or if the soluble sodium is more than 70%. Kanwar and Kanwar (1968) showed that residual sodium carbonate due to CO_3 is more hazardous than that due to HCO_3. The research information on the effect of associated anions is inadequate.

Generally the boron content of most waters is less than 2 ppm except in parts of Rajasthan where its concentration may be as high as 5 ppm. As the leaching of boron is difficult it accumulates much faster than other ions and may present a serious problem, but there have been very few reports of boron toxicity under field conditions.

References

Abrol, I. P., Bhargava, G. P., Sharma, R. C. and Singhla, S. K. 1972. Studies on ground water quality and soils of Mathura district, U.P. Report No. I. CSSRI, Karnal.
Asghar, A. G., Puri, A. N. and Taylor, E. M. 1936. Soil deterioration in irrigated areas of Punjab, Part I. *Punjab Irrig. Res. Inst.* Lahore, Pub. 4, No. 7.
Bapat, M. V. and Shukla, P. J. 1971. Geochemistry of surface waters and their utilization in arid regions with special reference to Kutch region of Gujarat State. Paper presented at Seminar on Water Resources on Rajasthan and Gujarat.
Bhumbla, D. R. 1969. Water quality and use of saline waters for crop production. Symposium on soil and water management, ICAR, pp. 87–108.
Bhumbla, D. R. and Abrol, I. P. 1972. Is your water suitable for irrigation? *Indian Fmg.* 22, 15–17.
Bhumbla, D. R., Kanwar, J. S., Mahajan, K. K. and Bhajan Singh. 1969. Effect of irrigation water with different sodium and salinity hazard on the growth of crops and the properties of the soil. Symposium on Arid Zones, Jodhpur.

Deo, R. and Ruhal, V. S. 1971. Effect of salinity on the yield and quality of rape (*Brassica campestris*). *Indian J. Agri. Sci.* **41**, 134–136.

Kanwar, J. S. 1961. Quality of irrigation water as an index of its suitability for irrigation purposes. *Potash Rev.* **24** (13).

Kanwar, J. S. and Kanwar, B. S. 1968. Quality of irrigation waters, *9th Intl. Cong. Soil Sci.* **1**, 391–402.

Paliwal, K. V. 1972. Irrigation with saline water. IARI, New Delhi.

Raghaw Reo, K. V., Raju, T. S. and Ramesan, V. 1969. Approximate estimate of the ground water potential of India. Symposium on soil and water management. ICAR, New Delhi, pp. 8–12.

Sharma, R. C., Abrol, I. P. and Bhumbla, D. R. 1975. Quality of ground waters in Jind district. *Haryana Bull.* No. 3. CSSRI, Karnal.

Singh, S. S. and Kanwar, J. S. 1963. Boron and some other characteristics of well waters and their effect on the boron content of the soil in Patti (Amritsar). *J. Indian Soc. Soil Sci.* **11**, 283–286.

Singh, B. and Bhumbla, D. R. 1968. Effect of quality of irrigation water on soil properties, *J. Res., PAU.* **5**, 166–169.

DISCUSSION AND CONCLUSIONS

compiled by I. SZABOLCS

(Budapest)

Soil as an essential part of the natural environment is basic to irrigated agriculture. Soil changes due to irrigation influence not only the irrigated fields, but also the agricultural area and the whole biosphere of the region concerned. Consequently the conservation and increase of soil fertility are of paramount importance: the soil conditions determine the necessity, possibility and efficiency of irrigation, as well as the possibilities of a proper soil-water management.

Irrigation is of vital importance in the agriculture of arid and semi-arid regions supplying crops cultivated on fertile soil with adequate water. But arid land irrigation exists in only a few cases without also producing undesirable side-effects on the soil. Among these water-logging, salinity and alkalinity are the most widespread and harmful. The papers discussed here analysed these problems in relation to irrigation and drainage, and considered methods of predicting and preventing harmful processes, including numerical methods for forecasting, modelling and controlling them. The balance of water and salts in soils, closely related to the subjects of Section III, and the influence of irrigation on the natural environment, are constantly emphasized.

Following the printed papers, Prof. K. R. Ramanathan (India) described the problems of water resource management in the arid and semi-arid areas of north-west India where drought conditions occur on average once every two or three years. Surface water resources are limited during the dry season, and the main water source (80% of the total) is ground-water which is saline to varying degrees. Active ameliorative measures are being taken to increase the supply of freshwater from the rivers and to recharge the depleting underground aquifers. Steps are also being taken to reduce contamination of the rivers passing through large cities.

In the lively discussions which pointed up the many-sided soil aspects of arid land irrigation in developing countries, the following took part: S. M. Abusadda (Kuwait); S. Alsouli (Saudi Arabia); A. Arar (FAO—Egypt); A. M. Balba (Egypt); J. E. Blom (Netherlands); M. A. Farid (WHO); K. E. Hansson (United Nations); A. F. El Kashef (Egypt); J. Khouri (Egypt); M. B. A. Malik (FAO—Egypt); I. Massoud (FAO); S. Mohamed (Egypt); A. Osman (ACSAD—Syria); and R. E. Quelenec (UNESCO).

The main topics of the discussion were the unfavourable side-effects of arid land irrigation, namely secondary salinization, alkalization and waterlogging, the efficiency of irrigation in the developing countries of the arid zone, the soil–plant–water relationships, and the present and future problems of proper land and water management. Most questions arose from Kovda's introductory paper and El Gabaly's report on water and land use problems in the 23 countries of the Middle and Near East Region.

The general conclusion was that the improvement of operation and maintenance practices in existing and future irrigation projects has vital importance, not only from the viewpoint of a more rational use of water resources, but also as a precondition of the prevention, regulation, or at least moderation of the hazard of waterlogging, salinity and/or alkalinity. The improvement of irrigation efficiency involves precise flow regulation, proper irrigation methods, steps to exclude or limit water losses (evaporation, filtration, seepage), and to prevent salinization of surface waters by groundwater infiltration. Better water management necessitates better land management, also more rational use of land resources and more efficient agricultural practices. Outstanding problems of modern agricultural development included land fragmentation, old-fashioned farming traditions, inadequate land use, cropping patterns and intensities, silting of canals, and salting of water sources and reservoirs. To increase irrigation efficiency at farm level it seems necessary to make the farmers pay for the water, giving them technical guidance and demonstrations.

The evaluation of soil and water resources needs a world-wide land and water survey, as an integrated approach by hydrologists, engineers, soil scientists and agronomists:

> to elaborate a comprehensive prognosis system for salinity, alkalinity and waterlogging;
> to analyse more thoroughly the many-sided plant–soil–water relationships, including such problems as quality of irrigation waters and salt tolerance of crops;
> to evaluate the influence of irrigation on the soils' salt balances, exchangeable cations and physical properties;
> to elaborate possibilities for diminishing filtration and evaporation losses;
> to determine the conditions of using saline and waste waters for irrigation purposes, and preventing salinization and alkalization from the groundwater.

On the basis of the papers and the discussions the following conclusions were drawn and recommendations made. Firstly, a world-wide analysis and mapping is necessary for the evaluation of the present situation and for forecasting the probable changes due to irrigation. In this respect such World-Projects as the World Soil Map, World Map of Salt Affected Soils, World Map of Desertification and the World Map of Soil Deterioration—initiated and realized by various international organizations, UNESCO, FAO, UNEP, ISSS, SCOPE, MAB, etc.—have to be encouraged, completed and made available to their users for various arid land irrigation projects.

Having a general idea of the existing and predicted processes and phenomena, a detailed soil and hydrologic survey prognosis and monitoring system needs to be elaborated which can be an exact scientific basis for planning and execution of irrigation systems. The steps taken in this respect were summarized during the FAO Expert Consultation (Rome, June 1975) on the "Prognosis of Salinity and Alkalinity," and the elaboration of tentative guidelines was accepted. The first version of these guidelines was presented at the meeting of the Subcommission on Salt Affected Soils of ISSS at Texas, USA, in August 1976 on "Saline water irrigation."

Session IV of the Symposium concluded that the methods of survey, field and laboratory analysis, processing data, mapping, monitoring, and the advisory service for recommendations of a proper salinity–alkalinity control, proper water and soil management, need to be standardized. International efforts are needed for the complex analyses of the changes in the hydrologic cycle and soil properties due to irrigation.

International standards need to be adapted and elaborated for these methods in the various regions. In this respect special graduate and post-graduate courses and international meetings have special importance. Even though there are some basic books available on these topics, new books containing information and tentative guidelines are much needed.

The FAO–UNDP Regional Land and Water Use Project for the Near East and North Africa is the first attempt to solve problems of land and water use by an integrated approach based on pilot studies which incorporate a number of special practices. In addition, this will review and appraise present and potential resources and identify constraints to better use. In the light of the findings from this project, the principles and practices can be extended to other areas.

SECTION V

Effects of Irrigation on Biological Balances

LA PRODUCTION DE POISSONS DE CONSOMMATION DANS LES ECOSYSTEMES IRRIGUES

par JACQUES DAGET

Professeur au Muséum d'Histoire Naturelle de Paris

Résumé

L'auteur rappelle brièvement les conditions écologiques que les poissons rencontrent dans les écosystèmes irrigués. Il fait le point des connaissances actuelles sur la rizipisciculture. Concernant l'utilisation des canaux d'irrigation, il suggère de faire des essais d'élevage en cages flottantes avec *Clarias lazera*. Enfin il met en garde contre l'emploi à des fins phytosanitaires de certains produits particulièrement nocifs pour les poissons (Endrin et Lindane notamment).

Summary

Ecological conditions faced by fish in irrigated ecosystems are briefly recalled by the author. He then sums up recent advances in rice fish farming. With regard to the utilization of irrigation watercourses, he suggests that *Clarias lazera* rearing experiments could be made by means of floating cages. Warnings are given against the phytosanitary use of some substances (namely Endrin and Lindane) very harmful for fish.

Introduction

Dans son rapport final (1971, projet 4), le Conseil International de Coordination du Programme MAB déclarait: "Il convient d'accorder une attention particulière aux sols de la zone aride qui ne peuvent être exploités sur une grande échelle que par irrigation ... L'irrigation pénètrera de plus en plus dans les zones semi-arides et même dans les zones semi-humides, permettant d'obtenir des rendements élevés dans des cultures telles que celles du blé, du riz, du maïs ... Il convient par conséquent d'envisager les écosystèmes irrigués dans une perspective interdisciplinaire et d'examiner leurs aspects biosociaux aussi bien que leurs aspects biologiques."

Or, jusqu'à présent, si les retenues d'eau, lacs de barrage et réservoirs de toutes sortes, destinés à alimenter les systèmes d'irrigation ont fait l'objet d'études poussées de la part des écologistes et des aménagistes piscicoles, peu d'attention a été portée aux canaux d'irrigation eux-mêmes et à leur utilisation possible pour la production de protéines animales sous forme de poissons. Une mention particulière doit cependant être faite de la rizipisciculture, connue et pratiquée de longue date dans un certain nombre de pays. Elle constitue en fait un cas particulier puisqu'elle utilise non pas les canaux d'irrigation mais les soles cultivées, noyées sous 10 à 50 cm d'eau durant tout le cycle de végétation du riz.

Dans un écosystème irrigué, l'eau amenée par un canal de dérivation circule dans un ensemble de ramifications primaires, les répartiteurs, elles-mêmes subdivisées en ramifications secondaires, les distributeurs, avant de s'infiltrer dans le sol s'il s'agit de cultures sèches (blé, maïs, coton, canne à sucre) ou de s'étendre sur le sol s'il s'agit de cultures noyées (riz). Les canaux, de section de plus en plus faible, sont tronçonnés par des vannes qui peuvent être ouvertes ou fermées à la demande suivant les besoins des cultures situées en aval. L'eau est donc tantôt courante et tantôt stagnante. Le niveau varie peu, au moins dans le canal de dérivation et dans les répartiteurs qui doivent rester en charge pour que l'eau puisse circuler facilement par gravité dans les distributeurs. La profondeur est variable mais faible dans l'ensemble, de sorte que la température de l'eau est susceptible de s'élever notablement dans la journée sous l'effet de l'insolation, d'autant plus que les digues qui bordent les canaux et délimitent les parcelles ne portent aucun arbre pouvant donner de l'ombre. Les canaux et leurs rives sont éventuellement pourvus d'une végétation aquatique ou semi-aquatique souvent jugée indésirable parcequ'elle gêne l'écoulement, diminue le débit et constitue des gîtes favorables à la prolifération des vecteurs de certaines maladies telles que le paludisme et la bilharziose.

D'après l'expérience personnelle de l'auteur qui a séjourné 12 ans au Mali, à proximité immédiate de l'Office du Niger, les canaux d'irrigation et les rizières ne constituent pas pour les poissons des milieux écologiques notablement différents de ceux rencontrés dans les plaines inondées naturellement par la crue des fleuves locaux. Les associations spécifiques que l'on y observe sont comparables mais le nombre des espèces présentes et la taille des individus diminuent en même temps que l'importance et la profondeur des canaux. Il est donc certain qu'un écosystème irrigué, alimenté par une retenue d'eau ayant un peuplement de poissons suffisamment riche et diversifié, ce qui est le cas des peuplements naturels équilibrés, est immédiatement colonisé par les espèces locales déjà préadaptées à vivre et à se reproduire dans ce milieu particulier. Les alevins et les petites espèces, qui se laissent facilement entraîner par le courant, pénètrent dans les rizières avec l'eau d'alimentation. L'expérience a toutefois montré que cet empoissonnement naturel est en définitive très peu productif, les rendements étant de l'ordre de quelques kilos à l'hectare pour les 3 à 4 mois que dure le cycle de végétation du riz. La production est plus élevée dans les canaux profonds qui restent en charge toute l'année. Elle pourrait y être équivalente à celle des retenues d'eau, généralement estimée entre 50 et 100 kg/ha/an.

Rizipisciculture

La rizipisciculture consiste à empoissonner, avec des alevins d'espèces et de tailles sélectionnées, des rizières spécialement aménagées en vue d'obtenir une production de poissons qui vienne s'ajouter au bénéfice que le cultivateur retire de sa récolte du paddy. Elle est pratiquée au Japon, en Indonésie, aux Philippines, en Malaisie, au Viet-Nam, en Italie, aux Indes, en Tanzanie et à Madagascar. Les techniques varient naturellement avec les pays, les climats et les conditions locales. Une abondante littérature existe sur ce sujet, à laquelle on pourra se reporter pour de plus amples détails. Nous nous bornerons ici à rappeler les aspects écologiques essentiels du problème.

Les espèces les plus souvent utilisées en rizipisciculture sont:

— la Carpe commune, *Cyprinus carpio*, à régime omnivore, dont on peut faire doubler le poids en deux mois environ, à condition de la nourrir;
— *Trichogaster pectoralis*, à régime planctonophage, qui possède des organes de respiration aérienne lui permettant de supporter une certaine désoxygénation dans les eaux peu profondes et surchauffées;
— diverses espèces du genre *Tilapia*, à régime également planctonophage, résistantes et bien adaptées aux milieux tropicaux.

Ces poissons sont couramment utilisés pour la pisciculture en étangs, de sorte que les techniques pour en obtenir la reproduction, pour les nourrir, les stocker, les transporter, sont parfaitement connues. Dans les rizières, ils trouvent un milieu particulièrement favorable à leur développement pour les raisons suivantes:

— les terres irriguées destinées à la culture du riz sont en principe assez fertiles, sinon naturellement au moins du fait des engrais, de la fumure et des soins culturaux qu'elles reçoivent. Sont extrêmement importants à ce point de vue l'assec annuel suivi d'un labour et d'un hersage qui ont l'avantage d'aérer le sol et de provoquer la minéralisation de la vase, ainsi que les sarclages qui remettent en circulation et en solution dans l'eau une certaine quantité de sels nutritifs;
— le phytoplacton, le zooplancton, le périphyton ainsi que les têtards et les insectes aquatiques sont souvent plus abondants dans les rizières que dans les étangs. Tous ces organismes utilisent une partie des éléments nutritifs et concurrencent de ce fait le riz sans aucun profit pour l'homme. Bien que les biomasses soient faibles, la production peut devenir assez élevée en raison d'un turn over très rapide. Par exemple, les larves de Chironomides sont souvent présentes dans le sol des rizières avec de fortes densités. Elles se nourrissent de matières végétales et jouent un rôle important dans la décomposition rapide de la paille enfouie dans le sol des rizières. De récentes recherches (Dejoux, 1974) ont montré qu'en milieu tropical le cycle de vie larvaire de beaucoup d'espèces est de l'ordre de 15 à 20 jours et le rapport P/B journalier de l'ordre de 0,24, soit un P/B atteignant 29 pour une durée de végétation du riz de 120 jours. Il y a évidemment tout intérêt à transformer en poisson bon pour la consommation humaine toute la production de matière vivante animale ou végétale autre que le riz.

Les poissons nettoient la rizière en la débarrassant du phytoplancton, de la végétation adventice, des insectes etc. Les excréments qui tombent au fond sont rapidement décomposés par les bactéries et les sels nutritifs qui en résultent vont enrichir le sol. C'est par ces mécanismes que l'on explique les majorations de rendement en paddy de 5 à 15% qui ont été parfois observés dans des rizières empoissonnées par rapport à des rizières témoins sans poissons.

La présence de poissons sélectionnés dans une rizière est donc doublement bénéfique. Le riz en profite directement. En outre, une partie des éléments nutritifs que le riz ne peut utiliser et qui seraient perdus autrement, sont valorisés par les poissons qui les transforment en protéines consommables par l'homme. A Madagascar, où des essais ont été entrepris depuis 1960 pour mettre au point une technique rationnelle de rizipisciculture, on estime que l'on peu compter sur les rendements moyens suivants: 20 à 30 kg/ha/120 jours sans fumure ni distribution de nourriture, 80 à 200 kg/ha/120 jours avec une bonne fumure mais sans nourriture, 200 à 400 kg/ha/120 jours avec une bonne fumure et distribution de nourriture complémentaire sous forme de rations journalières équivalant à 1/10e environ du poids total du poisson existant dans la rizière (Moreau, 1972).

En contrepartie de son intérêt, la pratique de la rizipisciculture implique un nombre de sujétions qui pourraient même dans certains cas être considérées comme des inconvénients dirimants.

1) Les premiers jours après le repiquage, les plants mal enracinés peuvent être déchaussés par les poissons qui fouillent plus ou moins le sol pour y rechercher leur nourriture et en déloger des larves d'insectes. Ce type de comportement devient sans inconvénient dès que les plants sont bien enracinés. Il équivaut à de petites façons culturales superficielles et favorise plutôt le tallage du riz. Il est cependant recommandé d'attendre une huitaine de jours après le repiquage pour empoissonner un rizière.

2) Il est indispensable que chaque parcelle soit munie d'un drain périphérique et d'un trou-refuge situé au point le plus bas lorsque la parcelle n'a pas une planéité parfaite. Le drain périphérique est un fossé de 20 à 30 cm de large, de 30 à 40 cm de profondeur, qui entoure la parcelle au pied des diguettes. Ces dernières peuvent d'ailleurs être construites ou renforcées par la terre provenant du creusement du drain. Le trou-refuge peut avoir 80 cm de diamètre et autant de profondeur. Il doit communiquer avec le drain. Le but de ces dispositifs est de fournir un refuge aux poissons durant les heures chaudes de la journée et de leur permettre, lors des assecs temporaires ou de la vidange finale, de se rassembler dans le drain ou le trou-refuge où leur capture est plus facile.

3) Il faut prévoir un système de grillage à mailles fines pour empêcher les alevins d'espèces sauvages de pénétrer dans la rizière avec l'eau d'alimentation. Ces espèces sont en effet de nature à concurrencer avec succès, pour l'utilisation de la nourriture, les espèces sélectionnées dont on veut faire l'élevage mais sans présenter les mêmes avantages au point de vue de la production. De même, il faut prévoir une grille à la sortie pour empêcher les poissons de s'échapper lorsqu'on baisse le niveau de l'eau ou que l'on vidange les parcelles.

4) Après le repiquage et durant quelques jours, on maintient en général une mince lame d'eau, parfois moins de 5 cm, dans la rizière. Or, pour empoissonner, il faut attendre que l'épaisseur d'eau soit au moins de 10 cm. Si la profondeur est trop faible, l'eau risque de s'échauffer dans la journée, ce qui oblige les poissons à se réfugier dans le drain périphérique et le trou-refuge où les effets de l'insolation se font sentir de façon moins sévère. Pour obtenir une production intéressante, il est préférable d'augmenter progressivement la hauteur d'eau jusqu'à 30 ou 40 cm, au fur et à mesure que les poissons grossissent.

5) Il est bon de disposer d'étangs de stockage, situés à proximité des rizières, afin d'y mettre les poissons qui, au moment de la récolte du paddy, n'ont pas atteint une taille jugée suffisante pour la consommation. Ces poissons peuvent être gardés ainsi en attente jusqu'à ce qu'on puisse les remettre dans une parcelle en culture pour y bénéficier d'une seconde période de croissance. Lorsque les cultures de riz sont annuelles, un assec des parcelles d'une durée d'un mois est généralement suffisant. Il en résulte qu'entre une récolte de paddy et la mise en culture suivante, on peut inonder la parcelle qui joue alors le rôle d'étang et dans laquelle on peut remettre les poissons.

6) Enfin, et c'est souvent le plus difficile à réaliser, il est indispensable de disposer d'un nombre suffisant d'alevins d'espèces et de tailles convenables pour empoissonner les rizières dès que la hauteur d'eau y est suffisante afin de ne rien perdre de la durée du cycle de végétation du riz pour la croissance des poissons.

Utilisation des canaux d'irrigation

Les canaux d'irrigation étant tronçonnés par des vannes, on pourrait penser qu'une portion de canal située entre deux vannes puisse être utilisée comme un bassin de pisciculture. Or, il n'en est rien car les canaux, qui doivent rester en charge, ne peuvent être vidangés pour la récolte du poisson comme le sont les étangs d'élevage. De plus, il est impossible d'empêcher les espèces sauvages de s'y implanter et de s'y multiplier. Si en effet on munissait les vannes de grillages pour empêcher les alevins de passer, il faudrait que les mailles soient très étroites, ce qui réduirait le débit, d'autant plus que ces grillages se colmateraient très vite. L'adjonction de grilles ou de grillages aux vannes pour contrôler l'entrée et la sortie des poissons apparaît donc incompatible avec la nécessité d'assurer une bonne circulation de l'eau, de maintenir un débit régulier et de respecter les plans d'utilisation de l'eau, toujours très stricts dans les systèmes irrigués.

Jusqu'à présent, les canaux étaient donc considérés comme des milieux sans intérêt piscicole particulier bien que les pêcheurs locaux y capturent des poissons comme dans les rivières et les mares naturelles. Une technique relativement récente, l'élevage des poissons en cages flottantes permettrait peut-être d'utiliser plus rationnellement l'eau qui circule dans les canaux ayant une certaine profondeur. Les essais dont l'auteur a connaissance ont tous été faits dans des lacs ou des étangs. Ils mériteraient d'être repris dans des écosystèmes irrigués. Les premiers, datant de 1968, concernaient la Carpe commune. Les cages avaient une capacité de 6,5 m^3. Elles recevaient une charge de 200, 400 ou 600 alevins au m^3. La température moyenne de l'eau au cours des essais était de 24°,2 C. Les poissons nourris artificiellement sont passés de 28 à 216 g en 116 jours ou de 50 à 268 g en 93 jours. La production moyenne avec une densité de 600 poissons au m^3, ce qui est très supérieur aux rendements atteints avec les méthodes de pisciculture habituelles (Steffens, 1969). Des essais avec la Carpe commune ont été récemment repris en Israël avec des cages de 1 m^3 et 200 poissons par cage. Les résultats sont très satisfaisants à condition d'assurer une bonne oxygénation de l'eau et une nourriture abondante (Shiloh et Viola, 1973).

Aux Etats-Unis, on s'est adressé au "channel catfish." Les premiers essais effectués au lac White Oak ont donné de bons résultats (Collins, 1970, 1971). Des études récentes ont été faites avec des cages de 1 m^3 et une charge de 300 alevins. Ceux-ci sont passés de 25 à 450 g en 117 jours. La température de l'eau oscillait entre 24 et 32°C et les poissons recevaient chaque jour une quantité appropriée de nourriture sous forme de granulés flottants. La mortalité était faible, 90 à 98% des alevins ayant survécu.

Dans cette technique d'élevage en cages flottantes, l'eau sert de véhicule à l'oxygène nécessaire pour la respiration des poissons. Elle sert également à l'évacuation des déchets de nourriture et des excréments. Il est nécessaire de nourrir régulièrement une fois par jour et de surveiller attentivement les cages pour enlever les individus morts ou malades de façon à éviter les risques d'épidémie. Le principal danger, avec les fortes densités au m^3 utilisées, est le manque d'oxygène surtout lorsque les élevages sont pratiqués en eau stagnante. Or, dans des canaux d'irrigation, ce danger serait minimisé puisque l'eau y circule, sauf quand toutes les vannes sont fermées, ce qui n'arrive guère dans le canal de dérivation et rarement dans les grands répartiteurs, les seuls qui se prêteraient à l'installation de cages de 1 m^3. Des essais devraient d'ailleurs être faits avec *Clarias lazera* qui possède des organes de respiration aérienne et peut, de ce fait, résister à une désoxygénation du milieu. Cette espèce a déjà retenu l'attention des pisciculteurs des zones tropicales. On connaît ses habitudes alimentaires (Imam, Roushoy, Philisteen, 1970) et si la production d'alevins en grandes quantitiés n'est pas encore définitivement résolue, elle ne devrait pas tarder à l'être.

Incidence des traitements phytosanitaires

Les traitements phytosanitaires par insecticides et herbicides sont absolument indispensables dans les écosystèmes irrigués car il est impératif d'obtenir des cultures le rendement le plus élevé possible. Or ces traitements ne peuvent qu'aller à l'encontre d'une augmentation de la production piscicole. Puisqu'ils ont pour effet de détruire les insectes et la flore adventice, ils réduisent nécessairement la quantité globale de nourriture naturelle disponible pour les poissons, surtout dans les rizières. Cet inconvénient apparaît cependant minime dans la mesure où, comme on l'a vu plus haut, il est nécessaire de distribuer un appoint de nourriture artificielle pour obtenir des rendements intéressants.

Les risques d'une action directe sur les poissons sont au contraire loin d'être négligeables et les essais de laboratoire ne sont pas toujours suffisants pour apprécier l'importance exacte du danger. Sur le terrain en effet, les doses qui atteignent effectivement les poissons sont quelquefois supérieures à celles théoriquement prévues et l'action nocive du produit employé peut être renforcée et aggravée par celle de la température, du manque d'oxygène ou de tout autre facteur synergique. Welcomme (1971) a effectué pour le compte de la F.A.O. des essais de toxicité concernant 4 formulations d'insecticides et 1 d'herbicide en usage dans les rizières du Dahomey. Ses conclusions peuvent se résumer ainsi:

— l'Endrin, en raison de sa toxicité propre et de sa persistence dans le milieu aquatique, est très dangereux pour la faune. Son utilisation doit être prohibée dans les zones irriguées si l'on désire y maintenir des poissons;
— le Lindane à 5%, sous forme de granulés, à la dose de 16 kg/ha, est moins nocif mais cependant à déconseiller;
— le Synexa 50 à la dose de 2,5 kg/ha et le Synexa 25 à la dose de 20 kg/ha présentent moins de danger et peuvent être utilisés sous contrôle strict;
— l'herbicide TOK, destiné à débarrasser les canaux de la végétation aquatique flottante, ne présente au contraire aucun inconvénient pour les poissons.

Des observations concordantes sur les effets nocifs du Lindane ont été faites à Madagascar (Moreau, 1972). Ce produit est épandu dans les rizières pour lutter contre les chenilles du "borer" blanc (*Maliarpha separatella*). Il est utilisé sous forme de granulés contenant 6% de Lindane et répartis sur le sol des rizières à la dose de 50 kg/ha. Lorsque l'épandage, comme c'est souvent le cas, est fait uniquement sur la sole cultivée au cours d'un assec de quelques heures, le traitement ne cause aucun dommage aux poissons qui se sont momentanément réfugiés dans le drain périphérique et le trou-refuge. Au contraire, lorsque l'épandage a lieu en pleine eau, on contate des mortalités atteignant parfois 7 à 8% avec la Carpe commune et davantage avec les diverses espèces de *Tilapia*.

Enfin, il convient de rappeler que les produits non biodégradables sont parfois stockés en concentrations importantes dans certains organes des poissons ce qui n'est pas sans présenter des inconvénients pour les consommateurs éventuels.

Conclusions

Pour un aménagiste piscicole, l'eau qui circule dans un écosystème irrigué, en plus de son rôle primordial vis-à-vis des cultures, est susceptibles de produire une quantité de protéines animales, consommables par l'homme sous forme de poisson, qui est loin d'être négligeable. Néanmoins, pour tirer parti de cette potentialité, il est indispensable que certains aménagements soient prévus sur le terrain et que certaines précautions soient prises dans les modalités d'irrigation et les traitements phytosanitaires. Il est donc nécessaire, pour arriver à une solution assurant la rentabilité maximale de l'écosystème global, que les responsables envisagent les problèmes d'aménagement des zones irriguées sous tous leurs aspects. Il ne faut surtout pas négliger l'intérêt du cultivateur puisqu'en faisant intervenir le poisson de consommation dans l'équilibre biologique des zones irriguées, on cherche essentiellement à améliorer sa ration alimentaire et son budget familial.

References

Collins, R. A. 1970. Cage culture of catfish in reservoir lakes. Proc. 24th Annual Conf. S.E. Association Game and Fish Commissioners. Atlanta, 489–496.

Collins, R. A. 1971. Cage culture of catfish in reservoirs. *Resour. Publ. U.S. Bur. Sport Fish Wildl.* **102**, 115–123.

Dejoux, C. 1974. Synécologie des Chironomides du lac Tchad (Diptères, Nématocères). Thèse Doct. Etat, Paris. 171 pp. multigr.

Iman, A. E., Roushoy, H. M. and Philisteen, A. 1970. Feeding of catfish *Clarias lazera* in experimental ponds. *Bull. Inst. Oceanogr. Fish U.A.R.* **1**, 205–222.

Moreau, J. 1972. Perspectives offertes par la rizipisciculture à Madagascar. *Terre malgache, Tany Malagasy*, **14**, 227–242.

Shiloh, S. and Viola, S. 1973. Experiments on the nutrition of carp growing in cages. *Bamidgeh* **25** (1), 17–31.

Steffens, W. *et al.* 1969. Results of cage rearing on carp in cool water from power stations. *Z. Fisch.* **17**, 45–77. 77.

Welcomme, R. L. 1971. The toxicity of four insecticides and a herbicide under tropical conditions. *Afr. J. Trop. Hydrobiol. Fish* **1**, 107–114.

INLAND FISHERIES IN ARID ZONES

by R. L. WELCOMME

Fishery Resources and Environment Division, Department of Fisheries, FAO

Summary

Inland fisheries are non-consumptive with respect to the aquatic resource, but are very sensitive to changes within the ecosystem. Various other uses of water can act upon fisheries usually to their detriment. Proper management techniques including aquaculture can maintain at least a proportion of the fish production from waters in arid zones thereby ensuring the continued supply of a valuable source of protein.

Résumé

La pêche dans les eaux continentales ne détruit pas les ressources en eau mais est très sensible aux changements qui se produisent dans les écosystèmes. L'eau peut être utilisée à bien d'autres fins que la pêche, mais généralement au détriment de celle-ci. Les techniques d'aménagement, y compris l'aquaculture, peuvent permettre au moins le maintien d'un certain niveau de production piscicole des eaux des zones arides, assurant ainsi un apport continu de protéïnes de bonne valeur.

Introduction

At first sight a discussion of fisheries in the context of arid zone agriculture appears somewhat paradoxical. However, many of the world's more important inland fisheries are situated adjacent to arid zones and the water which supports them is required for a variety of agricultural and pastoral activities other than fisheries. Multi-purpose use of this kind reacts upon the various elements of the user complex. From the point of view of the individual elements the effects resulting from the interactions can either be positive, where more than one element benefits, or negative, where benefits to one element are achieved at the cost of others. Fish and the fisheries which depend on them are essentially non-consumptive with respect to the aquatic resource, but because they live entirely within the aquatic system and are behaviourally and physiologically sensitive to even slight variations in their environment, fisheries react rapidly to changes in use. Unfortunately such changes usually act to the detriment of fish stocks and the fisheries which depend on them. There is therefore a need to seek strategies whereby the least loss of a valuable source of protein occurs. This depends on a clear understanding of the importance of the various elements and the ways in which they react and this paper discusses some of the positive and negative aspects of fisheries *vis-à-vis* the management of arid zone waters for other purposes.

303

Effects of other uses on fisheries

Positive effects. Positive effects on a fishery may be regarded as those which increase the potential catch either in quality or quantity so as to make a greater amount of acceptable fish available for consumption.

Creation of artificial water bodies. One solution to the problems of those arid zones which suffer from an uneven distribution of rainfall in time is the creation of various types of impoundments to conserve water for irrigation and for domestic or animal consumption. Such water bodies range in size from small cattle dams to major reservoirs. Many of these are quite adequate for supporting fish populations, although in those dams which dry up completely annual stocking of fry is necessary. In some countries stocking programmes to utilize fully such dams already exist and in others projects are being planned to do so. However, a large proportion of such impounded waters remain unused and the stocking and harvesting of fish in these should be encouraged. Maximum yields, particularly from smaller reservoirs are obtained by strict management practices whereby the species living in the lake are rigorously controlled. Additional yield may be obtained from reservoirs not destined for drinking water by artificial feeding or fertilization and even with only 6–8 months of water appreciable fish production can be achieved. In the larger reservoirs naturally breeding populations arise and local fish species usually cannot be prevented from colonizing the water body. Desirable species may also have to be introduced for the full achievement of potential, but subsequent stockings are usually non-productive.

"Modulos." A special type of water management system is being tried in Venezuela where the floodplain of the rivers of the Llanos region remain dry for several months of the year, but are inundated for a short period of time. Here the problem is the lack of grass for grazing through the long dry season. To surmount this a system of dykes is being constructed. These retain either rain water or overspill water from the river for several months and as the enclosed water bodies shrink a constant fresh zone of grass is available for grazing. Such areas are suitable not only for the rearing and capture of those fish which migrate into the enclosures during the flood but also for stocking with additional fry.

Aquaculture. Rearing additional fish by means of intensive or semi-intensive aquaculture presents great possibilities when associated with irrigation systems. Classic fish culture stations consisting of groups of ponds can easily be placed between the irrigation source and the irrigated land, but in tropical arid areas this involves tremendous loss of water through evaporation from pond surfaces and bottom seepage. An upstream location for such large installations is advisable, especially in areas where intensive agriculture is practised. Here the increase in nutrients, insecticides or herbicides in the water at downstream sites can lower the water quality so as to make it unsuitable for fish. In areas where chemical interventions are minimal individual ponds can be integrated into the regular irrigated system providing a type of small-scale rural aquaculture. Fish can also be raised in pens or cages constructed in the irrigation channels themselves. These of course must be confined to those main channels which are constantly filled with water.

Negative effects. Drawdown in ponds and reservoirs. The lowering of level consequent upon the withdrawal of water for diverse purposes usually decreases the fish populations inhabiting reservoirs. The decrease results from several factors. Firstly the diminution in area concentrates fish increasing competition for food and predation. Secondly, the water may become deoxygenated and overheated at the time of least area, producing massive fish kills. Thirdly, the changes in level themselves may disrupt the breeding patterns of those species which build nests at or near the water edge (e.g. *Tilapia*). As a result, the productivity of those dammed lakes and ponds with extensive drawdown tends to be less than that of the equivalent lake with less fluctuation in level. These factors may render smaller dams unsuitable for stocking, or means that they have to be stocked every year and harvested before the lowest water level is reached.

Changes in rivers. Much of the water required for irrigation is extracted from rivers which have to be modified for the purpose. Modifications include canalization, diversion or damming and each of these inevitably has an effect on the fish stock, which may be positive or negative. Fisheries ecology in tropical river systems is closely linked to the annual cycle of flood seasons which overspill the river to inundate large areas of the adjacent low-lying ground. Fish yields are optimal in any given system where both high and low water areas are maximal. Changes in the flood regime, which may be brought about by drainage or upstream withdrawal for irrigation, and which reduce either the area or the duration of the flood also tend to reduce catch. Variations in the rate of change also affect the fish population, and abrupt drainage of the swamp systems at the end of the floods leaves many fish stranded in pools which later desiccate.

Many migratory species depend on certain flow conditions to trigger their migrations. Most of the major runs of fish in temperate and tropical rivers are thus correlated with individual spates in a general flood condition. Such spates appear to act physiologically upon the fish, as well as to fill the river channel with oxygenated water to ease their passage. The timing of the spate and the physiological readiness of the fish to respond appears to be a finely tuned mechanism. Therefore if spates are suppressed by regulation of the flow, or if they occur at the wrong time of the year, reproduction does not occur.

Control of flooding has further effects in that it tends to simplify the specific diversity of the population by favouring certain species over others. Less diverse populations are generally less stable than more diversified ones, and extreme fluctuations in catch may be set up which would not otherwise occur. A lessening of the number of species in the stock is also produced by simple canalization.

Low flow conditions within the river can result in partial deoxygenation of the waters, particularly if, as happens in many tropical systems, the channel breaks up into a series of small pools. This tendency is aggravated where organic pollutants are present which by their oxygen demand can rapidly deplete the water of dissolved oxygen. Lessened flow within the river also increase siltation and colonization by emergent aquatic vegetation with a consequent loss of habitat diversity.

Effects of fisheries on other uses

Generally speaking fisheries only affect other uses in a positive way. In other words if a system is to be managed for maximal fisheries it is essential that certain considerations of flow or volume be maintained which may detract from the amount of water available for other purposes. Equally, should a healthy fish population be desired the use of insecticides and fertilizers should be restricted or avoided completely.

Fish do occasionally react more generally with other uses. Properly selected species have been used to control weed growth in canals, mollusc-eating fish may help to control the snail hosts of bilharzia, and many species of fish eat mosquitoes at some stage in their life history. Equally important interactions have been noted in rice culture areas where fish have been accused of wholesale destruction of the rice plants. On the other hand rice/fish culture is widely practised in many areas of the world and gives excellent yields of both fish and rice. Some fish in fact help control stem-boring insects and rice yields in fields where such fish are stocked are higher than in fishless fields. Various intensive aquaculture practices may introduce organic BOD into the water of ponds or streams or may favour algal bloom conditions, occasionally creating problems for other users.

Management of arid zone aquatic systems for fisheries

Production of fish from natural waters or by aquaculture is both feasible and highly desirable in arid zones. However, as may be seen from the foregoing, conflicts may arise between fisheries and other uses of the resource which can act to the detriment of one or other of the uses. Management of the waters for fisheries therefore requires the inclusion of fish production as an objective in establishing the management plan with an appropriate weighting in the whole scheme of multi-purpose use.

In general, management considerations need to be directed at two areas. Firstly at the source of supply of the water, where the withdrawal of water for irrigation or other purposes should be modified so that at least minimal conditions for the maintenance of the fish population are conserved. Thus by a controlled minimum pool size in dams and reservoirs a breeding stock of fish can survive the dry season obviating the need to restock such waters annually. In river systems appropriate water release regimes based on studies of the biology of the fish species should be established so as to permit the breeding and survival of at least a portion of the fish population. Secondly at the irrigated system itself where various types of aquaculture can be used alongside irrigation systems or within the river channels. Existing aquaculture practices may not be sufficient to take full advantage of the possibilities presented by such systems and new types of semi-intensive and extensive practices are being investigated. The use of small dams to maintain flooding in the depression of floodplains, for instance, which may serve to counterbalance in part the loss of fish due to a curtailed or lessened flood.

There is no doubt that through proper application of these and other techniques fisheries can be integrated into the general scheme of water use in arid and semi-arid areas ensuring a supply of fish protein from most waters which are also used for other purposes.

A COMPARISON OF MOSQUITO POPULATIONS IN IRRIGATED AND NON-IRRIGATED AREAS OF THE KANO PLAINS, NYANZA PROVINCE, KENYA

by M. N. HILL, J. A. CHANDLER and R. B. HIGHTON

Medical Research Council Project, P.O. Box 1971, Kisumu, Kenya

Summary

Studies on the mosquitoes of the irrigated and non-irrigated areas of the Kano Plains, Nyanza Province, Kenya, have been carried out in connection with work on the changes in the epidemiology of insect-borne diseases associated with irrigation. Irrigation effects have reduced the diversity of mosquito species. Temporary water breeders, e.g. *Anopheles gambiae* s.l. and *An. funestus* have replaced permanent water species, e.g. *Mansonia uniformis* and *M. africana*, as the predominant house-frequenting mosquitoes in the irrigated areas, and mosquito populations overall have increased four-fold in the irrigated regions. The epidemiological implications of these changes are discussed.

Résumé

Des études sur les moustiques des régions irriguées et non-irriguées des Plaines de Kano, Nyanza Province, Kenya, ont été faites en connexion avec des études sur les changements dans l'épidemiologie des maladies transmises par l'eau associées avec l'irrigation. Les effets de l'irrigation ont réduit la diversité des espèces de moustiques. Ceux qui habitent les gîtes temporaires, par example, *A. gambiae* s.l. et *A. funestus*, ont remplacé les espèces vivant dans les gîtes permanents de l'eau, e.g. *M. uniformis* et *M. africana*, comme les moustiques en predominance dans les maisons des terrains irrigués, et le nombre de moustiques en général a quadruplé dans les regions irriguées. Les implications épidemiologiques de ces changements sont discutés.

Introduction

Plans for the irrigation of the Kano Plains, a semi-arid area on the N.E. shores of Lake Victoria have long been in existence. A pilot irrigation scheme of rice production was completed in 1968, and 2 crops of rice per annum have been produced regularly since 1969, using water from the Nyando River (Fig. 1) to supplement the natural precipitation in 2 wet seasons (November–December and March–June). A second pilot scheme is being constructed at the Kano II site situated near the Lake-shore swamps, and drawing water directly from the Lake. Fifty percent of the irrigated area has been allocated for sugar cane

307

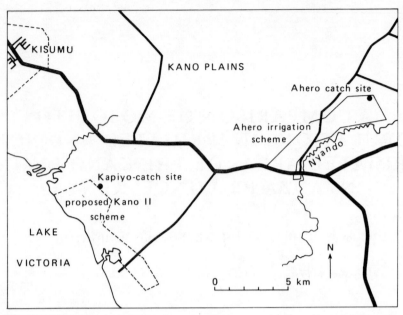

FIG. 1. Map of the Kano Plains, Nyanza Province, Kenya, showing sites of irrigated areas, and mosquito collection
      sites.

production, and the first crops were planted during mid-1975. The remaining area will
be used for rice production, and planting is scheduled for early 1976.

The Medical Research Council (UK) set up a research project in 1971 at Kisumu to
study the effects of irrigation on the health of the population in the area, with special
reference to vector-borne diseases, notably arboviruses, malaria and bilharzia. The project
has carried out medical and demographic studies in the populations living on and near
the schemes to establish basic health parameters. In addition long-term entomological,
veterinary, rodent, malacological and ornithological studies have been carried out to establish basic biological parameters which might be expected to affect the status of vector-borne
diseases.

The purpose of this paper is to report on some of the findings with regard to the mosquito
populations.

The study area

The Kano Plain lies on the N.E. shore of the Winam Gulf of Lake Victoria, and consists
of an area 50 km long varying from 5 km wide in the West, to 30 km in the East. The Plain
is covered with fertile alluvial deposits from the Nandi Escarpment in the North and the
Kisii Hills in the East (altitude 750 m). The town of Kisumu (lat. 0° 05'S long. 34° 48'E)
lies on a rocky promontory at the East of the Kano Plain and contains approximately
23,000 inhabitants. The Plain is crossed by the Nyando and Luanda Rivers which are permanent, and several parched rivers also cross the Plain, causing extensive flooding during
the wet seasons. The Lake shore swamps are thickly vegetated with Papyrus and other

Cyperaceae. The predominant plain vegetation is thorn tree, Euphorbia and Eucalyptus stands.

The major ethnic group of plains dwellers are Luo (a Nilotic tribe) who are essentially subsistence farmers and fishermen. The rural population has a density of approximately 300 persons per square kilometre (Ominde and Ojany 1969). The Luo cultivate staple crops of maize and sorghum, and maintain extensive herds of cattle, goats and sheep resulting in heavy overgrazing of the area. Extensive commercial sugar farms are located on the foothills of the escarpments.

The climate of Kisumu has been described by Haddow (1942) with special reference to mosquitoes. Rainfall averages 124 cm a year at Kisumu, the main wet season being March to May, with an erratic short rain season during November and December. The average daily maximum temperature ranges from 27°C in June and July to 29°C in January. Relative humidities at 14.00 hr vary between 41 and 57% and rise to over 90% after dark.

TABLE 1. *Showing total catch of all mosquito species in CDC-type light traps operated in irrigated and non-irrigated areas of the Kano Plains, Kenya, January–December 1973.*

Spp	Irrigated (83 trap nights)	Non-Irrigated (86 trap nights)
An. gambiae s.l.	12,734	457
An funestus	1,309	210
An. ziemanni	122	37
An. pharoensis	16	5
M. uniformis	73	1,462
M. africana	14	447
M. fuscopennata	3	2
M. karandalaensis	—	2
M. metallica	—	1
C. antennatus	73	47
C.p. fatigans	128	—
C. poicilipes	2	17
C. univittatus	12	79
C. tigripes	2	—
C. rima group	—	6
C. ethiopicus	1	—
C. aurantapex	—	1
Ae. lineatopennis	—	2
Ae. circumluteolus	6	21
Ae. ochraceous	—	3
Ae. sudanensis	2	—
Ae. africana	3	1
F. splendens	—	2
F. mediolineata	—	29
Total	14,500	2,831
Total species	16	20
Index of diversity	1.74	2.6

Methods

Mosquitoes were collected by the use of modified CDC-type light traps (Sudia and Chamberlain 1962) which were locally manufactured from metal sheeting. The traps were operated in two areas: (a) on the Ahero Irrigation Scheme and (b) a site near the swamp edge of the non-irrigated plains at Kapiyo. Two houses were randomly selected at each site, and light traps were operated inside near the eaves spaces, from dusk to dawn on 4 consecutive nights each month from January to December 1973. The mosquito collections were transported to the Kisumu laboratory alive, killed by freezing in an ultra-deep freeze and identified.

Results

The aggregated result for the full year's collections are shown in Table 1. In the irrigated collection sites, a total of 14,500 individuals comprising 16 species were collected, the majority (87.8%) were *An. gambiae* s.l. while *An. funestus* formed 9.0%, *An. ziemanni* 0.8%, *M. uniformis* 0.5%, *M. africana* 0.1%, *Culex antennatus* 0.5%, *C.p. fatigans* 0.9% and

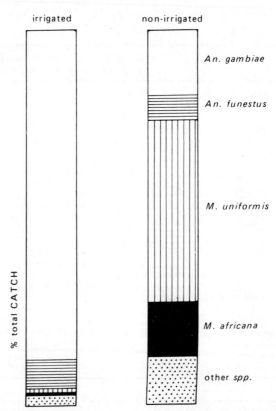

Fig. 2. Percentage catch of major mosquito species in CDC-type light traps in irrigated and non-irrigated area of the Kano Plains, Kenya (1973).

TABLE 2. *Showing monthly Williams Means (MW) of collections of four of the commonest mosquito species taken in CDC-type light traps operated in irrigated and non-irrigated areas of the Kano Plains, Kenya (January–December 1973).*

Spp.	J	F	M	A	M	J	J	A	S	O	N	D
				Irrigated area								
An. gambiae s.l.	37.4	0.4	6.4	26.2	154.4	294.7	66.3	29.7	5.6	8.1	8.5	34.4
An. funestus	55.9	4.2	3.5	1.1	3.7	9.4	17.3	35.0	0.8	2.1	0.7	3.0
M. uniformis	1.7	3.9	0.2	0.3	0.3	1.0	0.9	0.7	0.1	0.7	0.1	1.0
M. africana	0.4	0.1	0.1	0.3	0.0	0.3	0.1	0.2	0.1	0.0	0.0	0.1
				Non-irrigated area								
An. gambiae s.l.	1.8	0.4	0.6	0.5	2.3	10.7	1.9	0.8	8.2	1.5	1.4	7.8
An. funestus	12.0	0.4	0.5	0.3	0.0	0.4	1.3	0.3	3.6	0.9	1.3	0.7
M. uniformis	20.0	53.5	51.4	18.2	1.6	7.1	5.7	3.8	5.0	4.5	2.6	2.1
M. africana	3.4	9.9	4.3	3.0	0.4	0.8	0.5	0.6	1.2	4.0	2.0	1.0

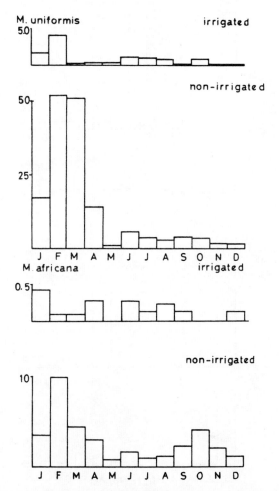

FIG. 3. Monthly Williams means of collections of *An. gambiae* s.l. and *An. funestus* taken in CDC-type traps (8 trap nights per month) operated in irrigated and non-irrigated areas of the Kano Plains, Kenya (1973).

C. univittatus 0.1%. the remaining species making up 0.2% of the total catch. The Index of Diversity for the irrigated area catches was 1.74 (Fisher Corbet and Williams 1943).

The sites on the non-irrigated areas yielded 2831 specimens, comprising 20 species, the majority of which (51.6%) were *M. uniformis* while *An. gambiae* s.l. formed 16.1%, *M. africana* (15.8%), *An. funestus* (7.4%), *An. ziemanni* (1.3%), *C. antennatus* (1.7%) and *C. univittatus* 2.8%, the remaining species making up 3.3% of the total catch. The Index of Diversity for the non-irrigated area catch was calculated at 2.6. The percentages of the major species in each of the areas are represented in Fig. 2.

In Table 2 the monthly Williams Means (MW) for 4 of the common mosquitoes in the irrigated and non-irrigated areas are presented, and these results are expressed as histograms in Figs. 3 and 4. In both habitat types, populations of *An. gambiae* s.l. and *An. funestus*

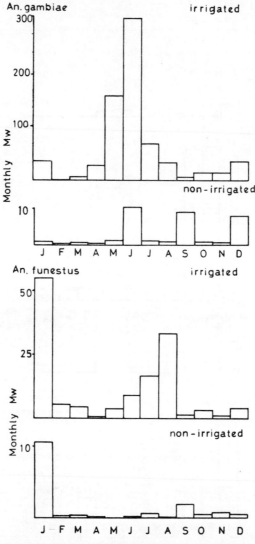

FIG. 4. Monthly Williams means of collections of *M. uniformis* and *M. africana* taken in CDC-type traps (8 trap nights per month) operated in irrigated and non-irrigated areas of the Kano Plains, Kenya (1973).

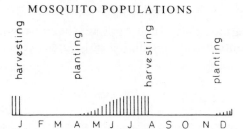

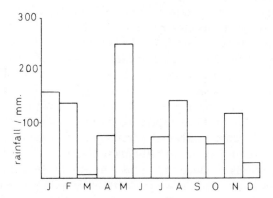

Fig. 5. Diagrammatic presentation of ricefield cultivation at the Ahero Irrigation Scheme, together with monthly amounts of rainfall in 1973.

show similar trends (Fig. 3). Peak populations of *An. gambiae* occur in May (MW 154.4) and June (MW 294.7) on the irrigated area, and minor peaks occur in January (MW 37.4) and December (MW 34.4). In the non-irrigated area, the peak population of *An. gambiae* s.l. was recorded in June (MW 10.7) while a secondary peak occurred in December (MW 7.8). On the irrigation scheme, high populations of *An. funestus* were recorded in January (MW 55.9) and in July and August (MW 17.3 and 35.0). Similarly, in the non-irrigated area, the population peaks of *An. funestus* occurred in January (MW 12.0) and July and September (MW 1.3 and 3.6).

M. *uniformis* and M. *africana* showed similar trends in population in both areas with highest populations of both species occurring in January, February and March (Fig. 4).

Rainfall figures from the Ahero Irrigation Scheme are shown in Table 3, while Fig. 5 shows monthly rainfall histograms and a diagrammatic representation of rice cultivation cycles near the study site during 1973. Highest rainfall occurred in May (250.5 mm) with substantial amounts in January (153.7 mm), February (132.9 mm) and August (138.4 mm). The cultivation cycle in 1973 commenced with a harvest of the previous year's crop in

TABLE 3. *Showing monthly rainfall totals/mm at the Ahero Research Station, Kano Plains, Kenya.*

Month	J	F	M	A	M	J	J	A	S	O	N	D
Rainfall amount/mm	153.7	132.9	4.1	76.1	250.6	49.2	71.1	138.4	70.1	56.7	113.0	28.4

January. Transplanting took place in mid-April, and the crop was harvested in early August, with the next transplanting occurring during early December.

Discussion

A review article by Surtees (1970) suggested 4 major ecological changes resulting from irrigation practices which could be expected to affect mosquito populations. These are simplifications of habitat, changes in water flow and water table levels, increased areas of above groundwater, and associated urbanization. The results presented above largely support this thesis. At the Ahero ricefield the simplification of habitat has led to a reduction of the number of species from 20 on the unchanged plain to 16, with a corresponding change in the Index of Diversity from 2.6 to 1.74. The changes in water flow and water table levels have resulted in a switch in species composition in the mosquitoes. Thus species, such as *M. uniformis, M. africana, C. antennatus, C. univittatus* and *F. mediolineata*, which are essentially permanent water breeders have given way to temporary water breeders such as *An. gambiae* s.l. and *An. funestus*, the former species occurring in the shallow waters of the ricefields, while *An. funestus* occurs largely in the drainage and irrigation canals. Chandler and Highton (1975) have described a cycle of breeding in the ricefields at Ahero. After flooding and transplanting, the fields produce enormous numbers of *An. gambiae* s.l. but later on, as the rice plants grow larger, and the water clears and develops a characteristic flora and fauna, *An. pharoensis, An. ziemanni* and *C. poicilipes* are the predominant species. Later still, prior to harvesting, *C. antennatus* predominates.

The increased area of above groundwater has clearly enhanced the population of mosquitoes in the ricefield areas, and the house entry rate for all species was approximately four times greater than in the non-irrigated regions.

Finally, consideration must be given to the increased urbanization occurring at the periphery of the rice-growing areas. On the Kano Plains the Luo people live in extended family groups and do not congregate into villages, thus dwellings are scattered throughout the area. The 2500 occupants of the ricefield scheme have been concentrated into 7 villages at the periphery of the scheme. Sewerage facilities in the villages are inadequate, and it seems likely that the enhanced populations of *C.p. fatigans*, an urban species, result from this.

As regards the seasonal distribution of mosquitoes in the two types of habitat, the data are difficult to interpret, owing to the fact that ricefield cultivation is associated with the rainy seasons in view of the high evaporation rate during the dry seasons. As a consequence, population densities of *An. gambiae* s.l. and *An. funestus* show similar seasonal variation on the ricefields and non-irrigated areas, except that periods of peak population density are longer lived in the ricefield areas. *M. uniformis* and *M. africana* are dependent on water table (Krafsur 1972), and consequently in the Lake basin population peaks have been shown to be highly correlated with Lake levels, which have been falling throughout 1973 and 1974.

To conclude, the epidemiological implications of these changes should be considered. The Kano Plains is an area already holoendemic for malaria. At the present time, the infant malaria rate in the ricefield villages is only slightly higher than on the non-irrigated plains (D. H. Smith, Pers. Comm.). This is probably due to the stable nature of endemic malaria which results in infection rates being less affected by changes in the factors which cause transmission than other forms of malaria (Macdonald 1957).

Arbovirus activity on the Kano Plains was at a low level, although Bowen *et al.* (1973) estimated human infection rates with Onyong nyong virus of 10% per annum. The mosquito vectors of this virus are *An. gambiae* and *An. funestus*, and consequently transmission may be enhanced on the ricefields.

Clearly in an area when the ecosystem is changing drastically the possibility of new mosquito, animal and human association giving rise to new disease systems is always present, and it is necessary for these changes to be monitored if epidemics of diseases in man and domestic animals are to be avoided.

References

Bowen, E. T. W., Simpson, D. I. H., Platt, G. S., Way, H., Bright, W. F., Day, J., Achapa, S., and Roberts, J. M. D. 1973. Large-scale irrigation and arbovirus epidemiology, Kano Plain, Kenya. II. Preliminary serological survey. *Trans. R. Soc. trop. Med. Hyg.* **67**, 702–709.

Chandler, J. A., and Highton, R. B. 1975. The succession of mosquito species in ricefields in the Kisumu area of Kenya, and their possible control. *Bull. ent. Res.* **65** (2).

Fisher, R. A., Corbet, A. S., and Williams, C. B. 1943. The relation between the number of species and the number of individuals in a random sample of an animal population. *J. anim. Ecol.* **12**, 42–58.

Haddow, A. J. 1942. The mosquito fauna and climate of native huts at Kisumu, Kenya. *Bull. ent. Res.* **33**, 91–142.

Krafsur, E. S. 1972. Observations on the bionomics of *M. africana* and *M. uniformis* in Gambola, Illubabar Province, Ethiopia. *Mosq. News* **32**, 73–79.

MacDonald, G. 1957. The epidemiology and control of malaria. Academic Press: London.

Ominde, S. H. and Ojany, F. F. 1969. *African Scientist* **1**, 7–11.

Sudia, W. D. and Chamberlain, R. W. 1962. Battery-operated light trap an improved model. *Mosq. News* **22**, 126–129.

Surtees, G. 1970. Effects of irrigation on mosquito populations and mosquito-borne diseases in man with particular reference to ricefield extension. *Int. J. Envir. Stud.* **1**, 35–42.

WATER WEED PROBLEMS IN IRRIGATION SYSTEMS

by D. S. MITCHELL

Biological Sciences, University of Rhodesia.

Summary

Many aquatic plants are troublesome weeds in irrigation canals and reservoirs and may cover extensive areas. In natural conditions, the extent or even the presence of suitable aquatic habitats is often seasonal and many of these plants have evolved to grow rapidly when good growth conditions exist. Other plants are particularly well adapted to relatively constant aquatic situations such as in a marsh or swamp and may form the persistent, almost monospecific stands of vegetation typical of these areas. Both types of aquatic plants can be regarded as primary colonizers in an ecological succession from aquatic to terrestrial conditions and are well suited to take advantage of various man-manipulated water regimes. Consequently, it is desirable to examine the ecological factors which promote the growth of these plants so that, hopefully, the environment can be managed to minimize the problems they cause. Before an aquatic plant population can become established, an adequate and constant supply of water is necessary for at least the duration of most, if not all, the main growing season. Aquatic plants are especially susceptible to desiccation and will be inhibited by extensive and rapid fluctuations in water level. Other important factors are the nature of the substrate, the regularity and rate of water flow, the availability of nutrients and, for submerged species, sufficient light penetration to ensure adequate photosynthesis. Irrigation canals and shallow reservoirs with earthern sides and bottoms generally provide suitable conditions for the growth of aquatic plants, which may therefore become a serious problem and interfere with the planned utilization of the water. Among the problems caused by aquatic plants are increased water loss through evapotranspiration (though this is difficult to measure accurately and, in arid areas may not be as serious as was first thought), interference with flow, and the provision of suitable habitats for the vectors of diseases, such as malaria and schistosomiasis. The role of aquatic plants in promoting the incidence and transmission of these diseases depends upon the extent to which the habitat is rendered more favourable for the host and disease organisms and this is dependant on complex ecological inter-relationships which now require urgent study. Mechanical, chemical and biological methods have been used for the control of aquatic weeds but more thought should also be given to the manipulation of environmental factors such as water level and the integration of different control measures. A rational approach to the problems of water weeds calls for inter-disciplinary liaison and suitable provisions to deal with these problems should be considered at the planning and subsequent stages of schemes to manipulate water resources.

Résumé

De nombreuses plantes aquatiques constituent des adventices néfastes dans les canaux et les retenues d'irrigation et peuvent couvrir des superficies étendues. Dans les conditions naturelles, l'extension, ou même la présence, d'habitats aquatiques convenables, a souvent un caractère saisonnier et nombre de ces plantes sont susceptibles de se développer rapidement lorsque les conditions de croissance sont favorables. D'autres plantes sont particulièrement bien adaptées à des conditions aquatiques relativement constantes, comme par exemple dans les marais ou marécages, et peuvent constituer les peuplements résistants, à peu près monospécifiques, de la végétation typique de ces milieux. Ces deux types de végétaux aquatiques peuvent être considérés comme des colonisateurs primaires

318 D. S. MITCHELL

dans une succession écologique allant du milieu aquatique au milieu terrestre. Ils peuvent tirer profit des variations de régime hydrique dûes à l'homme. En conséquence, il est souhaitable d'étudier les facteurs écologiques qui stimulent la croissance de ces végétaux, de sorte qu'on puisse espérer réaliser un aménagement de l'environnement qui permettrait de minimiser les problèmes soulevés. Pour pouvoir s'installer, une population de plantes aquatiques doit bénéficier d'une alimentation en eau adéquate et constante pendant au moins une grande partie, sinon la totalité de la durée de la période principale de croissance. Les plantes aquatiques sont particulièrement sensibles à la dessiccation et seront freinées dans leur développement par des variations importantes et rapides du niveau de l'eau. Les autres facteurs importants sont: la nature du substrat, la régularité et la vitesse de circulation de l'eau, la disponibilité des éléments nutritifs et, en ce qui concerne les espèces submergées, une pénétration suffisante de la lumière pour permettre une photosynthèse convenable. Les canaux d'irrigation et les réservoirs peu profonds, dont le fond et les parois sont en terre, offrent des conditions favorables à la croissance des plantes aquatiques, qui peuvent alors poser de sérieux problèmes et influer sur une utilisation rationnelle des eaux. Parmi les problèmes soulevés par les plantes aquatiques figurent: l'accroissement des pertes d'eau par évapotranspiration (bien qu'il soit difficile de mesurer cet accroissement avec précision et qu'il ne soit pas aussi grave qu'on ne l'ait pensé au premier abord dans les pays arides), l'incidence sur le débit, et la création d'habitats favorables aux vecteurs de maladies telles que le paludisme et la schistosomiase. L'importance du rôle des plantes aquatiques dans l'apparition et la transmission de ces maladies est fonction de l'existence d'un habitat plus favorable à l'hôte et aux vecteurs, ce qui dépend d'un ensemble d'interrelations écologiques complexes dont l'étude devrait maintenant être entreprise d'urgence. Des méthodes mécaniques, chimiques et biologiques ont été utilisées pour lutter contre les mauvaises herbes aquatiques, mais il conviendrait également de s'attacher davantage à l'intervention sur les facteurs de l'environnement, tels que le niveau de l'eau et à l'intégration des différents moyens de lutte. Une approche rationnelle des problèmes posés par les adventices aquatiques nécessite le recours à la multidisciplinarité, et il conviendrait de réserver les sommes et moyens nécessaires à l'étude de ces problèmes lors de la planification et du déroulement des phases successives des projets d'utilisation des ressources en eau.

Introduction

Aquatic plants growing in or along the edges of irrigation canals may impede flow, encourage silting, increase loss of water through evapotranspiration, and provide a habitat for the vectors of diseases, such as malaria and schistosomiasis. Troublesome growths in ditches draining irrigated fields restrict flow and cause soil to become waterlogged. Aquatic weeds may also cause problems in man-made lakes associated with irrigation schemes. Through all these effects, aquatic plants can interfere with the controlled supply of water in such a way as to make the limitation or removal of the plants essential, thus increasing both capital and recurrent costs to an unforeseeable extent and upsetting the careful calculations of both engineers and economists.

Water weeds often infest extensive areas. For example, Timmons (1960) estimated that of the 232,000 km of canals in the 17 western states of the United States of America, 76,000 km were infested with submerged aquatic plants, and nearly 36,800 km by emergent aquatic plants. A further 32,000 km were infested by algae. Of the 213,183 ha of ditchbanks, 159,176 were infested by plants. In 1957, the total cost of weed control was $8,113,297 and the resulting savings in losses due to weeds was estimated to total $15,860,026. The productive value of the water lost in any case was estimated at $27,574,920.

Almost all the drains and waterlogged areas in the Chambal Irrigation system, India, are infested with emergent weeds which occupy an area of about 10,000 ha, while submerged weeds affect many of the canals and occupy a total of about 1500 ha (Mehta *et al.* 1973). Consequent weed control has been expensive and, even then, designed water discharges have not been fulfilled.

Up to now, most attempts to control aquatic weeds in man-manipulated water regimes have involved the use of herbicides, or the mechanical or manual removal of the plants. There are relatively few examples of the application of ecological principles, such as the employment of biological control agents. Possibilities also exist for the manipulation of the hydrological regime so as to make growing conditions unfavourable for the problem plant. Such measures demand an understanding of the biology and ecology of weeds. A knowledge of the structure and function of the whole ecosystem concerned and of the role of the plants in it is also desirable.

A rational approach to aquatic weed problems therefore depends upon an adequate ecological knowledge of the plants and ecosystems concerned. It is here that research is now most urgent and that synthesis of knowledge is essential, if a foundation is to be laid for further progress beyond the trial-and-error approach that has been utilized so often in the past. Hopefully, such progress could lead, on the one hand, to modifications in engineering designs which would minimize the problems caused by aquatic weeds or, on the other, to a reasonably accurate estimate of the cost of controlling them.

The ecology of aquatic plants

Aquatic macrophytes may be divided broadly into two ecological types. There are those that are adapted to survive in situations where there is a marked seasonal difference in growing conditions, as in temporary pools or seasonally flooding rivers, and there are those that are adapted to take advantage of relatively constant aquatic conditions, as in a marsh, or swamp. The former have to contend with adverse conditions for part of every annual cycle and consequently are adapted to grow very rapidly during the time of the year when conditions are favourable and survive through the adverse season, either as small populations of reduced plants, or as resistant propagules, such as seeds or rhizomes.

Both types of aquatic plants cause problems. Most man-manipulated systems of water management provide permanent and relatively stable hydrological conditions. On the one hand, this removes the adverse period for aquatic plants and, at the onset of stable conditions, native or introduced species grow rapidly from initially small numbers of plants or resistant propagules to form very large populations, which persist until some form of ecological instability recurs (Mitchell and Thomas 1972). On the other hand, the presence of constant shallow water and waterlogged soils may result in the relatively slower but more persistent growth of dense stands of swamp plants.

Aquatic plants have different life-forms, namely emergent (e.g. *Typha*), submerged with floating leaves (e.g. *Nymphaea*), submerged (e.g. *Potamogeton*) and free floating (e.g. *Salvinia*), each best adapted to different depths of water. Around the edge of the water body, phreatophytes (plants which grow in high water tables) may also occur. Aquatic plant communities often exhibit a zonation of life-forms from terrestrial to aquatic conditions and it is important to realize that this represents stages in an ecological succession. Many aquatic plants, in a sense, can be considered as primary colonizers in a successional series as borne out by their "opportunist" characteristics of rapid growth, an ability to adapt to widely differing conditions, and high reproductive capacity. Removal of these plants therefore merely creates the conditions which they are most adapted to exploit, and such an exercise will thus normally require frequent repetition.

The general environmental requirements for aquatic plant growth are summarized by Mitchell (1974). The essential requirement is obviously an adequate and constant supply of water for at least the duration of most, if not all, the main growing season. Aquatic plants are especially susceptible to desiccation and are usually rapidly killed when completely exposed to dry air. For this reason, rapid and extensive lowering in water level will markedly inhibit the establishment of populations of these plants, and Bowmaker (1973) has suggested that, in Lake Kariba, annual fluctuations exceeding 2 metres, or a rate of fluctuation greater than 0.6 metres per month, would be detrimental to the development of littoral hydrophytes.

If water is drained from a canal, there is bound to be an adverse effect on any plants which may be growing there. The extent of this effect will depend upon the length of time during which the canal is left dry and upon the extent to which water is retained within the aquatic plant colonies and in the mud in which the plants are growing.

Certain aquatic plants are able to adapt to the presence or absence of water by changes in growth form (Sculthorpe 1967), while others survive adverse dry conditions in the form of resistant propagules as described by Misra (1973) for the Upper Gangetic plain in India. Certain plants with no resistant stages, such as *Salvinia molesta*, survive in the water bodies that remain in the dry season, but are killed in waters that dry up completely.

Any aquatic plants which are native to arid regions will be confined to the few isolated water bodies that may occur there. As it is highly probable that these water bodies will be seasonal in nature, it is also likely that the plants will only be able to survive if they have some form of resistant stage in their life cycle. The creation of permanent water conditions in the reservoirs and canals associated with irrigation of such areas provides the opportunity to these plants for rapid and sustained growth. In addition other plants which previously were unable to survive in the area are likely to invade and establish themselves.

Other important factors which affect the establishment and growth of aquatic vegetation are temperature, the nature of the substrate, light, and the availability of plant nutrients. Warm temperatures, likely to promote the growth of aquatic weeds, are generally present in arid areas, and earthen sides and bottoms of irrigation canals provide a suitable substrate for aquatic plant growth. Light is an essential for the growth of all green plants. It is only likely to be limiting where water is excessively turbid, too deep or shaded in some way. Plant nutrients are almost inevitable in the ditches draining irrigated fields and may often escape from fertilized fields into adjacent irrigation canals and reservoirs. Excessive growths of aquatic plants in water systems are only possible if sufficient nutrients are available and the general eutrophication of many man-manipulated water systems undoubtedly contributes substantially to this problem.

The rate and frequency of water flow down a canal can be important. Aeration of the water associated with this movement can encourage the growth of attached submerged species, but too rapid a flow will have an adverse effect on floating plants such as *Eichhornia crassipes* and *Salvinia molesta*. These plants tend to be carried along the canal until prevented from further movement by an obstacle of some sort. In this situation they may build up in numbers and cause a serious blockage or interfere with operation of machinery.

Natural aquatic ecosystems are complex and exhibit a web of inter-relationships between living organisms and environmental factors. Plants play an important role in the stability of the system. They are the primary producers on which the secondary producers depend. In addition they provide habitats for other plants and for animals. They absorb nutrients from the system and release nutrients into it when they decay. They stabilize the substrate and may also stabilize the water flow. However, it is unusual for the presence and function of aquatic vegetation to be considered in man-manipulated water systems. It should be borne in mind that aquatic vegetation has positive as well as negative effects, and the normal role of the vegetation in the water should be examined before extensive control measures are instituted.

Effects of aquatic weeds in irrigation systems

Increased water loss through evapotranspiration of aquatic vegetation is often assumed to be considerable, but, as it is difficult to measure this accurately, the situation is far from clear. Indeed some investigations indicate that, under certain conditions, the opposite may be true, especially in arid areas. Linacre *et al.* (1970) compared water loss from a *Typha* infested swamp and a lake in New South Wales, Australia. The water bodies were 16 km apart and were both surrounded by gently rising, low hills and relatively dry areas. They showed that the evapotranspiration from the swamp (E_S) was less than that from the lake (E_W) during dry periods $(E_S:E_W=$ about $1:1.6$ and $1:3)$, but was more or less the same following rainfall which saturated the surrounding countryside. Rijks (1969) obtained similar results from an African papyrus swamp where $E_S:E_W$ ratios reached about $1:0.6$ and ranged from $1:0.38–0.81$. These results contrasted with earlier expectations and with results obtained in more temperate areas. For example, Rudescu, Niculescu and Chivu (1965) reported that the annual evapotranspiration from a dense *Phragmites* stand in Romania to be 1.0 to 1.5 metres. Guscio *et al.* (1965) reported that phreatophytes cover about 6.5×10^6 ha in the seventeen western states of the United States of America and annually lose 30.65 km^3 water. They reported that *Typha* uses 211 to 254 cm of water every year in comparison with evaporation rates of 127 to 190 cm from open water surfaces. More recently Smid (1975) has measured evaporation from a reed swamp in Czechoslovakia, using an energy-budget approach. He measured daily totals of evaporation of 5.6, 6.9, 5.5 and 1.4 mm respectively on four days when the weather was constant and relatively undisturbed. These values were compared with evaporation from a 3 m^2 pan anchored in the open water area of a nearby fish pond where water losses on the equivalent days were 5.4, 3.7, 5.5 and 1.6 mm respectively.

Loss of water from floating vegetation, or vegetation with floating leaves, does not appear to have been measured in field conditions. However, Timmer and Weldon (1967) carried out measurements with *Eichhornia crassipes* growing in tanks during March in Florida. The plants were grown until leaves had reached a length of 75 cm and the loss of water from these containers was then compared with the loss of water from similar tanks containing water only. They obtained an average evaporation to evapotranspiration ratio of $1:3.7$. Little (1967) and Mitchell (1970) carried out measurements of water lost from dishes containing *E. crassipes* and *Salvinia molesta* in greenhouse conditions. Little showed that the ratio of evaporation to evapotranspiration was $1:4–5$ for *E. crassipes* and $1:1–2$ for *Salvinia molesta*. Mitchell obtained a mean ratio of $1:1.14$ for *Salvinia molesta* and $1:1.62$ for *E. crassipes*. However, humidities were generally high during the latter experiment.

Brezny, Mehta and Sharma (1973) measured the loss of water from cement tanks in which they had established healthy stands of two emergent weeds, *Typha latifolia* and *Cyperus rotundus*; three floating weeds, *Eichhornia crassipes*, *Pistia stratiotes* and *Ipomea aquatica*; and a floating leaved weed, *Trapa natans* var. *bispinosa*. These water losses were compared with water losses from similar tanks without plants. The experiments were carried out in Rajasthan, India, in order to assess the effect of these plants on water loss from the Chambal irrigation scheme. Average ratios of evaporation to evapotranspiration were $1.00:1.26$ for *E. crassipes*, $1.00:0.93$ for *P. stratiotes*, $1.00:0.99$ for *T. natans*, $1.00:1.13$ for *I. aquatica*, $1.00:1.52$ for *T. latifolia* and $1.00:2.41$ for *C. rotundus*.

The general lack of agreement between the results cited above indicate the urgent need for detailed and accurate quantitative measurements of water loss from different types of aquatic vegetation under a range of field conditions, especially in arid areas. However, it should be noted that such measurements are difficult to perform accurately with current methods. For example, Smid (1975) considered that his measurements had an error of $\pm 20\%$ in the case of instantaneous values and about 10% in that of daily sums of evaporative water loss.

Submerged aquatic weeds decrease the cross-section of a canal or river and may reduce the design flow rate for an artificial canal by as much as 97% (Stephens *et al.* 1962). Floating species restrict flow to a lesser extent but, when massed in a thick mat, create friction losses similar to those provided by the sides and bottom of the canal. For example, in Florida a dense mat of *Eichhornia crassipes* reduced the efficiency of large canals by 60% and small canals by 80% (Bogart 1949). Emergent species also interfere with flow and in addition act as a trap for silt, causing sedimentation, which is impossible to deal with while the plants remain. The decrease in flow is brought about by an increase in the retardance coefficient of a canal which may be more than doubled by floating weeds and may be increased twentyfold by submerged weeds (Gupta 1973). Mehta and Sharma (1973) carried out a systematic study to measure the effect of aquatic weeds on the supply of water from the Chambal Irrigation System using Manning's formula to calculate coefficient of roughness. Comparison of design values and values measured before and after weed control showed that weeds reduced flows by about 50–70% in the main canal and by 83% and 77% in distributaries.

The role of aquatic plants in the incidence and transmission of human and livestock diseases is complex. Perhaps the most serious disease and the one most likely to be favoured by the development of irrigation systems is schistosomiasis (bilharziasis), but the nature of the relationship between the presence of aquatic vegetation and the incidence of the disease is far from clear. The relationship depends upon the degree to which the plants render the aquatic environment suitable for the transmission of the disease and build-up of large populations of the aquatic snails, *Bulinus*, *Biomphalaria* and *Oncomelania*. Species of another aquatic snail, *Limnaea*, the intermediate host for the liver fluke, *Fasciola*, which causes fascioliasis in man and other mammals (especially ungulates), has similar habitat requirements to the intermediate snail hosts for schistosomiasis (van Someren 1946, Hira 1969). Thus studies of the habitat requirements for *Limnaea*, while of economic importance themselves, are also relevant to an understanding of the ecology of bilharzial snails.

The scientific literature on the ecology of aquatic snails contains many general observations that the development of aquatic snail populations is favoured by the presence of aquatic plants, but there is very little hard data to substantiate these observations. For example, in a pamphlet issued by the World Health Organisation (1959) the following statement is made:

"Although young snails feed mainly on plankton and are not dependant on higher plants for food, the presence of aquatic weeds in irrigation canals favours the establishment of snail colonies for a variety of reasons. The plants slow down the current, oxygenate the water and provide shade for excessive sunlight, shelter from predators, food for adult snails and suitable surfaces on which the snails can crawl and deposit their egg masses. Furthermore, the leaves and stems are usually covered with algae and the decaying leaves themselves provide additional food and humus."

Abdel-Malek (1958) made a similar statement, but also pointed out that, while the presence of vegetation makes the habitat more favourable for the subsistence and breeding of bilharzial snails, these animals can occur in smaller numbers in the absence of weeds.

Boycott (1936) found no direct relationship in England between species of weeds and species of aquatic snails, while El-Gindy (1962) found such a relationship to exist in Egypt, and other studies (Dawood *et al.* 1965), also in Egypt, indicated that bilharzial snails prefer *Potamogeton crispus*, followed by *Eichhornia crassipes* and then *Panicum repens*. Similarly, Airey (1973) showed that the ventral surfaces of floating *Nymphaea* leaves were the preferred sites for oviposition of *Limnaea natalensis* and the young snails were found in largest numbers there.

Bulinus truncatus rohlfsi the intermediate host for *Schistosoma haematobium*, was found by Paperna (1968) to be closely associated with the submerged aquatic weed, *Ceratophyllum demersum*, in the lower Volta in Ghana. Large populations of the snail were found on the plant which was liable to seasonal fluctuation in extent. The snail numbers were highest when the plant was abundant in January and declined gradually until both plant and snail disappeared from May up to late October, or even November. *Biomphalaria pfeifferi*, the intermediate host for *Schistosoma mansoni*, was associated with the marginal emergent plants and bank vegetation and its population numbers were liable to much smaller seasonal fluctuations.

Hira (1969) observed young *Biomphalaria* snails clinging to the submerged portions of *Salvinia* plants in Lake Kariba, and I have seen the same to occur with *Eichhornia crassipes* in Lake McIlwaine, Rhodesia. Thus these plants are important agents in distributing the snails around a lake. The reported effect on *Salvinia* of mollusciding the areas where both plant and snail occurred caused Wild and Mitchell (1970) to perform experiments which showed that *Salvinia* was more sensitive to Bayluscide, the chemical used, than other aquatic plants. The one chemical could therefore be used to control both pests.

Aquatic vegetation also assists snails to resist desiccation during periods when water bodies may dry out. The aestivation of the snails during these periods seemed to be favoured where weeds were present (Abdel-Malik 1958). Even though *Limnaea* snails were highly susceptible to desiccation and had little burrowing ability, Airey (1973) observed that young snails were able to move into the mud along the roots of rooted aquatic plants and in this way escape from the immediate effects of the drying out of a water body.

Schiff (1974) has shown that light and temperature affect the location of host snails by the miracidia which hatch from the *Schistosoma* eggs passed out by infected people. During summer, the miracidia tend to concentrate on the bottom and those parts of the surface which are shaded by floating vegetation. During winter, the miracidia move to the surface only and, although caged sentinel snails are infected in open water at this time, the number of infections are much greater in the shaded areas. The presence of dense stands of marginal vegetation also encourage alfresco defaecation and urination by possibly infected persons adjacent to or into the water.

The relationship between aquatic vegetation and schistosomiasis is complex, and it is difficult to define causal links (Jordan and Webbe 1969). A considerable number of observations, however, indicate that the presence of aquatic plants favours the development of snail populations and the transmission of the disease, but more research is urgently required to clarify the situation.

The other group of diseases which cause most concern are those carried by mosquitoes, and the importance of aquatic plants to the incidence of these diseases depends on the ecological relationships between the plants and the aquatic stages of the mosquito's life cycle. Malaria is carried by species of the *Anopheles* mosquito. In Africa, mosquitoes of the *An. gambiae* species-complex breed in small sunlight pools and are not dependant on aquatic vegetation, but the larvae of the *A. fenestus* complex apparently require vegetation for shelter. Rozeboom and Hess (1944) showed that the breeding of *A. quadrimaculatus* in the Tennessee Valley, United States, was related to the extent that plants intersect the water surface. Plants with fine stems and many leaves at the water surface have a high value for this "intersection line" and provide conditions which are especially conducive to mosquito breeding.

Gupta (1973) and Williams (1956) both point out that *Mansonia* mosquitoes which are responsible for transmitting rural filariasis, mainly breed in association with *Pistia stratiotes* in India and Sri Lanka, although Gupta states that *Eichhornia crassipes* and *Myriophyllum spicatum* are also preferred breeding sites. In the United States, *Mansonia* is responsible for the transmission of encephalitis. The larvae of these insects are equipped with specialized breathing trachea which allow them to obtain air supplies from the air canals within submerged plant stems. Consequently, these mosquitoes are not open to normal means of control and can only be limited by removal of the aquatic vegetation (Guscio *et al.* 1965).

Bruns and Kelley (1974) have shown the importance of irrigated water in disseminating weed seeds. Screenings of water from a canal infested with an abundant growth of ditchbank weeds yielded seeds of 149 species of plants compared with 85 species from a canal where weeds were being partly controlled. If water and seeds were to be evenly distributed throughout the year, the former would deliver 15,467 seeds per annum to each hectare of land it serves and the latter 1,680.

Control of aquatic weeds

Perhaps more has been written on this subject than on any other aspect of the problems caused by water weeds. For this reason only a brief summation will be given here.

Problem growth of aquatic plants may be controlled by appropriate *"preventative"* measures to avoid providing the environmental conditions most conducive to growth, or by *chemical, mechanical* or *biological* methods. These different procedures have been recently reviewed by Mitchell (1974), Blackburn (1974), Robson (1974) and Bennett (1974) respectively. Frequently the most satisfactory results are achieved by various combinations of methods, and *integrated* control, based on an interdisciplinary approach, is receiving increasing consideration.

Preventative measures require a knowledge of the biology and ecology of the problem plants and often call for a compromise between the ideal manipulation of the water resource and the ideal procedure for the control of the weed. Prior to construction of an irrigation system or of a man-made lake, the catchment area should be diligently surveyed for the presence of potentially dangerous plants. Special attention should be paid to areas in which the environmental conditions are most like those that are likely to be experienced in the proposed system. Manipulation of water level has proved a powerful tool in the control of undesirable aquatic plant growth in reservoirs, especially when integrated with other methods (Penfound *et al.* 1945, Manning and Johnson 1975, Manning and Saunders 1975). Canals should be constructed with steep banks and, where serious weed problems are predictable, consideration should be given to the use of piping. The additional expense of such measures, however, can only be justified on the basis of quantified predictions of likely overall savings.

Fairly simple preventative procedures may be very effective. Malhotra (1972) has described how a procedure of draining canals and letting them dry for five days prevented weed problems developing for a subsequent six months in the canals of the Bhakra Irrigation System in India. Sainty (1974) advocates the same sort of procedure for the control of *Elodea canadensis* in Australia. Such procedures, however, will be unlikely to succeed against a species such as *Hydrilla* which produces resistant turions and may have to be combined with the application of suitable herbicides to the exposed sides and bottom of the canal.

Chemical methods of control are among the most widely and successfully used but care must be taken to avoid injury to plants subsequently irrigated with the water and pollution of associated aquatic systems (Bowmaker 1973). Extensive studies on the effect on crops of the chemicals used for weed control, the rate of dissipation of the chemicals and of the nature and effect of residues left after this break-down have been carried out in the United States (Bruns 1967, 1969, Bruns *et al.* 1964, 1974, Frank *et al.* 1970, Demint *et al.* 1970). Bartley and Gangstad (1974) considered the environmental effects of the use of herbicides for the control of aquatic plants and conclude that, when properly selected and carefully used, herbicides may be employed in potable water and water used for irrigation of crop plants. The range of chemicals that have been successfully used, and their application, has been described by Blackburn (1974).

Mechanical and manual methods of controlling excessive plant growth have also been widely employed. Malhotra (1972) records that, prior to the procedures described earlier, submerged weeds in the Bhakra system were removed manually, which required that channels be closed for seven days every two months. In the Chambal Irrigation System, *Typha* populations may be reduced by about 80% for a year by three underwater cuttings at monthly intervals during the rainy season when the plant is flowering. Different times of cutting in relation to the growth phases of the plant have different effects (Robson 1974) and these should be investigated when formulating a programme of weed control. Weed cutting machines of various types have been described by Robson. These are generally more efficient and cheaper than manual methods but more expensive than chemical methods.

Biological control can take several forms. The most common is the introduction of an animal (often an insect) which feeds specifically on the problem plant. The introduction of pathogens and of other competing plants with more desirable characteristics has also been employed. Biological control has the benefit of providing inexpensive perpetual control with the minimum of detrimental side-effects (Bennett 1974), but is difficult to integrate with other methods of control. The successful control of *Alternanthera philoxeroides* by the beetle *Agascicles hygrophila* in the United States provides perhaps the best example of this method (Maddox *et al.* 1971). Recently much attention has been given to the use of a herbivorous fish *Ctenopharyngodon idella*, the Chinese grass carp or White Amur (Michewicz *et al.* 1972), and Mehta, Krishna and Taunk (1973) describe experiments to investigate the control of submerged weeds by this fish in the Chambal Irrigation System in India.

The introduction of competing plants has been discussed by Yeo and Fisher (1970), but there are relatively few instances of this method being successfully employed. Bowmer (1973) reports that the introduction of low growing species of *Eleocharis*, which will not obstruct water flow severely, is being actively investigated in Australia. In the Chambal Irrigation System, *Brachiaria mutica*, which is a valuable fodder grass in India, has replaced *Typha* within six months when planted after the *Typha* leaves have been cut. However, the grass may obstruct the flow of water in a canal, and regular cutting is essential. The method is not appropriate in sections where flow must be unobstructed at all seasons but is useful in drains and other areas where the flow must be unobstructed in the rainy season only, as, between the rains, the area may be cropped and productively used (Mehta 1975).

The integration of different methods of control offers attractive possibilities and is often appropriate against a particular aquatic weed, as for example in the case of *Typha* (Timmons *et al.* 1963).

Conclusion

The problem of aquatic weeds in irrigation systems should be seen as an ecological response of the environment to the manipulation of hydrological resources by mankind. It is a problem which requires an inter-disciplinary approach if it is to be rationally and satisfactorily solved. Of paramount importance is an appreciation of the environmental inter-relationships which are being affected by the weed and which will be affected by any control measures. The present state of knowledge of aquatic ecosystems is in most cases inadequate and there is little alternative to a trial-and-error approach based on appropriate previous experience. The need for co-operation between aquatic biologists, hydrologists, engineers, economists and agriculturalists at all stages of planning, implementation and management of irrigation systems, is essential, and methods to deal with potential or actual aquatic problems must be incorporated in all planning and management proposals. Such proposals must be appropriate to local circumstances of climate as well as social and economic structure. Thus, what may be appropriate in a developed country with a low density of rural population may be quite unsuitable in an undeveloped country with a relatively dense rural population and a low level of employment. Similarly, measures which have proved successful in one developing country may not be applicable in another.

The introduction of irrigation systems in developing countries may be divided into three phases: feasibility studies, period of design and decision, management of introduced scheme. In each of these phases it is necessary to include consideration of water weeds. Feasibility studies should include identification of potential problem species, evaluation of probable cost of adverse effects as well as of cost of control. The critical period of decision making is one in which different ways of handling the weed problems are evaluated and presented to the decision makers. One of the ways presented should be that of doing nothing, with the consequences that are likely to follow. Areas of uncertainty and speculation should not be concealed from the decision maker. Once the scheme is introduced, it has to be managed in the short term and the long term. This involves monitoring which should be based on ecological understanding of the plants concerned so that important factors are not excluded but unnecessary work is avoided.

It is in respect of the areas of feasibility study, decision making and monitoring that synthesis of what we now know is essential and that research is most urgent if ecologically based and economic programmes of water weed management in irrigation systems are to be developed.

References

Abdel-Malik, E. 1958. Factors conditioning the habitat of bilharziasis intermediate hosts of the family Planorbidae. *Bull. Wld Hlth. Org.* **18**, 785–818.

Airey, J. D. 1973. The biology of *Limnaea natalensis* (Krauss) in Rhodesia, M.Phil. Thesis, University of London.
Barley, T. R. and Gangstad, E. O. 1974. Environmental aspects of aquatic plant control. *J. Irrig. Drain. Div. Am. Soc. Civ. Engrs.* **100**, 231–244.
Bennett, F. D. 1974. Biological control. In: *Aquatic vegetation and its use and control*, pp. 99–106, ed. Mitchell, D. S., Paris, UNESCO.
Blackburn, R. D. 1974. Chemical control. In: *Aquatic vegetation and its use and control*, pp. 85–98, ed. Mitchell, D. S., Paris, UNESCO.
Bogart, D. B. 1949. The effect of aquatic weeds on flow in Everglades Canal. *Proc. Soil Sci. Soc. Fla.*, 9 Clewston, Fla., USA, 1948–1949, pp. 32–52.
Bowmaker, A. P. 1973. Hydrophyte dynamics in Mwenda Bay, Lake Kariba, Kariba Studies, No. 3. Salisbury, Rhodesia, Trustees National Monuments and Museums.
Bowmer, K. H. 1973. Aquatic weeds, herbicide use and water quality. Seminar on pollution problems of the River Murray, 27–29 Nov. Wentworth, Australia.
Boycott, A. E. 1936. The habitats of fresh water Mollusca in Britain. *J. Anim. Ecol.* **5**, 116–186.
Brezny, O., Mehta, I. and Sharma, R. K. 1973. Studies on evapotranspiration of some aquatic weeds, *Weed Sci.* **21**, 197–204.
Bruns, V. F. 1967. Submerged aquatic weed control in an irrigation system with acrolein. Abstr. Meeting Weed Soc. America, Feb. 13–16, Washington, DC.
Bruns, V. F. 1969. Response of sugar beets, soy beans and corn to acrolein in irrigation water, Abstr. Weed Control: why and how. Washington State Weed Conf. Nov. 5–7, Yakima, Washington.
Bruns, V. F., Demint, R. J., Frank, P. A., Kelley, A. D. and Pringle, J. L. 1974. Responses and residues in six crops irrigated with water containing 2.4-D. Coll. of Agric. Res. Centre, Washington State University.
Bruns, V. F. and Kelley, A. D. 1974. Weed seeds in irrigation water, Abstr. Meeting Weed Soc. America, Feb. 12–14, pp. 33–34.
Bruns, V. F., Yeo, R. R. and Arle, H. F. 1964. Tolerance of certain crops to several aquatic herbicides in irrigation water. Tech. Bull. 1299 Agric. Res. Serv. US Dept. Agric., Washington, DC.
Dawood, K. J., Farooq, M., Dazo, B. C., Miguel, L. C. and Unrau, G. O. 1965. Herbicide trials in the snail habitats of the Egypt 49 project. *Bull. Wld Hlth. Org.* **32**, 269–287.
Demint, R. J., Frank, P. A. and Comes, R. D. 1970. Amitrol residues and rate of dissipation in irrigated water. *Weed Sci.* **18**, 439–442.
El-Gindy, H. I. 1962. Ecology of snail vectors of bilharziasis. *Proc. 1st Int. Symp. on Bilharziasis*, pp. 305–318, Cairo, Govt. Printer.
Frank, P. A., Demint, R. J. and Comes, R. D. 1970. Herbicides in irrigation water following canal-bank treatment for weed control. *Weed Sci.* **18**, 687–692.
Gupta, O. P. 1973. Aquatic weed control for efficient water use. Rajasthan College of Agriculture, Udaipur Tech. Bull. 2.
Guscio, F. J., Bartley, T. R. and Beck, A. N. 1965. Water resources problems generated by obnoxious plants. *J. Watways Harb. Civ. Am. Soc. Div. Engrs.* **10**, 47–60.
Hira, P. R. 1969. Transmission of schistosomiasis in Lake Kariba, Zambia. *Nature Lond.* **224**, 670–672.
Jordan, P. and Webbe, G. 1969. Human schistosomiasis. London. Heineman Medical Books Ltd.
Linacre, E. T., Hicks, B. B., Sainty, G. R. and Grauze, G. 1970. The evaporation from a swamp. *Agr. Meteorol.* **7**, 375–386.
Little, E. C. S. 1967. Progress report on transpiration of some tropical water weeds. *Pestic. Abstr.* **13**, 127–132.
Maddox, D. M., Andres, L. A., Hennessey, R. D., Blackburn, R. D. and Spencer, N. R. 1971. Insects to control alligator-weed: an invader of aquatic ecosystems in the United States. *Bioscience*, **21**, 985–991.
Malhotra, S. P. 1972. Remedy for aquatic weeds in Bhakra canals. Chandigarh, Roxana Printers.
Manning, J. H. and Johnson, R. E. 1975. Water level fluctuation and herbicide application: an integrated control method for *Hydrilla* in a Louisiana reservoir. *Hyacinth Contr. J.* **13**, 11–17.
Manning, J. H. and Saunders, D. R. 1975. Effects of water level fluctuation on vegetation in Black Lake, Louisiana. *Hyacinth Contr. J.* **13**, 17–21.
Mehta, J, Krishna, R. and Taunk, A. P. 1973. The aquatic weed problem in the Chambal Irrigation Area and its control using Grass Carp fish, Abstr. Reg. Semin. Noxious aquatic vegetation in tropics and sub-tropics. 12–17 Dec., New Delhi, pp. 48–49, New Delhi, India Printers.
Mehta, I. 1975. Asst. Plant Physiologist, Soil and Water Management Station, Kota, Rajasthan, India. Personal communication.
Mehta, I. and Sharma, R. K. 1973. Effect of weeds on the flow capacity of Chambal Irrigation System in Kota Rajasthan, Abstr. Reg. Semin. Noxious aquatic vegetation in tropics and sub-tropics, 12–17 Dec., New Delhi, pp. 7–8, New Delhi, India Printers.
Michewicz, J. E., Sutton, D. L. and Blackburn, R. D. 1972. The White Amur for aquatic weed control. *Weed Sci.* **20**, 106–110.
Misra, R. 1973. Seasonal dynamics and aquatic weeds of the low-lying lands of the Upper Gangetic Plain, Abstr.

328 D. S. MITCHELL

Reg. Semin. Noxious aquatic vegetation in tropics and sub-tropics, 12–17 Dec., New Delhi, pp. 24–26, New Delhi, India Printers.

Mitchell, D. S. 1970. Autecological studies of *Salvinia auriculata*, Ph.D. Thesis, University of London.

Mitchell, D. S. 1974a. Water weeds. In: *Aquatic vegetation and its use and control*, pp. 13–22, Paris, UNESCO.

Mitchell, D. S. 1974b. Environmental management in relation to aquatic weed problems. In: *Aquatic vegetation and its use and control*, pp. 57–71. Paris, UNESCO.

Mitchell, D. S. and Thomas, P. A. 1972. Ecology of water weeds in the neotropics, Paris, UNESCO (Technical papers in hydrology, no. 12).

Paperna, I. 1968. Studies on the transmission of schistosomiasis in Ghana, III. Notes on the ecology and distribution of *Bulinus truncatus rohlfsi* and *Biomphalaria pfeifferi* in the Lower Volta Basin. *Ghana Med.* 7, pp. 139–145.

Penfound, W. T., Hall, T. F. and Hess, A. D. 1945. The spring phenology of plants in and around the reservoirs in North Alabama with particular reference to malaria control. *Ecology*, 26, 332–352.

Rijks, D. A. 1969. Evaporation from a papyrus swamp. *Q.J.R. Met. Soc.* 95, 643–649.

Robson, T. O. 1974. Mechanical control. In: *Aquatic vegetation and its use and control*, pp. 72–84, ed. Mitchell, D. S., Paris, UNESCO.

Rozeboom, L. E. and Hess, A. D. 1944. The relation of the intersection line to the production of *Anopheles quadrimaculatus*. *J. Natn. Malar. Soc.* 3, 169–181.

Rudescu, L., Niculescu, C. and Chivu, I. P. 1965. Monografia stufului den delta Dunarii, Romania. Editura Academiei Republicii Socialiste.

Sainty, G. 1974. *Elodea canadensis*, Fmrs.' Newsl., Lecton, 123, 10–13.

Schiff, C. J. 1974. Seasonal factors influencing the location of *Bulinus* (*Physopsis*) *globosus* by miracidia of *Schistosoma haematobium* in nature. *J. Parasit.* 60, 578–583.

Sculthorpe, C. D. 1967. The biology of aquatic vascular plants. London, Edward Arnold.

Smid, P. 1975. Evaporation from a reedswamp. *J. Ecol.* 63, 299–309.

Stephens, J. C., Blackburn, R. D., Seaman, D. E. and Weldon, L. W. 1962. Flow retardance by channel weeds and their control. *J. Irrig. Drain. Div. Am. Soc. Civ. Engrs*, 89, 31–53.

Timmons, F. L. 1960. Weed control in western irrigation drainage systems. Washington DC, Agric. Research Service, US Dept. of Agric. and Bur. of Reclam., US Dept. of the Interior, joint report. (ARS 34–14).

Timmer, C. E. and Weldon, L. W. 1967. Evapotranspiration and pollution of water by water hyacinth. *Hyacinth Contr. J.* 6, 34–37.

Timmons, F. L., Bruns, V. F., Lee, W. O., Yeo, R. R., Hodgson, J. M., Weldon, L. W. and Comes, R. D. 1963. Studies on the control of common cattail in drainage canals and ditches. Washington DC, Agric. Research Service, US Dept. of Agric. and Bur. of Reclam., US Dept. of the Interior (Techn. Bull. no. 1286).

van Someren, V. D. 1946. The habitats and tolerance ranges of *Lymnaea* (*Radix*) *caillaudi*, the intermediate snail host of Liver fluke in East Africa. *J. Anim. Ecol.* 15, 170–197.

Wild, H. and Mitchell, D. S. 1970. Der Einfluss von Bayluscid auf den Wasserfarn *Salvinia auriculata* und andere Wasserpflanzen, Höfchenbr. Bayer PflSchutz-Nachr. 23, 112–117.

Williams, R. H. 1956. *Salvinia auriculata* Aublet: the chemical eradication of a serious aquatic weed in Ceylon. *Trop. Agric. Trin.* 33, 145–157.

World Health Organisation. 1959. International work in bilharziasis, Geneva, WHO.

Yeo, R. R. and Fisher, T. W. 1970. Progress and potential for biological weed control with fish, pathogens, competitive plants, and snails, Rome (Italy), FAO Int. Conf. Weed Control.

CHANGES IN TERRESTRIAL BIOTA

EFFECT OF IRRIGATION ON SOIL FAUNA by S. I. Ghabbour
(University of Cairo)

Soil fauna in desert soils are very sparse but consist of specially adapted species which can withstand drought and can take advantage of minimal gains in soil moisture from rainfall or dew to feed and to breed. These species are mainly: harvester and carpenter ants which feed on seeds and on dead decaying wood; sand roaches which have the unique ability of absorbing water vapour from the soil atmosphere with relative humidities of 82%; a host of tenebrionid beetles which feed on decaying litter; and a number of predators such as carabid beetles, ant lions and centipedes. These animals usually have a daily rhythm of vertical migration from the soil surface to deeper layers governed primarily by the changing temperature and moisture gradients which are inversed twice daily at sunset and at sunrise. They usually concentrate under desert shrubs where litter accumulates, places where protection is provided against excessive heat, evaporation and predators. The population densities of these animals are closely correlated with seasonal variations in the moisture and temperature regimes, as has been found in deserts both in Egypt and in the United States.

Earthworms, on the other hand, are the main constituent of the soil fauna in irrigated soils, together with a host of collembola, mites, nematodes and other minor animal groups. The population densities of these animals are also dependent upon variations in soil moisture which vary according to the frequency of applying irrigation water. They also depend on soil texture which determines the level of water available to the organisms. Studies on survival of earthworms in sandy and clayey soils of Egypt when irrigation is stopped have shown that earthworms cannot remain active for long in sandy soils (of 28% water-holding capacity) but remain active indefinitely in clayey soils (of 50% water-holding capacity).

When sandy soils are irrigated, levels of moisture content rise from the natural range of 0–7% to an artificial range of 15–22%. Under such conditions earthworms, nematodes and other groups of soil animals invade the newly irrigated sandy soils and their populations build up gradually at the expense of the original groups of soil fauna which cannot withstand the new high moisture levels. Populations of earthworms in newly irrigated sandy soils in Egypt reach from 32,000 to 1,000,000 individuals per acre. If the plot is used for afforestation, they may reach 3,000,000 individuals/acre. These figures are attained in favourable conditions, which are permanent under orchards and artificial forests and transient under field crops, namely: an undisturbed soil (ploughing is very harmful for earthworms), a regular and adequate water supply, a fine soil texture (to raise water availability), and a regular and adequate supply of organic matter.

The replacement of the original xeric soil fauna by ruderal species does not necessarily mean that all autochthonous species are wiped out. On the contrary, some species of namatodes, bugs, etc., originally present, may become pests of the introduced crops — as has happened, for example, at Wadi Natrun in Egypt.

Because of differences in degree of water availability between sandy and clayey soils, sandy soils of 16–27% moisture content are optimal for species of earthworms which thrive in clayey soils of 25–45% moisture content. The determining factor is the pF and the relative humidity of the soil atmosphere, which is a function of pF.

Earthworms have been deliberately introduced in the southern parts of USSR to accelerate reclamation of newly irrigated sandy desert soils by improving their properties and this has met with considerable success. The nitrogen excretion of earthworms is one of their contributions to soil fertility. Under experimental conditions, *Allolobophora caliginosa*, the common earthworm in Egypt, produces 6.5% of its fresh weight as urine each day. The average fresh weight of a normal population in sandy soils might be 30 kg/acre, which produces about 2 litres of urine per day. This urine will contain 0.06 g ammonia and 0.8 g urea; the mucus produced will contain about 20 g protein; and further quantities of ammonia and urea are also contained in the casts. With increase in soil moisture levels, more ammonia and less urea are excreted by earthworms.

Remarkable changes have been observed also in the vertebrate fauna associated with the soil in newly irrigated areas of arid regions. For example, rodent outbreaks are recorded from Ismailia, Faiyum and Kom Ombo in Egypt. Under natural arid conditions, rodent populations are kept under control by scarcity of food and predator pressure, but both constraints are removed by irrigation. This is also the case with birds such as the desert sand-grouse (*Pterocles*) which is a pest in Nubia and *Passer hispaniolensis* on the northwestern coast of Egypt.

EFFECT OF IRRIGATION ON WEED FLORA IN EGYPT by M. Imam
(University of Cairo)

A five-year survey of plant species directly connected with the irrigation system showed the following changes:

(a) Changes related to the shift from basin to perennial irrigation: many species have not been observed in most of the localities mentioned in old literature. Among these are *Caldesia reniformis* and *Alisma plantago-aquatica*, which, it seems, need to be inundated for at least part of the year.

(b) Changes related to cessation of annual flood: *Glinus lotoides* and *Riccia* sp. (a liverwort), which used to cover Nile banks after the recession of flood water, have disappeared. *Glinus lotoides* appeared on the shores of Lake Nasser where it is mixed with local desert plants.

(c) Changes in the rice-field weeds: a survey showed that fourteen species recorded in old literature are now rarely encountered. Most of these disappeared owing to the lower salinity of the soil especially in the northern Delta.

(d) The change from "Nili" rice (3–4 month growing season) to the Summer rice "Seifi"

(6-month growing season) in the Faiyum province probably accounts for ten new records of rice weed species. Some of these weeds are chronically harmful to the crop.

(e) Changes in the aquatic weeds in the Nile and associated irrigation system have followed construction of the High Dam. The annual flood used to wash away most of the aquatic weed growth, in particular floating weeds. In 1964, the year when the flood stopped, water hyacinth suddenly became a serious problem. In 1965 the Ministry of Irrigation started a three-year crash control programme. This helped for a short time, then the problem reappeared. Another programme was started in 1975. The confined water between the Old Aswan Reservoir and the New High Dam is now blocked with submerged weeds. Lake Nasser waters support submerged weeds but not, as yet, floating plants. Preventive means need to be developed for this body of water.

IMPACT OF CHANGING IRRIGATION ON AGRICULTURAL PESTS AND WILDLIFE
IN EGYPT by A. Mahir Ali (University of Assiut)

(a) An interesting study is provided by the impact of the High Dam and associated change in the irrigation system from basin to perennial on the agricultural pests and wildlife in the area. For instance, there is a noticeable increase in the number of crocodiles (*Crocodilus niloticus*) in lake Nasser, and gazelle herds (*Gazelle leptocera* and *G. dorcas*). Ibex (*Capra nubica*) and barbary sheep (*Ammotragus lervia*) are reappearing and plant growth is also increasing.

(b) On the other hand there are a few unpleasant changes. There is a change in the type and density of weeds in parts of the area; some of the perennial weeds such as *Cynodon dactylon*, *Cyperus rotundus*, and *Convolvulus arvensis* are now very common.

(c) There is an increase in some economic insect pests, and this increase seems to be moving southward. The grape moth (*Polychrosis botrana*) which was not known in Upper Egypt has become common. The corn stalk borer (*Chilo agamemnon*) is extending southward.

(d) The cotton leafworm (*Spodoptera littoralis*) is increasing in density. The following table gives the numbers of egg masses in millions per year as collected throughout the whole season in all cultivated lands.

Governorate	1967	1968	1969	1970	1971	1972	1973	1974	1975
Beni Suef	18.8	22.7	84.2	9.1	22.7	27.5	49.9	101.2	55.6
Minia	2.7	4.4	26.4	18.3	19.7	8.2	42.1	430.4	166.4
Assiut	0.3	1.6	30.9	5.7	6.6	3.9	5.5	96.2	34.7
Sohag	0.5	1.7	3.5	3.7	5.2	2.7	3.5	45.7	14.0
Quena	0.1	0.1	1.4	0.7	0.9	2.4	2.1	12.6	0.8

(e) The Nile grass rat (*Arvichanthis niloticus*) is now invading barns and villages; apparently it has changed its habits from being a purely field rat to a commensal rat. *Rattus* spp. used to invade dwellings and Nile boats at the beginning of rising water during flood time. Now, because the water level in the Nile is almost constant and the rat burrows are no more submerged under flood water, this seasonal invasion has stopped.

(f) Another area with a more diversified ecology than the Nile Valley is Sinai and the Canal Zone. This includes vast desert, cultivated land and wetland. Agriculture will depend on open irrigation and water sprinkling with a part of the wetlands to be turned into cultivated land. Such projects are expected to bring about some drastic changes in wildlife, and these will have ecological and economic impacts on the development of the area.

EFFECT OF IRRIGATION ON SOIL MICRO-ORGANISMS by M. S. El-Abyad
(University of Cairo)

The effect of irrigation on the microbial populations of desert soils can best be illustrated through studies of a series of profiles made at Tahrir Province where soils of different reclamation history are found. A description of the localities in which profiles were made is as follows:

Profile No.	Date of sampling	Locality	Description
1	23 Aug. 71	Northern Sector (annual rainfall 138 mm)	Yellow calcareous soil cultivated for 8 years with monthly irrigation.
2	23 Aug. 71	„ „	Uncultivated calcareous soil dominated by *Mesembryanthemum nodiflorum* and *Salsola inermis*.
3	6 Feb. 72	Southern Sector	Sandy soil reclaimed 12 years ago, cultivated with *Manginfera indica*.
4	6 Feb. 72	„ „	Sandy soil reclaimed 8 years ago, cultivated with *Pisum sativum*.
5	6 Feb. 72	„ „	Red sand and gravels, reclaimed and cultivated with *Arachis hypogaea* 2 years before sampling.
6	6 Feb. 72	Desert Wadi	Sand and gravels, plant cover dominated by *Artemisia monosperma*.

Each profile was sampled at 4 depths: 0, 0–10, 10–20 and 20 cm, and for purposes of comparison, mean values of the results obtained at the 4 depths are presented below. The microbial counts were obtained by using the "soil dilution plate".

Prof. No.	Moisture content (%)	Org.C (%)	Total carbonate (%)	T.S.S. (%)	Loss-on-ignition (%)	pH	F ($\times 10^2$)	B & A ($\times 10^3$)
1	21.42	0.279	31.172	0.388	7.06	8.01	17.2	1080.4
2	7.79	0.192	23.624	2.500	10.61	9.32	5.2	26.2
3	2.32	0.408	1.714	0.139	2.04	8.49	51.6	165.7
4	2.79	0.025	2.620	0.115	0.55	8.59	10.9	46.5
5	3.09	0.055	2.810	0.096	3.033	8.95	17.4	99.3
6	0.66	0.014	0.192	0.157	0.546	9.37	1.5	11.7

F = Fungi; B & A = Bacteria and Actinomycetes.

These results indicate that the counts of fungi, bacteria and actinomycetes were higher in the reclaimed soils (profiles 1, 3, 4 and 5) than in the non-reclaimed soils (profile 2 and 6). However, among the last two profiles, the calcareous profile (No. 2) supported greater number of micro-organisms than the sandy one (No. 6). This may indicate that the total population of soil micro-organisms is greatly influenced by soil fertility, and this is affected by the type of soil, moisture content, organic matter content, total soluble salts, total carbonates, and whether the soil is regularly cultivated with crop plants or not. Generally the counts of bacteria and actinomycetes are far greater than those of fungi. This is expected because of the high alkalinity of the studied soils. When the level of alkalinity (as determined by the total carbonates) decreases coupled with an increase in the percentage of organic carbon, the fungal counts increase (profile 3). The results also indicate that when a calcareous and a sandy soil have the same reclamation history, the former soil supports a larger microbial count than the latter (Profiles 1 and 4 respectively). Thus it may be concluded that although soil water is very important in desert reclamation, yet it must be considered as only one factor among many that induce fertility.

The most common genera of fungi (as determined by the soil plate method) isolated from the different profiles are as follows:

Profile No.	1	2	3	4	5	6
Alternaria	+	+	○	+	+	+
Ambylosporium	○	+	○	○	○	○
Aspergillus	+	+	+	+	+	+
Chaetomium	○	+	○	○	○	+
Cladosporium	○	+	○	○	○	○
Cladotrichum	+	+	○	○	○	○
Cunninghamella	+	+	+	○	○	+
Curvularia	○	○	+	+	+	+
Fusarium	+	+	+	+	+	+
Helminthosporium	○	○	○	+	○	○
Hormiscium	○	○	○	+	○	○
Mucor	+	+	○	○	○	○
Mycelia sterilia	○	+	○	○	○	+
Nigrospora	○	○	○	○	○	+
Paecilomyces	○	○	+	+	○	○
Penicillium	+	+	+	+	+	+
Rhizopus	+	+	+	+	+	+
Stachybotrys	○	+	○	○	○	+
Stemphylium	○	+	○	+	○	○
Syncephalastrum	○	○	○	○	○	+
Torula	○	○	○	○	+	○
Total	8	14	8	9	7	12

It is interesting to notice that both the unreclaimed profiles (2 and 6) supported the largest numbers of genera (14 and 12 respectively). The rest of the reclaimed profiles support practically equal numbers. This is probably attributed to the fact that reclamation and its effects on increasing soil-moisture content and organic content through amendment by manures and excretions by the roots of plants create selective conditions to the fungal genera. However, the number of genera is not always correlated with the fungal counts, for a lower number of genera may give high counts and the reverse is also true. Also a small number of genera may account for a larger number of species. In this respect, identification up to the species level is useful for such comparisons.

Generally the soil samples were dominated by four genera: *Aspergillus, Fusarium, Penicillium* and *Rhizopus. Alternaria, Cunninghamella* and *Curvularia* are also very well represented.

DISCUSSION AND CONCLUSIONS

Compiled by M. KASSAS

University of Cairo

Irrigation is part of a complex of practices comprising management of available water resources, controlled distribution of this water over cultivated land, and withdrawal of excessive water through drainage. Ecological consequences of this complex include: (1) the creation of new ecological systems related to water bodies such as reservoirs, irrigation canals and drainage ditches; (2) radical modification of ecological systems of the terrestrial habitat, related to practices of irrigated agriculture such as ploughing, introduction of crop plants and fertilizing.

Some of these ecological changes may be conceived as change from an ephemeral situation to a perennial situation. Water bodies in the form of pools and ponds of various sizes and running water in torrential wadis are ephemeral features that follow the incidents of cloudburst rainfall in arid lands. These provide short-lived habitats for aquatic life. But as irrigation schemes are established, perennial water bodies provide habitats for types of aquatic life that are alien to the arid lands.

Other ecological changes are related to the intensive management. The sub-humid and arid ecosystems are inherently unstable and exhibit a potential for dramatic changes which are readily triggered by the sudden appearance of extensive areas of irrigated crops; for example, sudden increases in insect or bird populations, or weed flora. Moreover, natural dry-land ecosystems have a limited capacity for assimilating, withstanding and responding to inputs of water, chemicals and energy that are associated with an irrigated agro-ecosystem. Farming practices extend this capacity through changing the character of soil and vegetation. Problems of salinization and other forms of fertility deterioration are in fact symptoms of manipulation exceeding the natural capacity of the ecosystem.

Irrigation changes the native ecological set up and creates new productive systems on land such as crop fields, orchards, wooded areas, irrigated pasture and housegardens, and in water it produces fisheries and aquaculture. These changes in the physical and biological environment lead to radical changes in the ecology.

Human ecology is likewise drastically disturbed, for the farmers of irrigation schemes often comprise settled populations of otherwise nomadic indigenous peoples, and the relocation of communities, families and individuals from other probably distant parts of the territory. There are sociological interactions and problems including socio-medical consequences that need to be carefully managed. An example is the Khashm-el-Girba irrigation scheme on the Atbara River in the Sudan where populations of three ethnic groups were

335

brought to live together: the Nubia river-terrace farmers, the Hadendowa from the Red Sea coastal mountain country who were mostly nomadic pastorals, and the Shukrya of the Butana plains that had a mixed rain-fed farming and stock-breeding occupation. Such mixed populations associated with new irrigation schemes have welcome positive socio-political results, but they also have problems.

Of the papers in this Section dealing with biological balances two are concerned with new productivity potentials, namely fisheries and aquaculture, and two with environmental hazards, namely vectors of disease and aquatic weed growth. The purely agricultural aspect of ecological change, which already has a huge literature, is not specifically covered; but the four short papers on changes to terrestrial biota, including soil fauna and flora, weeds and wildlife, introduce topics of importance to agriculture which have been somewhat neglected. One of the rather striking conclusions, whatever the subject—fish or fungi— is that, whereas irrigation obviously increases the potential of biological productivity, it generally results in a reduction in the diversity of plant and animal life. The more uniform habitat created by regular application of water allows some species to wax tremendously, while others disappear under this artificial form of "natural" selection. Unfortunately the species which become dominant are not always those which are desirable from the social and economic viewpoints.

At the symposium in February 1976 there was a discussion around these papers and other matters of biological balance. The contributions, with a mention of the scientists who made them, are summarized below under three headings.

Inland fisheries and aquaculture

The two papers by Drs. Daget and Welcomme stimulated the following contributions from C. Storsbergen and F. L. Hotes respectively.

Experiences in the Far East have shown that rice farming is controversial to fish farming management. Modern rice varieties need water only 5–10 cm deep, which is the minimum for efficient fish farming, and fish in paddy fields are an easy prey for birds. Fish and rice have different development cycles and draining the fields to harvest fish may harm the rice. Irrigation engineers object to fish cages and similar obstructions in irrigation canals as they hamper the water flow. In Indonesia and the Philippines fish culture is always kept separate from irrigation systems; but even so, discharge problems occur during floods, as the rice field discharge is hampered by extensive areas of fish ponds. A combination of rice and fish farming is only possible in places like Bangladesh with rice varieties accepting deeper water (floating rice).

The World Bank is prepared to finance inland fisheries as part of irrigation schemes if economically and technically feasible. Last year the World Bank agreed to finance a US $50 million irrigation-flood control project in the lower Sâo Francisco River Valley in northeast Brazil which included an initial $2 million for fisheries resources development. This comprised three productive components: (1) a few commercial fish ponds, using primarily local species; (2) stocking of small ponds in the flood plain which are filled with water for at least six months of the year and where hybrid monosex *Tilapia* stocking is proposed; (3) stocking small farm ponds (0.5 ha or less, adjacent to rice fields).

Dr. Daget commented: "Il est certain que l'élèvege des poissons est parfois incompatible avec certaines pratiques culturales ou certains aménagements dans les systèmes irrigués.

Mais on ne saurait en conclure que les cultures irriguées et l'élévage des poissons doivent toujours être séparés. En effet, dans certains cas leur combinaison harmonieuse, sous forme de rizipisciculture, a donné de bons résultats. La confrontation des résultats obtenus dans des conditions et des contrées différentes est toujours intéressante. Or en Afrique les techniques de pisciculture et de riziculture sont souvent moins perfectionnées qu'en Indonésie ou en Extrême Orient. Tenant compte des expériences déjà faites, des essais devraient être tentés dans divers pays pour savoir dans quelles conditions la production de poissons de consommation est ou n'est pas compatible avec les exigences de la culture dans les systèmes irrigués."

Disease vectors in water bodies

The paper by N. M. Hill *et al.* was discussed by M. A. Amin, J. A. Coumbaras, A. M. Farid, A. A. Idris, L. Obeng, R. S. Odingo and S. Pels, who raised the following points.

Is the spread of *An. gambiae* in the irrigation schemes in the Kano Plains due to irrigation itself or mainly due to misuse and lack of discipline in irrigation practices? To which A. J. Chandler replied that the effect of irrigation and rice cultivation has been to magnify enormously the normal breeding cycle of this mosquito. Under normal conditions *An. gambiae* peak populations follow seasonal rainfall. Two crops of rice are grown each year, and the effect of this is to increase and extend breeding populations. Approximately 70% of the breeding of *An. gambiae* takes place in rice fields, in particular those recently transplanted, whilst the remainder occur in flood and waterlogged areas near drains and seepage areas.

Conclusions drawn from studying the ecology of *An. gambiae* in Kenya apply to the local mosquito population of that area. When this species invaded Egypt and was exposed to a seasonal climate it bred during the winter in vegetation in the canals as well as in vegetation mats along the Nile. It cannot usually fly more than 1 km so houses built beyond this range from canals and rice fields might avoid mosquito invasion. To this N. M. Hill replied that each case needs individual study. Siting the homes of workers away from irrigated areas should be considered during the planning of any irrigation scheme, but it is difficult to house the population outside the environs of a very large scheme, neither is it easy to persuade land owners whose homes are adjacent to irrigation schemes to move away when areas with an indigenous human population are irrigated.

Under conditions of arid land irrigation do mosquitoes breed where the water quality tends to have a high salt content? To which N. M. Hill replied that many mosquito species are capable of breeding in saline and brackish waters, for example *An. merus* of the *An. gambiae* complex, and many *Aëdes* species are also saltwater breeders.

On the Kano Plains the increase in malaria is important, but the problem may not be so serious in an area where this disease is already endemic. The implication that schistosomiasis is also prevalent in the irrigated area may, however, be questioned. Research by Dr. Kimoti on the same irrigation scheme has established that *Bulinus* and *Biomphalaria* snails which play such a significant part in the transmission of schistosomiasis do not like the black cotton soil areas but proliferate more where soils are sandy and nearer the escarpments. Also, why are the vectors of schistosomiasis not present on the AHERO Pilot Scheme? To this A. J. Chandler replied by emphasizing that the Kano Plains is not a truly arid area, and that conditions exist for the maintenance of endemic malaria and schistosomiasis in this heavily populated region. The irrigation schemes are essentially land reclamation from swamp and a redistribution of available water. In this context malaria, while it is undoubtedly a serious problem in the region, does not seem greatly affected by irrigation. It should be stressed that if general malaria control measures were to be applied on the Kano Plains, particular vigilance would be required to ensure adequate protection of the populations on the irrigated areas. Similarly while the schistosomiasis vector (*Bulinus*) has not been recorded on the irrigations schemes, the vector snail and cases of the disease are found in the Kano Plains region. In the AHERO Pilot Scheme since its inception in 1968 the main inlet canal and major subcanals have been treated with the molluscicide "frescon"; this pilot scheme is an important example where chemical control of vector snails has been successful.

Why is rice hand-planted in the Kenya irrigation project? This method was traditionally developed in regions where the period of natural flooding was too short for the growing period of rice. With adequate artificial control of irrigation water, pregerminated rice seed can be sown directly into flooded fields, and this allows more adequate weed control and also assists the establishment of rice on salt-affected land. The average yield of rice grown in this way in Australia is 7 tons per hectare. Also, does slowly moving water reduce the number of mosquitoes derived from rice fields? In Australia rice is commonly grown in bays with a continuous small discharge into the drainage system and hence in moving water. To this N. M. Hill replied that hand-planted rice is a labour-intensive crop requiring low capital investment. Maintaining a water flow in rice field plots as is done in Australia would probably reduce mosquito breeding. Examination of the distribution of mosquitoes in rice field plots with water inlets and outlets has shown that mosquitoes become concentrated near the corners of the field away from the inlet, and local application of insecticides has been proposed.

Irrigation provides habitats for organisms (including vectors) which have not existed in the area prior to irrigation, for example snails have spread from the Nile to the Rahad Irrigation Project in the Sudan. Furthermore, the people coming to live in the irrigated area bring with them vectors and organisms uncommon in the area.

There is an obvious need for a multi-disciplinary approach to the interrelated processes and problems associated with irrigation schemes. Thus before making the designs it is important to consult the health authorities; before introducing fish it is important to consider the environmental and public health hazards and their possible means of biological control; before pouring in insecticides or molluscicides it is important to consider the environment and whether the objective can be attained by other means.

The authors of the paper (Hill and Chandler) made the following points:

(a) Irrigation may result in a considerable simplification of natural habitat, with a corresponding reduction in species diversity.

(b) Changes in water flow and water table levels result in changes in species composition. Thus species such as *Mansonia uniformis* and *M. africana* which breed essentially in permanent waters have given way to *Anopheles gambiae* and *An. funestus* which breed mainly in temporary waters.

(c) Increase in areas of above-ground water leads to increase in the total mosquito population; in Kisumu rice fields the night biting rate upon humans increased four times. Changes in faunal composition cause changes in mosquito feeding habits: on Kano Plains 70% *An. gambiae* feed on cattle, 30% on man, the reverse in rice growing areas.

(d) Human populations tend to increase near irrigated areas, and this brings greater numbers of people into closer contact with mosquito populations. Urbanization with inadequate sewage facilities may lead to increase in other vectors, such as *Culex pipiens fatigans*, a vector of filariasis, and *Aëdes aegypti*, a domestic species and vector of yellow fever. Large numbers of mosquitoes constitute a nuisance which easily justifies control measures. In areas with stable malaria, irrigation practices are unlikely to have a serious effect on prevalence of the disease, but in areas with unstable malaria irrigation may lead to epidemic infection, as occurred in Portugal in 1930 and in the Sudan in more recent years.

(e) The connection between arbovirus infections and irrigation practices has been well documented for many viruses. Examples include Japanese encephalitis in the Far East, West Nile encephalitis in Europe, and equine encephalitis in the USA. All these diseases have occurred in epidemic proportions as a result of complex interrelationships between mosquito fauna, avian fauna and man.

(f) It is not intended to stress the negative aspects of irrigation schemes, but it is important to emphasize that in areas where the ecosystem is changing, the possibility of different relationships between mosquitoes, man and other animals, giving rise to new disease systems is always present. This situation needs constant monitoring if outbreaks of disease in man and domestic animals are to be avoided.

Water weeds

The main paper by D. S. Mitchell summing up the ecological relationships and environmental impact of aquatic plants, and ways of controlling water weeds, was commented on by M. Abu Zeid, H. Said, C. H. Swan and E. B. Worthington.

The very high cost of weed clearance is well known to anyone who has been responsible for operating irrigation schemes. Great contributions can be made by engineers at the design stage if they know the problems likely to arise. From this two points emerge: (a) pilot projects are very desirable where schemes are for development in new areas, and (b) the cost of weed clearance is often not sufficiently taken into account when assessing the desirability of lining canals.

Aquatic weeds are one of the most serious problems facing water management in Egypt. One can hardly find a drain or canal which is free from weeds. The Ministry of Irrigation has been introducing weed control measures but it is a very costly process and complicated to handle. In Egypt irrigation canals are unlined; irrigation rotations are applied where main canals run continuously, but branch canals and laterals receive water for a period of 4–6 days every 8–18 days. Every year during December and January canals are closed and drained for about three weeks for maintenance. Since the High Dam came into operation the absence of silt in irrigation water has increased weed growth. Last year two Egyptian programmes were launched: the first is a country-wide manual and chemical weed control scheme, costing about 6 million dollars; the second is a joint Egyptian–Dutch agreement for weed control studies concerning mechanical and biological control.

Asked whether water weeds could be transformed into useful materials, D. S. Mitchell replied that their utilization has been reviewed by Boyd (1974). The conversion of excessive aquatic vegetation to animal and human food is possible and has been investigated. The main problem is that 95% of the plants is water and that often their use is uneconomic.

On the biological control of weeds mention was made of coypu (*Myopotamus coypu*, a large South American rodent from which "nutria" fur comes). This was accidentally introduced into L. Naivasha in Kenya about 1973, and by 1975 practically every water lily had gone and the emergent vegetation was also affected. This was a disaster from the viewpoint of conservation and tourism because the birds and other fauna depended on a super-abundance of water weeds in that lake. D. S. Mitchell commented that this case is a salutary reminder of the need for research on the use of biological control organisms; such research is expensive but is important and must be done. We must express the need for engineers and biologists to cooperate at the planning stage to minimize weed problems. The cost may be justified in terms of savings in water weed control.

SECTION VI

The Efficiency of Irrigation Schemes

AN OVERVIEW

by MILOS HOLY

President of the Czechoslovak National Committee of ICID

Summary

Supplies of fresh water in the hydrosphere are restricted and unevenly distributed in space and time. The biological function of water as the basic component of the biomass cannot be substituted by any other substance. Growing demands on water resources caused by population growth, as well as by growing economic activity of the developing countries, industry and its need of energy and demands on water for personal consumption, are common phenomena. This requires that a systems approach be taken to attain rational use of water resources including multipurpose water conservancy systems to which irrigation systems are connected.

In arid and semi-arid regions the main demand placed on multipurpose water conservation is the safeguarding of water supplies which, in view of fluctuating discharge in rivers, is often accomplished by retention and storage. In such cases the water conservancy will consist of a system of reservoirs of which the high capital cost requires that their use be maximally efficient. Considering the conflicting demands placed on multipurpose systems this efficient use by different sectors of the national economy must be assured. In deciding on the different requirements placed on the system the criterial technical and economic function may be applied in the form for an equation (p. 346).

Only such demands placed on the system should be met whose fulfilment guarantees efficient use of water from the economic and social point of view. This criterion should be applied in the assessment of irrigation efficiency as only those schemes which prove efficient have hope for future priority in water supply.

Efficiency in irrigation requires that water be conveyed to the system and distributed with minimum losses in such a way as to secure maximum efficiency of water use as determined by the ratio of the amount of water used by plants to the amount of water withdrawn from the system. Water losses comprise losses by evaporation, seepage and losses arising in the actual operation of the system. They may be prevented by various technical measures, e.g. closed pipe conduits, canal lining, the use of different irrigation methods, good operation of the system. Technical measures securing high irrigation efficiency are costly. In evaluating each project a wide decision-making process should be applied comprising economic and social criteria which consider the environmental aspects of the project and its impacts on society; but these are not easy to assess in economic terms.

Résumé

Les stocks d'eau douce de l'hydrosphère sont limités et répartis de façon inégale dans le temps et l'espace. La fonction biologique de l'eau, en tant que composante fondamentale de la biomasse, est irremplaçable. La demande accrue en eau due à l'accroissement de la population globale aussi bien qu'à l'essor économique des pays en voie de développement, à l'essor industriel et à son besoin en énergie et à l'élévation de la consommation personnelle liée à celle du niveau de vie, devient particulièrement forte. Ce fait nécessite une approche systémique pour promouvoir une utilisation rationnelle des ressources en eau, y compris des plans conservateurs d'utilisation de l'eau à des fins multiples auxquels les plans d'irrigation sont liés.

Dans les régions arides et semi-arides, l'impératif principal de l'utilisation de l'eau à des fins multiples est le stockage de l'eau pour faire face à une distribution aléatoire des régimes hydriques. Dans ce cas, un système de conservation de l'eau s'identifie à un système de réservoirs. Le coût de ces structures exige qu'elles possèdent une efficacité maximale. Etant donné les pressions contradictoires que subit un système hydrique ayant des fins multiples, l'efficience de l'utilisation de l'eau pour les différents secteurs de l'économie doit être vérifiée. Pour juger des demandes les critères d'une fonction socio-économique peuvent être exprimés par une équation (p. 346).

On ne peut donner suite à des demandes que si elles offrent toutes garanties d'utilisation efficace de l'eau à des fins économiques et sociales. Ce critère devrait être appliqué à l'évaluation de l'efficience de l'irrigation car seuls les projets qui présentent des caractères d'efficacité ont des chances de recevoir de l'eau dans le futur.

L'efficience de l'irrigation nécessite que l'eau parvienne aux périmètres irrigués et soit distribuée avec des déperditions minimes afin d'assurer une efficacité maximale à l'eau, que l'on pourrait déterminer par le rapport de la quantité d'eau utilisée par les plantes à celle gardée à l'intérieur du périmètre. Les pertes en eau comprennent les pertes par évaporation et infiltration ainsi que celles qui se produisent au cours de son utilisation. Elles peuvent être contrôlées par diverses techniques telles les conduits fermés, les canaux ouverts recouverts d'un revêtement, de bonnes opérations d'irrigation. Les techniques assurant à l'irrigation une grande efficacité sont coûteuses. Dans l'évaluation de chaque projet, les processus les plus larges de prise de décision devraient être appliqués, y compris la prise en considération des aspects du projet relatifs à l'environnement et à son impact sur le développement de la société, fait qu'il est difficile d'évaleur en termes économiques.

The uses of water

The importance of water in the biosphere is growing with the advance of society. Water is the most frequently occurring substance on Earth, estimated at $1400 \, km^3$, yet only 1% of the total supply of water in the hydrosphere is fresh and therefore of major importance to mankind. According to Dooge et al. (1973) all this water other than that usable only for fishing and navigation could be used economically by present-day technology.

Supplies of fresh water are not only limited but are also unevenly distributed in space and time. In the course of movement of water within the global circulation it is indestructible and cannot be induced on a large scale. The water cannot be physically consumed, but its properties can undergo transformation of chemical composition, colour, temperature, etc.

As one of the principal resources of the biosphere, water fulfils the following functions in its service to man: domestic purposes, agricultural and industrial production, fisheries, power production, recreation, and river and maritime navigation. In the fulfilment of all these functions the two basic parameters of quantity (distribution of water in space and time) and quality (physical, chemical and biological properties) are always important.

In its biological function as the essential component of the biomass water cannot be substituted for by any other substance. Its significance for vegetation consists in the solution of plant nutrients, the transportation of substances from resource to consumer, participation in cell formation, the hydration of enzymes, its heat regulation, and participation in photosynthesis. The requirements for water of plants cultivated for human nutrition are considerable. Not even the most progressive agrotechnical achievement can alter the fact that 400–500 litres of water or more are necessary for the production of one kilogram of organic (dry) matter.

The United Nations World Food Conference of November 1974 in Rome showed that millions of people starved even in periods of highest food production. World food may be sufficient in total, but this does not solve the problem for it cannot be distributed in the required manner due to political, economic and social problems. Food production must be safeguarded in the area of consumption, and this raises demands on water resources in arid and semi-arid regions. The growth of demand ensues not only from population growth and increasing food consumption, but also from economic activity, including the development of industry and need for energy of all kinds, as well as from growing demands on domestic consumption of water.

For effective development of arid areas it will be necessary to harmonize economic development with the natural conditions. Science will have to help resolve competition in demands on the use of natural resources, particularly of water resources. From the point of view of the national economy, primarily the hydrological cycle is the source of the water supply, secondarily it is the recipient of an acceptable proportion of waste products of the development process.

Water conservancy

Arid areas usually have a mean annual rainfall below 250 mm occurring irregularly; in many desert areas there may be no rain for several years at a time, but occasionally heavy rainfall causes considerable damage. The uneven distribution of discharges in rivers is characteristic for the tropic and subtropic zones, where deviations from mean precipitation are most pronounced. With low precipitation the supplies of groundwater are also small and usually suffice to cover only local subsistence requirements. For efficient use it is therefore necessary to retain surface water in suitable areas and to transport it to places where it is needed, sometimes to considerable distances. For the future the possible desalinization of saline and sea water should also be considered.

Problems of water quality frequently arise, and in arid zones are assessed mainly with regard to the salt content and bacterial pollution. Special attention needs to be devoted also to the re-use of urban and industrial waste waters for irrigation purposes.

It is evident from the above that growing demands on water, of which demands for irrigation water in arid zones form a significant part, require a systems approach to attain rational use. This approach may be applied in multipurpose water conservancy systems to which irrigation would be connected. The multipurpose conservancy is a system of reservoirs and other constructions and measures which, on the basis of prevailing natural conditions and by means of technical, biological, economic, social and other measures, meet the demands placed by society in general. The most frequent demands are securing water for primary use, control of runoff, improvement of water quality, improvement of environment.

In arid and semi-arid zones the principal demand will be the guarantee of adequate water supplies which, in view of fluctuating discharges caused by the uneven precipitation, may be acquired by storage. In areas with heavy rainfall the control of runoff and floods may become dominant to water storage. The management of the system will in either case be extremely exacting, modern techniques will have to be used, and capital costs will be considerable.

A technical and economic model of a multipurpose water conservancy system must be formulated to attain an effective use of investments, with a view to attaining optimum balance between aims and means. In arid and semi-arid zones these aims are likely to include intensive development of irrigation and runoff control, hydropower and improvement of the environment.

The demands may become competitive as not every system is capable of meeting them all. A decision on which demands are to be given priority, and to what extent, needs a criterial function comprising three groups of variables, namely unit costs for equipment for the individual purposes (N^+), unit contribution of water supply to the individual purposes (P^+), unit losses for failure to supply water for the individual purposes (Z^+). If the water conservancy system comprises k reservoirs and is to fulfil i purposes and N^+, P^+,

Z^+ are variable as related to the time development of the system, the time factor t must also be considered.

J. Ríha (1975) formulated a model of such a system by a general criterial technical and economic function expressed in the formula

$$\Phi = \sum_{k=1}^{m} \sum_{i=1}^{n} \sum_{t=1} f(N_{k.i}^+; P_{k.i.t}^+; Z_{i.t}^+) \tag{1}$$

The solution of this function is extremely difficult. It may be simplified by the optimization of only one parameter while considering the yield coefficient, the reproduction coefficient, maturity by profit, etc. Of late non-economic criteria are being applied in connection with the complex problems of the human environment, which makes solution even more difficult.

In optimizing the withdrawal of water from the conservancy system the model must be elaborated for a system having an adequate number of consumers with requirements spread in time. The model may be static or dynamic. The static model considers constant withdrawals and the constant capacity of the reservoirs without regard to future development. It considers the relations of capital and running costs, the price of water, losses caused by limited water supply and the value of profit gained. The result is the optimal distribution of withdrawals assuming constant input. The dynamic model seeks to optimize the time needed for building new storage reservoirs and other works with a view to the assumed development of the whole system as related to the progress of development. The result of the mathematical elaboration of a number of objectives is the elaboration of a rule for operations, determining the storage quantity and the distribution of water among consumers, including irrigation systems.

While it is not the purpose of this paper to deal in detail with the criterial function of multipurpose water conservancy systems, it is evident that through such systems conflicting demands can be given priority which will guarantee the effective use of water from the economic and social points of view. This must be borne in mind when considering irrigation schemes because, with continuously rising demands for water, only efficient schemes will in the future have hope of priority supply.

Losses from seepage and evaporation

The use of water will be the decisive factor in considering the efficiency of an irrigation scheme, given the distribution of irrigation in time and space as related to optimal yields of crops. Water therefore needs to be transported and distributed with minimum losses. The type and arrangement of supply as well as the irrigation methods are arranged to attain this goal.

Losses of water are basically due to evaporation and seepage and losses brought about in operation, as shown by Bos (p. 351). Losses by evaporation from open water surfaces are considerable, for in arid zones potential evaporation is significantly higher than rainfall. Direct solar radiation supplies approximately 60% of the total energy reaching the surface in the course of the day with albedo at 0.11 (Yaron et al. 1973). The high level of evaporation has been brought out by data collected and published by ICID (1967). For instance, in Guiana the annual evaporation loss from open surfaces reaches 139.7 cm, in Burma 114–152 cm, in some regions of India up to 300 cm, in Egypt the Aswan Reservoir has an annual evaporation of 250 cm; in Sudan in the vicinity of Khartoum daily evaporation reaches 7.5 mm, near Wadi Halfa 7.9 mm.

Such values show that surface supply of irrigation water considerably reduces its effective use. In order to reduce the losses the water surface exposed to the sun should be as small as possible, which basically means designing and building deeper narrow canals and screening them by vegetation. Good results have been attained by monolayers forming a film on the open surface preventing evaporation. The National Academy of Science in Washington (1974) suggests aliphatic molecular monolayers which do not restrict the transmission of oxygen to the water and are non-toxic. But in irrigation canals with permanent discharge and varying withdrawals there is the problem of maintaining a continuous film on the water surface, so this method will probably only be applied in reservoirs with a slow-moving water surface. In order completely to prevent water losses by evaporation water must be conveyed through closed conduits, usually placed underground.

For economic reasons most arid countries build conveyance and distribution canals of earth, and in view of the usually high porosity of the soil in these regions, very often sandy, great losses occur by seepage. In unlined canals these losses are related to the infiltration capacity of the soil, the depth of the groundwater table in the neighbourhood of the canals, discharge and velocity of water, length of canal, withdrawal timetable and other factors.

Some data on water losses by seepage from canals are given in ICID (1967). Thus, for instance, of the Upper Bari Doab Canal in India losses range from 13.1% to 19.15% in distributaries and minors. In the alluvial plains of Uttar Pradesh and Punjab in India the transit losses due to seepage in canals are as large as 36% of the supply entering at canal head. In Egypt, during the critical summer months, the main canals draw on an average of about 15,000 million m^3 and losses by seepage are estimated at about 1500 million m^3, i.e. 10%. Observations of the Kara-Kum Canal in the USSR showed that in the first year on average the seepage losses amounted to 43% of the overall discharge at the head, while within several months they had risen to 60%. Seepage losses in Algeria reach about 40% from canals in sandy soils.

Canal linings and closed conduits

The values of seepage from canals vary considerably, depending on climatic and soil conditions and on the density and length of the conveyance and distribution systems. Yet it is evident that losses are large, and to prevent them canals have to be lined, the type of lining selected with a view to the required effect and with regard to costs. Lining techniques are assessed in ICID (1967). Of the less costly techniques colmation is a favourable method used in sealing porous soils, such as sands. It consists in sealing the pores by smaller soil grains which enter the pores during water infiltration. Of chemical lining techniques there is artificial salination by introducing sodium salts into the soil; the soil saturated by exchangeable sodium swells and becomes impervious. The development of plastics makes it possible to line canals with membranes, which may be made on the spot by spreading asphalt, or laid as an industrially manufactured foil. Thermoplastic foils have been used in recent years. Very reliable is concrete lining, which is watertight if joints are sealed properly; the best sealing is by concrete on-site lining. In some arid and semi-arid regions prefabricated concrete flume canals are used.

Losses by evaporation and by seepage are not the only water losses. Others are from canal leakage, damaged dams, runoff from canals by uneconomic operation and discharge on completion of irrigation in intermittent operation. Losses from canals and distributaries may be prevented almost completely by introducing closed pipe conduits. In arid and semi-arid regions these are not being widely built mainly due to high capital cost and because they require a certain level of mechanization in building. One must also bear in mind that the closed pipe conduit can only be used for small flows because costs are very high for big discharges. Therefore, the supply of large quantities of water will always have to be by open canals which will, however, have to be lined. Open canals are useful also for other purposes than irrigation, namely for recreation and navigation.

In considering the economic aspects of closed conduits many factors will have to be considered—the available amount of water, costs of storage and costs of building and operating open canals in areas with restricted water resources, the possibility of enlarging irrigation systems and thus providing food for a larger proportion of the population, reducing the dependence of the area on imports, the use of the water for other sectors of the national economy, etc. Vast savings of water would be attained by using pipe conduits which would certainly be important when considering the planned expansion of irrigation from the present 200 million ha to 500 million ha of land which is suitable for irrigation in different parts of the world.

Factors to be considered should also include social and health factors which cannot as yet be expressed in economic terms. They are aptly described by H. A. Rafajtah of WHO in a paper on the impact of irrigation and drainage on public health presented at the ICID Moscow Congress in 1975. Surface water offers suitable conditions for breeding and development of disease vectors and intermediate pests. It also serves as a carrier of causative organisms of intestinal diseases, such as dysentery, typhoid fever, paratyphoid and cholera. Surface conveyance of water in irrigation causes the spread of disease vectors and pests in irrigated areas and the rate of parasitic diseases in the agricultural population may rise alarmingly. Slow-moving irrigation canals offer suitable conditions where mosquitoes can breed, so malaria may be a serious threat to socio-economic development. The UN Economic Commission for Africa stated in 1970 that "the incidence of water-related diseases has considerable effect on the cost benefit relations of water resources development projects." All these negative phenomena are worsened by the discharge of wastes from human settlements into open canals.

In this connection the WHO representative spoke to ICID of the pipe conduits used in sprinkler irrigation. Such views are being voiced ever more frequently, which means that pipe conduits should be considered in a wide perspective and that social aspects will in some cases be decisive (see Holý and Říha 1973 and Karadi 1975).

Other losses and their prevention

Irrigation efficiency is reduced by water losses on irrigated land, which include the loss of water in the soil profile, return flow, direct evaporation from soil and plants and in sprinkler irrigation losses in the atmosphere. Loss in the soil profile consists of water not used physiologically by plants and seepage into the subsoil. The amount of physiologically unused water is of consequence in heavy soils, but in light soils with low hygroscopicity is negligible for the water balance. Loss by seepage into subsoil below the root zone is affected by the soil class, soil moisture and the state of the vegetation. It is directly dependent on irrigation depth and distribution and on the irrigation method used. Great losses occur in surface and flood irrigation while in sprinkler and trickle irrigation using suitable irrigation intensity they are excluded.

Water losses by return flow from the irrigated area occur when more water is supplied than can infiltrate into the root zone. Water accumulates on the surface and runs off, often washing out nutrients and pesticides with which it may cause pollution. These losses do not occur in correctly performed sprinkler, trickle and furrow irrigation. Losses by evaporation from the surface of irrigated land depend on the method of irrigation. In flooding, border irrigation and sprinkler irrigation, evaporation occurs from the entire surface; in furrow irrigation, only from water in the furrows. Losses by evaporation in sprinkler irrigation depend on the temperature and relative humidity of the air, the saturation deficit, the velocity of wind, the size of the nozzle and the working pressure. It is very difficult to determine this loss, but various investigations have shown that it does not usually exceed 2% of the sprinkled water.

All these losses determine, together with the level of local management, the irrigation efficiency. They are usually expressed by the coefficient of farm efficiency, which depends to a considerable extent on the irrigation method applied. The most frequently quoted coefficients of irrigation efficiency are 0.85 for sprinkler irrigation, 0.75 for furrow irrigation, 0.65 for border irrigation, and 0.50 for flooding.

Of the methods prevailing in the arid zones subsoil irrigation and flooding are most frequent, using water fed by gravity. These methods involve low capital and running costs and low demand for power and machine equipment. Flooding is normally used for crops of high water requirement. That these methods are labour intensive is balanced by the great number of agricultural workers.

Sprinkler irrigation beyond doubt has certain advantages against other methods. It is technically the most efficient, making possible the accurate distribution of water, thus reducing water losses by runoff and infiltration below the root zone. It makes possible mechanization and automation. On the other hand it demands higher skills of the operating personnel and the investment is high (see Pilsburg and Degan 1968). Some specialists believe that surface irrigation can be as economical and as efficient as sprinkler, but in practice losses generally prove to be smaller with sprinkler systems. Moreover it eliminates the danger of propagating and spreading dangerous microbes and vectors of disease. The wide decision-making process, considering all economic and social aspects, may in many cases result in use of sprinkler irrigation in arid and semi-arid zones.

Trickle irrigation has been introduced more widely in recent years. It improves the efficiency of water use by good control of quantity and distribution, and it minimizes surface water where mosquitoes breed. Trickle irrigation is beyond doubt a promising technique for the future, but for wide application operational difficulties arising from clogging of trickles will have to be removed as well as salt accumulation near the margins of the wetted area and on the soil surface (see FAO 1973).

Irrigation efficiency is also affected by other factors, such as the size and shape of the area, its distance from the water resource, the species of cultivated plants. In the effort to increase irrigation efficiency ICID has given these problems detailed study in cooperation with the University of Agriculture and the International Institute for Land Reclamation and Improvement at Wageningen (see Bos and Nugteren 1974).

The advance of society, and with it the continuous exploitation of natural resources, requires that a truly scientific approach be taken to the solution of these problems, and especially to all new projects, anticipating their impacts on the environment and applying systems analysis consistently. This assumes that projects be considered not only as compact units but also as complexes composed of separate components, for this makes it possible to design optimally their structure, organization and functional behaviour. The attainment of irrigation efficiency will contribute considerably towards making comprehensive water conservancy systems purposeful and efficient.

References

Bos, M. G. and Nugteren, J. 1974. *On Irrigation Efficiencies*. ILRI, Wageningen.

Dooge, C. I., Costin, A. B. and Finkel, H. J. 1973. Man's influence on the hydrological cycle. Irrigation and Drainage paper 17. Rome, FAO.

FAO, 1973. Trickle irrigation. Irrigation and Drainage paper 14, Rome.

Holý, M., Kutílek, M. and Ríha, J. 1975. Evaluation of positive and negative phenomena of irrigation in the human environment in projects of multipurpose water conservancy systems. *Proceedings IX Congress ICID.* New Delhi.

Holý, M. and Ríha, J. 1973. Socio-economic problems of water resources development and approaches to their solution. *Proceedings 1st Congress IWRA.* Chicago.

Holý, M. 1975. *Irrigation Structures*. Prague, SNTL (in print).

ICID, 1963. *Economic Quantity of Irrigation Water*. Czechoslovak National Committee, Prague.

ICID, 1967. *Controlling Seepage Losses from Irrigation Canals*. New Delhi.

ICID, 1967. *World-wide Survey of Experiments and Results on the Prevention of Evaporation Losses from Reservoirs.* New Delhi.

Karadi, G. M. 1975. Environmental considerations in water resources planning. *Proceedings IX Congress ICID.* New Delhi.

National Academy of Sciences, Washington. 1974. *More water for Arid Lands.*

Pillsbury, A. F. and Degan, A. 1968. *Sprinkler Irrigation*. FAO, Rome.

Rafatjah, H. A. 1975. The impact of irrigation and drainage on public health. *Proceedings IX Congress ICID.* New Delhi.

Ríha, J. 1975. Evaluation of water in the dynamic system of human environment. Prague, Sc.D. Thesis.

Yaron, B., Dantors, E. and Vaadia, Y. 1973. *Arid Zone Irrigation*. Springer Verlag, Berlin–Heidelberg–New York.

SOME INFLUENCES OF PROJECT MANAGEMENT ON IRRIGATION EFFICIENCIES

by M. G. BOS

International Institute for Land Reclamation and Improvement (ILRI), Wageningen, Netherlands

Summary

There is an urgent need to improve the water utilization efficiency in irrigation projects. To obtain information that would allow guidelines on methods of water distribution to be set, a questionnaire was organized among the National Committees of the ICID. From the data collected, it became obvious that water losses due to the complexity of the project management far exceed losses due to seepage and evapotranspiration from the canals. In addition to recommending that canals be lined, being the generally accepted method of fighting seepage losses, a number of other guidelines could be formulated. The most important of these are: (1) avoid irrigation projects of less than 1000 ha; (2) divide large irrigation projects into lateral units of between 2000 and 6000 ha, depending on topography; (3) let each lateral unit contain a number of rotational units, the size of which should vary between 70 and 300 ha depending on local farm size and topography; (4) operate main, lateral and sublateral canals on a schedule of continuous flow; (5) within a rotational unit, organize the rotation of water supply to farm inlets or group inlets independently of the distribution in adjacent units; (6) on large irrigation projects (more than 10,000 ha) decentralize the project management so that each lateral unit has its own staff.

Résumé

Dans les projets d'irrigation, on a un grand besoin d'améliorer l'efficacité de l'utilisation d'eau. Pour obtenir des informations sur les méthodes de distribution de l'eau, on a fait une enquête parmi les comités nationaux de l'ICID. Jugeant des réponses, il est évident que les pertes en eau causées par la complexité des aménagements surpassent fréquemment cells qui sont causées par l'infiltration et l'évapotranspiration de l'eau des canaux. Outre les revêtements des canaux, méthode bien connue pour combattre les pertes par infiltration, on a formulé un certain nombre d'instructions. Les plus importantes de celles-ci sont: (1) ne pas installer des projets d'irrigation de moins de 1000 ha; (2) diviser les projets d'irrigation en unités de 2000 à 6000 ha en fonction de la topographie; (3) composer chaque unité latérale d'un nombre d'unités rotatives d'une taille de 70 et 300 ha en fonction de la topographie et de la dimension des fermes; (4) faire fonctionner les canaux principaux et sublatéraux de façon à ce qu'existe un écoulement continu; (5) organiser à l'intérieur d'une unité de rotation une rotation d'apport d'eau aux fermes ou groupes de fermes, indépendante de la distribution d'eau dans les unités adjacentes; (6) dans les grands projets d'irrigation (plus de 10,000 ha), décentraliser la direction du projet de manière à ce que chaque unité latérale ait sa propre personnalité.

Introduction

For centuries farmers have been practising irrigation and have arrived at certain standards for the operation of their irrigation systems. These empirical standards have only

351

regional significance and are aimed at either maximum crop production under the given conditions or at an acceptable amount of labour. Often the standards represent a compromise between the two. With more and more land being brought under irrigation, many of these standards were applied in newly developed regions with entirely different physical and social conditions.

The operational aspects of water distribution in irrigated areas that are still largely dominated by tradition usually do not reflect the efficient utilization of the available water as a primary objective. Because of the increasing pressure of population, water often becomes a limiting factor in countries where irrigated agriculture forms a basic element of food production. Thus there is an urgent need for a more efficient use of the water resources and for a more scientific approach to the problem of operating irrigation systems.

It was felt that if a large number of existing irrigation areas could be analyzed a number of guidelines might be produced that could be used with confidence in planning and designing irrigation systems. To obtain the necessary information, a questionnaire, covering no less than 93 items, was organized among all National Committees of ICID through its Central Office in New Delhi. At the closing date, 29 National Committees had submitted questionnaires covering a total of 91 irrigated areas. The work of processing the data obtained from the questionnaires was performed by ILRI and the Irrigation Department of the University of Agriculture, both at Wageningen, The Netherlands.

Grouping of data

Since it was understood that the results of the inquiry could only be of value if the basic climatic and socio-economic conditions were taken as the primary variables, it was decided to group the 91 investigated areas into four main categories:

Group I: Columbia, Egypt, India, Iran, Israel, Mexico, Rhodesia (a total of 28 areas). All areas of this group have a severe rain deficit so that crop growth is entirely dependent on irrigation. In general the farms are small and have cereals as their most important crop. Secondary crops, if any, are rice, cotton, or sugar cane.

Group II: Columbia, Guyana, Japan, South Korea, Malaysia, Malawi, Philippines, Taiwan, Thailand (a total of 22 areas). Although the economic structure of these countries is about the same as those of Group I (except Japan, see below), Group II differs in that the rain deficit is less and that the main crop in all the areas is rice.

Group III: Australia, Cyprus, France, Greece, Italy, Portugal, Spain, Turkey, United States of America (a total of 32 areas). In this group the irrigation season is usually somewhat shorter than in the first two groups, and the economic development, in general, is more advanced. Besides cereals, the most important cultivations are fodder crops, fruit and vegetables.

Group IV: Austria, Canada, German Federal Republic, The Netherlands, United Kingdom (a total of 10 areas). The areas of this group all have a cool, temperate climate and a relatively short irrigation season (3–4 months). Most of the soils irrigated are light textured and most of the irrigation is by sprinkler and has a supplementary character.

It should be noted that climatic indications only set broad outlines, facilitating the use of the data for comparable areas. It is beyond the scope of this paper to indicate summary areas on the world map to which the data of each group should be applied; here the reader must use his own judgement. Neither were specific indices used for a country's economic situation; Japan, for instance, was included in the second group for the sake of simplicity although it differs from the other countries in the group both as to climate and economic development.

Since data were collected under a promise of anonymity to their suppliers, we have given each irrigated area a three-figure code. The first figure stands for a geographical region, the second stands for a country and the third for an irrigated area or project. The first two figures are in Table 1.

TABLE 1. *Coding of countries*

11	Austria	32	Egypt	64	Philippines
12	Fed. Rep. of Germany	33	Iran	65	Taiwan
13	The Netherlands	34	Israel	66	Thailand
14	United Kingdom	35	Turkey	71	Australia
21	France	41	Malawi	81	Canada
22	Greece	42	Rhodesia	82	United States of America
23	Italy	51	India	91	Columbia
24	Portugal	61	Japan	92	Guyana
25	Spain	62	South Korea	93	Mexico
31	Cyprus	63	Malaysia		

Definitions of efficiencies

In an irrigation canal system, the distribution of water over the irrigated area can be split up into three successive stages:

conveyance by main, lateral, and sublateral canals to the farm inlet or group inlet;
conveyance by farm ditches to the field, or, if group inlets are used, conveyance by distributary and farm ditches to the field;
application to and distribution over the field from the field inlet onward.

The efficiency in the first stage is defined as the *Water Conveyance Efficiency*, e_c, and can be expressed as

$$e_c = \frac{V_f}{V_t}$$

where V_f is the volume of water delivered to all farm or group inlets in the area and V_t is the total quantity of water supplied to the area.

The efficiency in the second stage is defined as the *Farm Ditch Efficiency*, e_h, and can be expressed as

$$e_h = \frac{V_a}{V_f}$$

where V_a is the field application to the cropped area and V_f is the volume of water delivered to all farm inlets in the area.

TABLE 2. *Calculated (average) efficiencies*

Project code	e_p	e_f	e_d	e_a	e_b	e_c	Project code	e_p	e_f	e_d	e_a	e_b	e_c
111				0.75			517						
112	0.29		0.60	0.49	0.80	0.75	518	0.15	0.30	0.29	0.51	0.57	0.50
121	0.29		0.64	0.46	0.80	0.80	519						
122†	0.20		0.35	0.57	0.80	0.44	51(10)						
123†	0.07		0.30	0.23	0.80	0.38	51(11)						
124	0.60	0.63	0.75	0.81			51(12)						
131	0.57	0.88		0.88	0.80								
132	0.41			0.41			611	0.34	0.41	0.75	0.45	0.90	0.83
							612	0.22	0.23	0.85	0.26	0.90	0.94
211	0.31	0.33	0.79	0.39	0.85	0.94	613	0.11	0.12	0.80	0.14	0.87	0.92
212	0.44	0.69	0.63	0.71	0.97	0.64	614	0.25	0.26	0.92	0.27	0.95	0.97
213							615	0.19	0.20	0.87	0.22	0.90	0.97
214	0.28	0.67	0.40	0.70	0.94		621						
215	0.46	0.56	0.69	0.66	0.85	0.82	622		0.28	0.72	0.35	0.80	0.90
216		0.62					631	0.38	0.34	0.76	0.40	0.85	0.89
217							632		0.17	0.54	0.25	0.68	0.80
218		0.94					633	0.33	0.39	0.86	0.39	0.97	0.88
219		0.71					634						
221	0.36	0.37	0.48	0.75	0.50	0.96	635						
222	0.20	0.34	0.31	0.65	0.53	0.59	641				0.52		
223	0.30		0.51	0.59	0.60	0.85	642	0.39	0.43	0.87	0.45	0.95	0.92
224				0.63			651						
231							652	0.22	0.40	0.34	0.64	0.60	0.56
232	0.20	0.36	0.36	0.56	0.65	0.56	653	0.33	0.34	0.93	0.36	0.95	0.98
233	0.29	0.43	0.47	0.62	0.70	0.67	661				0.38		
241	0.34	0.43	0.46	0.72	0.60	0.77							
251	0.30	0.33	0.58	0.51	0.65	0.89	711				0.67		
							712						
311	0.41	0.51	0.78	0.52	0.96	0.81							
312				0.62			811	0.45					
313	0.39	0.44	0.74	0.52	0.84	0.88	821	0.26		0.66	0.40	0.80	0.83
321	0.30	0.46	0.46	0.66	0.70	0.66	822	0.33		0.70	0.58	0.80	0.88
331	0.29						823						
332				0.76			824	0.28	0.53	0.52	0.55	0.97	0.54
333							825						
334				0.50			826	0.33		0.50	0.59	0.80	0.63
341	0.51						827				0.71		
351	0.15	0.56	0.22	0.65	0.86	0.26							
352		0.61	0.37	0.70	0.87	0.42	911	0.20					
							912	0.33	0.38	0.78	0.42	0.90	0.87
411							913	0.11					
421	0.32	0.45	0.57	0.47		0.71	914	0.13					
422	0.49	0.86				0.56	915	0.13	0.25	0.33	0.38	0.65	0.51
							916	0.19					
511							921						
512	0.40	0.57	0.58	0.70	0.82	0.70	931	0.27	0.57	0.31	0.87	0.65	0.48
513	0.14	0.20	0.34	0.40	0.50	0.67	932	0.51	0.56	0.77	0.66	0.85	0.91
514	0.25	0.32	0.47	0.53	0.60	0.78	933	0.24	0.27	0.52	0.45	0.61	0.86
515	0.16	0.24	0.34	0.47	0.51	0.67	934	0.21	0.42	0.41	0.50	0.83	0.50
516													

Note: Italic values have 50% weight.
†Waste water disposal installations.

The efficiency in the third stage is defined as the *Field Application Efficiency*, e_a, and can be expressed as

$$e_a = \frac{V_n}{V_a}$$

where V_n is the rainfall deficit (i.e. the difference between the consumptive use and the effective rainfall over the cropped area) and V_a is the field application to the cropped area.

Apart from these three efficiencies, it was found necessary to define several other efficiencies. The reason for this was that not all the questionnaires had been completed in full detail and others contained answers whose reliability was doubtful because the questions had apparently been misunderstood. To allow a different approach in analysing these questionnaires, therefore, the following additional efficiencies were defined:

Farm Efficiency, e_f, which is the ratio between the quantity of water placed in the root zone (rainfall deficit) and the total quantity under the farmer's control, or

$$e_f = \frac{V_n}{V_f} = e_a e_b$$

Distribution Efficiency, e_d, which is the ratio between the quantity of water applied to the fields and the total quantity supplied to the irrigated area, or

$$e_d = \frac{V_a}{V_t} = e_b e_c$$

Overall (or Project) Efficiency, e_p, which is the ratio between the quantity of water placed in the root zone (rain deficit) and the total quantity supplied to the irrigated area, or

$$e_p = \frac{V_n}{V_t} = e_a e_b e_c = e_a e_d = e_f e_c$$

The overall (or project) efficiency represents the efficiency of the entire operation between diversion or source of flow and the root zone. By taking the complementary value, one can obtain the total percentage of losses. The values of V_n, V_a, V_f and V_t were derived from the questionnaires and totalled over the irrigation season. The calculated efficiencies are listed in Table 2.

Manageability of the irrigation project

It has long been recognized that the water losses that occur in conveying water to the group or individual farm inlets via main, lateral and sublateral canals are influenced by the quality of the management and the manageability of the irrigation system. The top echelon of an irrigation project usually consists of a manager and one or more specialized engineers who are responsible for the operation of the entire distribution system. The quality of the management, besides depending on local availability, also depends on the size of the project; the larger the project, the more funds can be made available to hire properly qualified personnel. The manageability of the project also depends on project size but, in addition, on what general schedule of continuous or intermittent flow the canals are being operated under, and on the method of water distribution applied.

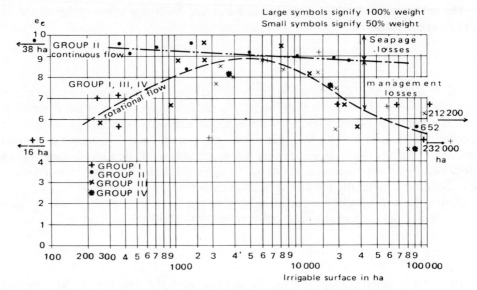

FIG. 1. The conveyance efficiency (e_c) values as a function of the irrigable area.

Conveyance efficiency versus average irrigable area. If the water conveyance efficiency of an irrigated area is plotted as a function of the size of the area where technical facilities are available for irrigation, we obtain Fig. 1. In this figure two curves are drawn showing mean e_c-values for areas in Group II and the combination of the Groups I, III and IV.

All areas in Group II have rice as their main or only crop and water is supplied continuously to the fields at an approximately constant flow through a system of canals and ditches. This procedure requires little or no adjustment of division or inlet structures and causes no organizational problems. It is only the increasing canal length related to a larger irrigable area that causes the conveyance efficiency to decrease slightly. We assume that most water lost can be attributed to seepage and to a lesser extent to evapotranspiration from the water surface and canal banks.

The Groups I, III and IV curve represents mean e_c-values for areas where either one main crop (other than rice) or a certain variety of crops is cultivated which may necessitate more or less frequent adjustment of the supply. The curve shows a maximum e_c-value with an average of about 0.88 for irrigable areas of between 3000 and 5000 ha. For smaller irrigable areas, e_c-values may be as low as 0.50, probably due to the reduction of the project management to one person who, besides handling the distribution of water, is engaged in agricultural extension work, maintenance, transport and marketing of crops, administration, etc. If the manager is to fulfil all his tasks satisfactorily, he must be highly skilled, but on small projects (less than 1000 ha) funds are not always available to hire such a person.

Also if the irrigable area is large (more than about 10,000 ha), the conveyance efficiency decreases sharply, probably due to the problems management faces in controlling the water supply to remote sub-areas. Large systems tend to be less flexible in adjusting the water supply because of the relatively long time it takes to transmit information on flow rates and water requirements to a central office and the long travel time for water in open canals. To avoid water deficits in downstream canal sections, there is often a tendency to increase the supply to the head of the canal system.

We assume that the difference between the Group II curve and Groups I, III and IV curves can be mainly attributed to management losses. This water will either be discharged into the drainage system or will inundate non-irrigated lands, creating a drainage problem as a harmful side-effect. In this context it is interesting to note that, in the only area (code 652) of Group II that has an e_c-value not fitting the mean curve, sweet potatoes, sugar cane and rice are cultivated and the supply to all these crops is on a schedule of rotational flow. It is also interesting to note that the relevant e_c-value corresponds well to the mean curve for irrigable areas in the Groups I, III and IV.

Conveyance efficiency versus size of rotational unit. At the headworks of many irrigation canal systems, water is diverted continuously throughout the irrigation season, its flow rate being adjusted to crop requirements only after periods that are long relative to the time the water travels through the canal system. Somewhere along the canal system, however, water is drawn continuously via a discharge measuring and regulating structure to serve an irrigation unit with internal rotation to the farms within it. Downstream of such a structure, the canals do not carry water continuously but function of some schedule of inter-mittent flow. The irrigation unit served by a canal system on intermittent flow is called a rotational unit. Within a rotational unit, the water distribution is organized independently of the overall conveyance and of the water distribution in neighbouring rotational units. It is based only on the farm water requirements in that unit. The size of the rotational unit influences the water conveyance efficiency markedly, as shown in Fig. 2.

This figure suggests that an optimum conveyance efficiency can be attained if the size of the rotational unit lies between 70 and 300 ha. If the unit is small (less than 40 ha) the conveyance efficiency decreases sharply because temporary deficiencies of water cannot be flattened out by managing the already low flow rate according to an alternative schedule. Because of unavoidable inaccuracies in the measurement of the flow rate, a tendency exists in small rotational units to set a safety margin above the actual amount required. If the

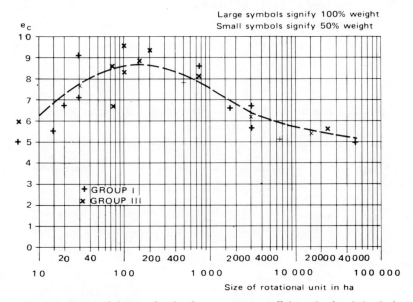

FIG. 2. Influence of size rotational unit on conveyance efficiency (surface irrigation).

rotational unit is large (more than 600 ha), rather long canals of large dimensions have to be filled and emptied after periods which are short relative to the time the water travels through the canal. Together with the organizational difficulties of correct timing, rotating the flow in large units causes the conveyance efficiency to decrease to values as low as 0.50. It may be noted that Fig. 2 does not include values for Groups II and IV since no irrigation is practised on a rotational schedule in these groups.

Farm ditch efficiency versus duration of delivery period. It has already been suggested that the use of a canal or ditch on a schedule of intermittent flow causes the conveyance efficiency to decrease because of operational losses. The influence upon the farm ditch efficiency of the period during which delivery lasts is illustrated by Fig. 3. On farms that have a water delivery period of less than 24 hours, we note that the water losses consist not only of seepage losses during the operation, but also of losses caused by the initial wetting of soil around the ditch perimeter. To improve the efficiency of farm ditches, we recommend that they

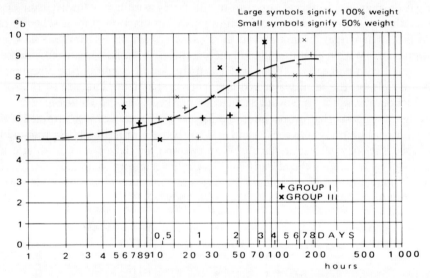

FIG. 3. Influence of average delivery period at farm inlet on farm ditch efficiency (surface irrigation).

TABLE 3. *Farm ditch efficiency of farm ditches flowing continuously*
(Group II)

Code	e_b	Average farm size (ha)	Code	e_b	Average farm size (ha)
611	0.90	0.05	632	0.68	0.08
612	0.90	0.03	633	0.97	1.6
613	0.87	0.1	641	—	2.8
614	0.95	0.05			
615	0.90	0.1	642	0.95	2.3
			653	0.95	0.85
622	0.80	1.5	661	—	<5
631	0.85	1.0			
Average	0.88		Average	0.88	

be lined, especially those ditches that have a low flow capacity and are used for short periods at a time.

The farm ditch efficiency of ditches that are used for at least 200 hours approaches about 0.88, which value is remarkably close to the average value of 0.88 for ditches carrying a continuous supply of water to rice fields (see Table 3).

Relation between conveyance and farm efficiency to methods of water supply to a farm. From a project management point of view, we may broadly distinguish four methods of water supply to a farm inlet or a group inlet for a number of farms:

A. Continuous supply, with only minor changes in flow size, generally used in conjunction with basin irrigation (rice). The conveyance system consists of a network of open canals, also flowing at a constant rate.

B. Rotational supply on a pre-determined schedule which depends mainly on the variable crop requirements and the availability of irrigation water at the head works. The schedule of rotational flow is decided by officials of the central irrigation service.

C. Similar to B, but now the schedule of rotational flow is based mainly on water volumes demanded in advance by the individual farmers. The water is conveyed to the farm inlet through a network of open canals.

D. Water is distributed through a system of pipe lines over the entire project, and farmers can draw water in accordance with their demands of the moment. All (6) questioned projects that have this distribution system use it in conjunction with overhead sprinkler irrigation.

The results of a calculation of the average farm, conveyance, and overall efficiencies for irrigated areas classified under these methods are shown in Table 4.

TABLE 4. *Average efficiencies for different distribution methods*

Method	No. of samples	e_f	e_c	e_p
A	12	0.27	0.91	0.25
B	20	0.41	0.70	0.29
C	6	0.53	0.53	0.28
D	6	0.70	0.73[1]	0.51

[1] Based on two values: 0.64 and 0.82.

If we consider the third column of this table, we note that the farm efficiency increases sharply from a low value of $e_f = 0.27$ for areas using Method A to a rather favourable value of $e_f = 0.70$ for areas using Method D. If we limit ourselves to open canal systems, this column alone would suggest that water can be saved by using Method C. If we consider the methods from the point of view of project management, we find that Methods A, B and C are successively more complicated to manage because flow rates in the canal system have to be adjusted at shorter and shorter intervals. From Table 4 we can see that this increasingly complicated management causes the conveyance efficiency to decrease from a high value of $e_c = 0.91$ for areas using Method A to a low value of $e_c = 0.53$ for areas using Method C. As a result the overall or project efficiencies of irrigated areas using Method A, B or C are very similar. This would suggest that if any measures are taken to increase the farm efficiency and these measures complicate the water conveyance in the irrigated area, then the sometimes tremendous efforts spent on improving the farm efficiency can be nullified by a decreasing conveyance efficiency.

Recommendations

The conclusion is drawn that the management of an irrigation project faces increasing difficulties in keeping the conveyance efficiency to an acceptably high level if canals have to flow intermittently or if frequent adjustments have to be made in the flow rates. In designing an irrigation system the engineer should strive for a distribution system which is simple and requires a minimum of management.

For irrigated areas where either a main crop (other than rice) or a combination of different crops is cultivated, a maximum conveyance efficiency can be obtained if the size of the irrigable area—or management unit—is between 3000 and 5000 ha. On small irrigation projects (less than 1000 ha) an acceptable conveyance efficiency can probably only be attained if a top-heavy project management is acceptable. On large irrigation projects we recommend a decentralization of the project management as follows:

(a) *General project management.* The general project management operates the dam site or diversion and main canal. The main canal should have a flow rate which can be adjusted to meet the water requirements of the various lateral units.

(b) *Local irrigation management.* Depending on topography and local conditions, the irrigation project should be divided into a number of lateral units, each having an area of between 2000 and 6000 ha (mean 4000 ha). Each lateral unit should receive its water at one point from the main canal and should have its own skilled local irrigation management staff who will be responsible for the water distribution within that lateral unit only.

We would further recommend that the main, lateral and sublateral canals be operated on a schedule of continuous flow and that the area not be divided into sub-rotational units. During the entire season the flow rate in each of these canals should be a function of the water requirement of the commanded area only. Each lateral unit should contain a number of rotational units whose size should be between 70 and 300 ha, depending on local farm size and topography. Within each rotational unit, the water distribution should be organized independently of the distribution in adjacent rotational units.

Within a rotational unit, each farm or group of farms (group inlet) should receive water during a sufficiently long period, preferably more than 3 days. If ditches are to be used for shorter periods but with briefer intervals in between, these ditches should be lined to prevent excessive seepage losses.

REUTILIZATION OF RESIDUAL WATER AND ITS EFFECTS ON AGRICULTURE

by ELOY URROZ

Director General of Water Use and Pollution Control, Ministry of Water Resources, Mexico

Summary

One of the limiting factors for urban and industrial growth in the Valley of Mexico is the availability of water. The aquifers below the old lake bed are being mined, even though it has been established that such a practice will accelerate the sinking of the downtown area. For the past 10 years all possible water supply sources have been under study by the authorities. All these sources combined could provide $126 \, \text{m}^3/\text{s}$ of water to the city; however, they have not been deemed the most economical solution, since water is still withdrawn from the lake-bed aquifer. The reuse of wastewater for industry and irrigation offers good prospects. Within Mexico City 20% of the municipal wastewaters are already being reused for watering parks and lakes. Since industry uses up about 28% of the water supply, the percentage of water reuse can easily be increased to about 50% as is the case in the city of Monterrey.

Irrigation District No. 03, where wastewater is used extensively, has been studied in detail. Dangerous concentrations of boron can be avoided by appropriate application of water, and the available data indicate that the use of raw sewage has not presented any ill effects on productivity from the high salt content nor from a bacteriological standpoint. However, the health effects of irrigation with raw wastewaters have not yet been determined.

Résumé

L'un des facteurs limitant de la croissance urbaine et industrielle dans la Vallée de Mexico est la disponibilité de l'eau. Les aquifères situés sous le lit de l'ancien lac sont en cours d'exploitation, bien qu'on sache que cette pratique risque de nuire à la zone centrale de la ville. Au cours des 10 dernières années, toutes les sources potentielles d'eau ont été étudiées par les autorités. Toutes ces sources combinées peuvent fournir $126 \, \text{m}^3/\text{s}$ d'eau à la ville. Cependant elles ne constituent pas la solution la plus économique puisque l'eau est toujours extraite de l'aquifère sous-lacustre. La ré-utilisation des eaux usées à usage industriel et pour l'irrigation ouvre de bonnes perspectives. A l'intérieur de la ville de Mexico, 20% des eaux usées municipales sont d'ores et déjà ré-utilisées pour fournir de l'eau aux parcs et aux lacs. Depuis que l'industrie utilise environ 28% des apports d'eau, le pourcentage d'eau ré-utilisée peut facilement s'accroître de 50% dans le cas de la Cité de Monterrey.

Le District d'Irrigation No. 03, où les eaux usées sont utilisées de manière extensive, a été étudié en détail. Les dangereuses concentrations en bore peuvent être évitées par une utilisation rationnelle de l'eau. Les données disponibles indiquent que l'utilisation des eaux d'égout n'a eu pour conséquence aucun effet néfaste sur la productivité en raison de leur très haute teneur en sel on de leur composition bactériologique. Cependant les effets sur la santé de l'irrigation avec des d'égout n'ont pas été déterminés.

Water availability in Mexico

Until the beginning of the sixteenth century, the Valley of Mexico was a closed basin with a relative abundance of water. Hydrological equilibrium was reached naturally

361

TABLE 1. *Present and future water supply systems for metropolitan Mexico City*

System	Q (m^3/s)	Wells (no.)	Aqueducts (km)	Elevation (m)
I. Valley of Mexico—basin				
A. Groundwater				
1. Urban area[1]	9.0			2400
2. Chalco[1]	3.0	30	40	2400
3. Apan[1]	3.0	40	245	2540
4. Cuautitlan[1]	1.8	40	100	2450
B. Surface water				
1. San Mateo-Nopala-Nadim[1]	1.5		85	2600
2. Guadalupe[1]	2.0		80	2600
II. Outside water basins				
A. Groundwater				
1. Oriental[1]	7.0	70	465	2750
B. Surface water				
1. Necaxa	20.5		250	
2. Upper Balsas (Tenancingo)	9.0		100	2020
3. Upper Balsas (Jojutla)	13.5		75	1544
C. Combined				
1. Tepeji-Cuautitlan[1]	2.5	40	135	2250
2. Alto Amacuzac	16.0		180	1544
3. Tecolutla	41.0		600	1000
4. Lerma[1]	8.0			2500
5. Alto Lerma[1]	14.0	40	140	2700

[1] Systems presently in operation.

TABLE 2. *Physical and chemical characteristics of the Mexico City wastewaters*

Characteristic	Concentration (mg/l)	Characteristic	Concentration (mg/l)
Temperature (°C)	18	Heavy metals:	
Conductivity (umho/cm)	1792	Lead	0.090
Total solids	1590	Mercury	0.0015
Total suspended solids	1150	Cadmium	0.027
Total coliform (MPN/100 ml)	63×10^7	Zinc	0.54
BOD	220	Copper	0.09
COD	500	Nickel	0.1
MBAS	8.2	Iron	2.40
Ammonia (as NH_3)	12.0	Manganese	0.17
Organic nitrogen	6.0	Chromium (total)	0.1
Nitrate—N (as NO_3)	0.2	Potassium	50.0
Total—P (as PO_4)	23.0	Sodium	308.0
Ortho—P (as PO_4)	10.0	Boron	1.40
Chlorides	182.0		
Sulphates	147.0		
Hardness (total)	483.0		
Alkalinity (total)	433.0		
pH	7.7		

through the increase of the lake areas during the wet years, slowly decreasing depending upon the length of the droughts. The risk of flooding has always been a problem, necessitating the construction of control works very early in its history.

The ancient great lake with no natural outlet was made up of the Zumpango, Xaltocan, Texcoco, Xochimilco and Chalco lakes that covered an area of approximately 150,000 ha. During prehispanic time one of the first flood control works to be constructed was a 16 km long dyke that divided the freshwater in the south-west from the brackishwater of the northeast. After the Spanish conquest, it was decided that the solution to flooding should be through the conveyance of the waters out of the Valley of Mexico. In 1607 a cut was made in the mountains to a depth of 60 metres to let the Cuautitlan River out of the valley. This resulted in the drainage and reduction of the lake area to about 15,000 ha. Thus the flood problem was solved with the generation of another: that of water shortage.

The lack of a separate storm sewer system with sufficient capacity to remove storm runoff led towards the end of the nineteenth century to the construction of the Gran Canal and the first tunnel of Tequisquiac. This system safeguarded the valley from flooding during the rainy season, but it speeded up the drying of the lakes. To make good the water shortage, large quantities of groundwater have been withdrawn from the aquifer below the old lake bed. The subsidence of the soil in this area has caused the sinking of a major portion of downtown Mexico City. At the start of the century, Lake Texcoco was 3 metres below the city; while, at the present time, the level of the lake bottom is 3 to 4 metres above the streets.

It is estimated (1975) that Mexico City has a population of about 8.5 million and that the City generates 50% of the industrial production of the whole country. By the year 2000 it should have between 16 to 25 million inhabitants, depending on whether the present growth rate is maintained or if it is reduced by about 20%. The population problem is further complicated by the fact that the greater metropolitan area of Mexico City has already a total population of about 11 million people.

The present and future sources of water supply are listed in Table 1, which shows that at least 25% of the present supply comes from the aquifer directly beneath the city and another 20% from the Lerma River system. All of the nearby sources are already under utilization, adding up to about $40 \, m^3$. Future growth of the city to the year 2000 will require bringing in water from a radius of 200 km to meet the demands, and the cost of such an enterprise shows that the reclamation of wastewater for industrial and agricultural purposes will become essential. The volume of sewage water generated within Mexico City amounts to about $30 \, m^3/s$ and average values for the main water quality constituents are given in Table 2. The dissolved solids, boron, MBAS and bacterial counts require special care before these wastewaters can be indiscriminately utilized for irrigation.

Water reuse

Water reuse is not a new topic in Mexico, for domestic wastewaters are utilized raw for the irrigation of farmland quite extensively throughout the country, 77% of which is arid or semi-arid. In Mexico City the watering of parks and recreational areas is provided from the disinfected effluent from three sewage treatment facilities, and the irrigation of nearby agricultural areas from the effluent of two other plants. The characteristics of these and seven other treatment plants that are located within the valley are summarized in Table 3.

The first of the plants run by the Mexico City Government was built in 1955 to provide water for Chapultepec Park, and since this was the first of its kind in Mexico, sludge treatment facilities were provided. However, difficulties in the operation of the anaerobic digester soon led to the solids being disposed of directly in the local sewer system. Other plants do not have sludge disposal facilities.

Chapultepec Park covers 400 ha with three artificial lakes; San Juan de Aragon has an artificial lake, sports fields and a large wooded area. The wooded area in both of these parks is watered during the dry season with treated wastewaters from the 160 and 500 l/s treatment facilities, respectively. The sports fields of the Ciudad Deportiva, as well as gardens and small parks in the surrounding area, are watered with the treated effluent from

TABLE 3. *Wastewater treatment plants in Mexico City*

Name	Capacity (l/S)	Type	Use of effluent
1. Chapultepec	160	Activated sludge	Parks and lakes
2. Cd. Deportiva	230	Conv. act. sludge	Parks and sports fields
3. San Juan de Aragon	500	Conv. act. sludge	Parks and lakes
4. Xochimilco	1250	Conv. act. sludge	Lakes and irrigation
5. Cerro de la Estrella	2500	Conv. act. sludge	Lakes and irrigation
6. Bosques de las Lomas	55	Extended aeration	Parks
7. Acueducto Guadalupe	80	Conv. act. sludge	Parks and cemetery
8. Unidad Ixtacalco	10	Act. sludge and nut. rem.	Lake and irrigation
9. San Juan Ixhuatepec	500	Conv. act. sludge	Industries
10. Club Campestre	60	Biological filters	Parks and golf course
11. Thermoelectric (CFE)	450	Conv. act. sludge	Industry
12. Plan Texcoco	2000	Not final	Parks, lakes and irrigation

a 230 l/s treatment plant. There are several country and golf clubs that re-use the effluent from small package treatment plants for watering the fairways. Most of the plants take wastewaters directly from the municipal sewer system as their needs make it necessary, returning to it the waste solids.

In many of the recreational lakes, the large amount of nutrients present in the treated wastewaters has given rise to nuisance growths, primarily algae and small aquatic plants. This problem reaches its peak during the hot months of the year, when mats of blue-green algae decompose on the water surface. In one treatment plant provision was made for nutrient removal by physical-chemical means, and a lake has thus been maintained free of nuisance conditions during its first year of operation.

The installation of wastewater treatment in order to reuse effluent for industrial purposes is also proceeding. A group of 11 industries has built a treatment plant to take raw waste-waters from the Rio de los Remedios. After secondary treatment, the disinfected effluent is distributed to the industries where it may receive some form of tertiary treatment in order to meet the requirements of each manufacturing process. In a similar fashion, the Thermo-electric Plant of the Valley of Mexico treats wastewaters from the Gran Canal to reuse as cooling water and this allows their freshwater allotment to be used for the generation of steam. This treatment plant began operating in 1963 at a capacity of 150 l/s, which was increased to 450 l/s in 1974.

One of the newest and largest treatment plants in Mexico City is the Cerro de la Estrella facility, with a capacity for 2500 l/s, which was installed specifically to provide treated waste-waters for the irrigation of a vast tract of land that used to make up the old lakebed. Much of the plant capacity will be available for exchange with freshwater that is presently being utilized for irrigation of farmland, and thus will increase the municipal water supply.

A large portion of the Mexico City wastewaters is now used for irrigation and this will probably be increased by a new trunk collector that has recently been put into operation. It is known as the Emisor Central, has a capacity of $200 \, m^3$, reaches depths of up to $230 \, m$ below the ground level and takes the municipal wastewaters out to the Irrigation District No. 03 in the Mezquital. This region is extremely arid and was almost totally sterile before it was provided with sewage for irrigation purposes.

The Lake Texcoco Plan now under study aims to rehabilitate the old lake bed. During the rainy season the lake bed is used to take runoff peaks and thus avoid flooding the downtown area, but it soon dries out, giving rise to dust storms which create a health hazard as well as a nuisance, since pathogenic organisms are easily transported by the fine detritus particles. In the old Texcoco lake bed, which has a surface area of about 14,500 ha, a series of five recreational lakes is planned, along with parks and agricultural zones, all to be watered from a treatment plant with a $2000 \, l/s$ capacity. The lakes will have progressively better quality water, with fishing and boating allowed in the last of the series.

Another important project getting under way is the establishment of Water Reuse Districts within the city. The Federal Government will install treatment facilities in certain areas of the city to treat the municipal sewage and sell treated effluent to industries located within each region. Along similar lines it is thought that treated wastewaters can be exchanged for freshwater from the farmers of the Apan, Chalco and Cuautitlan regions.

Wastewater for agriculture

The Ministry of Water Resources has funded a continuing research programme to look into water quality standards for the reclamation of wastewaters to be used in agriculture, industry and aquifer recharge. This project was initiated two years ago with a comprehensive study of an area that had been irrigated with raw sewage during the past 60 years, namely Irrigation District No. 03 located north of the Valley of Mexico in the Mezquital Valley (Fig. 1). This district affords the unique opportunity of evaluating the effects of irrigating with different water qualities on different types of soil with a variety of crops.

This district, which is run by the Ministry of Water Resources, has available a large amount of data on water use and agricultural yield. It is fed primarily by the Tula and Salado rivers which drain a basin of about $11,000 \, km^2$. Two reservoirs (Endho and Requena) are located in the upper portion of the Tula River, and control the flow of freshwater into

TABLE 4. *Water consumption in the Irrigation District No. 03*

Farming cycle	Yearly volume ($\times 10^6 \, m^3$)		
	Wastewater	Clean water	Total
1969–1970	760.7	177.8	938.5
1970–1971	700.0	176.6	876.6
1971–1972	731.4	247.8	979.2
Average	730	201	931

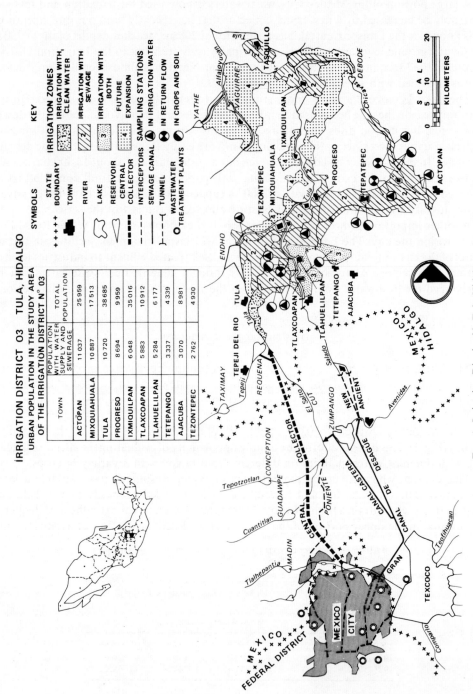

IRRIGATION DISTRICT O3 — TULA, HIDALGO

URBAN POPULATION IN THE STUDY AREA
OF THE IRRIGATION DISTRICT N° O3

TOWN	POPULATION WITH WATER SUPPLY AND SEWERAGE	TOTAL POPULATION
ACTOPAN	11 037	25 959
MIXQUIAHUALA	10 887	17 513
TULA	10 720	38 685
PROGRESO	8 694	9 959
IXMIQUILPAN	6 048	35 016
TLAXCOAPAN	5 883	10 912
TLAHUELILPAN	5 284	6 177
TETEPANGO	3 337	4 339
AJACUBA	3 070	8 981
TEZONTEPEC	2 762	4 930

SYMBOLS

+++++	STATE BOUNDARY
	TOWN
	RIVER
	LAKE
	RESERVOIR
▬▬▬	CENTRAL COLLECTOR
- - -	INTERCEPTORS
———	SEWAGE CANAL
⊢⊣	TUNNEL
○	WASTEWATER TREATMENT PLANTS

KEY

IRRIGATION ZONES

	IRRIGATION WITH CLEAN WATER
2	IRRIGATION WITH SEWAGE
3	IRRIGATION WITH BOTH
4	FUTURE EXPANSION

SAMPLING STATIONS

◐	IN IRRIGATION WATER
◑	IN RETURN FLOW
◒	IN CROPS AND SOIL

SCALE

0 5 10 20
KILOMETERS

FIG. 1. Irrigation District 03, Tula Hidalgo.

TABLE 5. *Urban population within the irrigation District No. 03*

Town	Sewered population	Total population
1. Actopan	11,037	25,959
2. Mixquiahuala	10,887	17,513
3. Tula	10,720	38,685
4. Progreso	8,694	9,959
5. Ixmiquilpan	6,048	35,016
6. Tlaxcozpan	5,883	10,912
7. Tlahuelilpan	5,284	6,177
8. Tetepango	3,337	4,339
9. Ajacuba	3,070	8,981
10. Tezontepec	2,762	4,930

the district. The Salado River receives the Mexico City municipal wastewaters, as these exit through the Tunnels of Tequisquiac. The water available for irrigation is 24% freshwater from the upper Tula River and 76% wastewaters from the Salado River, and insures the irrigation of about 42,000 ha, with a remainder of about 10,000 ha dependent upon rainfall. Several studies were conducted to determine what risks, if any, developed to human health, soil conservation and productivity, e.g. from boron in soils, salinity and pathogens.

Boron in soils.[1] Boron, an essential nutrient at trace concentrations, can be toxic in varying degrees to different types of plants, as shown in Table 6.

TABLE 6. *Plant tolerance to boron*

Plants	Boron in irrigation water (mg/l)
Highly sensitive	0.4
Sensitive	0.4–1.25
Semi-tolerant	0.7–2.50
Tolerant	1.6–4.0

Soil characteristics are also important in determining whether the plant will be at the lower or upper end of each group. Soils with a saturation extract between 0.7 and 1.5 kg/l are considered marginal and anything above 1.5 mg/l is contaminated. Since the Mexico City wastewaters contain 1.4 mg/l of boron on average, a careful look was necessary to determine any toxic effects. Hence a series of experiments was designed to test the boron desorption capabilities of three soil types—clay-loam, sandy-clay-loam, and clay—at three different depths and with four different amounts of water applications. The initial boron concentrations in saturation extracts of the soils were for clay-loam, 1.21–1.54 ppm; sandy-clay-loam, 0.92–1.60 ppm; and clayey soil, 2.05–2.3 ppm. The data obtained so far seem to indicate that the boron removal capacity for each applied water layer is dependent upon the soil texture and the depth of the soil stratum to be washed. It is easier to remove boron from the upper stratum since the concentration gradient between the water and soil is greater, decreasing as the water passes from one stratum to another.

[1] The original paper contained details of studies on boron, on which the author can be consulted.

Salinity. The Mexico City sewage falls within the C3–S2 classification, corresponding to highly saline water with a medium content of sodium. It is not suitable for application to soils with a poor drainability. When applied to soils with good drainability salinity should be controlled with the aid of chemicals or organic matter (Samplon 1968). It should not be used on fine textured soils and, when used on rough textured soils, the plants should be of varieties highly tolerant of salts.

Since the texture of the soils studied was fine and the salt concentration in the saturation extract low (1150 ppm), it is thought that the high content of organic matter (125 mg/l of BOD) in the water is helping to lessen the salinity effects. This situation seems to be responsible for the good growth of salt-sensitive crops, such as beans, radish, celery and fruit trees. They do not appear to be affected by salt contents of from 0 to 2500 ppm.

Nutrients. The principal sources of nutrients in the irrigation water are divisible into (a) natural and (b) induced by man. The natural sources, chiefly from precipitation, sedimentation of dust and the wash-out of non-cultivated soils, contribute nitrogen of 0.2–0.9 mg/l and phosphorus of 0.1 mg/l, 61% approximately from rain water. The sources induced by man consist of discharges of residual domestic and industrial waters and the wash-out of soils under cultivation. The waters of Irrigation District 03 present the following average concentrations of nutrients:

	Class of nitrogen (mg/l)			Phosphorus (mg/l)	
	N-NH$_3$	N-NO$_3$	N-Org.	Total PO$_4$	Ortho. PO$_4$
Clean	3.65	0.51	3.05	7.22	3.57
Sewage	10.13	0.17	6.90	26.30	9.10
Mixed	7.72	0.93	6.60	14.00	5.70

Considering that the annual volume of water received by the district is approximately 800 m^3, they contribute a large quantity of nutrients to which must be added commercial fertilizers, as well as the employment of adequate agricultural practices, which includes growing alfalfa and other plants which fix atmospheric nitrogen.

Characteristics of waters, soils and cultivations. In order to obtain the best sustained production it is of vital importance in waters of irrigation to know the magnitude and ranges of variation of the physico-chemical and bacteriological parameters. It is also necessary to determine the quality, structure and agricultural characteristics of the soils. In the Irrigation District 03 the characteristics are listed in Table 7. It may be noted that the mixed waters are more polluted than the sewage and clean waters, as indicated by coliforms, BOD and COD. The explanation of this lies in the local effluents of domestic waters and residual waters from stables in the vicinity of the irrigation canals.

Productivity. At first it was thought that the Mexico City sewage might contain substances toxic to crops, specifically salinity and boron. However, although the salt and boron content is high, a difference could not be seen between the productivity of plots watered with raw sewage and those watered with clean water. It seems that most of the crops grown in the irrigation district are semi-tolerant to salinity and boron, except for beans and alfalfa that

TABLE 7. *Characteristics of waters in Irrigation District No. 03*

Parameter	Clean waters	Mixed waters	Sewage
Electric conductivity EC	803.52	1473	1726
Total solids TS mg/l	880.28	1422	1564.6
Total suspended solids TSS mg/l	66.83	114.2	202.67
Total dissolved solids TDS mg/l	696.60	1218.0	1373.2
Faecal coliforms MPN/100 ml	3.7×10^4	47.24×10^8	8.38×10^8
BOD mg/l	48.64	184.0	171.6
COD mg/l	173.53	489	405.0
MBAS mg/l	89.91	5.75	8.36
NH_3–N mg/l	3.65	7.72	10.13
Organic N mg/l	3.05	6.60	6.90
NO_3–N mg/l	0.51	0.93	0.17
PO_4–P mg/l	7.22	14.0	26.34
PO_4–P ortho mg/l	3.57	5.67	9.07
Chlorides Cl^- mg/l	121.76	207.0	184.4
Sulphates SO_4 mg/l	70.89	153	210.35
Total hardness mg/l	282.83	302.0	472.3
Total alkalinity mg/l	319.95	459.0	450.2
pH	7.76	7.7	7.74
Lead mg/l Pb	0.053	0.132	0.130
Mercury mg/l Hg	0.0005	0.002	0.0013
Cadmium	0.013	0.04	0.020
Zinc	0.026	0.36	0.457

TABLE 8. *Nutrient input to the Irrigation District No. 03*

Type of water	Yearly nutrient input +(kg/ha-yr)		
	N	P–PO_4	K
Clean water	42	42	208
Combined water	303	199	810
Sewage	320	490	913

TABLE 9. *Nutrient requirements by different crops*

Crop	Optimum N:P:K relationship (kg/ha-yr)		
	N	P–PO_4	K
Alfalfa	20	200	300
Wheat	420	39	24
Corn	145	221	95
Barley	70	47	34

are considered tolerant. It is also significant that these soils within a semi-arid zone give yields comparable to those of fertile zones (86 ton/ha of alfalfa), explained by the fact that these soils have been under irrigation with sewage for more than 60 years. During that time the organic matter in the wastewaters has been able to condition the soils, as well as adding nutrients, which to a great extent precludes the use of fertilizers. Table 8 presents the yearly contribution of nitrogen, phosphorus and potassium to the irrigation district from the municipal wastewaters. The clean waters require the use of fertilizers wherever they are the only source of water. Table 9 illustrates the nutrient requirements of some of the most widely grown crops in the study area.

The differences between mean productivities for each crop with different types of irrigation water were subjected to a null hypothesis with a significance level of 1%. The results of this analysis did not reject the hypothesis in any one of the cases studied, thus the observed differences in productivity for alfalfa, barley, oats, tomato and corn are not significant.

References

Aguilar Alvarez, I. 1975. *Los Recursos Hidràulicos en America Latina al Ano 2000, Una Imagén de lo Inaceptable.* Ediciòn Especial, Colegio de Ingenieros Civiles de México.

Anon. 1971. *Standard Methods for the Examination of Water and Waste Water,* 13th ed. APHA, AWWA and WPCF, Washington, DC.

Anon. 1972. *Plantas de Tratamiento: Xochimilco, Cd. Deportiva, Cerro de la Estrella, San Juan de Aragòn, Chapultepec.* Informe Técnico, Oficina de Plantas de Tratamiento, del Departamento del Distrito Federal.

Anon. 1973. *El Plan Texcoco, Solucion a un problema ancestral.* Obras, September.

Anon. 1973. *Informe Descriptivo de la Planta de Tratamiento de aguas negras.* Informe Técnico, Aguas de San Juan Ixhuatepec, Sociedad de Usuarios.

Benftez Zenteno, R. and Cabrera, G. 1966. *Proyecciones de la poblaciòn de México, 1960–1980.* Banco de México, S.A., México.

Bernard, Ostle. 1970. *Estadistica aplicada.* Limusa–Wiley, México.

Lamartine, Paul. 1965. *El Desarrollo regional de México.* Banco de México, S.A., México.

Lees, N. D. 1965. *Localizaciòn de industrias en México.* Banco de México, S.A., México.

Orozco, J. V. 1964. La Comisiòn Hidrològica de la cuenca del Valle de México, su origen, finalidad y resultados hasta 1963. *Proceedings: Tercer Seminario Latino-americano de Irrigaciòn en México.*

Paz Sànchez, A. 1973. *La utilizaciòn de agua negra tratada.* Obras.

Ramos Magaña, J. and Ortiz Santos, G. 1969. *Comparaciòn de los diversos anteproyectos estudiados para abastecimiento de agua a la cuidad de México.* Ingenieria Hidràulica en México.

Richards, L. A., Ed. 1954. *Diagnosis and Improvement of Saline and Alkali Soils.* US Salinity Lab., USDA. Washington, DC.

Samplòn, S. *et al.* 1968. *Diez temas sobre suelos.* Ministerio de Agriculture, Madrid.

Ordoñez, B. R. and Salcedo, N. 1974. Personal communication of unpublished data.

THE EFFICIENCY OF WATER USE
IN IRRIGATION IN EGYPT

by I. Z. KINAWY

Chairman of the Egyptian Committee of ICID

Background

Egypt has a population of about 38 million and a total land area of about one million km^2. Nearly 96% of this area is barren desert, and the remaining 4%, concentrated mostly in the Nile valley and the Delta, is intensely cultivated. These irrigated areas support the population and have a density of 1320 persons per km^2. The area under cultivation amounts to about 6.5 million feddans[1] with an average of 1.67 crops per year for each feddan.

The agricultural sector is dominant in the economy of Egypt and provides employment for over 60% of the population. Moreover, it contributes about 27% of the total income, as against 21% for industry. About 80% of exports are agricultural products, but Egypt imports large quantities of wheat and flour, with lesser quantities of maize. In the development planning the agricultural programmes are given priority, with particular emphasis on increasing production of food crops in order to reduce the imports.

Perennial irrigation was introduced in Egypt in very early times. However, only in 1902 was this system used on a comparatively large scale. It necessitated placing the water at the disposal of every farmer. Consequently, many canals were constructed (over 30,000 km at present) to convey water to all areas. Although this system has contributed favourably to the extension of the actual irrigated area, yet it has also contributed to the rise of the groundwater table which has persistently deteriorated the fertility of the land and decreased the productivity of the soil. Free-of-charge irrigation, in addition to a noticeable percentage of lands using gravity irrigation, has led to the unwise use of irrigation water. It is important to note that the lands of Egypt which are composed of clayey loam have a relatively low degree of permeability and thus do not allow for a rapid movement of sub-soil water.

The Nile is the main water resource of Egypt. Rain is scarce and the quantity of groundwater is limited at present. The safe yield of the groundwater in the Delta aquifer is estimated at 370 million m^3/yr. Hence the groundwater as a source is hardly competitive

---

[1] One feddan = 4200 m^2 = 0.42 hectare.

with the Nile water. Nevertheless, 130 productive pump stations are distributed over the southern part of the Delta to feed the tails of the irrigation canals as an auxiliary source of water in the summer dry season.

Until 1902 lands during the summer season relied solely upon the natural supply of the river. In certain years, the supply was far short of the water requirements, causing disasterous results to the crops. Hence annual storage, i.e. storing water in seasons of plenty for use in the next season of shortage, has become a necessity.

Thus the old Aswan Dam was erected with a reservoir capacity of 970 million m^3. In order to make the best of the stored water, more barrages across the Nile were erected simultaneously with the construction of the Dam. Early in 1912 the Dam was heightened to increase the impounded water to 2.5 milliard m^3 (1 milliard = 1000 million). For the second time, the Dam was reheightened in 1933, and the total capacity of the reservoir became 5 milliards. Further in the same year, Gebel Awlia Dam in the Sudan was erected to the benefit of Egypt to store an additional 2.5 milliards.

By and by, the need for more water became imperative, because the annual storage system has failed to cover the increasing water requirements necessary for the extensive programme of land development planned to face the rocketing growth of population. There was no alternative except to move to the over year storage by the Aswan High Dam in order to provide, in the shortest possible time, extra water for further development of Egypt and the Sudan.

The Aswan High Dam gives an increase in the guarantee draft of 22 milliards, of which 14.5 milliards are for the Sudan and 7.5 milliards for Egypt. Ultimately the quota of Egypt will be 55.5 milliards and that of the Sudan is 18.5 milliards, while 10 milliards will be left for compensation of evaporation and seepage losses in the reservoir zone. These figures are based on the average annual supply of the natural river during the present century (84 milliard m^3/year). The records show fluctuation between 120 and 45 milliard m^3/year.

Before the construction of the Aswan High Dam, the higher ranges were not reached. The water had to be discharged right away so as not to threaten the safety of the country. At present all the Nile waters are utilized by Man, discharge into the Mediterranean being limited to drainage water drawn off the irrigated lands, and some spillage during three weeks closed irrigation season in January. The Sudan, under its extensive programme of development, will very soon use its share of the Nile waters. In the absence of the Aswan High Dam, this would not have been possible without Egypt suffering. Egypt has to be careful in optimizing the use of her limited quota.

At the dawn of this century the Nile water was in abundance. On the average, one man had 25 m^3/day, while today this average has fallen to 4 m^3/day. During the same period, the cultivated area has increased from 5.1 to 6.5 million feddans, and the crop index has changed from about 1.0 to 1.67. The industrial and domestic requirements have also risen rapidly, so that the provision of adequate supplies has become difficult.

Agriculture in Egypt

Since ancient times the Nile valley and its Delta have been, and still are, one of the most famous agricultural lands in the world. Thanks to the rich soils, the ample water supply and the favourable climate, most of the lands carry more than one crop annually.

The main crops are cotton, rice and maize during the summer season, wheat and clover during the winter. Other grains as well as vegetables and fruits are also cultivated to a large extent all the year round. In recent years certain changes in agricultural patterns have been observed. In 1952, the cultivated area in Egypt during the summer season was no more than 50% of the total area (3.120 million feddans), while in 1972, this area increased to 80% (5.331 million feddans).

Maize was normally cultivated in July, but subsequent to the construction of the Aswan High Dam the cultivation season has been changed to early summer during April–May. In 1952, the maize areas in summer and autumn were 27,000 and 1.677 million feddans, while in 1972, these figures were changed to 1.210 million and 321,000 feddans respectively. Moreover in 1952 the cultivated rice area was only 350,000 feddans, whereas in 1972 it became 1.200 million feddans. Sugar cane has increased from 92,000 feddans in 1952 to 202,000 feddans in 1972.

Table 1 shows the crop pattern as surveyed in 1972. The main crops in Egypt are wheat which covers 19% of the cultivated area, cotton 24%, maize 23.5% and rice 17.5%. The water requirements in this table are according to the results of experiments carried out in pilot areas which have been selected so as to cover the whole country. They are not identical to those given by Criddle and Blaney. On the basis of their formulae, water requirements for wheat, beans and onions are respectively 2%, 10% and 22% higher, whereas for cotton it is 28% less. The experimental results are therefore chosen for representing the irrigation for the country.

These rapid changes in the season and area of cultivation should correspond to a simultaneous change in the design of irrigation rotations, canal sections and above all the discharges of the Aswan High Dam. In other words very careful water management is now exceedingly necessary. If the crop pattern is left without strict control, Egypt may soon find itself faced with a water shortage. Thus the less water-demanding crops must be given priority over the more water-demanding ones such as rice (10,000 m^3 of water per feddan) and sugar cane (17,800 m^3 per feddan), whatever the value of such crops may be.

The irrigation system

Free-of-charge flush irrigation is the system prevailing in Egypt. In many cases pump stations commissioned by the government are used to ensure gravity irrigation in certain zones. Three hundred and eighty governmental pumping stations are now serving 1.9 million feddans or 35% of the total cultivated area. This system makes for extravagance in water consumption. Sprinkler irrigation has recently been used over very small areas. Furthermore, lift irrigation is becoming more frequent, aiming at keeping the water table at its lowest. The discharges of the Aswan High Dam vary from one month to another with the water requirements, together with the climatic conditions prevailing over the country (see Table 2).

For about three weeks in January, all irrigation canals are closed and the discharges of Aswan are at their lowest, enough only to secure navigation in the course of the river and to cover the industrial, electrical and domestic requirements. Normally the discharge during this period is 80 million m^3/day, but owing to the increasing demand for hydropower the discharge has now been raised to 90 million m^3/day. The draft during the winter closure is practically left to spill into the Mediterranean sea, but this lost discharge (2.0

TABLE 1. *Water requirements in the area based on the flush system of irrigation dominating in Egypt*

	Lower Egypt (the Delta)			Middle Upper Egypt			Uppermost Egypt			Total cropped area in Egypt ×1000 fed.	Total water requirement for Egypt, mill. m³
	Cropped area ×1000 fed.	Water requirements per fed. m³	Total requirements mill. m³	Cropped area ×1000 fed.	Water requirements per fed. m³	Total water requirements mill. m³	Cropped area ×1000 fed.	Water requirements per fed. m⁴	Total water requirements mill. m³		
Winter crops:											
Wheat	725	1600	1160	217	1700	370	297	2100	624	1239	2154
Beans	135	1350	183	153	1470	225	77	1600	123	365	531
Barley	73	1400	102	6.7	1500	10	11.3	1700	19.5	91	31.5
Fenugreek	1.8	1050	2	11.7	1170	14	15.5	1290	20	29	36
Lentils							67	1210	81	67	81
Flax	32.5	1070	35	0.5	1190	0.6				33	35.6
Onions	2.5	2250	5.5	6.5	2500	16	22	2700	59.5	31	81
Lupine	6	1000	6	2.5	1120	3	1.5	1230	2	10	11
Chick-peas	3.5	1020	3.5	.3	1130	0.4	6.2	1240	7.5	10	11.4
Egyptian clover (Bersim)	1166	3100	3600	274	3650	1000	125	4000	500	1565	5100
Temporary (Bersim)	903	1910	1730	226	2120	480	125	2330	292	1254	2502
Garlic	2.2	1600	3.5	6.6	1780	117	.2	1950	0.5	9	121
Vegetables	45	2220	100	129.5	2470	320	4.5	2720	12	179	432
Others	11.5	1900	22	1.5	2110	3	16	2320	37	29	62
Summer crops:											
Cotton	1000	3400	3400	334	3900	1300	218	4700	1030	1552	5730
Rice	1140	8800	10,100	6	9920	59				1146	10159
Corn				92	2800	258	392	3080	228	284	1468
Maize	1140	2700	3100	322	3000	966	69	3300	228	1531	4294
Sugar cane	12	14,760	117	25	16,400	410	165	17,800	2940	202	3527
Earth nuts	18	3160	57	11	3510	39	5	3860	19	34	115
Sesame	6	2000	12	6	2220	13.5	30	2440	73	42	98.5
Vegetables	98	3780	377.5	129	4200	544	335	4620	1550	562	2471.5
Others	60	3420	205	18	3800	68.5	42	4180	175	120	448.5
Orchards	188	6850	1290	44.2	7600	322	20.8	8370	174	253	1786
Total	6769		25671	2023		6539	2045		9177	10837	41387

Water requirements shown in this table are according to extensive experiments carried out by the Ministry of Irrigation in pilot areas.
Total cropped area is 10,837 million feddans as actually surveyed in 1972.

milliards) should be made use of by storing the water either in the northern lakes bordering the sea, or in some other suitable reservoirs such as Natrun depression in the Western Desert. In the first case, the reservoir, through its artificial head, would safeguard against the intrusion of sea water in the northern lands of the Delta, which is becoming a problem at present, and at the same time it would allow for the irrigation of the cultivable lands neighbouring the lakes. In the second case the reservoir, whose capacity is 3.5 milliards, would help in the reclamation of more lands and might be necessary in event of the hydro-electric power scheme at Qattara, 90 km to the west of Alexandria, being executed and the sea waters allowed to discharge into the Qattara depression.

In Egypt there are two harvest seasons; in April–May and in September–October. In the first, being a transitional season between the winter and summer croppings, the irrigation water is badly needed in order to cultivate the rice and maize; whereas in the second season, water is to cover the needs of winter crops and to allow for navigation. In the latter season, the water slopes either in the Nile or in the navigable canals tend to be flat, and more water is allowed to discharge into them in order to secure the safe minimum water depth for navigation, especially just downstream of the barrages. Thus it seems that for purposes of navigation, Egypt loses about 3.5 milliard m^3/year, or 6% of her quota, to which we believe that more attention should be paid. At present, the cost for storing one milliard m^3 of water is about LE 70 million or 150 million US dollars.

Industrial and domestic water use

While the dominant use of water in Egypt is for irrigation purposes, the demands for domestic and industrial purposes, though comparatively moderate today, will very soon increase. A decision has been taken to build 10 new industrial towns around Cairo, the water requirements of which will be of the order of 4 million m^3/day or 1.5 milliard m^3/year. Currently, 3.0 milliard m^3/year are estimated to cover the industrial and domestic water use, and this figure will rise to about 4.5 milliards by the turn of this century. Waters used for industrial and domestic purposes are regarded in Egypt as losses from the point of view of irrigation management.

Drainage in cultivated area

In Egypt, the main object of drainage is to control the rise of groundwater within the root zone, so that air can penetrate through the soil pores more easily and become available to the roots of plants. Moreover, providing adequate drainage helps to prevent salt accumulation in the soil.

The mixing of drainage water with irrigation water is not a common practice in the Delta. It is hardly utilized unless a serious deficit in the water supply occurs. This is due to the salinity and unchecked industrial pollution of drain waters as well as to untreated sewage in the drains. But in Upper Egypt the drainage water is pumped either into the river or into the canals. As a reasonable estimate, 4.5 milliards of drainage water are re-used in irrigation. Re-use of drainage water in future seems inevitable in order to meet the great development of land reclamation in the north of the Delta, either directly if the salinity is within safe limits, or by mixing with irrigation water if the salinity has to be diluted.

Water distribution

Water is distributed among the irrigation canals in rotation according to the season, locality and the crop under cultivation. The water is allowed to flow in a group of canals for a period long enough to have the area irrigated. The supply period is followed by closing the head regulators and the turn is given to another group of canals.

Distribution among the holders of land is through outlet pipes, the diameter of which is determined by the property of the holder. The smallest diameter is 15 cm (6 inches), with increase of 1 inch or 2 inches in the larger pipes. Such a system of water distribution is not fair, and some oversize outlet pipes lead to the misuse of water. As a rough estimate 2.0 milliard m³/year spill into the drains without being efficiently used for irrigation purposes.

Water losses

Losses include seepage, infiltration and evaporation from the Nile as well as from the irrigation canals. Subsequent to construction of the Aswan High Dam the Nile is no longer higher than the land levels except in some locations upstream of the barrages. However, there is a serious seepage from the canals. For example, Ismailia canal loses about 800,000 m³ annually because of its unlined sandy section and its high water levels. In some stretches, water is more than 4.5 m above the ground level. We believe that special care should be given to this problem which is common also to a considerable number of canals constructed in the reclaimed desert areas.

Farmers are used to re-irrigate their fields as long as their supply rotation is on. This is very common in rice cultivation. It seems that the habit was formed prior to the construction of the Aswan High Dam, when the summer rotations were not very regular due to the insufficiency of water supply. Moreover, when the rotation is on, head regulators are opened to let water discharge into the canals day and night. Farmers are supposed to serve their lands continuously during the rotation period, but some work only in the course of the daylight, with the result that a high percentage of water is left to spill into the drains.

Over-irrigation is another form of extravagance. It is recorded that some farms in the west of Delta have drawn double as much as the average amount of water required with consequent deterioration of fertility and fall of productivity.

Water balance in Egypt

The water discharges from the Aswan Dam are shown in Table 2.

This volume is intended for a total cultivated area of 6.5 million feddans, which means an average of 8500 m³/feddan/year (= 20.000 m³/hectare) or 5100 m³ per feddan of the cropped area per year (= 12.000 m³/hectare).

Table 3 shows the water balance in 1972. The overall efficiency has been improved since 1966 when the efficiency was 66%, though the water capital (55.5 milliards) stood unchanged. It seems that the inclination to more water-demanding crops has led to this result. The increase in rice and sugar cane areas has necessitated the utilization of 5 milliard m³ more

TABLE 2. *The Aswan High Dam discharges during 1972*
(milliard m³)

			BF 27.200
Jan.	3.350	July	6.950
Feb.	4.000	Aug.	6.200
March	4,200	Sep.	4,300
April	4.000	Oct.	3.850
May	5.200	Nov.	3.650
June	5.450	Dec.	3.000

BFW 27.200 milliards	Total 55.150 milliards

TABLE 3. *Balance of water budget in Egypt based on 1972 records*
(milliard m³)

Total discharges of the Aswan High Dam		55.150
Industrial and domestic abstractions	− 3.000	
Upper Egypt (UE) irrigation requirements (Table 1)	− 15.720	
Reuse of drainage water in UE	+ 4.570	
Losses in UE	− 4.420	18.570
Water passing Cairo (measured)		36.580
Delta irrigation requirements (Table 1)	− 25.700	
Spillage into the Mediterranean (measured)	− 3.500	
Losses in the Delta	− 7.380	− 36.580
Overall efficiency of irrigation water	41.420	= 75%
	55.150	

of water. On the one hand, this trend increases the apparent efficiency; but on the other, it calls for a much wiser use of water and control of crop areas.

From the above table one can conclude that the accumulated losses in the water income are 13.880 milliard m³ classified as follows:

		% quota
Industrial and domestic use	3.000	5.5
Excess discharges to secure navigation, and to cover the hydro-power requirements, which finally spill into the sea	3.500	6.5
Losses due to conveyance, seepage and extravagance in water utilization	9.800	17.5
	18.300	33.0%
Recovered through the reuse of drainage water	− 4.880	− 8.0
Net losses	13.880	25.0%

Conclusion

Should the population of Egypt continue to grow at the present rate of 2.5%, 250,000 feddans must be reclaimed annually just to keep the current standards as they are. This necessitates looking for a further 2 milliard m³ of water every year.

In the south of the Sudan, the Jonglei diversion canal project is designed to channel the water of the Upper Nile in a confined artificial canal instead of flowing in the swamp region. Ultimately the water benefit from this project will be 3.6 milliard m³, which means that the new supply will be absorbed within two years. It seems essential therefore carefully to plan and control the use of all available water.

In this respect the Ministry of Agriculture on the basis of preliminary soil survey has listed an area of 1,174,000 feddans subject to possible reclamation and irrigation, of which 156,400 feddans are in Upper Egypt and 1.018,500 feddans in the North. It is understood that, pending the outcome of further investigations only a portion of these areas will be finally brought under cultivation. However, based on the gross listing it is estimated that the new lands would require about 10 milliard m³ of water annually. In addition to these lands an estimated reclamation of 765,000 feddans in the Sinai Peninsula will require 6.9 milliards more.

Ten milliard m³ is left to compensate for the evaporation and seepage losses in the Aswan High Dam Reservoir, but owing to the urgent need for conserving as much water as possible for further development, it is vital to encourage researches aiming at suppressing the huge evaporation losses.

Moreover since the available supply under the 1959 treaty is already taken up, the trend should be to look to this and other sources of loss for the fulfilment of the projected demands. Water-serving devices, improvement in irrigation practices, elimination of waste, minimizing the conveyance losses and reuse of drainage water, must take priority to satisfy the ever-increasing demand.

Irrigation projects should be viewed as parts of one great framework of the water balance for all Egypt. They should be studied with a view to the available water supply for optimum crop pattern. Cost benefit studies should be made and particularly for those projects serving newly reclaimed lands.

The present crop pattern in conjunction with the irrigation system should be reviewed to plan for a higher utilization of the limited water resources. A principal aim should be to save irrigation water for more crops by raising the crop index, and that seems possible, especially in the newly reclaimed lands.

DISCUSSION

Compiled by MILOS HOLY

The ICID investigation

Discussion on the efficiency of irrigation schemes was based mainly on the results of investigations carried out by ICID into the influences of project management, as presented by M. G. Bos (p. 351). In answer to questions M. G. Bos stated that the 91 projects from which data were acquired represented 5 million ha of irrigated land. While not fully representative, they give a relatively good idea of the problems related to efficiency. He outlined research now proceeding in order to improve efficiency, and indicated that several aspects which came to light in this Symposium will be included in a second questionnaire. W. Imrie stated that the examples presented had convinced him that data provided by the individual ICID National Committees for the evaluation of efficiency are reliable.

Factors in irrigation efficiency

The efficiency of irrigation is significantly affected by the level of training of users of irrigation water and it was recommended that information be acquired on the level of users of the individual irrigation systems. It was stated that in Pakistan, for instance, even the biggest efforts made by engineers at an efficient use of water are nullified by the whims, mood and traditions of the farmers. The view was voiced that charging the farmer to pay for water used would increase efficiency: the farmer himself benefits from the water, and funds thus raised could be used to employ qualified engineers to manage the irrigation project in a proper manner.

ICID research showed, however, that charging the farmer at a relatively low level for the water he uses does not necessarily improve efficiency of water use. The level of the fee should be studied and adjusted in such a manner as to increase efficient use. Some speakers indicated that recent measurements carried out in the Middle East showed that the efficiency of water use, including wastage in the farming processes, is no more than 30–40%, in some countries as low as 25%.

The efficiency of water conveyance may be increased substantially by canal lining. Economic techniques must, however, be sought. Experience gained in Egypt and India shows that in sandy and sandy-loam desert areas it is useful to tile the canals on the canal bottom only, not at the sides, as water losses through the canal bottom comprise a large proportion of the overall loss. India's experience shows that gravel packing to a thickness of approximately 1 foot is cheaper than tiling, allows the water to spread sideways, and makes possible the growing of shade and fruit trees on either bank of the canal.

Efficiency in use of irrigation water may be attained only with adequate measurements

379

of the amount of water flowing at one moment at each gauging point in the distribution system. This implies that every irrigation system should be equipped with a sufficient number of measuring devices. Insofar as the developing countries are concerned there are two obstacles to this, namely the devices are too expensive and personnel trained in their operation and maintenance are too few. In consequence measuring devices are not part of many irrigation systems, and farmers in the upper section of the system use most of the water, over-irrigating their fields, while farmers in the lower sections do not receive any water at all in times of shortage. In some countries up to 30% of the area developed for irrigation is abandoned for this reason.

Water control at the farm gate is the decisive factor in irrigation efficiency. Water should be delivered in the amount and at the frequency needed for optimal crop production, neither more nor less. A simple measuring device for farm and field outlets is urgently needed; it should not be sensitive to varying boundary conditions, should be resistant to rough handling, sediment and debris, ought to be inexpensive, all of which attributes would make it operable by farmers without training. Next to such devices attention should be devoted to the use of existing irrigation structures, such as gates, drops or diversions, for purposes of measurement. Simple and accurate formulae are available to convert the depths at gates into discharges.

The advantages and disadvantages were discussed of the individual irrigation methods. It was stated that sprinkler irrigation is very suitable for soils with a high infiltration rate, for rough topography and for shallow soils, but has the disadvantages of high capital and maintenance costs, danger of mechanical failure, the problems of efficient operation and organization of maintenance, the burning of crop foilage in hot climates, and losses by evaporation. Consideration is needed also of other highly efficient methods such as advanced surface irrigation, low-pressure pipe irrigation, proper land grading, and last but not least, the training of farmers in improved surface irrigation.

In arid lands with a scarcity of good water it is important to give full consideration to the selection of land areas and crops to secure the highest returns per unit of water, and also to the danger of the accumulation of salts when subsoil methods are applied.

Great emphasis was put on close cooperation between irrigation engineers and agronomists in the preparation of irrigation projects. The fact that irrigation and agricultural engineers are trained and work as separate specializations causes many problems in many countries.

Participants in this discussion were: A. Arar, A. M. Balba, A. M. Baligh, M. G. Bos, R. C. Carpenter, S. J. Chimonides, J. De Forges, M. El Gabaly, A. Farmi El-Kashef, G. Garbrecht, P. Karakatsoulis, I. Z. Kinawi, J. C. Mongelard, and K. R. Ramanathan.

Use of saline water

A further discussion showed the great interest which exists in irrigation using water containing some salt. The possibility of using sea water mixed with drain water or with any other water of low salt content was discussed. Experience in Egypt shows that drain water may be used in emergency conditions for irrigation, for leaching saline soils, for drip irrigation on sandy soils, on relatively salt-tolerant crops, and for domestic and industrial uses to reduce the consumption of fresh water. In Saudi Arabia no drop in alfalfa yields was observed when irrigated with drain water containing 4 mhs/cm, but with a content of 8 mhs/cm yields went down by 21%.

Important data were presented in the discussion on the effects of the Aswan High Dam on discharge, suspended load, water quality and salinity. The fertile land of Egypt will require some time to become accommodated to the new conditions. New irrigation projects in arid zones should be studied in the light of previous experience elsewhere and intensive research should be conducted before projects are launched. Detailed information of these subjects is available in "Environmental Control for Irrigation and Flood Control Projects," Proceedings of the IX Congress ICID, Moscow 1975.

Participants in this discussion were F. Ahmed Mohammed, A. Arar, A. M. Balba, A. A. Kamal, M. Kansoh, H. Mohammed Al Haskim and M. Shaker Hanna.

SECTION VII

Human Problems in Irrigation Areas

SANTE ET IRRIGATION[1]

by ALEXIS COUMBARAS

Chargé d'Enseignement à la Faculté de Médecine
de l'Université Paris VII, Paris, France

Summary

The wide-scale irrigation programmes in the arid zones of developing countries bring about some serious dangers and sometimes catastrophic situations in the field of health. The health problems that arise are of course those of water-transmissible diseases, such as schistosomiasis and malaria, but also those connected with the immigration of new populations, the creation of new rural communities, new habitats and housing. All these phenomena cause a potential aggravation to the environment, but may also result in making it healthier. They may permit to solve such problems as drinking water supplies, excreta disposal, improvement of sanitary installations and of habitat, a better organization of basic medical preventive and curative structures. Mass health education may become possible as well as a new mode of life in better accordance with health standards.

To achieve this target, action in the field of health should begin as early as the planning phase of an irrigation project, in view of the evaluation of the situation, the forecast of the dangers, the discussions and the choice of the measures to be taken. This action has to be continued during the implementation phase, during the execution of the project and long after the irrigation system has been built and put into operation, so as to assume the surveillance of the health situation and the maintenance and the improvement of the results that have already been achieved.

The health action can be efficient only if integrated in the project as part of a whole, financially, administratively, and only if sincerely accepted by those responsible for the programme as part of their duties. Only a multidisciplinary approach to the problems that arise from irrigation can provide us with solutions which would be satisfactory from every point of view, including health.

Résumé

La pratique de l'irrigation sur une vaste échelle dans les zones arides des pays en voie de développement, s'accompagne sur le plan de la santé, de dangers graves et parfois de situations catastrophiques. Les problèmes sanitaires posés par l'irrigation sont évidemment ceux des maladies transmissibles par l'eau comme le schistosomiase et le paludisme, mais également ceux liés à l'immigration de populations nouvelles, à l'implantation de nouvelles communautés rurales, à la création d'un habitat nouveau. Ces différents phénomènes constituent une aggravation potentielle pour la santé, mais ils pourraient également être l'occasion d'un assainissement: permettre la solution de problèmes de l'eau potable, de l'élimination des excréta, l'amélioration des installations sanitaires et de l'habitat, une meilleure structuration médicale curative et préventive, une éducation sanitaire des masses et un mode de vie plus conforme aux normes de l'hygiène.

Pour parvenir à ce résultat, l'action sanitaire doit intervenir dès la planification du project d'irrigation en vue de l'évaluation de la situation, de la prévision des dangers, de la discussion et du choix des mesures à prendre. Elle doit se poursuivre par une phase d'exécution tout au long de la réalisation du project. Elle doit continuer

[1] Préparé à la demande de l'Organisation mondiale de la Santé.

bien au delà de l'achèvement des travaux pour assurer la surveillance de la situation sanitaire et le maintien et l'amélioration des résultats acquis.

L'action sanitaire ne saurait être efficace que si elle est intégrée dans l'ensemble du projet, tant sur le plan administratif et budgétaire, que dans l'esprit des responsables du programme. Seule une approche pluridisciplinaire des différents problèmes posés par l'irrigation pourra leur apporter des solutions satisfaisantes à tout point de vue, y compris celui de la santé.

Les zones arides couvrent près du tiers des terres émergées de la surface de notre planète et, depuis la plus haute antiquité, l'homme a essayé de les rendre viables et de les exploiter à des fins agricoles. Les zones arides sont caractérisées par une pluviométrie faible et souvent irrégulièrement répartie au cours de lannée. L'homme a donc cherché à suppléer à ces défaillances de l'eau du ciel par des aménagements des eaux de surface et des eaux souterraines. Il a foré des puits, branché sur des cours d'eau des canaux de dérivation; il a construit des barrages pour constituer des réservoirs de stockage et de régulation d'écoulement.

Il nous est particulièrement agréable, alors même que nous sommes réunis ici, sur cette terre d'Egypte, pour parler d'irrigation, de signaler que le plus ancien barrage du monde découvert à ce jour se trouve dans ce pays. Il remonte à la troisième ou quatrième dynastie donc à plus de 4500 ans. Il s'agit du Sadd el Kafara ("barrage des païens" des auteurs islamiques), dont les ruines s'étendent près de Helwan, à une quarantaine de km au sud du Caire.

Si donc les préoccupations des hommes n'ont guère changé avec les siècles quant à l'irrigation des terres arides, leurs besoins liés à la démographie et à leur incessante recherche du progrès se sont considérablement accrus, peut-être plus ces dernières décades qu'au cours des millénaires passés. Les moyens techniques dont ils disposent, allant de pair avec leurs besoins, les hommes ont pu réaliser en matière de bassins de rétention des lacs artificiels énormes. Il est vrai qu'à côté des besoins agricoles, un autre besoin a souvent présidé à ces créations : celui de l'énergie hydroélectrique. Avec l'accélération du développement, l'accroissement des besoins d'énergie et son enchérissement, on peut prévoir pour les années à venir un accroissement encore plus frénétique du rythme de ces réalisations.

D'autres vous ont dit les bienfaits de ces créations. Il nous appartient la tâche ingrate d'en montrer surtout les dangers. Bienfaits et méfaits atteignent ici parfois les dimensions gigantesques des ouvrages qui leur ont donné naissance.

Parmi ces méfaits il en est un de taille: celui qui porte sur un bien de l'homme parmi les plus précieux et que curieusement nous gaspillons souvent avec la plus grande insouciance, la santé. Mais il est encourageant de voir de plus en plus, comme dans cette réunion par exemple, mais sur le terrain aussi, s'amorcer le dialogue entre les hommes du sol, de l'eau et de l'agriculture et ceux de la santé. C'est de cette coopération que naîtra une meilleure compréhension des problèmes et qu'apparaîtront les solutions. Car il serait dérisoire et tragique que l'homme, dans le combat qu'il mène pour son bien-être, y mine la condition première de celui-ci: sa santé.

Les collègues ici présents traiteront tout à l'heure plus en détail tel ou tel problème particulièrement important relevant de leur compétence. Je voudrais ici définir, classer, délimiter ces problèmes et énoncer quelques idées générales quant à la manière de les aborder et de leur chercher des solutions.

A. *Et tout d'abord, quels sont ces problèmes?*

Les problèmes de santé liés à l'irrigation sont nombreux et évidemment très différents d'un pays à l'autre, d'un schéma d'irrigation à l'autre. Ils peuvent cependant presque tous être classés sous les trois rubriques suivantes:

I. *Maladies transmises par l'eau.* Maladies don't l'agent pathogène, l'hôte intermédiaire ou le vecteur séjourne dans l'eau, et qui ont évidemment d'autant plus tendance à se répandre en étendue, fréquence et gravité, que les plans d'eau s'étendent, que les réseaux d'irrigation se multiplient et que le contact de l'homme avec l'eau s'accroît.

La schistosomiase (bilharziose) véhiculée par un mollusque aquatique, *le paludisme* transmis par certains moustiques dont les larves vivent dans l'eau, sont des grands fléaux des pays en voie de développement. Leur expansion catastrophique par les pratiques d'irrigation dans les territoires arides n'a été que trop fréquente pour qu'il soit nécessaire de donner des exemples. Leur importance est telle que je ne saurais les aborder dans cet exposé très général. Deux interventions séparées leur seront consacrées.

II. *Les problèmes d'assainissement du milieu et de l'habitat.* Parmi ceux-ci:
(1) *Le problème de l'eau potable* revêt une importance capitale en raison de nombreuses maladies bactériennes, virales et parasitaires véhiculées par l'eau de boisson. L'établissement d'un réseau d'irrigation constitue cependant ici en général un bienfait, du moins en ce qui concerne les territoires arides. Tout simplement parce qu'il met l'eau, potable ou non, mais en quantité suffisante, à la portée des habitants. Elle est aussi en général plus propre que celle dont la population disposait avant la création du réseau d'irrigation. L'installation de ce réseau peut d'ailleurs souvent être l'occasion de résoudre de manière satisfaisante et à peu de frais le problème de l'eau potable. C'est le cas notamment des programmes d'irrigation où l'eau est apportée par des puits de forage. C'est là souvent une eau bactériologiquement pure à son émergence, mais qui ne tarde pas à être souillée dès sa sortie. Il suffirait bien des fois d'assurer l'étanchéité au point de sortie et d'aménager très simplement l'adduction et la prise à des fins domestiques d'une infime partie de l'eau destinée à l'irrigation, pour que le problème de l'eau potable soit résolu. C'est là un exemple typique, presque caricatural, de ce qui peut ou pas se passer si on prend ou non en considération les besoins de santé au cours d'un projet d'irrigation. Dans le premier cas on offre à peu de frais de l'eau potable à la communauté; dans l'autre, non seulement on la prive de ce besoin élémentaire, mais pour peu qu'on laisse l'eau se répandre inconsidérément sur le sol, on risque de créer des foyers de transmission de paludisme, de schistosomiase et de bien d'autres maladies encore.
(2) *Le problème des maladies transmises par les matières fécales* est parfois directement lié au nouvel apport d'eau et d'humidité dû au nouveau réseau d'irrigation. C'est le cas de l'ankylostomiase, de l'anguillulose. Mais c'est surtout par des voies indirectes que l'irrigation favorise la transmission de des maladies. L'implantation de populations nouvelles, la création de nouvelles agglomérations rurales, l'instauration d'une vie communautaire plus étroite, sont favorables, faute d'hygiène fécale, à la diffusion de ces maladies. La création de nouveaux villages aura ainsi contribué à l'expansion de maladies graves ou pouvant le devenir: amibiase, ascaridiase, ankylostomiase, anguillulose, la schistosomiase sous sa forme intestinale, pour ne citer que les principales, mais aussi, quelle occasion nous est offerte de combattre avec plus d'efficacité ces maladies par une politique de construction de

latrines et d'éducation sanitaire! Tous ceux qui se sont occupés de la question savent combien cette action est relativement plus facile en agglomération plutôt qu'auprès de cultivateurs dispersés. Et l'augmentation des revenus et celle du niveau de vie ne doitelle pas s'accompagner d'une amélioration de l'hygiène et de la santé?

(3) *L'implantation de structures médicales curatives et préventives* de base est également facilitée par le regroupement de la population le long du réseau d'irrigation. Il est évident qu'un dispensaire ou même un petit hôpital sont plus efficaces quand ils sont établis au sein d'une agglomération qu'au milieu d'une communauté rurale très dispersée.

(4) *Planification des implantations rurales nouvelles et de l'habitat.* Nous concevons parfaitement que le solution de ces problèmes importants relève de bien de facteurs autres que la santé, parfois impératifs. C'est ainsi que les habitations ne sauraient être implantées loin des terres à cultiver et à irriguer, et ceci nonobstant l'argument sanitaire qui viserait souvent à éloigner l'habitation de la proximité immédiate des foyers hydriques d'infection. Il serait cependant normal que l'argument sanitaire soit *également* pris en considération, à côté de tant d'autres, quand plusieurs options s'offrent pour le choix de l'emplacement et de la structure d'une implantation nouvelle.

Quand l'autorité chargée d'élaborer le projet d'irrigation s'occupe aussi (et il est fort souhaitable qu'elle le fasse) de la construction du nouvel habitat, il est capital que l'aspect sanitaire des nouvelles maisons n'échappe pas à ses préoccupations. Il faut que les pièces d'habitation, les cuisines, les installations sanitaires, notamment les latrines, répondent à des exigences d'hygiène et de confort minima. Il faut aussi que le choix des matériaux de construction, la disposition et la dimension des locaux, répondent certes aux possibilités économiques mais aussi à un souci d'amélioration par rapport à la situation précédente. Il faut toutefois également que ces innovations puissent être acceptées par la population transplantée, habituée à une certaine manière de vivre et d'habiter, qui est d'ailleurs souvent parfaitement justifiée. Il faut se méfier d'un modernisme intempestif est insensé. Il nous est arrivé de voir des villages entiers nouvellement construits en béton, rester inhabités parce qu'il régnait dans les maisons une chaleur infernale alors que la construction traditionnelle, moins chère, en boue séchée offrait une température beaucoup plus agréable. Des latrines inconfortables, malodarantes, difficiles à entretenir, ne respectant pas l'intimité et la pudeur des gens, ne seront pas utilisées par la population qui continuera à se soulager dans la nature. Aussi, on n'hésitera pas à interroger un grand nombre d'habitants sur leur désirs, on demandera leur avis sur toute innovation projetée en matière d'habitat. On tiendra compte des capacités de la population d'accepter telle ou telle nouveauté. Les habitudes séculaires relevant de l'habitat, du mode de vie domestique, les habitudes sanitaires aussi, sont profondément enracinées dans la population. Tout changement, si minime soit-il, implique de la part de l'individu un effort psychologique considérable. Si l'on veut qu'il le fasse, il convient de lui faciliter la voie et être modéré dans ses exigences. Mieux vaut, à set égard, s'assurer d'une amélioration relative et progressive que d'élaborer un changement radical, mais qui ne sera pas accepté.

III. *Education sanitaire de masse.* La nécessité d'un échange de vue entre la population et les responsables du projet a déjà été évoquée à propos de questions d'eau potable, d'élimination hygiénique de matières fécales, de questions relatives à l'habitat. Nous nous sommes attachés à montrer le nécessité de s'informer auprès de la population sur ses desiderata et d'apprécier sa réceptivité envers les améliorations à apporter. Il est évident que ce courant de pensée doit se faire aussi dans le sens inverse. Il faut que la population soit persuadée de l'utilité d'une eau pure, de l'usage des latrines et autres installations sanitaires du nouvel habitat. Faute de cet effort de propagande, nous n'avons que trop de fois constaté l'abandon par les habitants des nouvelles installations qu'on leur avait fournies, et leur retour, pour l'eau de boisson, au marigot ou au cours d'eau ancestral.

Il est certain que l'éducation sanitaire, telle qu'elle est souvent pratiquée par un personnel étranger ou même national, mais étranger à la région et à ses problèmes propres, est très peu efficace. Les méthodes de facilité habituelles, telles que la distribution de prospectus (souvent à des illétrés!), les affiches, les messages à la radio et à la télévision, les conférences et les films, n'auront que peu d'effet s'ils sont artificiellement apportés de l'extérieur sans une propagande en profondeur assurée par des gens qui entretiennent un contact permanent avec les habitants. Ces gens sont les chefs et notables traditionnels ainsi que les responsables du projet vivant en étroite collaboration avec la population. Ce sont:
— les agents de santé ruraux, les maîtres d'école, les agents de vulgarisation agricole du projet, les chefs traditionnels et religieux des villages, les chefs politiques locaux, et même les sorciers et guérisseurs traditionnels locaux qui ont encore bien souvent une audience étendue parmi la population locale, et qui peuvent être "récupérés" en vue d'une collaboration des plus fructueuses.

Un programme d'éducation sanitaire de masse ne peut jamais réussir si une large fraction de la population ne se sent pas personnellement concernée, si elle demeure une activité isolée de quelques spécialistes qui se bornent à donner de sages conseils aux habitants. Ces spécialistes ne doivent être que les premisers chaînons et les animateurs d'une armée de propagandistes sanitaires, aussi nombreux et variés que possible, choisis parmi les membres les plus actifs, les plus écoutés et les plus au contact de la population. Il n'y a que ce procédé qui peut mobiliser suffisamment l'opinion de la communauté pour modifier son comportement et assurer le succès d'un programme d'éducation sanitaire.

B. *Les problèmes qui viennent d'être évoqués sont définis par certains facteurs dont quelques-uns sont inhérents à l'homme et d'autres au milieu. Quels sont ces facteurs et quels sont nos moyens d'action?*

1. *Facteurs hydriques.* Un système d'irrigation comprend principalement un réseau de canaux, les uns en matériaux artificiels, les autres en terre, des bassins, souvent un lac artificiel de rétention. C'est là un système écologique nouveau dont le volet épidémiologique qui intéresse l'homme n'est que l'un des aspects. Cet écosystème est cependant inséré au sein d'une région donnée, et c'est par conséquent avant tout les régions limitrophes qui vont imprimer au nouvel écosystème son épidémiologie. La nouvelle unité épidémiologique sera donc faite de ses données hydrologiques propres et de ce que lui apporteront les régions environnantes. Et de fait, il s'est presque toujours avéré illusoire d'essayer d'empêcher par quelque dispositif technique ou par un effort continu, la pathologie environnante, schistosomiase, plaudisme par exemple, d'envahir le périmètre d'irrigation si toutefois les conditions de vie y sont favorables au vecteur. L'homme se bornera donc à limiter autant que possible le phénomène et à se mettre à l'abri.

Si donc les conditions écologiques sont propices à l'installation de mollusques ou de moustiques, ils s'installeront coûte que coûte dans les nouveaux gîtes qui leur sont offerts. Le lac artificiel et ses abords immédiats deviennent à cet égard une réserve intarissable de vecteurs, notamment en ce qui concerne les mollusques hôtes intermédiaires de la schistosomiase. C'est dire qu'il est illusoire de chercher à tuer jusqu'au dernier mollusque du système d'irrigation, car quand bien-même on y parviendrait, la réinfestation venue de l'extérieur ne tarderait pas à se produire.

L'homme est donc condamné à subir les incidences épidémiologiques de l'écosystème qu'il a créé. Toutefois il peut en atténuer considérablement les effets et s'y soustraire dans une grande mesure par des actions focales et limitées aux points de contact infectants. Le mollusque dangereux pour l'homme est celui qui vit, se multiplie, s'infecte et transmet la maladie là où l'homme entre à son contact : points de baignade, lavoirs, passages à gué, embarcadères. C'est là que doivent porter les efforts de lutte contre les mollusques et ceux tendant à nous soustraire de leur contact infectant. Le mollusque qui vit à l'autre bout du lac n'a que fort peu de chances d'approcher l'homme et de devenir dangereux. Il est inutile de chercher à l'exterminer en engloutissant un budget considérable pour un résultat éphémère et inconstant.

Il est à noter ici que parmi les points de contact infectants, les plus dangereux sont souvent ceux créés et entretenus par l'homme par une mauvaise pratique de l'irrigation, et qui nuit à l'exercice même de l'irrigation autant qu'à la santé. C'est ce qui se produit quand le réseau d'irrigation est mal entretenu, quand il s'y produit des fuites, quand le réseau de drainage est insuffisant ou qu'on le laisse se colmater et être envahi par la végétation. Une discipline rigoureuse dans l'utilisation de l'eau rendrait là un service aussi important à la pratique de l'irrigation qu'au maintien de la santé.

2. *Facteurs humains.* Ces facteurs agissent eux aussi d'une manière aggravante sur l'épidémiologie de la nouvelle zone d'irrigation. Dans bien des cas l'aménagement, grâce à l'irrigation de nouvelles terres cultivables, constitue une attraction pour les populations voisines ou lointaines qui viennent s'y installer. C'est d'ailleurs souvent l'effet recherché par les gouvernements. Ces migrants apportent parfois avec eux une pathologie nouvelle pour la région. D'autres fois, la densification de la population dans le périmètre irrigué ou aux abords du lac facilite et accélère la transmission d'une maladie qui y existait déjà et, de ce fait, la rend plus fréquente, qu'en cas de population moins nombreuse et plus clairsemée.

Avec la création d'un complexe d'irrigation, de nouvelles occupations professionelles peuvent apparaître ou se développer. Tel est le cas des pêcheries qui se développent dans les nouveaux lacs artificiels. La pêche implique un contact plus étroit avec l'eau et peut résulter en une expansion de la schistosomiase, comme cela s'est déjà vu dans biendes projets.

Aussi, convient-il d'exercer un contrôle sanitaire rigoureux sur les immigrants, portant principalement sur les maladies transmissibles pouvant mettre en danger l'ensemble de la nouvelle communauté et aggraver la situation épidémiologique de la région. Ce contrôle doit être exercé autant que possible avant l'arrivée des immigrants sur le nouveau territoire; sinon, le plus vite possible, après. Il est à noter toutefois que ce contrôle se heurte dans la pratique des grands projets à des difficultés considérables et c'est bien rare qu'il soit mené à bien de manière satisfaisante. Il faut se souvenir que le nomadisme et semi-nomadisme sont courants dans les zones arides, et si la réalisation des projets d'irrigation tend justement à fixer les nomades, le nomadisme se poursuit néanmoins, en général pendant de longues années, et souvent indéfiniment à travers le périmètre irrigué, ce qui implique les échanges de pathologies que l'on devine. Dans une certaine mesure, le projet d'irrigation augmente même ce mouvement d'export-import de maladies avec l'appel périodique de main-d'œuvre agricole extérieure, rythmé par la périodicité des pratiques agricoles.

3. *Considerations administratives*. Nous voyons donc que si la création d'un schéma d'irrigation s'accompagne souvent d'un renforcement des structures sanitaires de base de la région, les problèmes sanitaires qu'il crée sont d'une telle ampleur et complexité que les services de santé risquent d'être débordés. C'est ce qui arrive régulièrement quand aucune considération administrative particulière n'est accordée à la composante sanitaire du projet d'irrigation, et que l'appréciation des problèmes de santé et leur solution sont tout simplement confiées aux services réguliers du Ministère de la Santé du pays en question. Ces services, même quant ils ont la compétence théorique adéquate, manquent de budget et de personnel pour faire face à des problèmes nouveaux, énormes et difficiles. Aussi, n'ont-ils que trop tendance à les traiter de manière routinière, comme ils le feraient pour n'importe quel secteur de leur activité. Dans ces conditions, leur action ne peut se solder que par un échec.

La composante sanitaire d'un projet d'irrigation doit, dès le départ, faire partie intégrante de l'ensemble du projet, relever d'un budget qui lui est propre et faisant partie du budget général du projet. Elle doit relever d'un personnel médical, technique et auxiliaire dont toute l'activité professionnelle serait consacrée uniquement au projet. Si celui-ci relève d'une autorité spéciale créée pour la durée du projet, le personnel doit relever de cette autorité, et de cette autorité seulement. Ceci n'exclut d'ailleurs pas une collaboration, toujours bénéfique, avec les services réguliers du Ministère de la Santé, appelés un jour à gérer de manière routinière les problèmes sanitaires du projet, mais seulement quand ceux-ci seront devenus routiniers.

Les problèmes soulevés par ces projets forment un tout et relèvent d'une même cause: savoir précisément l'implantation d'un schéma d'irrigation. Ils appellent des solutions d'ensemble impliquant une intégration dans le temps et dans l'espace des différentes activités et cela n'est possible que sous une seule autorité et responsabilité. Si l'action est menée par le Ministère de la Santé, cette intégration ne sera apparente qu'administrativement. Dans la pratique, nous l'avons bien des fois constaté, les différentes activités seront dispersées à travers les différentes sections techniques de ce Ministère, avec l'habituel manque de coordination et d'approche d'ensemble et une dilution de responsabilité et d'autorité.

Si les problèmes sanitaires soulevés par de tels projets sont liés entre eux et doivent être traités en tant que tels, ils sont aussi étroitement liés aux différentes autres activités du projet: hydrologiques, agricoles, sociales, et doivent donc être considérés et traités en étroite collaboration avec les responsables de ces différentes formes d'activité. Or, là encore, l'expérience prouve que l'effort d'intégration des activités sanitaires à l'ensemble du projet ne peut réussir que lorsque, sur le plan administratif, elles ne relèvent que d'un même organisme.

4. *Echelonnement des problèmes de santé dans le temps*. Les différents problèmes médicaux et sanitaires se posent lors des phases successives de la planification et de l'execution d'un projet d'irrigation. Ils doivent être traités *à temps* si l'on veut leur trouver une solution satisfaisante.

(a) *Lors de la planification du projet:* Nous venons de dire que les problèmes de santé doivent être considérés comme formant une composante à part entière de l'ensemble du projet. Comme tel, le chapitre santé doit bénéficier au sein du projet de dispositions budgétaires, administratives et techniques au même titre que toute autre branche d'activité, et ceci dès le début de la planification. Comme pour les autres champs d'activité du projet, la réussite de la composante santé dépend, pour une grande part, d'une vision correcte des problèmes et des dangers, d'une prévision et d'une planification judicieuse de l'action *avant* le commencement des travaux. L'expertise et les conseils de l'épidémiologiste doivent être demandés dès que la décision d'exécution du projet est prise et, en tout cas, autant que possible, avant le début des travaux. Il nous faut malheureusement constater que cette façon de faire, qui devrait être la règle, est, de nos jours encore l'exception et que les avis concernant la santé sont demandés alors que toutes les options sont prises, ou déjà même exécutées, qu'aucune prévision budgétaire n'est prévue pour les problèmes sanitaires, et souvent seulement à l'apparition des premiers phénomènes épidémiologiques. Le point de vue sanitaire doit être pris en considération et discuté lors de la prise des différentes options concernant les problèmes de déplacement de populations, d'implantation de nouveaux villages, de l'habitat, d'assainissement du milieu, d'eau potable, et de bien d'autres. L'argument sanitaire peut appuyer tel ou tel choix en ce qui concerne l'ingéniérie même du dispositif d'irrigation. S'il est vrai que l'irrigation est souvent source de dangers sur le plan de la santé, il est faux, comme on se l'imagine habituellement, que les recommandations inspirées de considérations sanitaires vont en géréral à l'encontre d'impératifs économiques et techniques. Au contraire, très souvent, il peut se dégager d'un tel dialogue pluri-disciplinaires des solutions satisfaisantes pour tous. C'est ainsi qu'il nous est arrivé de conseiller pour certains réseaux d'irrigation des conduites bétonnées fermées ou enterrées. Celles-ci constituent évidemment une excellente mesure pour la suppression de gîtes à mollusques et à larves de moustiques.

Elles sont d'un prix de revient plus élevé au départ mais les économistes du projet ont pu calculer qu'on s'y retrouvait largement en diminuant l'entretien des réseaux, en éliminant les pertes d'eau par évaporation et infiltration, en augmentant la vitesse d'écoulement de l'eau (ce qui permet d'étrendre le périmètre irrigué et la surface cultivable) en facilitant le tracé des routes et le transport, en éliminant enfin les mesures antilarvaires et antimollusques qu'il aurait été nécessaire d'appliquer sur des canaux ouverts.

D'une manière générale, tout avis formulé pour des raisons sanitaires, prônant une certaine discipline d'utilisation de l'eau pour éviter son gaspillage, son écoulement anarchique et sa stagnation grâce à des aménagements appropriés (drainage efficace, entretien rigoureux des réseaux d'irrigation et de drainage) constitue un conseil qui rejoint les intérêts fondamentaux de la pratique de l'irrigation et apporte un argument sup-plémentaire pour une utilisation rationnelle du réseau.

(b) *Pendant l'exécution du projet:* il y aura lieu évidemment d'assurer un service médical de protection et de traitement à la main-d'œuvre plus ou moins nombreuse selon l'importance du chantier. L'implantation d'un ou plusieurs dispensaires, éventuellement d'un petit hôpital seront parfois nécessaires. Ces structures pourront d'ailleurs demeurer à titre permanent, au service de la population qui s'installera dans la région.

C'est à ce moment là aussi que commenceront à se poser les problèmes inhérents au déplacement et à la réinstallation des habitants. Nous avons déjà évoqué certains de ces problèms qui surviennent à un moment où la population présente justement une résistance moindre du fait d'une certaine désorganisation de ses capacités professionnelles et financières, qui accompagne toujours un déplacement. C'est aussi le moment privilégié pour effectuer quelques contrôles sanitaires, faire de l'éducation sanitaire et apporter à la population réinstallée, en même temps qu'un nouvel habitat, un apprentissage du moeurs sanitaires nouvelles.

Lors de travaux de grande envergure, le déplacement des populations provoque parfois une rupture des activités agricoles pendant une ou deux saisons. Dans ce cas, le gouvernement doit prévoir une aide massive pour aider la population, non seulement à se réinstaller, mais aussi à subsister pendant cette période critique. C'est là surtout un problème d'économie et de planification, mais nous ne pouvons pas ne pas souligner les effets catastrophiques que peut entraîner sur la santé physique et morale une transplantation anarchique d'une population, momentanément privée de moyens de subsistance.

L'action sanitaire se poursuivra pendant toute la durée des travaux, en quelque sorte greffée sur la progression de l'ensemble du programme. Le dialogue sera maintenu entre les hommes de santé et les autres responsables du projet. Les difficultés seront examinées au fur et à mesure qu'elles se présenteront et les solutions satisfaisant au mieux toutes les parties seront recherchées dans cet effort de compréhension mutuelle.

La surveillance des phénomènes épidémiologiques, le dépistage d'éventuelles erreurs dans la mise en place des dispositifs d'assainissement seront assurés de la manière la plus étroite à cette période, car le remède sera d'autant plus facile à appliquer que le mal sera dépisté à temps.

(c) *Une fois les travaux achevés:* il importe que l'action sanitaire se poursuive et ne connaisse pas une période d'interruption, comme cela ne se voit que trop souvent sous prétexte que la réalisation technique du programme est achevée, que l'on passe à présente à une phase d'exploitation routinière, ou que l'on change de Ministère responsable. Bien souvent les problèmes épidémiologiques n'apparaissent justement qu'après la mise en eau du système d'irrigation. A l'inverse de réalisations techniques, les problèmes de santé impliquent une surveillance à assurer, des recommandations à faire et des mesures à prendre longtemps après l'achèvement de l'ouvrage et la mise en fonctionnement du système d'irrigation. Il est par aileurs malheureusement certain qu'un périmèter d'irrigation situé dans un pays en voie de développement, sous un climat aride et chaud, demeurera toujours un point hautement vulnérable sur le plan sanitaire et exigera une vigilance accrue quant à son fonctionnement et à son entretien.

5. *L'approche pluridisciplinaire des problèmes d'irrigation.* De tout ce qui précède, il se dégage une idée capitale: la nécessité d'intégrer les problèmes de santé dans tout programme d'irrigation dont ils forment une composante importante et qui ne saurait être traitée ni en parente pauvre, ni comme une sorte d'activité marginale, indépendante du reste du projet et traitée de manière annexe, si possible "après," par des spécialistes qui ne s'occuperont que "de ce qui les regarde," à qui on demandera de créer la santé dans une région qu'on aura tout fait pour rendre insalubre, surtout "sans toucher à rien" et sur un mini-budget qu'on aura rafistolé in extremis en râclant les fonds de tiroirs. Tant que les hommes de santé ne seront considérés que comme des gêneurs ou des corps étrangers, rien de constructif ne sera fait pour résoudre les graves problèmes de santé posés par l'irrigation de terres arides. En contrepartie, nous devons admettre que bien des médecins et des scientifiques, bien des experts de grande valeur parfois, n'ont pas toujours voulu ou su adopter un point de vue réaliste, tendant à faire s'accorder les nécessités sanitaires et les impératifs techniques, économiques, hydrauliques, agricoles d'un projet d'irrigation. Certains d'entre nous prétendent proposer des solutions idéales aux problèmes de santé et conseillent des mesures incompatibles avec les possibilités budgétaires ou avec les impératifs techniques du projet. D'autres jugent qu'ils n'ont pas à entrer dans des considérations qui ne sont pas de leur compétence et les ignorent superbement alors que, sur ces considérations, repose la faisabilité même du projet. Comment alors s'étonner que leurs recommandations ne sont pas appliquées?

Les problèmes d'ingéniérie hydraulique et agricole en rapport avec l'irrigation sont faciles à comprendre, du moins dans leur application pratique en rapport avec la santé. Tout médecin qui veut bien s'en donner le peine en découvrira la relative facilité et l'intérêt. Il découvrira aussi combien son dialogue avec les techniciens de l'irrigation sera rendu plus ais´ et combien plus fructueuse et agréable la collaboration avec des hommes. De même il s'attachera à leur expliquer, très simplement, les problèmes sanitaires qui le préoccuppent et qui n'ont rien d'incompréhensible pour des non-médecins. J'ai souvent été frappé par l'intérêt et le changement d'attitude qu'on peut ainsi éveiller chez les responsables d'un projet agricole en comprenant tant soit peu leurs problèmes, en leur parlant de béton et de canaux, mais en les introduisant aussi dans notre propre domaine, en leur expliquant nos propres préoccupations. De ce langage commun, de cette volonté de comprendre les difficultés de l'autre, naît la véritable collaboration qui, elle seule, permet de dégager des solutions parfois imparfaites mais réalisables dans la pratique et acceptables par tous. L'approche pluridisciplinaire des problèmes d'irrigation, l'intégration des problèmes de santé n'est pas seulement une affaire de structure administrative et de budget. C'est aussi la recherche d'un but et d'un langage commun et une coopération d'hommes.

SANTE PUBLIQUE ET PROJETS D'IRRIGATION: ROLE DE L'ASSAINISSEMENT ET DE L'HYGIENE DU MILIEU

par MICHEL LARIVIERE

Professeur de parasitologie, Université Paris VII
France Expert de l'O.M.S.

Quand les épidémiologistes sont consultés sur les implications sanitaires des projets d'irrigation, ils mettent essentiellement l'accent sur les dangers des modifications écologiques susceptibles de favoriser le développement d'hôtes intermédiaires (ex. mollusques) ou de vecteurs (ex. Anophèles, Simulies). Leur crainte légitime de voir ainsi s'installer ou s'étendre des foyers de schistosomiases, de paludisme ou d'onchocercose devient chez eux une préoccupation majeure au point qu'elle semole résumer tous les aspects sanitaires péjoratifs de l'irrigation.

Certes, les retentissements sur la santé de telles parasitoses sont lourds et les exemples passés de flambées épidémiques catastrophiques résultant de l'inexistence ou de l'insuffisance d'études épidémiologiques préalables justifient que leur survenue demeure une crainte réflexe des responsables de la prospective sanitaire.

Mais en contre-partie on entretient une vision figée des relations "santé-irrigation" dans laquelle les hôtes intermédiaires invertébrés s'identifient à l'environnement tandis que l'homme n'y apparait plus qu'en acteur de second plan. On en arrive alors à masquer l'ensemble de ses besoins fondamentaux de santé et à oublier que leur satisfaction, qui incluent la lutte contre les maladies vectorielles ou par hôtes intermédiaires, dépendent pour l'essentiel de son pouvoir de mobilisation.

En réalité les problèmes sanitaires, soulevés par l'irrigation sont d'abord ceux du sous-développement, communs à toutes les populations de faible niveau de vie vivant dans le tiers monde, et préexistent à l'irrigation. La manière dont celle-ci sera abordée et conduite rondra plus aigue la situation épidémiologique de départ, introduira des maladies nouvelles (soit par la voie de l'immigration de main d'œuvre véhiculant de nouveaux agents pathogènes, soit en créant les conditions de prolifération d'hôtes intermédiaires ou de vecteurs venant de régions avoisinantes, ou bien harmonisera les rapports des hommes avec le milieu par l'action conjointe de l'assainissement et de l'hygiène et d'une organisation économique et sociale de qualité.

"Il n'y a pas de vie sans eau," dit la Charte européenne de l'eau. Il résulte de cette évidence que fournir de l'eau c'est créer une vie nouvelle et que toute décision d'irrigation de terres arides engage les responsabilités de la prise en charge d'une naissance.

Après avoir envisagé les données épidémiologiques de base du sous-développement, nous soulignerons la nécessité d'une action sanitaire intégrée dans le cadre général d'une politique d'assainissement et d'hygiène du milieu.

Les données épidémiologiques de base du sous-développement. Lorsque l'on compare les niveaux sanitaires des pays nantis et des populations du tiers monde, on constate que les différences portent essentiellement sur la fréquence des carences nutritionnelles et la haute prévalence des maladies transmissibles qui stigmatisent la situation de sous-développement.

La solution de ces problèmes est de ce fait une exigence prioritaire et conditionne l'aptitude des populations frappées par ces fléaux à progresser vers le mieux-vivre.

Les problèmes nutritionnels: La satisfaction quantitative et qualitative des besoins nutritionnels de l'homme est la base de toute politique économique et sanitaire. Elle constitue le capital primaire indispensable au développement harmonieux de l'individu et à la capacité de travail de la collectivité. Elle est un atout majeur de la résistance des populations aux agents d'agression présents dans l'environnement, "Le traitement du paludisme est dans la marmite", affirmait un adage français. Sans être aussi catégoriques sur les vertus curatives de l'alimentation, la plupart des médecins s'accordent pour reconnaitre que l'état d'équilibre nutritionnel des individus est une condition importante de leur capacité de défense contre les infections. Aussi le premier objectif sanitaire de la mise en valeur des sols par l'irrigation doit-il être de couvrir les besoins alimentaires de la population et d'organiser les structures d'une éducation nutritionnelle adaptée.

Les maladies transmissibles: Les problèmes sanitaires posés dans la plupart des projets d'irrigation sont, nous l'avons vu, presque toujours limités aux affections par hôtes intermédiaires ou par vecteurs. C'est nous semble-t-il une vue par trop limitative qui procède de deux attitudes qu'il convient de dénoncer:

l'une est que l'action sanitaire demeure presque partout dans le monde du seul ressort de techniciens à qui appartiennent les décisions qui sont ensuite projetées sur la population dont le rôle se résume à les subir. Il en résulte une mise en condition passive des collectivités devenue si réelle que les rapports hôtes-parasites ne sont plus perçus qu'à travers l'identité du parasite. Par exemple l'endémie ankylostomienne est bien due à la présence d'ankylostomes adultes dans l'intestin de l'homme. La nécessité de les détruire apparait donc comme une évidence qui conduit par automatisme au choix d'une chimiothérapie, dont on connait pourtant les limites, plutôt qu'à programmer une campagne de constructions sanitaires visant la neutralisation des excretas, donc de l'ensemble du réservoir de virus. Mais une telle action sur le milieu n'est plus de la seule compétence du technicien de la santé et réclame une participation active des usagers. Elle remet donc en question le conceptualisme des professions de santé. De même la bilharziose intestinale se maintient à l'état endémique en raison de la présence de mollusques qui en sont les hôtes intermédiaires. Leur destruction mobilise tous les efforts des épidémiologistes qui en oublient que la construction de latrines serait d'un grand poids dans la prophylaxie en neutralisant les excrétas humains qui finalement sont bien le point de départ du cycle biologique du parasite.

l'autre attitude est la permanence d'une mentalité de santé fermée à la globalisation des problèmes et de leur solutions. C'est en quelque sorte la rançon du progrès scientifique qui nous impregnant d'une connaissance de plus en plus spécialisée, finit par modeler notre esprit sur la perception des différences. Cette approche analytique des maladies et de leur étiologie est indispensable en pratique médicale individuelle. Mais la santé du plus grand nombre exige la démarche inverse de recherche des analogies.

Un vibrion cholérique par exemple et une amibe ont très peu de points communs: leur morphologie, leur biologie, leur rôle pathogène, leur sensibilité aux médicaments sont totalement différents. Pourtant, nous le verrons, ils possèdent des points de convergence épidémiologique susceptibles d'être exploités en de communes mesures de prophylaxie.

Malgré leurs diversités en effet, les germes pathogènes s'inscrivent, pour la plupart, dans trois types de chaines épidémiologiques que nous caractérisons en

Cycle direct court
Cycle direct long
Cycle indirect.

Cycle direct court: il concerne tous les germes susceptibles de passer directement, sans changement d'un individu contaminé, dénommé réservoir de virus (RV) à un hôte non contaminé (HNC) et qui par cette voie le deviendra à son tour.

C'est la chaine épidémiologique la plus simple, que l'on peut énoncer comme suit:

$$RV \rightarrow HNC$$

Une très grande diversité d'agents pathogènes se regroupent dans ce cycle.
— certains sont transmis par voie respiratoire, tels les germes de la tuberculose, de la rougeole, de la coqueluche...
— d'autres par voie cutanéo-muqueuse: ceux de la lèpre, du tétanos des trépanématoses (pian, syphilis endémique, syphilis vénérienne), de la trichomonose vaginale...
— d'autres enfin sont disséminées par les *matières fécales*: virus de la poliomyelite, bacilles typhiques et paratyphiques, shigelles, vibrion cholérique, amibes, flagellés intestinaux...

Cycle direct long: il concerne essentiellement des parasites à *dissémination fécale* tels que trichocéphales, Ascaris, Ankylostomes, Anguillules, dont le pouvoir infectant est différé du fait de la nécessité pour les formes parasitaires excrétées avec les fèces de subir une maturation plus ou moins longue dans le milieu extérieur, avant d'acquérir leur capacité de contamination. Ainsi les kystes d'amibes (cycle direct court) rejetés dans les selles du réservoir de virus sont directement contagieux pour toute personne qui absorberait par exemple de l'eau souillée de matières fécales. A l'inverse, un porteur d'Ascaris (cycle direct long) élimine dans ses seeles des oeufs non embryonnés, donc non infectants. Ceux-ci doivent trouver un milieu extérieur favorable pour poursuivre leur maturation jusqu'au stade d'œufs embryonnés, seuls capables d'infecter un sujet neuf.

C'est dire l'importance du milieu extérieur qui conditionne l'endémicité de ces parasites. Les conditions les plus favorables sont réunies dans les zones tropicales humides, mais la plupart d'entre eux sont très cosmopolites et l'irrigation en particulier peut être un facteur important de leur extension. Leur cycle épidémiologique se schématise ainsi:

$$RV \rightarrow \text{milieu extérieur} \rightarrow HNC$$

Cycle indirect: c'est le plus complexe car il fait intervenir dans la chaine interhumaine de transmission un et parfois plusieurs *hôtes intermédiaires* invertébrés ou vertébrés dans lesquels les parasites émis par le réservoir de virus doivent obligatoirement séjourner et se transformer pour devenir infectants. Il répond au schéma suivant:

$$RV \rightarrow HI \rightarrow HNC$$

La présence ou l'absence dans le milieu extérieur de tels hôtes intermédiaires déterminent l'existence ou non d'une endémicité. Ainsi il ne peut y avoir de bilharzioses sans la présence de mollusques hôtes intermédiaires, de paludisme sans anophèles, d'onchocercose sans simulies.

La prophylaxie des maladies transmises par hôtes intermédiaires apparait donc a priori dépendre des moyens mis en œuvre pour les détruire et le rôle des populations vivant dans les zones d'endémie semblerait devoir se limiter à l'attente passive de procédés d'extermination plus efficaces.

Nous verrons cependant que l'effort collectif de l'homme peut, dans certains cas, briser cette apparente sujetion.

La lutte contre les maladies transmissibles. Le regroupement des maladies transmissibles au sein de trois grands types de chaines épidémiologiques ne procède pas d'une négation des caractères originaux de chacune d'elles, dont il est au contraire indispensable de tenir compte dans des mesures de prophylaxie. Mais nous pensons qu'il n'est pas de couverture sanitaire d'une population en partant d'actions ponctuelles, sélectionnées sur des motivations événementielles, mais au contraire en créant des structures générales de convergence sur lesquelles les actions spécifiques trouveront leurs points d'articulation.

Ces structures de convergence ressortent de la synthèse des différents moyens de prophylaxie dont nous disposons contre les maladies transmissibles.

Les vaccinations: Il existe des procédés de vaccination efficaces surtout contre les germes transmis par voie respiratoire ainsi que contre certains agents pathogènes à dissémination fécale relevant du cycle direct court. L'effort de santé publique doit donc avoir pour objectif de réaliser des vaccinations collectives, d'en assurer le controle et le renouvellement régulier selon les schémas éprouvés pour chacune d'entre elles. La réalisation des plans de vaccinations n'exige pas l'édification de structure opérationnelles lourdes, mais doit s'intégrer dans des services de santé de base polyvalents et d'éventuelles formations itinérantes que peuvent nécessiter certaines situations géographiques et démographiques particulières.

L'action thérapeutique: Nous disposons actuellement d'un arsenal thérapeutique efficace pour combattre la plupart des maladies transmissibles. Mais le traitement de masse est le "pis aller" des solutions de la santé publique puisqu'il n'intervient que sur des sujets contaminés alors que le véritable objectif de la santé est de prévenir l'installation de l'infection.

Néanmoins la plupart des affections à transmission cutanéomuqueuse (à l'exception du tétanos) ne peuvent être à ce jour contrôlées par des mesures vaccinales. Seule la chimiothérapie permet la neutralisation du réservoir de virus, mais elle n'a de valeur en santé publique que si elle s'exerce sur la totalité des sujets infectés. Le depistage systématique des porteurs de germes est donc primordial et son succès dépend des degrés d'efficiences des formations sanitaires de base, du niveau de perception des responsabilités par la population et de sa participation active.

Il en est de même d'une forme voisine d'intervention médicamenteuse, la chimioprophylaxie préventive, qui demeure par exemple un des moyens clés de la lutte antipaludique. Celle-ci en effet n'a de chances de succes qu'en s'appuyant sur des services de santé fondés sur la participation active des populations et en s'intégrant dans le plan général d'assainissement et d'hygiène du milieu.

Assainissement et hygiène du milieu: Assainir le milieu c'est se donner pour objectif l'élimination ou la neutralisation des agents d'agression qui entravent le développement harmonieux de l'homme dans son milieu. L'hygiène est l'ensemble des moyens mis en œuvre pour maintenir l'état d'assainissement et éviter toute pollution nouvelle.

La lutte contre la dissémination fécale est étrangement, la grande laissée pour compte des plans de santé publique. Et pourtant, un très grand nombre de maladies transmissibles a pour origine la contamination fécale du milieu, dont le sol et l'eau sont les deux constituants essentiels.

L'assainissement du milieu est donc, singulièrement dans le cadre de projets d'irrigation, l'impératif de base de toute politique de santé. Il doit comporter.

a) *la neutralisation des excreta:* un réseau d'égouts terminé par des stations d'épuration constitue le meilleur modèle actuel d'élimination des déchets. Il s'impose dans les villes dès qu'une certaine densité de peuplement est atteinte. Mais il nécessite des investissements considérables qui ne peuvent être programmés que sur de nombreuses années. Aussi, l'assainissement surtout en zones rurales d'habitats dispersés demeure-t-il dans la plupart des cas, la solution qui s'impose. Les fosses septiques sont le type même des installations sanitaires familiales ou des petites collectivités, mais leur coût et les exigences d'entretien que nécessite leur fonctionnement obligent souvent à recourir, tout au moins dans un premier stade, à l'usage de *latrines à fond perdu.*

b) *la mise à la disposition de la population d'eau potable* est le deuxième volet indispensable des mesures d'assainissement. Mais l'eau est aussi pour l'homme d'un usage externe et l'aménagement de points d'eau pour le lavage, les baignades etc... est une nécessité d'hygiène publique d'autant plus que ces eaux de surface peuvent aussi constituer des gites de prolifération de mollusques ou de larves de moustiques. Nous touchons là un exemple de lutte contre des hôtes intermédiaires ou des vecteurs, articulée sur des prestations sanitaires dont le point de départ est l'obligation de satisfaire les besoins quantitatifs et qualitatifs fondamentaux de l'homme en eau.

c) *le contrôle sanitaire du circuit économique alimentaire:* à l'exception des Ankylostomes, des Anguillules, des Schistosomes dont la pénétration chez l'homme se fait par voie cutanée (marche pieds-nus, immersion de toute ou partie du corps dans l'eau contaminée) tous les autres germes à dissémination fécale sont transmis par voie orale. Le rôle joué par les mains sales n'est pas négligeable, mais c'est aussi au niveau des différents maillons de la chaîne économique alimentaire qu'il faut chercher à enrayer les risques de contamination.

Nous venons de voir la priorité qu'il faut accorder à la protection de *l'eau de consommation.* Mais la seconde source usuelle de l'infection digestive est l'ingestion d'aliments absorbés crus, qui pose tout d'abord le problème de l'hygiène des cultures maraîchères. Le choix de leurs emplacements, la constitution de périmétres de protection vis-à-vis de la souillure fécale humaine ou animale sont, parmi d'autres, des mesures d'assainissement indispensables.

Dans certains pays les excreta humains et animaux sont utilisés comme engrais. Ils possèdent des propriétés fertilisantes parfois irremplaçables et plutôt qu'en interdire l'utilisation, il faudra tenter d'en neutraliser le contenu pathogène.

Mais la pollution peut se poursuivre au delà des points de production. Au cours de la manutention et au niveau des points de commercialisation. L'hygiène des marchés est, dans bien des pays, inexistante. L'étalage des légumes et des fruits se fait le plus souvent au niveau même du sol. L'entretien de leur "fraîcheur" est assuré par un trempage renouvelé dans une eau dont l'origine est souvent douteuse.

L'entassement anarchique des ordures permet la prolifération de mouches dont on a demontré qu'elles étaient susceptibles de transporter à distance des germes pathogènes. Les dangers que font courir ces insectes sont augmentés du fait que de nombreau marchés sont dépourvus d'installations sanitaires ou en possèdent hors d'état d'utilisation.

Au niveau des points de consommation, l'hygiène alimentaire s'inscrit dans le cadre plus général de l'hygiène de l'habitat, de la santé familiale. On y retrouve l'exigence de la qualité de l'eau à usage domestique à laquelle s'ajoutent la necessité d'un stockage éfficace des aliments avant leur consommation et l'adoption d'une hygiène corporelle stricte.

L'éducation sanitaire. La nécessite d'insérer et de développer des actions éducatives dans tout programme de santé n'est plus à démontrer. Les sceptiques ou les détracteurs de jadis en sont souvent même devenus les plus ardents défenseurs. La contre-partie péjorative de ce grand mouvement d'adhésions est que l'éducation sanitaire a dans bien des pays une tendance fâcheuse à la sacralisation, avec ses temples (services, bureaux, associations éducatives...), ses ministres (présidents, vice-présidents, éducateurs specialisés...), ses oraisons diffusées a grand renfort de mass media (films, télévision, radio...). Les "apôtres" ont cédé le pas aux bâtisseurs de temples, la foi au dogme, les discours aux litanies, voire aux patenôtres. L'éducation sanitaire n'est pas une religion. Elle n'est pas, non plus "une tarte à la crême" offerte à la population en écran de solutions concrètes à venir que l'on ne prépare souvent pas.

L'éducation sanitaire est une technique de santé au service d'une action programmée. C'est le préalable et le complément indispensable au succès de tout engagement en matière de santé. Elle doit devancer, accompagner et prolonger toute activité. Elle doit être une prise de conscience et une participation réelle de toute la population concernée.

Conclusions: Les problèmes sanitaires que posent les projets d'irrigation s'inscrivent dans le cadre général de la pathologie de tous les pays à faible developpement économique et à bas niveau d'hygiène.

Certes il n'existe nulle part de situations épidemiologiques absolument identiques et il serait érroné et lourd parfois de conséquences d'aborder les actions de prophylaxie avec des plans-types conçus pour être applicables en toutes circonstances.

Une erreur également serait de cristalliser toute l'action sanitaire sur un phénomène épidemiologique dominant qui restirait de l'étude du milieu, préa lable indispensable à toute programmation d'une politique de santé. Mais s'il existe partout des facteurs endémiques spécifiques, ils ne sont jamais qu'une majoration de la prévalence d'un ou de plusieurs constituants des deux grands fleaux qui stigmatisent l'état de "sous-développement": la malnutrition et les maladies transmissibles.

La lutte contre la malnutrition se pose en termes économiques et éducatifs. Il s'agit donc d'une part de couvrir les besoins quantitatifs et qualitatifs alimentaires de la population et d'organiser d'autre part les structures d'une éducation nutritionnelle adaptée qu'il faut intégrer dans le cadre général de l'éducation sanitaire.

La lutte contre les maladies transmissibles pourrait paraître à priori comme une succession d'actes ponctuels, tant sont différants les agents pathogènes responsables.

En fait, nous avons vu que malgré cette diversité il existe entre eux des points de convergence épidemiologiques qui permettent de les regrouper dans des mesures communes de prophylaxie tout en ménageant les possibilités d'articulation d'actions personnalisées en fonction de leur importance épidemiologique relative. La première structure de convergence est *l'assainissement et l'hygiène du milieu* dont les grandes lignes opérationnelles doivent nous semble-t-il être axées sur la lutte contre *la pollution fécale du milieu*. Les raisons qui justifient ce choix prioritaire sont d'abord le grand nombre d'agents pathogènes à dissemination fécale, leur incidence directe sur la mortalité et la morbidité générales des populations atteintes et la plasticité des structures de lutte qu'ils requièrent permettant au maximum l'articulation de moyens plus spécifiques lorsque ceux-ci s'imposent.

La neutralisation des excréta par le moyen d'installations sanitaires choisies en fonction des possibilites économiques et sociales est une exigence de base.

La mise à la disposition de la collectivite d'eau potable est un devoir non moins fondamental mais l'eau n'est pas qu'un aliment pour l'homme.

Elle est nécessaire à l'hygiène corporelle dont les répercussions sur la santé ne sont pas négligeables, elle est la source indispensable du développement économique... Son captage, sa circulation, son stockage, nécessaires à l'approvisionnement quantitatif et qualitatif de l'homme sont aussi autant de points de la vie de l'eau sur lesquels peuvent s'articuler des contingences épidémiologiques spécifiques, telles que la multiplication où l'introduction d'hôtes intermédiaires ou de vecteurs de maladies transmissibles.

Mais le contrôle de la pollution fécale du milieu doit se prolonger tout le long du circuit économique alimentaire (au niveau de l'organisation des zones de cultures maraîchères, des marchés, de la planification des conditions de manutention, de la consommation familiale ou collective... Quittant alors son rôle mobilisateur, le contrôle de la pollution fécale s'intégrera cetta fois dans des structures de sante de base polyvalentes tels que les centres de santé familiale, l'hygiène scolaire ou des collectivités...au même titre queles vaccinations, les actions thérapeutiques éventuelles et l'éducation nutritionnelle.

Certes un certain nombre de maladies transmissibles justifieront peut-être longtemps encore les recours à des organismes techniques très specialisés. Mais là encore, leurs possibilités d'intervention seront d'autant plus faciles que la population se sentira concernée.

Car il n'est pas de succès possibles en santé publique sans la participation active de la population.

Mais, comme le souligne O. Adeniyi-Jones (chronique OMS 30:8–10) (1976): "Le role de la collectivité en matière de soins de santé ne se limite pas á une simple participation à des campagnes d'assainissement, ou à la construction de puits, de latrines et de centres de santé. Il suppose un partage des responsabilités et une collaboration active à la planification et à l'organisation appropriée de ces services."

Mais pour ce faire il importe déjà de modifier profondément la mentalité des propriétaires traditionnels de la santé: les médecins et le personnel médical. Il convient également de mettre un terme à la querelle qui oppose encore trop souvent dans la recherche du développement, les radicaux de la voie économique et ceux de la voie sanitaire. La maîtrise de l'environnement pathogène de l'homme ne peut aboutir réellement que dans le cadre d'une intégration harmonieuse du décollage économique et de la politique de santé.

SCHISTOSOMIASIS—THE ENVIRONMENTAL APPROACH

by LETITIA OBENG

United Nations Environmental Programme

Schistosomiasis has been estimated to affect about 200 million people in Asia, Africa, the Caribbean and Latin America. It is an old disease and although the life cycle and pathology of the parasite have long been known, chemotherapy is still unsatisfactory, drugs being toxic or controversial; prophylaxis and immunization are far from being available. Although the involvement of the host snail and water in the spread of the disease are well understood, we have yet to achieve large-scale control. The snail vectors of schistosomiasis are associated with sluggish waters and these may either be natural or man-made. Water development projects, such as dams and irrigation schemes contribute very much to the spread of the disease; but such projects are vital for development. It is therefore important to understand the health hazards and other undesirable environmental effects which they cause.

For many years now, various organizations have tried to bring schistosomiasis under control: UN Agencies, and particularly WHO, private Foundations like the Rockefeller Foundation and the Edna McConnel Clark Foundation, as well as many national institutions. Numerous professional and technical individuals have, with diligent effort, worked on the disease and its vectors. In a further attempt to this end, an International Conference on Schistosomiasis was organized by the Ministry of Health in Egypt in association with a Joint Working Group for Medical Cooperation set up by the USA, the Arab Republic of Egypt, WHO and UNEP, which met in Cairo during October 1975. My task is to outline the outcome of that Conference, particularly with reference to the ecological and habitat management of schistosomiasis.

All the major aspects of the schistosomiasis problem were discussed. There were five committees to deal with the epidemiology and socio-economic aspects, chemotherapy, control of vector snails with molluscicides, immunological aspects, and the ecological and habitat control. UNEP and WHO were responsible for organizing the last subject.

As background information UNEP's Governing Council as early as its second meeting expressed concern over the threat posed by pests to man and animals, and governments requested the Executive Director to pursue the search for methods of control alternative to the use of chemicals. In its priority programme on integrated pest management UNEP is giving attention to schistosomiasis, malaria and diseases caused or transmitted by cotton pests, which were identified by the government as of major importance. At the Cairo Conference, therefore, we considered a plan for the control which will stress the bio-environmental approach within an integrated programme.

Concerning other aspects of schistosomiasis the main speakers at the Conference on socio-economics indicated that it was difficult, except in a few instances, to cost the economic benefits of control, and further studies were recommended to relate control to productivity. As regards epidemiology, it was urged that schistosomiasis should be considered in terms of the total health problems of the areas affected: it was recommended among other more clinical studies to establish the relationship between egg output of the worm parasite, infection intensity, infectivity of transmission foci, and the need to base remedial action on local conditions. Discussions on chemotherapy stressed the urgent need for safer drugs because of possible mutagenicity and the long-term toxicity of some drugs now in use. The merits and demerits of hycanthone in particular were hotly debated and opinion was divided as to its continued use because of its reported carcinogenic effect in mice. It was recommended that hycanthone niridazol and other schistosomicidal drugs, including new agents, should be rigorously tested for mutagenic and carcinogenic effects. An urgent need was noted for clinical pharmacologists with expertise in parasitic diseases who can correctly design, conduct and analyse drug trials in different populations so as to obtain accurate estimates of their therapeutic properties and side effects. The matter of mass treatment versus selective treatment also received attention.

For control of the vector snails, molluscicides were recognized as the most effective method and their use should continue in conjunction with other measures. It was recommended that molluscicide application should be based on studies which include snail biology and water management of irrigation practices, weed clearance, and covered canals and drains. Also further study is needed of toxicity and pathogenicity of the molluscicides in man, and on the long-term effects of snail resistance to molluscicides and their cumulative effect on other biota.

Discussions on immunology showed that research has made progress, but there are still gaps in knowledge of its immunopathology and therefore full understanding of the disease process and its management are lacking. Immuno-diagnosis would assist effectively in better identification of patients in infected populations.

UNEP, as an organization concerned about the environmental impact of existing methods of controlling diseases, was most closely associated with the ecological and habitat control of schistosomiasis. Although molluscicides provide the best-known method of breaking the disease cycle and they are used sparingly, they still have adverse side effects and are expensive, beyond the means of most developing countries where the disease is endemic. UNEP seeks to encourage the development of alternative and additional methods, for no one method will be completely successful in controlling schistosomiasis transmission. UNEP advocates a carefully planned strategy based on effective mobilization of all available resources and methods in an integrated action-oriented programme which views health as an essential element of the overall development process. The main components of such a programme would be:

Destruction of the parasite in the human host or interference with its egg producing mechanism and capacity through the use of chemotherapy.
Prevention of contact between the eggs and the snail breeding sites. This would depend on satisfactory sanitation and waste disposal systems.

Prevention of contact between man and infected waters. This would require safe water supply systems.

Destruction of the snails and their habitats by various means which are environmentally safe so as to make them ecologically unsuitable for further proliferation of the parasite.

All except the first of these components imply effective water management, especially in connection with development projects such as dams and irrigation schemes.

It is clear that a concerted programme should, in addition to using environmentally sound methods to break the life cycle of the parasite, recognize the need for an equally sound economic and social infrastructure to support control programmes. Public awareness of the serious implications of the disease and the existing methods for its control, popular participation in strategies to combat the disease, and the political will on the part of the international community to co-operate in eliminating or at least reducing the factors which spread the disease, are pre-requisites for successful plans to relieve mankind of schistosomiasis.

At the Conference an action plan of demonstration projects for field control was recommended. The methods proposed are simple and could contribute also in other public health problems: to prevent contact of man with cercarial-infested waters, the provision of safe and convenient water supply and the creation where feasible of physical barriers; for the reduction of snail populations, construction methods, aquatic weed control and the modification and channelling of rivers and streams to destroy snail habitats. With particular reference to irrigated areas recommended measures include drainage, stream canalization, lining of canals, land levelling and filling to eliminate low spots and borrow pits, seepage control, piped or covered canals and drains, weed control and improved water management. It was also noted that the institution of techniques such as sprinkler and drip irrigation and covered canals will contribute significantly. An essential factor which was underscored was the need for public involvement, training and information dissemination in any activities which may be launched.

Recommendations were also made for further investigation, including the role of latrines in preventing transmission of the disease by reducing the contamination of surface waters, and further research on the biological and ecological aspects of the snail vectors.

I should like to stress again that UNEP seeks to emphasize the environmental, ecological and habitat management approach in an integrated system for controlling schistosomiasis. Some people tend to consider bio-environmental control methods as unrealistic, but with an integrated programme some headway may be achieved. The Peoples Republic of China has successfully controlled schistosomiasis, by a similar approach, to the point where it is no longer of public health importance. It is true that the vector in China, *Oncomelania*, has different ecology from *Bulinus* and *Biomphalaria*, but the components of control such as water supply and sanitation are applicable anywhere. The Chinese experience confirms that what is really important in the implementation of control measure of the type recommended is the political and personal will to do what is environmentally right and safe. Let us entertain the hope that through international communication like the Cairo Conference and this Alexandria irrigation Symposium, international effort will be able to make advances in the control of schistosomiasis without adverse environmental effects, as we strive to use land and water in the process of development.

PROBLEMS AND EFFECTS OF SCHISTOSOMIASIS IN IRRIGATION SCHEMES IN THE SUDAN

by MUTAMAD A. AMIN

London Khartoum Bilharzia Project, P.O. Box 2371, Khartoum, Sudan

Summary

The establishment of irrigation schemes in the Sudan has led to great modifications in the environment which favoured the spread and multiplication of snails and produced dramatic increase in the prevalence of schistosomiasis. In one irrigation scheme, Gezira, the prevalance of the disease is over 80% in children and up to 60% in the general population.

The economic loss as a result of absenteeism due to ill-health was estimated to be 30 million Sudanese pounds per year. Further, mollusciciding will cost over a million pounds every year to control the disease and nearly half a million to mass treat the infected people.

Résumé

La construction de canaux en vue de l'irrigation a provoqué des bouleversements dans l'environnement de la région. Ceux-ci ont facilité la propagation et la prolifération des mollusques. Il s'en est suivi un accroissement dramatique du nombre de cas de bilharziose. C'est ainsi que dans une plantation de la Gézira 80 pour cent des enfants et 60 pour cent de la population entière sont atteints.

La détérioration de la santé des travailleurs provoque un absentéisme important qui se traduit en une perte économique évaluée à trente millions de livres soudanaises par an. On estime que le coût de la lutte contre les mollusques s'élèvera à un million de livres et le traitement de masse des malades à un demi-million par an.

Schistosomiasis, or bilharziasis as it is also known, is one of the most important parasitic diseases of man. It is caused by infection with parasitic blood flukes, schistosomes, transmitted by aquatic and semi-aquatic snails. The disease is intrinsically linked with the social and economic problems of the community and it has been defined as a man-made disease. Infected as well as uninfected people are brought together to work and live on the irrigated land. Water contact activities whether for work, domestic or recreational needs usually make use of canal water. Human waste is often excreted or washed into the same water where snails are abundant. This deadly combination is all that is needed for the rapid spread of the disease.

The Sudan is an agricultural and cattle-breeding country with colossal opportunities for agricultural development. Naturally, several dams were built for the conservation and supply of water into canalization systems. These are Sennar, Jebel Awlia, Khashm-el-Girba and Roseires dams. Even Aswan High Dam has created a lake one-third of which is inside the Sudan. In addition many agricultural development schemes were established along the Niles. The present irrigated land is about 3 million acres, and more than one million acres are under construction.

The development of water resouces in the Sudan has led to great modifications in the environment which favoured the spread and multiplication of snails and produced dramatic increases in the prevalence of schistosomiasis. This paper gives an account of the problems and effects of the disease in two major irrigation schemes, Gezira and Khashm-el-Girba.

The Gezira Scheme

The Gezira constitutes that area which lies between the Blue and White Niles. Before irrigation the area was a semi-arid clay plain with scanty seasonal rainfall and schistosomiasis was non-existent. In 1910–1911, cotton was planted on a small experimental scale irrigated by pumps on the Blue Nile. After cotton production proved successful, it was decided to construct a dam at Sennar, and this was completed in 1925. After 1950 the irrigated area was greatly expanded and the present growth area of the scheme is nearly 2 million acres. Before irrigation by canalization, the population was approximately 135,000 (El Nagar 1958). Following irrigation a rapid increase in the population density began, today over one million people are living in the irrigated area. In addition a floating population of seasonal labourers, estimated at one-third to half a million from Western Sudan and from West Africa, visit the area to take part in cultivation and cotton-picking.

The Gezira irrigation system contains three main types of distributive canals, the main, major and minor canals, and the field channels known as Abu eshreens and Abu Sittas. Of these the minor canals are the most important foci in the transmission of schistosomiasis since snail populations there are large and man–water contact is high. The function of both the main and major canals is to transport water rather than to irrigate land. The minor canals are positioned at right angles to the major canals at intervals of 1.42 km. They are, however, fed from the major canals in pairs so that the offtakes from the major canals are 2.84 km apart. The function of the minor canal is not to transport water but to command land with sufficient head. The water is taken from the minor canal through field outlet pipes into Abu eshreens and Abu Sittas and thence to the crops.

Khashm-el-Girba Scheme

Khashm-el-Girba scheme lies on the Atbara River, a tributary of the Nile, 400 km east of Khartoum. The scheme was established for the resettlement of the population of the flooded lands at Wadi Halfa after the establishment of Aswan High Dam, and also for the settlement of the indigenous nomadic and semi-nomadic tribes of the area.

The scheme was executed in five phases. The first phase was mainly for the resettlement of Wadi Halfa people and this was completed in 1964. Following 1964 the area was expanded and the present total area is about 182,000 acres. The irrigation system and methods of irrigation is more or less similar to that of the Gezira scheme. No pumps are used and the water is carried by gravity. Crops include cotton, wheat, sorgum, vegetables, ground nuts and sugar cane.

Schistosomiasis: Past and present situations

Archibald (1933) thought that schistosomiasis in the Sudan started as far back as 2600 B.C., and had been brought by Ancient Egyptian raids and trade missions. During the Mahdist Revolution (1880) the Sudan was invaded by Egyptian troops, mainly recruited from peasants, and after the reconquest of the Sudan in 1898 Egyptians were stationed in several districts. In 1918, Egyptian labourers were imported in large numbers for irrigation schemes in the Northern Province. Christopherson (1919) stated that schistosomiasis is endemic in all provinces of the Sudan. This statement seems to be based on cases reporting to hospitals and not on epidemiological surveys.

Interest in schistosomiasis has been aroused and intensified by the building of Sennar Dam and successful treatment by antimony tartrate as a result of the work of Christopherson in 1918 in Khartoum following the elucidation of the life cycle of the worm by Leiper in 1915.

TABLE 1. *Prevalence of* Schistosoma mansoni *infection in the Gezira Irrigated Area*

Age	Males		Females	
	Number examined	Percentage infected	Number examined	Percentage infected
0–4	120	8.3%	128	2.3%
5–9	163	45.4%	177	45.2%
10–14	160	76.3%	137	79.6%
15–19	65	82.2%	121	79.3%
20–24	25	68.0%	81	51.9%
25–29	27	55.6%	72	45.8%
30–34	25	64.0%	67	44.8%
35–39	27	51.7%	56	48.2%
40–44	31	32.3%	38	21.1%

In the Gezira Irrigated Area of the Sudan, schistosomiasis is considered to be one of the most important public health problems. This belief is based mainly on the known prevalence of the disease which has apparently increased dramatically during the past 20 years. From the start of the scheme in 1925 until 1944, routine urine examinations detected *Schistosoma haematobium* in less than 1% of samples tested. However, between 1942 and 1944 while infection rates in dispensaries were reported as 0.9% and 1.67% and 2.43% respectively, Stephenson (1947) carried out independent village surveys and discovered an overall *S. haematobium* prevalence of 21% in adults and 45% in children with infection rates varying widely from village to village. Studying *Schistosoma mansoni*, Stephenson found only 5% infected in villages (compared to the reported 1.32% from dispensaries), but he stated his belief that *S. mansoni* was probably more widespread than he had been able to detect.

A few years later Greany (1952a) examined 81,000 people from some 300 villages and discovered 8.86% with *S. haematobium* and 8.77% with *S. mansoni*. In the 5–9 and 10–14-year-olds *S. mansoni* infection rates were 12.6% and 13.1% and *S. haematobium* rates were 15.0% and 18.5% respectively. The same author (1952b) considered *S. mansoni* to be an important disease since a 3-year follow-up of 77 infected individuals between the ages of 2 and 45 showed that 9 (all under 21) had died, 4 were bedridden, while 37 had enlarged livers.

Amin (1972) using the digestion method of stool examination (in which 1 g of faecal material is emulsified in 10 ml of 2% sodium hydroxide and left to stand for at least 2 hours before 0.1 ml aliquots are examined), found infection rates of up to 60% *S. mansoni*. In 1973, Teesdale (pers. comm.) used the Kato method and estimated the overall *S. mansoni* infection rate in 3 villages at over 70%. From 1200 children in 6 schools the infection rates in 7-year-olds was 77%, 8-year-olds 86% and 9-year-olds 86% which is too much higher than the 1952 levels to be accounted for by the improved diagnostic methods.

TABLE 2. *Snail species present on the Gezira Scheme*

Snail species	Medical and veterinary importance	Relative frequency
Biomphalaria pfeifferi	Intermediate hosts of: *S. mansoni*	Abundant
B. sudanica	*S. mansoni*	Common
B. alexandrina	*S. mansoni*	Common
Bulinus truncatus	Intermediate host of *S. haematobium* and also *S. bovis.*	Abundant
B. forskalii	Intermediate host of *S. bovis* (experimental)	Common
Lymnaea natalensis	Intermediate host of *Fasciola gigantica*	Common
Melanoides sp.	Nil	Abundant
Cleopatra sp.	Nil	Abundant

Table 1 shows up-to-date age prevalence of *S. mansoni*, and in Table 2 types of the parasite, snails species and their relative abundance in relation to the different types of water courses are recorded. Before irrigation Khashm-el-Girba was a schistosomiasis free area. Now both types of the disease are prevalent amongst the population. The urinary type was mainly introduced by Wadi Halfa people and the intestinal form was thought to be imported from the Gezira and the White Nile.

Schistosomiasis as a constraint to economic development

The relationship between schistosomiasis and economic development can be looked upon from the following angles:

(a) Morbidity resulting in temporary or permanent disability, and hence absenteeism or reduction in the work capacity.
(b) Cost of medical care.
(c) Cost of control.

Since schistosomiasis is a chronic disease which runs a long insidious course, the first of these considerations is difficult to assess. However, the clinical manifestations such as anaemia, hepatosplenomegaly and ascites are certainly capable of incapacitating the agricultural worker. However, these occur as late sequelae of the disease when its history has run for many years.

Farooq (1967) estimated that the infection in Egypt cost $560 million annually and working in a sugar estate in Tanzania, Fenwick (1972) showed that savings of about US $1400 per year could result if treatment of infected workers raised productivity by 5%.

In the Sudan it was estimated that about 9% of the output of agricultural productivity is lost as a result of absenteeism due to ill-health or during treatment which often requires hospitalization. In terms of the gross national product from agriculture it has been calculated that the total economic losses due to this factor alone would amount to 30 million Sudanese pounds per year. This is a low estimate since traditionally the whole family goes with the patient when he has treatment.

The costs of treatment and control are extremely high. It has been estimated that L.S. 500,000 pounds are needed to treat the infected people in the Gezira scheme. The annual cost of molluscicide is nearly one million pounds. Even if the money is available there are many problems as regards the choice of drug and its administration; choice of molluscicide and its application. The Gezira schistosomiasis control team has conducted a research programme with a view to recommending control procedures. Various molluscicides and methods of applications such as drip-feed, ground and aerial spraying, were studied. The project team has now the knowledge to treat the whole of the Gezira and would expect a continuous 99% reduction in overall snail population to result and hence a marked reduction in the incidence of the disease to a level whereby it will no longer be a public health problem.

The economic justifications for the development of irrigation schemes are beyond argument. However, there is a need for consultation among the engineers responsible for the design, the funding agency and the health authorities with a view to minimize schistosomal risks.

References

Amin, M. A. 1972. Large-scale assessment of the molluscicides copper sulphate and N-trityl morpholine (Frescon) in the Northern Group of the Gezira Irrigated Area. *J. trop. Med. Hyg.* **75**, 169–175.

Archibald, R. G. 1933. The endemiology and epidemiology of schistosomiasis in the Sudan. *J. trop. Med. Hyg.* **36**, 345–348.

Christopherson, J. B. 1919. Laboratory and other notes on seventy cases of bilharzia treated at the Khartoum Civil Hospital by intravenous injections of antimony tartrate. *J. trop. Med. Hyg.* **22**, 129–144.

El Nagar, H. 1958. Control of Schistosomiasis in the Gezira, Sudan. *J. trop. Med. Hyg.* **61**, 231–234.

Farooq, M. 1967. Medical and economic importance of Schistosomiasis. *J. trop. Med. Hyg.* **67**, 105–112.

Fenwick, A. 1972. The cost and cost-benefit analysis of a *S. mansoni* control programme on an irrigated sugar estate in Northern Tanzania. *Bull. Wld. Hlth. Org.* **47**, 573–578.

Greany, W. H. 1952a. Schistosomiasis in the Gezira irrigated area of the Anglo-Egyptian Sudan. I.—Public health and field aspects. *Ann. trop. Med. Parasit.* **46**, 250–267.

Greany, W. H. 1952b. Schistosomiasis in the Gezira irrigated area of the Anglo-Egyptian Sudan. II.—Clinical study of *Schistosomiasis mansoni*. *Ann. trop. Med. Parasit.* **46**, 298–310.

Stephenson, R. W. 1947. Bilharziasis in the Gezira Irrigated Area of the Sudan. *Trans. Roy. Soc. Trop. Med. Hyg.* **40**, 479–494.

IRRIGATION AND MALARIA IN ARID LANDS

by MOHYEDDIN A. FARID

Formerly of Division of Malaria Eradication, World Health Organization

Summary

The association of malaria with irrigation in arid lands has been known in ancient and recent history. The deleterious effects of this disease are not only related to its direct, harmful influence on the health of the victims but also to its indirect adverse effect on economy, education, and welfare of the people. The extent of the malaria problem connected with irrigation in arid zones does not depend as much on the climatic condition or the potentiality of the vector in establishing malaria transmission as on man-made disturbance of the ecological balance through defective planning and water mismanagement, thereby providing havens for vector breeding, and through temporary or permanent labour aggregation increasing the spread of the disease. A knowledge of the degree of malaria hazard in the various climatic regions of the arid zones, gives an idea of the malaria expertise needed to tackle the problem before, while, and after an irrigation project is undertaken in any potentially malarious areas. As the problem is a man-made one, there is an urgent need for mutual understanding and active cooperation among those who apply their skills in the various sciences, those who apply them in engineering, and those who occupy the irrigated lands. In irrigation schemes, which benefit neighbouring countries. there is also a great need for inter-country cooperation in preventing the malaria hazard as the disease does not respect national boundaries. The specialized agencies of the United Nations have a major role in contributing to the promotion of such international cooperation and in providing the governments in arid zones with the necessary guidance to avert or minimize the malaria hazard through comprehensive planning of irrigation projects.

Résumé

L'association du paludisme avec l'irrigation des terres arides est connue depuis la haute antiquité. Les effets néfastes de cette maladie ne sont pas seulement liés à une action nocive directe sur la santé de ses victimes, mais également à ses effets indirects fâcheux pour l'économie, l'éducation et le bien-être de la population. L'extension du paludisme liée à l'irrigation des zones arides ne relève pas tellement des conditions climatiques ou des potentialités de transmission du vecteur, mais beaucoup plus de la perturbation par l'homme de l'équilibre biologique par une planification et utilisation défectueuse de l'eau, ce qui crée des gîtes nouveaux pour la prolifération du vecteur et également de la concentration permanente ou temporaire de la main d'oeuvre, qui augmente les chances de diffusion de la maladie. La connaissance du danger potentiel de l'expansion du paludisme dans les régions climatiques variées d'une zone aride est indispensable, tant avant qu'après l'implantation d'un projet d'irrigation dans toute zone impaludée ou qui peut le devenir. Etant donné que le problème du paludisme est dans tous ces cas de création humaine, une compréhension mutuelle et une coopération active sont indispensables à établir entre tous ceux qui exercent leur science dans les différents aspects du génie rural, et ceux qui occupent les territoires irrigués. Une coopération inter-pays pour la prévention du danger paludique dans les schémas d'irrigation intéressant plusieurs pays voisins est aussi indispensable, car la maladie ne respecte pas les frontières des états. Les agences spécialisées des Nations Unies ont un rôle majeur à jouer pour établir une telle coopération internationale et pour procurer aux gouvernements des pays arides intéressés des conseils en vue d'éviter ou de minimiser le danger paludéen lors d'une planification d'ensemble des projets d'irrigation.

Introduction

The Holy Koran states: "And We created from water every living thing," and in another chapter, "And He has set up the balance in order that ye may not transgress balance."

Arid lands cover almost one-third of the terrestrial area of our globe and although generally characterized by the absence, or scarcity and irregularity of rainfall, they offer a habitable environment to man, animals, insects and plants, depending on the various interactions of climate, groundwater, existence of rivers, land form, proximity to seas or oceans, type of soil and plant cover. History tells us that since time immemorial man inhabited many of these natural environments as a nomadic shepherd or as a settled cultivator. One can assume that such primitive people, comparatively isolated, and possessing few tools, could not disturb the natural balance and were able to prosper without great health hazards. The ancient civilizations which flourished along large rivers, such as Euphrates and the Nile, led to the development of irrigation to increase the acreage of arable lands and thus meet the people's needs for food and fibre. Such irrigation, executed through huge labour forces with a rather primitive technology, created an imbalance between the water supply and the water need and led to the spread of malaria as recorded by inscriptions on temples (AAT disease in Denderah temple of Egypt) and by frequent reference to intermittent fevers in Babylo-Assyrian medical lore.

The relationship between undrained stagnant water and malaria was noted by Hippocrates 400 B.C., and in his treatise on "Airs, Waters, and Places" he wrote, "Should be rivers in the land which drain off from the ground, the stagnant water and the rain water, then (the people) will be healthy and bright. But if there be no rivers, and the water that the people drink be marshy, stagnant, and fenny, the physique of the people must show protruding bellies and enlarged spleens." (Russell 1955.)

With the advent of our era of modern science and technology, associated with an accelerated rate of population growth, it is irrelevant to talk about a return to the natural balance, or to lead a new life without technology, as this will condemn the majority of the world population to starvation and death.

To enable man to prosper under the difficult arid conditions, he has to rely on modern technology to tap and utilize all available water resources, but, in doing so he has to prevent, or minimize the health hazards connected with irrigation as only the latter can contribute to the well-being and socio-economic progress of the people in the arid zones.

Malaria and its epidemiology in the arid zones

In these zones malaria is caused by infections with three different species of *Plasmodium* (*P. falciparum, P. vivax,* and *P. malariae*), and is characterized by bouts of intermittent fevers which incapacitate the victim for about ten days and, if not treated properly, periodical relapses of the same occur over one or more years, leading ultimately to severe anaemia and enlarged spleens. This lowers the victim's vitality and level of immunity to intercurrent infections, apart from the fatal issues caused by the malignant form of the disease due to *P. falciparum.* Where malaria is endemic the disease smoulders insiduously, reducing the vitality of the people, lowering their work output, destroying their ambitions and decimating their children. Due to its insiduous nature, the fear that endemic malaria arouses is not proportionate to the damage it inflicts on the community. It is only during epidemics, particularly when *P. falciparum* infections predominate, as occurs cyclically in some sub-tropical areas of the arid zones, or when certain malariogenic factors prevail, that the devastating ravages of the disease become obvious. During the malaria epidemics in Upper Egypt (1942 and 1943) which were caused by the invasion of the vicious *A. gambiae* from the south, 130,000 malaria deaths occurred in the blighted area and the whole economic life was paralysed (Soper 1970). The people were bed-ridden with fevers and could not attend to the harvesting or to the watering of their fields or take care of their animals. The crops were left in the fields to rot, domestic animals and equines died of starvation and thirst, schools were closed and the railway service was interrupted due to absenteeism of local staff.

The epidemiology of malaria revolves around the dynamics of its transmission from malaria victims to healthy persons through the intermediary of certain species of anopheline mosquitos. Various physico-biological and socio-economic factors interplay in maintaining the transmission and the endemicity of this disease in various areas of the arid zones. The argument by some malariologists that malaria is simply a function of vector density, longevity, and human biting rate, ignores the fact that man with his modern science and technology has become the dominant factor in the epidemiology of malaria, not only as a passive carrier or a victim of the disease but "as an occasionally rational being who could eradicate malaria if he would" (Russell *et al.* 1946). The fact that the disease cannot withstand the tides of progress has been corroborated by the fact that many countries situated in arid zones, or with extensive areas therein, have succeeded in recent history in eradicating the disease, i.e. Australia, Israel, Lebanon, USA and USSR. These countries, however, as well as those that are advanced in their control or eradication programmes in arid areas of North Africa and the Eastern Mediterranean region, are still exposed to malaria risk as the local malaria vectors still exist in their normal or pre-eradication density although the malaria parasites have been eliminated from the whole or the bulk of the population. This condition of anophelism without malaria can be disrupted leading to malaria epidemics or resurgence of the disease if no effective and continuous disease surveillance exist, or if no serious measures are taken to prevent the malaria hazard connected with mass movement of the population from or to malarious areas. The same catastrophy may follow the invasion of a potent and exogenous malaria vector through major engineering schemes in border areas (e.g. High Aswan Dam project) or through modern and rapid transport systems.

The knowledge of distribution, ecology and habits of the malaria vectors in arid zones is fundamental in understanding the epidemiology of malaria and in planning its control or eradication. At least 17 vectors belonging to 7 malaria epidemiological zones (out of the world's 12 zones) are known to transmit malaria in arid zones (Macdonald 1957). The main vectors in the arid areas of these 7 zones are the following:

(a) North American zone (South California, Arizona, New Mexico of USA and Mexico: *A. pseudopunctipennis* and *A. freeborni*.
(b) European–Asian zone (South-East USSR and South-West China): *A. messae* and *A. sinensis*.
(c) Mediterranean zone (North-West Africa and countries extending from Levant to Aral Sea): *A. l. labranchiae*, *A. m. maculipennis*, *A. superpictus*, *A. sacharovi*, *A. claviger*.
(d) Desert zone (oases in the Sahara, Jordan and north Saudi Arabia): *A. sergenti*.
(e) Ethiopian zone (subtropical arid zone of Africa and West Coast of Arabian Peninsula: *A. pharoensis* (Nile Valley and Delta Region in Egypt), *A. gambiae* (arid areas of

Senegal, Mauritania, Mali, Niger, Upper Volta, Chad, Sudan—recently drought stricken countries—Yemen, West Coast of Saudi Arabia).

(f) Indo-Persian zones (countries bordering Persian Gulf, Afghanistan, Rajasthan desert in West India): *A. culicifacies, A. stephensi, A. fluviatilis, A. pulcherrimus.*

(g) Australasian zone (North Australia): *A. farauti.*

Each of these vectors has its own habits of breeding, feeding, day-time resting and range of flight and dispersal. They differ also in their susceptibility to infection with malaria parasites, and this factor together with the human-biting rate, longevity of the vector, and the critical density at which the vector population can establish transmission, decide the quantum of malaria transmission under natural conditions.

When considering the epidemiology of malaria in arid zones, one has to realize the importance of the climatic conditions and the environmental factors that contribute to the malaria situation in any area. The most important climatic factors in arid zones are the temperature and the relative humidity which determine the length of the malaria transmission season. Summer isotherms above 16°C are essential for the development of the parasites inside the mosquito. In the instance of *P. falciparum*, the summer isotherms should be above 19°C. On the other hand, the longevity of the vectors depends on the relative humidity during the hot months, and monthly averages of 52% are required to permit the mosquito to live beyond the period the parasites take to develop inside it until they reach the infective stage. These climatic parameters together with the natural or introduced environmental factors within a socio-economic pattern give rise to varying types of malaria situations in the arid zones. These types can be outlined as follows:

(a) In arid regions situated in temperate climates (with summer isotherms of 20–25°C), the optimum temperatures are maintained for 1–3 months during the short summer season. There common infections are caused by *P. vivax* and the disease has a tenuous hold and can be easily controlled. Agricultural development through draining swamps, and animal husbandry have caused malaria to disappear in the absence of specific control measures.

(b) In arid regions in sub-tropical climates with cool winters and summer isotherms of above 25°C, the transmission of malaria occurs during the humid summer months when the average monthly relative humidity is maintained above 52%. Both malaria infections of *P. vivax* and *P. falciparum* occur with peaks in early summer and late summer respectively. In areas using water from rain, rivers or wells for dry farming and irrigation of winter crops, malaria may be absent or at low endemicity (hypoendemic) and does not constitute a serious health problem, unless man-made environmental changes such as introduction of perennial irrigation, wet cultivation and population movements set the stage for malaria epidemics and postepidemic hyperendemic situations to occur. This tragedy which has been repeated in many areas in the sub-tropical arid belt, is due to the lack of vision with regard to the serious malaria hazard connected with these undertakings. It is noteworthy to mention that once malaria is entrenched under such circumstances, it is most difficult to control or eradicate. The high malaria endemicity in the Delta region of Egypt, particularly where rice is grown, and the failure to interrupt malaria transmission in areas under perennial irrigation in Syria, Iraq, and the Khouzestan Province in Iran, in spite of eradication campaigns that have been conducted over the last two decades, bear witness to the fact. One may mention also that perennial irrigation to promote cotton cultivation, which requires huge amounts of agricultural pesticides, leads to the rapid development of the malaria vectors' resistance to the common insecticides.

(c) In the semi-arid areas of tropical climates where the climate is continuously warm and where a short season of marginal rainfall occurs (less than 250 mm per annum), malaria transmission is confined to the humid months which may extend from July to December. Most of the African countries stricken by the recent drought are situated in this belt. The main vector is *A. gambiae* and the dominant species of parasite is *P. falciparum*. Due to the habit of this vector to bite man almost exclusively, its prolific breeding in small puddles of rain water, untidy irrigation, dugout pits or hoof prints, its high infectivity rate, and very low critical density to establish malaria transmission (in the order of one mosquito per 30 persons), this vector has gained its reputation as the most efficient malaria vector (Macdonald 1957). With such a vector, any favourable malariogenic factor such as unusual rainfall, or man-made disturbance of the environment providing extensive breeding places, or any tropical aggregate of labour, is fraught with a serious malaria hazard in the form of fulminant epidemics with high malaria morbidity and mortality. If such malariogenic factors are maintained, a postepidemic hyperendemic situation sets in which will baffle health authorities to control it effectively even with the modern malaria armamentum that is used successfully in other parts of the world. The threat of such happening in connection with the recent construction of the High Aswan Dam and the possible infiltration of this vector into *gambiae*-free areas in Egypt has led Egypt to conclude an agreement with Sudan to keep a *gambiae*-free buffer zone south of Lake Nasser and to maintain continuous surveillance.

Irrigation in arid zones and man-made malaria

The arid zone, covering a surface area of 6400 million acres, is just over two and a half times as large as the presently cultivated area in the world. The population pressure has been forcing and will continue to force economic planners to increase the acreage of lands to be put under irrigation. One must stress here that it is not irrigation itself that gives rise to the malaria problem but mostly the lack of appreciation on the part of engineers, agronomists and administrators, as well as on the part of the people served by these schemes, of the principles of community hygiene and environmental sanitation. In the presence of natural unstable malaria in the arid zone, the introduction of favourable malariogenic factors by man in the form of extensive breeding places for the vectors, or by increasing the vector-man contact through aggregation of people in labour camps or new settlements, represent a malaria hazard. The scope of the malaria hazard depends largely on other epidemiological factors connected with climatic conditions, feeding habits of the vector and its potency, the housing conditions, the socio-economic status of the people, and above all, the consciousness and degree of cooperation of the planners and executives of the irrigation projects with the health authorities, to prevent or minimize malaria hazard. It may be relevant here to list some of the main attitudinal pitfalls among all those concerned with irrigation in the arid zone which are at the origin of such hazard.

The economic planner. When considering investments in an irrigation scheme in the arid zone, the economist should not be guided only by the material gains in crops in relation to expenditures, but should realize that in the last analysis the scheme is intended to contribute to the welfare of the people served and that this welfare is not to be assessed only on the basis of the annual *per capita* income of the family, but also on certain health indices, such as life expectancy which can reflect the people's enjoyment to live in a productive, salubrious and aesthetic environment. For this reason the economist has to consider the health implications of the project before, while, and after, an irrigation project is undertaken and to include adequate funds in the budget for this purpose.

It is gratifying to note that this concept is taking root, and one notes in the 1974 document of the "Permanent Inter-State Committee on Drought in the Sahel" the following statement: "The principal adverse health implications of irrigation may include the creation of havens for larvae, and parasitic and microbial pollution of water. This may mean a local increase in vector-borne diseases (malaria, onchocerciasis, etc), parasitic diseases (bilharzia, Guinea worm, etc.), and water-borne diseases (typhoid fever, etc.). It is therefore essential that the effect on health of the adoption of the project should be included in the study of the project so that both financing and solutions may be found before it is implemented."

The dam construction engineer. He should realize that his job is not only to plan and construct the dam, but he should realize the health hazards resulting from the storage of a huge expanse of water behind the dam, and be guided in the engineering design regarding water level fluctuation, the preparation of the impounding reservoir and the protection of the labour force by malariologists experienced in malaria control on impounded water.

The irrigation engineer. Whether the water for irrigation comes from rain, rivers, reservoirs, or underground, he should realize that in arid lands water is a valuable commodity, and that untidy irrigation by misplacing water through defective canals, absence of drainage leading to water logging, over-irrigation, miscellaneous excavations, multiply the breeding places of the malaria vectors and make the control of malaria expensive and difficult.

The agronomist. He should realize that the levelling of arable land is both beneficial to the irrigation of the crops and to the prevention of land depressions in which water stagnates and acts as breeding places of the vectors. In selecting the crops to be raised he has to realize the malaria implications connected with wet plants such as rice and sugar cane, or with such crops as cotton. He should anticipate that in clay soil, irrigation without draining will lead to waterlogging and should insist that proper drainage is attended to.

The administrator. He has to be guided in the selection of village sites, and in the design of housing structures for the settlers, by the technical advice of malariologists. He should see that a proper anti-malaria ordinance based on local conditions is operating in his area, and give his full support to those who are controlling the disease.

The sociologist. In studying the habits and living conditions of the inhabitants of irrigated areas in arid lands, he should study the means to bring about a community understanding of the seriousness of the malaria problem and the ways in which they can contribute to the prevention of this health hazard.

The public health administrator. Whether recruited as a team member in the feasibility study of an irrigation project, or as a public health officer in rural settlements, he should realize that the science of malariology is a sophisticated one, and requires the services of experienced malariologists, entomologists and sanitary engineers. Unless he seeks the technical advice of these experts, the malaria problem may get out of hand, and its consequent control will prove very expensive.

The urgent need for active cooperation and mutual understanding

The development of irrigation projects in arid lands may prove a blessing or a curse depending on whether they contribute to the health and well-being of the people. Arid zones are characterized by scarcity of water and variability of climate. To meet the needs of an accelerated rate of population growth in many countries within the arid belt, the utilization of all available water is essential to extend the areas under cultivation or to introduce multiple cropping. Modern technological knowledge is being applied to increase the water supply through rain-harvesting, construction of dams, drilling of wells, and even desalination of water from seas. Such hydrologic engineering schemes are mostly prompted by economic and social needs without regard to the environmental changes that they produce in upsetting the ecological balance, and in introducing serious health hazards, or increasing those that already exist. Malaria has been always associated with such engineering schemes, and its devastating effects have been obvious in many arid lands where vision and cooperation were lacking. This is the reason why a wedding of knowledge and a cultural balance have to be promoted among those who apply their skills in various sciences, those who apply them in engineering, and those who benefit from such irrigation schemes. The epidemiology of malaria in arid zones shows that the malaria problem is mainly man-made. As malaria does not respect national boundaries, the need for international cooperation is obvious. The specialized agencies of the United Nations have a great role in fostering such cooperation and in promoting comprehensive planning of such schemes as the restricted choices of an arid climate leave little margin for ignorance, inefficiency, lack of vision and lack of cooperation.

References

Macdonald, G. 1957. *The epidemiology and control of malaria*, Oxford University Press, p. 66.
Russell, P. F. 1955. *Man's mastery of malaria.* Oxford University Press, p. 8.
Soper, F. L. 1970. *Building the health bridge*, Indiana University Press, p. 301.
Russell, P. F., West, L. S. and Manwell, R. D. 1946. *Practical malariology*, W. B. Saunders Company, p. 343.

DISCUSSION AND CONCLUSIONS

Compiled by A. COUMBARAS

A lively discussion developed at the Symposium around the subject matter of the papers, and is summarized below. The following took part: A. H. M. Bakr, A. Coumbaras, M. M. Dieye, A. M. Farid, M. Hakimi, C. E. Houston, A. A. Idris, A. F. Kashef, M. Larivière, M. B. A. Malik, L. Obeng, R. S. Odingo and H. Said.

The irrigation engineer seems to bear a major responsibility for health deterioration due to irrigation. But have you thought of the financial provisions for preventing health hazards, to be included in the budget of an irrigation project? Is there legislation to force irrigation authorities to submit their plans to health authorities for their comment and approval? Can we predict with certitude the health hazards of an irrigation project, and make an evaluation of their costs? (M. B. A. Malik)

Definitely, an integrated approach of the public health problems that arise directly and indirectly (M. Larivière, A. F. Kashef) from irrigation is desirable from a technical as well as from an administrative and financial point of view. The health component has to be included, when necessary and possible, in the irrigation project considered as a whole, with public health specialists under the same project authority (A. H. M. Bakr, A. Coumbaras, H. Said). Good legislation is of course of great help to define the cooperation between health and engineering, but it is not sufficient. A sincere cooperation, with a great deal of good will is necessary (A. A. Idris, A. M. Farid).

Good results may be achieved if this cooperation is established, provided the decision makers keep in mind, not only the presumed financial profit of a project, but also try to evaluate the objective to be achieved in terms of health and environment. The final target is the well-being of man, and not the crude income of a project (A. Coumbaras, M. Hakimi, L. Obeng, R. S. Odingo).

Some encouraging results have been achieved through that kind of inter-disciplinary approach. Let us mention the Rahud Sudan integrated project (A. A. Idris), the programme de développement intégré du Bassin du Fleuve Sénégal (M. M. Dieye), the FAO/WHO co-operation, particularly in consideration of public water supplies within irrigation water development projects (C. E. Houston).

Dr A. Coumbaras, the Convener of this section, concluded as follows: In this section which has been devoted to human impacts of irrigation, we have ascertained that, although irrigation is expected to, and effectively does, improve the people's health and well-being, by the increase of their resources, this favourable effect is often altered by an increase in the prevalence and gravity of water-transmissible diseases, particularly schistosomiasis and malaria, and also by the aggravations of other problems of hygiene and public health less directly connected to water, but which are essentially due to a disorganized rush of the populations towards the irrigated zone.

We have ascertained that the spread of these diseases may often be, if not stopped, at least considerably restrained by adequate hydrological engineering techniques, a better maintenance of the irrigation schemes, and a better discipline in the use of water. These improvements are desirable for a better practice of irrigation itself. With the establishment of irrigation and the subsequent development, there is opportunity for promoting a health policy, concerned specially with drinking water supplies, habitat improvement, faecal pollution control and sanitary education. Solutions that would fit everybody could be found only by a close collaboration of all the specialists and administrators that are concerned, and an administrative and financial integration of the health activities, to be considered as a component of the scheme as a whole. This integration has to be achieved as early as the planning of a project, and it has to be maintained during its implementation. The sanitary surveillance has to be continued indefinitely in the irrigation scheme, which will always remain fragile from the viewpoint of epidemiology.

We have also ascertained that the altered environment due to irrigation has to be corrected, also through environmental measures, as far as schistosomiasis, malaria, and faecal transmission of diseases are concerned. Specific methods of control, such as molluscicides, insecticides and mass treatment are certainly essential, but they cannot give their full effect and long-term results, unless they are applied to a corrected environment.

Bibliography

References to literature in this Section have been relatively few. The reason is that WHO have produced a bibliography on "Water Resources Development and Public Health" (document MPD/75.4, pp. 45). Copies of this document (of which a revised and expanded version will soon be issued, possibly before the end of 1976) are available to officially or professionally interested persons on application to the Division of Malaria and other Parisitic Diseases, WHO, 1211 Geneva 27, Switzerland.

SECTION VIII

The International Viewpoint

In this last Section viewpoints are expressed by the specialist agencies of the United Nations which co-sponsored the Symposium of February 1976. On the non-governmental side, a statement from ICID provides a link with the ICID conference on irrigation and drainage which took place in August 1975 at Moscow. The International Commission on Large Dams, as another international organization involved, did not contribute a separate statement because the consequences on the environment of building dams had been thoroughly discussed at previous meetings.

These statements were made at the beginning of the Symposium of February 1976, but for this book the overall conclusions and guidelines for the future are given prominence in Section I. There is no separate statement here from UNESCO, which was, however, intimately associated with the Symposium at all stages. UNESCO's regional meeting on arid land irrigation in the Near and Middle East, which was a stage in the prosecution of MAB project No. 4, immediately followed the Symposium in Alexandria and is the subject of a separate UNESCO report.

IRRIGATION DEVELOPMENT IN THE WORLD

by CLYDE E. HOUSTON

Chief of Water Resources, Development and Management Service,
Land and Water Development Division,
FAO, Rome

Summary

Large increases in food production are necessary for the world's increasing population. Much can be obtained by placing more lands under irrigation and by improving the operation and management of existing irrigation schemes. The greater portion of newly irrigated lands will come from the developing countries while improvement of existing facilities is necessary in developing and developed countries. For all practical purposes the total water supply of the planet is fixed, but man can manipulate the supply in regard to location and time of use. The greatest potential for new development of irrigated land is through use of groundwater. This should be in conjunctive use with surface water and should be strictly controlled to avoid waste, over-development and salination. Water resources surveys should be carried out where there is a dearth of information and where the end use is planned—not for the sake of making a water survey. New irrigation programmes before 1985 might come to 23 million hectares at 1975 prices of about $40 billion. Improved water management of existing irrigation schemes can probably do more towards increasing food supplies than any other agricultural activity. Such practices would include improvement and rehabilitation of existing irrigation schemes plus training farmers in improved irrigation practices. If this could be carried out on 50 million hectares the returns in increased production would be enormous and might cost as much as $25 billion.

Résumé

L'importance de la croissance démographique mondiale nécessite une augmentation considérable de la production alimentaire, qui pourrait être obtenue en grande partie par la mise en place de plus de terres irriguées et l'amélioration du fonctionnement et de la gestion des terres déjà irriguées. La plus grande partie des exploitations qui seront mises sous irrigation à l'avenir se situeront dans les pays en voie de développement, alors que l'amélioration des périmètres existants sera nécessaire à la fois dans les pays développés et les pays en voie de développement. Du point de vue de l'utilisation pratique, on peut considérer que les disponibilités totales en eau de la planète sont fixés; l'homme peut cependent gérer ces ressources dans l'espace et dans le temps. Le potentiel le plus important pour les développements futurs des terres réside dans l'utilisation des eaux souterraines. Celle-ci devra cependant être entreprise conjointement avec le développement des eaux de surface, et devra être strictement contrôlée afin d'éviter le gaspillage, le sur-développement et les problèmes de salinité. Les études des ressources en eau doivent être entreprises où il y a un manque d'information et où l'exploitation ultérieure a été planifiée, et non pas seulement dans le but de réaliser des études en tant que telles. D'ici 1985, le programme d'irrigation de nouveaux périmètres pourrait atteindre 23 millions d'ha, pour un coût estimé (prix 1975) à 40 milliards de dollars. Une gestion rationnelle des ressources en eau dans les périmètres irrigués existants pourrait probablement contribuer plus à l'augmentation de la production que toute autre activité agricole. Une telle pratique, comprenant à la fois l'amélioration et la remise en état des périmètres existants et la formation des paysans sur ces terres, et entreprise sur quelque 50 millions d'ha, signifierait une augmentation énorme de la production à un coût estimé à 25 milliards de dollars.

Introduction

As the world population increases, food demands increase. These are two facts with which all people and governments agree. Where opinions differ, they are usually in reference to the pros and cons of regulating the population increase and how to provide the needed increased amounts of food. The population problem I leave to others, while the increased food question will be discussed from the irrigation standpoint. This does not mean that we overlook the importance of rain-fed agriculture. On the contrary, we recognize that the largest portion of food production comes from rain-fed agriculture and there are numerous means known for improving production under these conditions. Also, we know of the importance of water to fisheries, forestry, industry and domestic purposes, but it is generally recognized that about 80% of the water consumed in the world is attributable to irrigation. It is also recognized that irrigation is a single input to crop production. Others, such as suitable soil, fertilizers, viable seed, credit, equipment, incentives, favourable land tenure, pest control, transportation, storage, extension and research, are also necessary.

Civilization first developed in an environment requiring irrigated agriculture. Only during the past 1500 years has the scene shifted from the dry areas to the more humid regions. It is in these humid areas that the great population centres of the world have developed today so that most of the arable soils of these areas are now used for crop production. To further increase food and fibre production for the world we are again returning to the arid and semi-arid lands that comprise about 50% of the earth's land surface. The majority of the developing countries are located in the arid or semi-arid zones and nearly one-half of the world's population depends upon food production under some degree of irrigation.

Data presented at the World Food Conference held in Rome in November 1974 indicated that there are about 200 million hectares presently irrigated in the world with nearly one-half being in the developing countries. Much of the needed increase in food production must come from the developing countries. To provide this it is estimated that new irrigation programmes should produce water for 23 million additional hectares by 1985. If double cropping were feasible on all these lands, the total increase in irrigated area would be equivalent to 46 million hectares. These figures indicate the intensification in use of the basic natural resources—land and water. Intensification will also be aimed at maximum production per unit of water and eventually into food value such as protein per unit of water.

The total fresh and sea water content of this planet is essentially fixed. Although man has been able to manipulate to a certain extent freshwater supplies in regard to location and time of use, availability has changed very little over the past thousands of years. This situation may change radically if and when the desalination of sea water decreases sufficiently in cost to make its use possible for irrigated agriculture, but at present there is no immediate prospect of a breakthrough in this respect.

The total fresh-water resources of the world may amount to some 75 million km^3. About three-quarters of these waters are frozen, and of the remainder about 99% is groundwater. Although some of this can be recovered, at a cost, much of it is at depths of over 1000 metres. Even the surface water resources, which are more easily accessible, are often poorly located in relation to man's needs; for example, the world's largest river, the Amazon, lies in one of the least inhabited parts of the globe.

In addition to variability in space there is the problem of variability in time. Man needs a constant supply of water but, apart from the ice caps, some great lakes and deep groundwater, the available water resources are in movement. This movement is highly variable;

even great rivers—for example, the Indus—may be in devastating flood at certain periods of the year and almost dry at others. Moreover, on top of seasonal variation must be added the variation from year to year, which can cause disastrous droughts as we have recently seen in Africa.

Water—the resource and its development

It is now accepted that all usable water is a single resource, and neither surface nor groundwater development can be considered in isolation. The scientific approach to the determination of the water resources of any natural unit involves the drawing up of an integrated water balance. In other words, the total precipitation on the watershed or unit must equal or be balanced by evapotranspiration, surface runoff, infiltration to the groundwater basin, and changes in surface and groundwater storage.

Successful management of all water in a unit or basin requires the conjunctive management of surface and groundwater. In this way, groundwater can be used to supplement a scarce surface water supply during dry periods. Surface water can be used mostly during medium and high runoff times to satisfy agricultural and other users and to recharge aquifers through percolation in natural or artificial basins and channels and wells.

A fully integrated water conservation and allocation system, with conjunctive use of surface and groundwater, may include a number of rivers or streams with their catchments, surface reservoirs, aquifers with underground water, recharge works, together with a network of wells, pipelines or canals supplying water to dispersed areas. Such a multiple system, although efficient from the standpoint of water conservation, may require large capital outlays for all the necessary facilities and structures. It will be appreciated, therefore, that the multitude of individual elements in systems of this kind, which not only require careful adjustment in relation to each other but also influence other growth aspects of regional development, make it necessary to introduce complex planning techniques.

In traditional irrigation development and even in fully modernized project design, preference is still often given to the sole use of surface water resources. But in several areas, and on an increasing scale, existing surface waters are already fully surveyed, known, developed or allocated for future development. Frequently, simple and cheap surface water developments have all been completed and only complicated and expensive schemes are left for implementation. There remain the groundwater resources, which, in many of these areas, are larger than generally recognized and available for rapid exploitation even if full development may be difficult and fraught with possible dangers of exhaustion of reserves, decline in water table levels and consequent water pollution by salination.

There has been a general tendency to ignore or overlook groundwater as a potential source for large irrigation development. The four major reasons for this neglect are (1) more accurate estimates of available supplies from calculations based on hydrological cycle studies have only recently become possible; (2) efficient pumps and low cost power are becoming generally available; (3) hitherto, agriculture relied on traditional surface sources for water supply; and (4) groundwater science and technology have been developed rapidly to a degree unknown before.

So, in the field of irrigated agriculture expansion, technical assistance aims at maximizing the available water resources by the combined and integrated use of both surface and underground supplies.

The term hydrogeology now embraces all aspects of hydrology and the use of geophysical, hydrochemical, drilling, aquifer testing, simulation models and other means of investigating and determining the groundwater resources of an area or region. The tremendous increase in the number, complexity and efficacy of the tools now available for groundwater investigations, development and management is of great potential value to the developing—as well as to the developed—countries of the world. However, groundwater lives and moves and has its being in a geologic environment, and a knowledge of this environment is basic to a knowledge of the supplies that can be safely extracted from it. So the inputs to, and answers received from, such tools as analog and mathematical models must be critically examined in the light of a sound understanding of their geological meaning.

Although the integrated use of groundwater with surface supplies is a basic principle, it may sometimes happen that the natural hydrogeological unit is a large basin although the groundwater is required for use in limited areas where soil surveys indicate optimum returns. In these cases any possible clash between regional and local interests can usually be solved by carrying out a reconnaissance-type survey of the whole groundwater basin, at the same time concentrating the investigation—and often pilot development—on the zone where development will take place. This enables an equitable apportionment of the total groundwater supplies to be made.

Continued improvement in the integration of surface and groundwater will require further hydrological, geological and water use research. Simple computerization and mathematical models are adequate for general relationships and planning, but the study of more complex systems will require the use of electrical analogy or other methods. The more complex systems must be analysed in depth, taking into consideration legal and economic limitations in addition to physical factors. Such analyses should include the types of organizations which may be best adapted to carrying out the planning, financing and, finally, the management and operation of the highly integrated system. The determination of the costs and how they should be distributed throughout the entire operation assume even greater importance than they had in the past.

This type of water development planning is a highly complex undertaking. It includes a long sequence of steps and involves a large number of widely differing activities. It is this complexity which makes timing important. Good planning of water development will have to create an optimal path of activities which will keep at a minimum the time for which non-productive investments have to be made; it will also have to specify measures which will speed up the process of reaching full utilization of the facilities constructed, thus enabling an early flow of benefits.

The effect of all of this on the natural environment must be studied more than it has been in the past. A new approach needs to be taken which embraces all environmental aspects and integrates them with other socio-economic and ecological considerations. Alternatives need to be explored and optimum solutions sought, using new methods of evaluation and analysis of the environmental elements involved. This new concept must be based on the correct understanding of the interrelationship between the development and management of water and the environment as a whole, taking into account the different phases of water occurrence in the atmosphere, on the land, in the soil and underground.

However, in the planning and even execution of projects it seems frequently to be assumed that the application of modern technology alone will suffice. The technology is, of course, valid for certain operations, but the problem more often lies in how to apply it in a given socio-economic situation than in the determination and processing of particular technical inputs. Far too much planning in resource development is unrelated to the human factor or to the overall needs and objectives of the country. The result is often the production of plans that never get off the shelf, or partial or complete failure because the plans, however sound technically, have not been related to the ultimate users—the farmers—and have not been conceived with the understanding of what the latter can or will do.

Clearly any development planning must be based on a realistic assessment of available physical resources. In most parts of the world much information is available, but it is surprising how seldom it is found to be systematically recorded or processed in such a way as to point towards important development potential.

Usually there is a mass of existing climatic records and often records of stream flows. Both may be much less than complete but may go back over a considerable period of years. Agricultural development requires accurate assessment of distribution and probability, neither of which can effectively be determined without reasonably long-term records. Inferences derived from plant or wider biological communities can sometimes be valuable as a basis for determining general climatic zones, in the absence of reliable meteorological data, in the particular area being studied.

Much development expenditure can be saved, especially in the early stages of exploration and planning, if examination of water resources—whatever their nature—is given equal or higher priority to the investigation of other physical resources. A country with a wide range of sunshine-hours could well direct its enquiries first to areas with the better radiation figures. Again, an assured rainfall in an area is a good reason for seeking to exploit it. Although it may be valuable—and academically satisfactory—to have a systematic coverage of a country by regional first-stage surveys, in practice time and funds are usually both in short supply. Usually some information is available about climate and the water resource so that it is possible to direct first-stage land resource surveys to regions where the water resource shows some promise.

Funding agencies have found through experience that resource surveys should not be undertaken as self-justifying and self-contained activities. There should always be a major end use in mind such as for specific agricultural, industrial or urban projects or for combinations of all of these. Major funding agencies and major operating organizations have also found that resource investigations should be undertaken by those organizations most closely allied with the major end user.

Several developing countries have already initiated action by launching water resource surveys, but such surveys are time-consuming and require professional competence which is often in very short supply and unlikely to increase in the near future. It is accordingly proposed that a World Survey of Water Resources and Irrigation Potential be undertaken as a cooperative project by the relevant international agencies and interested countries. The cost of starting the proposed survey could be in the region of $10 million. The work would include:

To identify objectives and establish guidelines for the survey.

In collaboration with the technical services of the countries involved, to collect data on water resources at national level and later to aggregate these at regional level.

In close collaboration with the national soil resource services, to correlate available water resources to land classification and land capability maps and thus determine the potential areas of irrigation development.

In close liaison with national and regional research institutes, to establish crop water

requirements which can be used for the purpose of making an assessment of the development potential of specific areas.

In close collaboration with the irrigation departments of the countries, to collect data
on the cost of irrigation development and capital outlays required as well as on the costs
of annual operation and maintenance for completed projects. These data can then be
used to develop the construction cost indices applicable to different types of irrigation
schemes.

The importance of new irrigation development may be realized from the following figures
presented at the World Food Conference.

The new irrigation programmes envisaged to be needed between now and 1985 might
be expected to encompass some 23 million hectares. In monetary terms, the total outlay
for the 10 years, 1975–1985, at 1975 prices may approach $40 billion. Many of the individual
projects to be included in such a programme would be large scale, slow-maturing and would
need long-term financial assistance on attractive terms. In the past, financial aid for irrigation development has already represented a fair share of total agricultural aid, but the needs
in the future will be even higher.

Water management

Improved water management is as important to successful irrigation as water development. In fact, irrigation and drainage workers in developing countries generally agree that
improved water management probably can do more towards immediately increasing food
supplies and agricultural income in irrigated areas than any other agricultural practice.

Such activities would include improvement and rehabilitation of existing irrigation
schemes which, through neglect, unawareness or both, are not now being fully utilized.
The objectives are to ensure that the projects, frequently long-established, are operating
at high efficiency and to provide the individual farmer or groups of farmers with knowledge
and means for more effective water use. A reasonable target for achievement between now
and 1985 might be to improve the use of one-quarter of the existing facilities or, say, 50
million hectares.

The returns obtainable by a thorough overhaul of the world's facilities for irrigated agriculture would be enormous. A large number of irrigation schemes are operating at less
than 50% efficiency and the doubling of staple food crop yields, such as cereals, with
improved management of the necessary inputs is, in many areas, perfectly feasible. If it
is assumed that 70 million hectares are used for cereal production, then an average improvement in crop yields of one ton per hectare, achieved in, say, five years would mean an increased production of 70 million tons or nearly 20% of the total estimated cereal output
of all the developing market-economy countries in 1969–1971. This would be the equivalent
of at least 18 million hectares of new, fully equipped, irrigation projects which would otherwise require a minimum capital outlay of the order of $30 billion.

For many irrigation schemes such improvement consists of renewing the existing installations which have deteriorated or have been reduced due to lack of proper management.
While such improvement works will require additions to the original capital investment,
it is very often the case that returns from investment in the improvement of existing schemes
are higher than could be expected from the same investment in new works.

Increased production in existing irrigated farms can be obtained by improving the following: water availability and storage; water distributing networks; on-farm facilities; operation and maintenance of installations; improvement of irrigation and drainage practices; field reshaping and grading; improved education and training and extension services.

The implementation of these programmes presupposes willingness on the part of developing countries' governments to accord such activities a higher priority than in the past. The principal obstacles to be overcome are shortage of trained manpower and lack of funds. Steps should be immediately taken in all countries to expand training facilities in the disciplines required for land and water development.

As to funds, the estimated total cost for improving approximately 50 million hectares would be about $25 billion, with possibly $10 billion in foreign exchange requirements.

The introduction of improved water management in agriculture is a complex problem involving the structure of rural society and the reform of agricultural administration. Such activities often have consequences which reach as far as the national legislative bodies which have to formulate the required changes in land and water laws.

It is clear that measures to accelerate benefits from water development projects for agriculture must necessarily be grouped around the farmer's field, since the success of investments depends ultimately on the effectiveness with which the water is used by the individual farmers. If water is to be effectively used by the farmer he must have adequate means to do so and there must be sufficient incentive in material terms for him to be productive and to sustain his effort. The main problem still remains to educate and assist farmers to use water properly, which constitutes a tremendous task as millions of farmers are involved.

Apart from improving crop yields and ensuring a more economic use of water, good water management also has a significant effect on the overall economy of a project. This is because low irrigation efficiencies result in demands for larger quantities of water which, in turn, call for greater storage capacity, larger water control structures and bigger canals and, therefore, heavier capital investment. Furthermore, water so wasted may considerably reduce the area that can be irrigated with the same supplies, thus raising the construction cost per unit area and reducing the overall project returns.

Degradation of land by waterlogging and salinity is a common by-product of irrigation. More than 70% of the 30 million hectares of irrigated land in Egypt, Iran, Iraq and Pakistan is moderately to seriously affected. Two hundred thousand hectares of newly irrigated land in Egypt is seriously threatened and in Pakistan it is reported that 100 hectares go out of production every day due to salinity and waterlogging. India has about 12 million hectares affected. Vast salty areas are found along the Senegal River, on the border of Lake Chad and in most countries of northern and central Africa. They also occur in the coastal valleys and plains of Chile, Peru, Argentina, Venezuela and Haiti, and the recent development of salinity is found in the Far East in traditional rice areas.

All of this does not present an encouraging picture for the future although we do know what to do about it from the scientific standpoint. We know how to reduce canal losses, hydraulic structural leakage and inefficient water application. We know that salinity can usually be overcome by proper drainage and leaching. We know many different drainage methods, but they must be tried under all the differing soil and water table and water quality conditions found in the world. Drainage, too, is expensive and sometimes the cost of removing water is as great as the cost of applying it to the land in the first place.

Of the nearly 150 active FAO field projects dealing with water, most are aimed at water development and improved water management at the farm level. The total contribution by the UNDP to date to FAO projects in which water is an important aspects amounts to about $100 million. The operations undertaken under these headings also include water resource surveys, the establishment of research and training institutes, the construction and operation of irrigation pilot projects and direct assistance to member governments in the field by qualified experts.

In addition to the programmes financed from other outside sources such as Funds-in-Trust, cooperative programmes like the FAO/World Bank Programme and the World Food Programme are interested in projects containing water development aspects. The World Bank has given loans of over $500 million during the last 10 years for irrigation, drainage and flood control schemes while WFP expenditure during the same period for projects with a water-use element has amounted to about $300 million. These figures give a striking indication of the vital role played by water in integrated development programmes.

Water problems are likely not only to intensify during the next decade, but also to demand urgent answers and action as an increased return, in terms of food, from irrigated agriculture becomes more and more a pre-condition for feeding the rising world population and improving the general standard of living. The importance which irrigation development and good water management practices will assume cannot be overstressed and unless continuous and vigorous action is taken the battle for maximum utilization of this indispensable resource is likely to be lost. A period of great uncertainty and readjustment lies ahead and it is strongly urged that the subject of water, its development and use, should be of permanent concern to all mankind.

THE INFLUENCE OF IRRIGATION ON MANKIND

by A. H. TABA, M.D.

Director, WHO Eastern Mediterranean Region, Alexandria

Summary

The extension of irrigation can be justified from the point of view of economic necessity and social betterment; however, it can also carry with it the increased threat of health hazards to mankind. Diseases transmitted by water are many, but two emerge as posing serious public health problems, namely, malaria and schistosomiasis. In order to make irrigation undertakings successful and fully beneficial to mankind, close collaboration between irrigation and health authorities should be established as early as possible during the projects' formulation stage.

Résumé

Si l'extension de l'irrigation peut se justifier pour des raisons économiques et de promotion sociale, elle n'est cependant pas dénuée d'effets qui risquent de mettre en danger la santé de l'homme. Les maladies liées à l'eau sont nombreuses. Deux d'entre elles, le paludisme et les schistosomiases sont toutefois les plus graves au regard de la santé publique. Si l'on veut que les systèmes d'irrigation soient à la fois un succès et une source de bien-être pour l'humanité, il faut que les responsables de l'irrigation et ceux de la santé publique établissent entre eux une étroite collaboration au stade même de l'élaboration de ces projets.

Man has learned from time immemorial that agriculture holds a sacred and essential place in the existence of man and animals, so it was not by accident that old civilizations have always been associated with water, without which agriculture cannot succeed. They grew and developed along rivers and water courses. The extension of irrigation systems nowadays can be readily justified in many areas in the world from the point of view of economic necessity and social betterment, since it permits the exploitation of otherwise useless land and the improvement of crop yields. However, the blessing of water is not devoid of untoward consequences; it may carry with it the increased threat of health hazards. But, have the potential health hazards of such developments been given due consideration and precautionary measures taken accordingly? The answers to these questions are unfortunately not always reassuring.

Leaving aside the question of chemical pollution, surface water intervenes in the spreading of communicable diseases in two distinct pathways. The first one is a short and direct way: if proper sanitary measures are not taken, water used for drinking and domestic purposes may be polluted by human and animal wastes and might contaminate the consumer directly. This is the case with viral infections such as poliomyelitis and hepatitis, bacterial infections such as typhoid and paratyphoid fevers, bacillary dysentery, cholera and protozoal infections such as amoebiasis and intestinal flagellates.

The second pathway is more complex as it involves an intermediary host and vectors which find in surface water suitable conditions for their breeding, multiplication and dispersion. It is also, compared with the first pathway of infection, the more difficult to deal with. Diseases thus transmitted are mainly parasitic. They are many, but two emerge as posing serious public health problems, namely malaria and schistosomiasis.

Malaria is transmitted by about sixty different species of *Anopheles* mosquitoes, all of which are aquatic breeders, although their preferences for specific types of water vary greatly. Malaria infection was highly prevalent in the entire tropical and sub-tropical belt and extended to temperate climates, but has been remarkably reduced through eradication/control efforts made in recent years. However, an upward trend has again been observed in certain localized areas, due primarily to the development of resistance in the vectors to the insecticides commonly used in the malaria eradication or control campaigns. Wherever the disease is endemic, it is considered to be one of the most serious public health problems. It also has a significant economic effect due to its debilitating nature. Instances of man-made epidemics of malaria through irrigation schemes are well known. For example, outbreaks occurred in 1934, 1939 and 1945 around artesian wells and their outlets in the cases of central and southern Tunisia.

Schistosomiasis is undoubtedly one of the main health hazards to be faced in the many large irrigation projects developed or in planning in Africa, in the Middle East and elsewhere. Human infections exist in seventy-one countries and the estimated number of people infected is over 200 million. The parasitic fluke of schistosomiasis lives in blood vessels in humans and produces large numbers of eggs which damage body tissues. The disease is most debilitating and its economic effect in areas where it is endemic is incalculable.

Thus an irrigation scheme designed to improve agricultural potential may have profound socio-economic and morbid effects on human populations by introducing or spreading parasitic diseases such as schistosomiasis and malaria. Appropriate measures must therefore be taken to control these health hazards of irrigation, so as to derive full benefit from such water resources development undertakings.

In the design of an irrigation project, every effort should be made to ensure that the problem of transmission of diseases is carefully considered. The Tennessee Valley development in the USA appears to be a good example to illustrate this principle in so far as malaria control is concerned. The building of dams in 1912 and 1927 was followed by violent epidemics of malaria, affecting everyone who lived within the flight range of the mosquito. The authorities solved this problem by fluctuating the water levels of their lakes and keeping the verges free from weeds, thus preventing the breeding of the local vector. Through such engineering measures, coupled with operational manipulation of water levels in the reservoirs, malaria was successfully put under control with a minimum application of insecticides.

Another important point is that many good irrigation practices help to avoid public health problems, and these should be followed. For instance, intermittent irrigation not only saves water but also greatly reduces the possibility of mosquito and snail breeding. The irrigation canals and drainage channels properly graded, maintained and cleared from weeds would discourage vector development, and also facilitate chemical control, if required. The lining of canals and the covering of canals and drains, though costly, might be justified under special circumstances, taking into account their advantages in reducing health hazards.

It is thus clear that in order to make irrigation undertakings fully beneficial to mankind, the irrigation engineers and the public health officials both have to contribute with respect to their specialized professional knowledge and experience. Close collaboration between irrigation and health authorities is essential for the success of any irrigation undertaking and needs to be established as early as possible during the project formulation stage.

THE VIEWPOINT OF UNEP

Representing the Director General of UNEP, Nairobi

From the time man shed his dependence on the gathering of chance fruit and nuts and evolved the art of cultivating crops, he brought on himself the uncertainty of adequate and regular supplies as he exposed himself to the mercy of environmental factors which control his food production. But in spectacular instances he did succeed in subjugating the undesirable elements of the environment and in employing beneficial ones. Indeed, civilization was cradled some 6000 years ago against a background of successful irrigation of the floodplains of the Tigris and the Euphrates and the valley of the Nile. Since then over the centuries man has indulged in irrigated agriculture. From the low figure of 20 million acres of irrigated land estimated at the beginning of the nineteenth century, there was a steep increase to about 460 million acres by 1970.

Man also widely used rain-fed land for agriculture and the breeding of domesticated stock and in the process not only voraciously consumed and destroyed vegetation but also exhausted the fertility of soils. The repercussion has been harsh. Today, of the total potentially arable land, which is about a quarter of the ice-free surface of the earth, only about 44% is under cultivation. The rest cannot readily be used, much of it because of man-induced soil problems.

Of all continents except Africa south of the Sahara much of the potentially arable land is already under cultivation although only about 2% is irrigated. Even so, following irrigation, large areas have become unusable and may continue to be so as a result of severe and direct side-effects which seem to accompany irrigation. However, under arid and semi-arid conditions, where climatic and soil deficiencies seriously hamper reliable food production, populations continue to grow; the number of mouths which need feeding increases and the best way we know of meeting the problem is through increased food production. Therefore we find ourselves considering further the possibilities of irrigated agriculture in the arid and semi-arid lands which now cover substantial areas of the earth's surface. The problem is how best to do this without aggravating the present situation.

Essentially, irrigation is interference with the soil regime by introducing moisture of a quantity and sometimes quality to which the soil is not accustomed, with consequent modification to the ecology of the environment. The resultant increased yield of crops is beneficial, but the ecological changes provide conditions for the establishment of foreign species, including the vectors of malaria and schistosomiasis, which carry with them new problems.

437

It is on record that the percentage of one population infected with schistosomiasis increased from 2% to 75% when perennial irrigation was introduced. It is relevant to mention that ecological and habitat management for the control of schistosomiasis was the subject of a meeting which took place in Cairo in October, 1975 (see p. 403).

Interference with the hydrologic cycle may cause underground flooding and waterlogging which, with the deposition of salts in the soil, may defeat the entire purpose of irrigation. Additionally, the irrigation of otherwise unusable arid land may create social and health problems as populations move into new areas. Fortunately, however, as may be said of most development projects which produce environmental hazards, timely precautions may minimize such undesirable effects.

A principle which is basic to the concept of the UNEP is the achievement of development without destruction, and in harmony with the environment. Although UNEP is not a funding agency, the Environment Fund which is administered by it encourages activities fulfilling two major roles: continued development towards a better life for all peoples, and support for critical components of development programmes which promote environmental safety.

It is a basic concept of the approach developed by UNEP that a systems approach to environmental considerations should be pursued. Development programmes tend to follow sectional interests, which need coordination in order to ensure a balanced and safe environment. The complications of arid land irrigation render this subject so important that the Governing Council of UNEP identified it as a priority area.

On a different level UNEP has been given the responsibility for the preparation of the UN Conference on Desertification to be held in 1977 following a resolution of the General Assembly calling for an understanding of the processes of desertification and the adoption of concerted action to arrest and, where possible, to reverse these processes through international cooperation. With the full involvement of appropriate UN agencies and experts, UNEP has initiated action for the analysis of existing knowledge, preparation of a map of desert-prone areas, and the formulation of a concerted plan of action.

Many centuries ago, irrigation in Egypt raised the standard of life of a people. In spite of problems that have resulted from irrigation, we are quite optimistic that it can be environmentally safe, but we require a better understanding of the processes which produce the adverse effects. We believe that then, at the policy and decision levels, adequate precautions will be taken in the design and management of irrigation schemes to ensure sustained yield. Thus can we contribute to the present major effort to combat the creation of arid lands and to restore degraded lands for the continued use of mankind on this our only one earth.

THE UNITED NATIONS WATER CONFERENCE, 1977

by KARL-ERIK HANSSON

*Assistant Director, Centre for Natural Resources,
Energy and Transport, UN, New York*

Water is expected to constitute the major resource problem in the decades ahead, and the United Nations has sounded an alert in the decision of its Economic and Social Council and General Assembly to hold a Water Conference from 7 to 18 March 1977. It will be held at Mar de Plata, Argentina, at the invitation of the Government of Argentina.

The Conference is to be preceded by a series of preparatory steps, including regional meetings organized by the regional economic commissions in Western Asia (June 1976 or later), Latin America (Lima, 30 August–3 September 1976), Asia and the Pacific (Bangkok, 20–27 July 1976) and Africa (Addis Ababa, 20–24 September 1976 immediately preceded by an ECA/UNESCO/WMO African hydrology meeting). The ECE Committee on Water is organizing the input for Europe. These meetings will be partly based on national reports and will be particularly important in developing recommendations for action at the national and international levels. It has been recommended that national committees be established for this and related purposes, and in several cases the formation of such committees has already served to bring together for the first time all the interests in development, management and use of water resources.

Non-governmental organizations which are expected to have an important place in preparations and at the Conference itself include, for example, IWRA (New Delhi, December 1975), IWLA (Caracas, February 1976) and COWAR (Alexandria, February 1976), all of which are relevant.

The Conference will have four major topics, and the preparatory meetings and documentation will be similarly organized, as likewise is the suggested outline for national reports. In addition, provision is made for "thematic papers" on subjects such as that considered in this volume. The four topics are: I. Resources and needs: assessment of the world water situation; II. The promise of technology: potential and limitations; III. Policy options; and IV. What to do: action proposals. A main coordinating consultant is helping with each topic, among them Prof. Gilbert White for topic I and Prof. M. Kassas for topic IV. All the relevant UN organizations are making coordinated inputs and close coordination is maintained with the United Nations Desertification Conference to be held later in 1977. These are all part of the series of major world conferences, on Environment (1972),

Population (1974), Food (1974), Women (1975), Habitat (1976), and will be followed later by another on science and technology.

The Conference itself will focus on topics III and IV, while documentation on topics I and II, which will be important in the preparatory process, will serve as background. More in-depth discussion is expected in parallel meetings on "thermatic papers." The Conference is seen as a high point in a continuing process, with follow-up action expected to be stimulated particularly at the national and local level but also in the scientific community.

THE INTERNATIONAL COMMISSION
ON IRRIGATION AND DRAINAGE

by MILOS HOLY

President of ICID

The International Commission on Irrigation and Drainage considers irrigation from engineering, economic and social aspects and is very much interested in environmental problems caused by irrigation projects.

The development of human society, especially in recent years, is characterized by growing economic activity manifested by the development of the power industry, the chemical and other industries and agriculture, and by other branches of the national economy. Investment activity is growing in all branches of the economy and if this is not controlled it is bound to bring about serious conflicts between society and the environment. A significant share in this investment activity is taken by water management projects, among them by the building of reservoirs, irrigation and drainage systems and flood control systems. Therefore, it is the task of all concerned to seek ways of designing these projects so as to avert conflicts with the environment, and in such a manner as to ensure that natural resources be used in harmony with the growth of the economy as well as in harmony with a healthy living environment.

For this reason the ICID included in its last Congress, held in August 1975 in Moscow, a session on "Environmental Control for Irrigation, Drainage and Flood Control Projects." We had papers from agencies of UN and from various countries, especially those which contain arid and semi-arid regions. On the basis of these papers we divided the impacts of irrigation into those which have positive or negative effects on the environment (see table on following page.)

After fruitful discussions, the session came to the following conclusions:

1. Irrigation, drainage and flood-control projects affect the environment not only as engineering projects but they also cause health, social, cultural, aesthetic and political impacts on the environment. These impacts are mostly positive, but negative impacts do also occur.
2. All impacts of the projects must be predicted and considered in advance. To achieve this, feasibility studies should be elaborated, in order to evaluate all future impacts on the basis of present knowledge. Experience gained from the construction and operation of existing projects should be considered.

441

	Positive	Negative
Engineering	Improvement of the water regime of irrigated soils. Improvement of microclimate. Possibility provided for use and disposal of waste waters. Retention of water in reservoirs and possibility of a multipurpose use.	Danger of waterlogging and salination of soils, rise in groundwater table. Changing water properties of water in reservoirs. Deforestation of area to be irrigated and with it a change of the water regime. Reservoir bank erosion.
Health	Ensuring increased agricultural production and thus improving the nutrition of the population. Recreation facilities in canals and reservoirs.	Possible spread of diseases ensuing from certain types of irrigation. The danger of pollution of water resources by return runoff. Possible infection by wastewater irrigation; new diseases caused by retention of water.
Social and cultural	Increasing the social and cultural level of the population. Interest in tourism to newly built reservoirs.	Displacement of population from reservoir area. Necessity of protecting cultural monuments in inundated areas. Problems of adjustment of people in irrigated areas.
Aesthetic	New artificial man-made lakes.	Project's architecture does not fit the environment.
Political	Increased self-sufficiency in food.	

3. Decisions to proceed with any project must be founded on knowledge of its complex impacts, including impacts on the environment. Traditional economic approaches do not allow for an evaluation of the social impacts. A broader decision-making method should, therefore, be elaborated involving the evaluation of all criteria, i.e. economic and social, of which the latter cannot as yet be fully expressed in economic terms.

4. The protection and creation of the living environment is carried out in the interest of society and must, therefore, be set above the interests of individuals or groups, and on the international scale above the interests of individual states. The protection and creation of the environment should, therefore, be put under State and International control, and the respective programmes of action must be drawn up and legislative bodies set up with a view to achieving this aim.

DISCUSSION

Compiled by E. B. WORTHINGTON

A world water survey

The world survey called for by Dr. Houston (p. 429) is of great importance for several reasons.

1. In a number of regions several countries compete for the same source of water, such as an international river.
2. The distribution of land and water on one hand, and of populations on the other, is extremely unequal between countries.
3. The infrastructure and availability of funds for irrigation development vary greatly and the systems of financial and technical aid, which depend on national needs, are largely uncoordinated.

Therefore a world water survey should be undertaken internationally, not merely as a patchwork of national studies. It should include rough estimates of the cost of possible developments, so that, while respecting national integrity, it would present a picture of world priorities. (E. J. Blom.)

Water management

Management of water is the key to irrigation efficiency and to many environmental effects. There are three stages—conveyance, distribution and application. Experience (for example in Iraq, Egypt, Palestine and India) shows that conveyance and distribution can be greatly improved by design, operation and maintenance. A programme of survey and study should be started in the 200 million hectares under irrigation in order to identify areas where and how improvement has been achieved. Such a survey, carried out perhaps in parallel with a world water survey, could do much to demonstrate how to increase food resources with quicker effect than new irrigation schemes. As an example, water management was much improved in a project in southern France by introducing a system of payment by farmers for water used in accordance with the needs of their crops. The socio-economic problems of water management are usually greater than the economic ones. (Y. Barrada, A. F. El Kashef, C. E. Houston.)

The population problem

Many papers presented to the Symposium imply a basic assumption that population must continue to increase at its present rate. More abundant food supplies and control of disease will continue to increase population, and this will lead to further demands on resources which have a limit in the long term. The process is apparent in Egypt and many other countries. The Symposium should draw attention to this overriding problem, although the need for population control is not a particular subject for our discussions. Evidence from several parts of the world suggests that there may be a turning point in the rate of population increase as the standard of living and of culture rises. This does not solve present problems, but may assist in reducing further problems in the future when resources may no longer be available to meet them. (H. B. Rofe, E. B. Worthington.)

APPENDIX 1

LIST OF PARTICIPANTS AT THE SYMPOSIUM

ALGERIA
HAKIMI, Moustapha — Ingénieur Agronome—Ecologiste, Ministère de l'Industrie et de l'Energie, 14 avenue de Pekin, Alger.

AUSTRALIA
MITCHELL, Searle David — University Lecturer, Zoology Department, University of Adelaide, North Terrace, Adelaide.

PELS, Simon — Geologist, PO Box 205, Deniliquin 2710, Sydney, NSW.

EGYPT
ABBAS, Mamdouh Hassib — Head of the Agriculture Research Branch, National Research Centre, Dokki, Cairo.

ABDEL GHAFFAR, Ahmed Sabry — Head of the Soil and Water Science Department, Faculty of Agriculture, University of Alexandria, Chatby, Alexandria.

ABDEL HAI, Abdel Motagally Mohamed — Agriculture Engineer, PO Box 786, Alexandria, Egypt 73/084.

ABDEL-HALIM, El Damaty — Professor of Soils, Ain-Shams University, Shubra, Cairo.

ABDEL-LATIF, Ibrahim Atia — National Research Centre, Soils Department, Dokki, Cairo.

ABDEL MONEIM, Balba — Professor of Soil Fertility, Department of Soil and Water Science, Faculty of Agriculture, Alexandria University.

ABDEL NABI, Ibrahim Said — Director of Land Water Researches, 81 Boustan Street, Cairo.

ABDEL SALAM, Abdel Kader Aly — Irrigation Director, 12 Mohamed Shawky Street, Kasr el Eini, Cairo.

ABDEL SAMIE, Ahmed Gamal — Vice-President of the Academy of Scientific Research and Technology, 101 Kasr el Eini Street, Cairo.

ABDOU, Ahmed Yehia — Director of Rural Environmental Health, Ministry of Health.

ABU ZEID, Mahmoud A. — Director, Water Management Institute, Ministry of Irrigation, 22 El Galaa Street, Boulak, Cairo.

AHMED, Mahmoud Fawzi — Assistant Director General, Department of International Organization, 17 Ismail Abou Fetouh, Dokki, Cairo.

AMER, Fathy — Professor, Department of Soil and Water Science, Faculty of Agriculture, Alexandria University.

AYYAD, Mohamed Abdel Gawad — Professor of Plant Ecology, Faculty of Science, Moharram Bey, Alexandria.

BADAWI, Saleh Abdel Aziz — Researcher, National Research Centre Botany Laboratory, El Tahrir Street, Dokki, Cairo.

BAHNA, Fawkia Labib — Researchers, National Research Centre, Dokki, Cairo.

BAKR, A. Helmy Mohamed — Associate Professor, College of Agriculture, University of Alexandria, Shatby, Alexandria.

*BALBA, A. M. — Professor of Soils, University of Alexandria.

BALIGH, Aly Mohamed — Professor of Irrigation, Faculty of Engineering, Cairo University, 62 El Orouba Street, Heliopolis, Cairo.

BEHAIRY, Awatef G. — National Research Centre, Dokki, Cairo.

EFFEICH, Nabil Elie — Research Department

EL ABD HESHAM, Sherif — Agricultural Engineer, PO Box 786, Alexandria.

EL ANSARY, Ahmed Ezzat — Assistant Professor of Hydraulics, Faculty of Engineering Alexandria.

* Not registered

445

EL ATTAR, Hatim A. — Associate Professor, Faculty of Agriculture, Alexandria.

EL AWADY, Mohamed Nabil — Associate Professor, College of Agriculture, Ain Shams University, Shoubra El Kheima, Cairo.

EL BAGOURI, Ismail Hamdi — Associate Professor, Soils Department, Desert Institute, Matariah, Cairo.

EL EBRASHY, Mohamed Fikry — Professor of Agronomy and Head, Department of Agronomy, Faculty of Agriculture, El Minia.

EL GALA, Abdel Moneim M. — Associate Professor, Faculty of Agriculture, Ain Shams University, Shubra el Khaima, Cairo.

EL GAMAL, Abdel Azim Shahin — Counterpart Expert, Soil and Drainage Section, FAO 73/048 Project, PO Box 786, Alexandria.

EL GHITANY, Mohamed Ali — Agricultural Engineer, 39 Moustapha Pacha Street, Sidi Gaber, Alexandria.

EL HABASHA, Kamal M. — National Research Centre, Dokki, Cairo.

EL HARIRI, Dardiri Mohamed — National Research Centre, Dokki, Cairo.

EL KASHEF, Ali Fahmi — Technical Consultant and Member of the Board of Directors, 5 Willcocks, Zamalek, Cairo.

EL KHOLI, Fouad Ahmed — Professor, Head of Agriculture Department, Atomic Energy Establishment, Cairo.

EL KOBBIA, Talaat Mohamed — Professor of Soils, Ain Shams University, Shubra, Cairo.

EL MAHDI, Mohamed Abdel Moneim — Chairman of Botanical Wealth Laboratory, Desert Institute, Mataria, Cairo.

EL RAFHI, Mohamed Salem — General Manager of the Research Department.

EL SAID, Farouk Anter — Associate Professor, Soil Laboratory, National Research Centre, Dokki, Cairo.

EL SHAFEI, Salah Aly — Department, Exer. Manager for Civil Projects, 4 Ahmed Morsi Street, Guiza, Cairo.

EL SHAZLY, Kamal Moh. Sabry — Assistant Co-Manager, Control of Water Loggings Salinity in Areas West of Nubaria. Canal ECY 73/048, 151 Tiba Street, Sporting, Alexandria.

EL SHERIF, Ahmed Fouad Mahmoud — Associate Professor, National Research Centre, Dokki, Cairo.

EZZAT, Mohamed Aly — Head of Research Department, General Organization for Rehabilitation and Development Projects, 8a El Sakhawy Street, Manchiet El Bakry, Cairo.

FAHMI, Sarwat H. — Ministry of Irrigation.

FATHY, Ahmed Mohamed — Lecturer, Faculty of Agriculture, Alexandria University, Soil and Water Department, Alexandria.

GEWAIFEL, Ismail Mohamed — Assistant Professor, Soil and Water Department, Faculty of Agriculture, University of Alexandria.

GHABBOUR, Samir Ibrahim — Lecturer, Animal Ecology, Institute of African Research and Studies, Cairo University, Giza, Cairo.

HABIB, Ibrahim Mohamed — Cairo University, Faculty of Agriculture, Giza, Cairo.

HAFEZ, Mohamed Badr — Director of Agricultural Department, Authority of Land Reclamation, Dokki, Cairo.

HANNA, Fayez Salib — Associate Professor, Soils Laboratory, National Research Centre, Dokki, Cairo.

HANNA, Luis Filib — Assistant Professor in Soil Science Department of El Minia Agriculture Faculty.

HANNA, Mikhail Shaker — Professor of Irrigation and Irrigation Design, Faculty of Engineering Alexandria University, Alexandria.

HASSAN, Hamdi — Professor of Soils, University of Ain Shams, Shubra, Cairo.

HASSAN, Mohamed Naguib — Professor, College of Agriculture, University of Alexandria.

HASSAN, Salah — Agriculture Division, National Research Centre, Dokki, Cairo.

HEFNI, Kamal Hussein — Director, Institute of Ground Water, Researches, Ministry of Irrigation, 15 Guize Street, Cairo.

HILAL, Mostafa Hassan — Head of Soils and Water Department, National Research Centre, Dokki, Cairo.

HUSSEIN, Mohamed Moursi — National Research Centre, Agriculture Division, Plant Physiology Department, Plant Water Relation Unit, Dokki, Cairo.

HUSSEIN, Touni Ali — Lecturer of Agronomy, Field Crop Production Department, Faculty of Agriculture, El Minia.

ISMAIL, Essam Khalil — Agriculture Engineer, 12 El Beroni, Sporting, rue El Horria, Alexandria.

ISMAIL, Hassan M.	Ex President of Cairo University, Head of Environment Com. of Academy, 26 Nahda Street, Dokki, Cairo.
KADR, Mostafa Mostafa	Professor of Pedology, College of Agriculture, Shatby, Alexandria.
KAMAL, Ali Ahmed	Chairman of National Committee for International Program of Hydrology, 21 Gamal El Din Abou El Mahasin, Garden City, Cairo.
KASSAS, Mohamed A. F.	Professor of Botany, University of Cairo.
KHALIFA, Ali Mohamed	President, Permanent Joint Technical Commission for Nile Waters, 13 Mourad Street, Guiza, Cairo.
KHALIL, Mohamed Bakr	Professor of Hydraulics, Faculty of Engineering, University of Assiout.
KINAWY, I. Z.	
KISHK, Mohamed Atia	Faculty of Agriculture, Minia.
KHOURI, J.	Arab Centre for Studies of Arid Zones and Dry Lands.
LASHIN, Abd El Hafiz Younes	Lecturer at El Minia Agriculture Faculty, Soil Science.
MAHER, Abdel Moneim Ali	Professor, Faculty of Agriculture, Assiout.
MAREI, Sherif M.	Lecturer, Faculty of Agriculture, Department of Soil and Water Science, Alexandria University, Alexandria.
MASSOUD, Massoud Ahmed	Lecturer at the Agricultural Department for Soil and Water, Atomic Energy Establishment, Cairo.
MAHMOUD, Rafik El Ghareeb	Lecturer of Botany, Alexandria University, Plant Ecology Botany Department, Faculty of Science, Moharrem Bey, Alexandria.
METWALLY, Yousef Salah El-Din	Professor of Soils, Ain-Shams University, Shubra, Cairo.
MOHAMED, Sami Soliman	Associate Professor, Desert Institute, Matariya, Cairo.
MOHAMED, Samir Ali	Associate Professor, Desert Institute, Cairo.
MORSY, Ahmed Abdel Aziz	Associate Professor, National Research Centre.
MORSY, Yousry Mohamed	Researcher, Soils and Water Laboratory, 6 Blokkat El Swarry, Darrassa, Cairo.
MOURSI, Hamdy Abdel Aziz	Researcher, National Research Centre, El Tahrir, Dokki, Cairo.
NICOLA, Farid	Ministry of Reconstruction.
NOUR, Saleh El Sayed	Head of Ground Water, Geophysical Department, 179 Nozha Street, Heliopolis, Cairo.
NOUR EL DIN, Nemat Abdel El-Aziz	Faculty of Agriculture, Ain Shams University, Shoubrah El Kheima, Cairo.
OMAR, Farouk Abdel Aziz	Lecturer in El Minia Agriculture Faculty.
QUENAMY, Ibrahim Zaki	Minister of Reconstruction and Housing, Egypt ICID National Committee.
RASLAN, Salah Abdel Azim	Director of Experiments, General Organization of Reclamation Projects and Agriculture Development.
REDA, Fatma	Associate Professor of Plant Physiology, National Research Centre, Dokki, Cairo.
SAID, Hussein	Academy of Scientific Research and Technology, Egyptian Department in ICSU.
SAID, Naguib Fahmi	Ministry of Reconstruction, ICID.
SEFAINE, William Naguib	General Director for Irrigation Department.
SHATA, Abdou Ali	Director, Hydrogeology Department, Desert Institute.
TALHA, Mahmoud	Associate Professor of Soils, Soils Department, Faculty of Agriculture, Ain Sham University.
TAWADROS, Anwar Ameen	Civil Engineer, General Organization for Rehabilitation Projects and Agricultural Development, 112 Hegay Street, Cairo.
YANSOUNI, Charles	Consultant, Egyptian Vineyards and Dist. Co., 122 rue El Shahid Galal El Dessouki.
YOUNAN, Nazeih Assaad	Assistant Professor, Faculty of Engineering, University of Alexandria, 191 Port Said Street, Sporting, Alexandria.
YOUNES, Hussein Ahmed	Researcher Pedologist, Soil and Water Use Laboratory, National Research Centre, Dokki, Cairo.
ZAKI, Mohamed Moheb	Supervisor General of the Agricultural Research Centre, Giza.
*ZEID, Mahmoud Abu	(see discussion, session III)

FEDERAL REPUBLIC OF GERMANY

EHLERS, Ekart	Professor of Geography, 355 Marburg-Geographisches Institut, Renthof 6.
FRITZ, Alfred	Agronomist, 6700 Ludwigshafen, BASF-6DGE/3.
GARBERCHT, Karl Heinz Gunter	Professor, Technical University, PO Box 3329, 33 Braunschweig.

FRANCE

Coumbaras, John Alexis	WHO Consultant, 11 rue de Vaugirard, 75006 Paris.
Daget, Jacques	Professor, 43 rue Cuvier, 75005 Paris.
De Forges, Jean	Professeur, 51 rue de Ranelagh, Paris.
Larivière, Michel	Professeur, Faculté de Médecine, 40 rue du Bac, 75007 Paris.
Person, Jacques	Charge de Mission BRGM, 2107 rue de la Source, 45.160 Olivet.
Roche, F. Marcel	Ingénieur en Chef, Chef du Département de la Recherche Fondamentale au Serv. Hydro. de l'ORSTOM, 19 rue Eugénie Carrière, 75018 Paris.

GREECE

Christodoulopoulos, Phottos Christodoulous	Agronomist, Thiseos 65 Kallithes, Athens.
Karakatsoulis, Georges Panayotis	Professeur d'Hydraulique Agricole de l'ENSA d'Athènes, Calinis 38, Zografou, Athènes.

HUNGARY

Kovács, Gyorgy	Civil Engineer, 11 Fo-utca 48–50, Budapest.
Varallyay, Gyorgy	Head of Department, 1022 Budapest Herman O.n. 15.

INDIA

Ramanathan, Ramakrishna Kalpathi	Professor (Emeritus), India Physical Research Laboratory, Ahmedabad 9.

IRAN

Adib, Mohamed	Engineer, Iran, Teheran, Varzesh Ave., Ministry of Energy, Water Division.
Roointan, Parviz	Engineer, Teheran, Iran, Amirabad Shermali No. 18 st. No. 39.

IRAQ

Abid, Awn Hussain	Director General of Irrigation, Irrigation Directorate, Baghdad, Iraq.
Al Bayati, Hussain Ali	Head of Design Division, Bureau of Soil Study of Design, Mansoon, Baghdad, Iraq.
El Samawi, Mohamed Ibrahim	Director General of Environment, In front of Medical City, Office of Human Environment, Baghdad, Iraq.
Hamami, Ali	Public Health Administrator, Alexandria, WHO.
Ibrahim, Ahmed Salman	Chief Engineer, Higher Agricultural Council, Baghdad, Iraq.
Ismail, Nashat Hamid	Head Soil Division, Ministry of Irrigation, Baghdad.
Kazzaz, Rashid Kamal	Chief Engineer of Irrigation Department, Sullamania.

JAPAN

Cho, Toshio	Director, Professor of Irrigation, Sand Dune Research Institute, 1390 Hamasaka, Tottori, Japan.

JORDAN

Abu Sa'da, Mohamed Said	Hydrogeologist, Water Gas Department, PO Box 12, Kuwait.

KENYA

Odingo, Richard Samson	Associate Professor of Geography, University of Nairobi, PO Box 30197, Nairobi, Kenya.

KUWAIT

Abusadda, Said M.	Water and gas Department.

MALI

Diakite, Samou	Ingénieur des Eaux et Fôrets, Directeur Station, Recherche sur les Plantations Irrigués, Mali.

MEXICO

HAUSER, Arturo — Assistant Director of Water Pollution Control, Ministry of Water Resources, Reforma 107, so Piso, Mexico 4, DF.

MAURITIUS

MONGELARD, Cyril Joseph — Plant Physiologist, 99–193 Aliea Heights Drive, AIEA, Hawai 96701.

MOROCCO

ISMAIL, Zeryouni — Chef du Service d'Hydrobiologie, Ministère des Travaux Publics, Rabbat.

KABBAJ, Abdellatif — Chef de la Direction des Ressources en Eau du Maroc, Direction de l'Hydraulique Casier Rabat Chellah, Maroc.

SERGHINI, Idrissi Mohamed — Chef du Service de l'Equipement à El Jadida.

NETHERLANDS

BLOM, E. Joop — 22 Beaulieustreat, Arnhem, The Netherlands.

BOS, Marinus G. — Civil Engineer, PO Box 45, Wageningen, The Netherlands.

ROSCHER, Klaas — Agronomist, Agricultural University, Nieuwe Kanaal II, Wageningen, The Netherlands.

STORSBERGEN, Cornelis — Head, Department of Irrigation Hydrology, DHV—Consulting Engineers, PO Box 85 Amersfoort, Holland.

SAUDI ARABIA

AL HASHIM, Mohamed Hashim — Irrigation Engineer, Research Department, Ministry of Agriculture and Water, Riyadh, Saudi Arabia.

ALSOULI, Saleh Sulaiman — Agronomist, Research and Development Department, Ministry of Agriculture, Riyadh, Saudi Arabia.

GHAZAL, Ahmed Ali — Agriculture Engineer, Hossa-Hofuf, PO Box 76, Saudi Arabia.

HLJAZI, Ali Abdul Khalek — Agronomist, Assistant Director for Agriculture and Water Affairs, Abha, Saudi Arabia.

SENEGAL

MOUHAMADOU, Makhtar Dieye — Directeur de l'Equipement Rural (Ministère du Développement Rural et Hydraulique), Senegal.

SUDAN

AMIN, Mutamad Ahmed — Faculty of Medicine, Khartoum, PO Box 2371.

IDRIS, Ahmed Ali — Director General Epidemiology, Ministry of Health, PO Box 303, Khartoum.

SYRIA

HAMIDEH, Taher — Engineer in the Design Department of the General Organization of the Euphrates Dam, GOED, Al Thawra Town, Syria.

KASSAB, Mouteb — Director of the Agricultural Research, GADEB, Agricultural Sector, Raqqa, Syria.

OSMAN, A. — ASCAD, Damascus.

OUESS, Id. — Director of the Design Department, GOED, EL Thawra, Syria.

SWEDEN

DANFORS, Erik — Senior Lecturer, Department of Land Improvement and Drainage, Institute of Technology, S 10044 Stockholm, Sweden.

EKLOF, Bertil — Manager Director, 47 Av. Grande Bretagne, Monte Carlo, Monaco.

FALKENMARK, Malin — Executive Secretary, Sveavagen 166, 15 tr, S 11346, Stockholm, Sweden.

THAILAND

KULAPONGSE, Precha — Irrigation Engineer, Royal Irrigation Department, Phitsanulok, Irrigation Project, Samsen Road, Bangkok.

SIRION, Sanan — Chief Design Engineer, Southern Thailand Royal Irrigation Department, Parkred, Nonthaburi.

UNITED KINGDOM

BELL, Peter John	Soil Hydrologist, Institute of Hydrology, Wallingford.
CARPENTER, Charles Raymond	Civil Engineer, c/o EPDC, Marlow House, 109 Station Road, Sidcup, Kent.
CARTEL, Gordon	PO Box 215, London SW1.
CHANDLER, Albert John	Medical Entomologist, PO Box 1974, Kisnum, Kenya.
DIXON, Chapman John	Director Soil Mechanics, c/o Foundation House, Eastern Road, Bracknell, Berks.
DOWNS, Barry	Chartered Civil Engineer, c/o Howard Humphreys and Sons, Kennet House, Kings Road, Reading.
HILL, Norman Michael	Government Scientist (Ecologist), c/o Microbiological Research Establishment, Porton, Salisbury, Wilts.
HILTON, H. Peter	Engineering Adviser, c/o Outward Mail Room, ODM, Bland House, Stag Place, London SW1.
IMRIE, William	Consulting Engineer, c/o Howard Humphreys and Sons, Kennet House, Kings Road, Reading, Berks.
LLEWELYN, D. A. B.	Technical Adviser, c/o British Petroleum, Britannic House, London EC2.
MIDDLETON, Allardyce Alexander	Consulting Engineer, Deneth House, Station Road, Cambridge.
ROFE, Henry Brian	
SWAN, Hugh Christopher	Civil Engineer, Standard House, London Street, Reading, Berkshire.
WAGER, Jonathan F.	Environmental Planner, Suez Canal Regional Plan, Development Advisory Group, c/o Ministry of Housing and Reconstruction, Cairo.
WHITING, Prichard Roger	Civil Engineer, c/o Scott Wilson Kirk Patrick & Partners, Scott House, Basingstoke, Hants.

USA

HESS, W. John	Professor of Engineering, 36 N 2nd Street, Clearfield, Pa. 14830.
McJUNKIN	
WHITE, F. Gilbert	Professor, University of Colombo, Boulder, Co.
WOLLMAN, Nathaniel	Professor of Economics, University of New Mexico, Albuquerque, NM 87131.

USSR

BORISSOV, Yuri	Representative of the USSR, Ministry of Reclamation and Water Resources in ARE, Office of the Economic Counsellor, Zamalek, Cairo.
POLINOV, Stanislav	Doctor of Technical Science, Tashkent, Saniiri, Yakuba, Kolossa 24.
ROZANOV, B. G.	Senior Programme Officer, UNEP, P.O. Box 30552, Nairobi, Kenya.
YOUSSENKOV, Eugeni	Doctor of Agricultural Science, Ministry of Amelioration of the USSR, Bassmanny, Jupik 6, Moscow.

SWITZERLAND

FARID, Ahmed Mohyeddin	Consultant, WHO, 15 François Lehmann, 1218 Grand Saconne, Genève.

UNITED NATIONS ORGANIZATIONS

UN SECRETARIAT

HANSSON, Karl-Erik	Assistant Director, United Nations Headquarters, New York 10017, USA.

FAO

HOUSTON, E. Clyde	Civil Engineer, Viala delle Termi di Caracella 00100. Rome, Italy.
MASSOUD, Ibrahim Fathy	Technical Officer—Soil Reclamation and Development, Land and Water Development Department, AGLS, FAO, Rome.

IAEA

BARRADA, Yehia   Head, Soils Irrigation and Crop Production Section, Joint FAO/IAEA Division of Atomic Energy in Food and Agriculture, IAEA, Vienna.

FAO—EGYPT

ARAR, Abdir-Razik Abdullah   Regional Land and Water Development Officer, PO Box 2223, FAO Regional Office, Cairo.

BARTLING, Olof Mats   Assistant Exp. Drainage EGY.048/73, PO Box 786, Alexandria.

CHIMONIDES, John Savvas   Senior Irrigation and Drainage Practices Engineer, PO Box 786, Alexandria.

*DASTANE, N. G.   FAO Agronomist, Egypt.

EL GABALY, M. M.   Regional Land and Water Project Director, FAO, Cairo.

EVERS, Alfred   Irrigation Engineer, PO Box 786, Alexandria.

GENET, William B.   Drainage Engineer, PO Box 786, Alexandria.

JONES, Peter H.   Project Manager, PO Box 786, Alexandria.

MALIK, Bashir Mohammed   Irrigation and Reclamation Engineer, PO Box 786, Alexandria.

PALMQUIST, N. Wilbur   Hydrogeologist, PO Box 786, Alexandria.

RE, Renaud Emile   Project Manager, UNDP/FAO New Valley Project, c/o UNDP, PO Box 982, Cairo.

FAO—LEBANON

VAN LEEUWEN, Nicholaas Hendrikus   Irrigation Engineer, 23 rue Du Fer à Cheval, 62000 Arras, France.

UNESCO

BLOEMENDAAL, Sander   UNESCO, Cairo

CISCHLER, Christian E.   UNESCO, Cairo.

DA COSTA, A. José   Senior Officer, Division of Water Sciences, Place de Fontenoy, Paris 75700, France.

REHEEM, Kamal   Director, UNESCO Regional Office, UNESCO, Garden City, Cairo.

WHO

DEOM, Jacques   Senior Scientist, Division of Malaria and other Parasitic Diseases, WHO, 1211 Geneva 27, Switzerland.

FARID, M. A.   Consultant. WHO, 15 François Lehmann, 1218 Grand Saconne, Genève.

WORLD BANK

HOTES, L. Frederick   Engineer, 8310 Botsford Ct., Springfield, Virginia 22152, USA.

UNEP

OBENG, Eva Letitia   United Nations Environment Programme, PO Box 30552, Nairobi.

ROZANOV, G. Boris   Senior Programme Officer, UNEP, PO Box 30552, Nairobi.

INTERGOVERNMENTAL ORGANIZATION

ACSAD

KHOURI, Jean   Director, Water Resources Division, ACSAD, PO Box 2440, Damascus.

OSMAN, Mustafa Ahmed   Director, Soil Sciences Division, ACSAD, PO Box 2440, Damascus.

NON-GOVERNMENTAL ORGANIZATIONS

IAHS

DOOGE, James   Professor and Senator, Department of Civil Engineering, University College, Upper Marrion Street, Dublin 2, Ireland.

ICID

HOLY, Milos   Professor, Prague 2, Karlovo No. 3, Czechoslovakia.

ICOLD

SCHMIDT, Gabriel

Directeur, Coyne & Helher, Bureau d'Ingénieurs-Conseils-Paris, 5 rue d'Héliopolis, 75017 Paris, France.

ISSS

SZABOLCS, Istvan

Director of Research, Institute for Soil Science and Agricultural Chemistry of Hungarian Academy of Sciences, Budapest 1022, Hermann O.U.15.

COWAR

WORTHINGTON, Edgar Barton

President of COWAR, Colin Godmans, Furners Green, nr. Uckfield, Sussex.

KELLER, Reiner

Secretary General of COWAR, Schwarzwoldstr. 18. D-7812 Bad Krozingen, Germany.

SERRA, Louis

Vice-President, 98 rue Xavier de Maistre, France.

SECRETARIAT

BOUERY BADR, Colette

Personal Assistant to the Head of UNESCO, Regional Office, Cairo.

DARELL-BROWN, Susan Elizabeth

Administrative Assistant, c/o Linnean Society, Burlington House, Piccadilly, London W1.

FOURNIER, Frederic

UNESCO Consultant, 2 rue Saint Charles, 78000 Versailles, France.

LE NAOUR, Jeannine

Secretary, 78 rue de la Garenne, Le Plessis-Robinson, 92350 Paris, France.

APPENDIX 2

GLOSSARY OF ACRONYMS

ACSAD	Arab Centre for the Study of Arid Zones and Dry Lands
ASCE	American Society of Civil Engineers
AWWA	American Water Works Association
COWAR	Committee on Water Research
DIP	Dez Irrigation Project
ECA	Economic Committee for Africa
ECE	Economic Committee for Europe
EPA	Environmental Protection Agency
FAO	Food and Agriculture Organization
GOED	General Organization of the Euphrates Dam
IAEA	International Atomic Energy Agency
IAH	International Association of Hydrogeologists
IAHS	International Association of Hydrological Science
IAWPR	International Association on Water Pollution Research
IBP	International Biological Programme
ICID	International Commission on Irrigation and Drainage
ICOLD	International Commission on Large Dams
IGU	International Geographical Union
IHD	International Hydrological Decade
IHP	International Hydrological Programme
ISSS	International Society of Soil Science
IUBS	International Union of Biological Sciences
IUCN	International Union for the Conservation of Nature and Natural Resources
IWRA	International Water Resources Association
MAB	Man and the Biosphere Programme
NAS	National Academy of Sciences
ORSTOM	Office de la Recherche Scientifique et Technique Outre-Mer
SCOPE	Scientific Committee on Problems of the Environment
SIL	International Association of Limnology
UATI	Union of International Engineering Organizations
UNDP	United Nations Development Programme
UNEP	United Nations Environment Programme
UNESCO	United Nations Educational, Scientific and Cultural Organization
USDA	United States Department of Agriculture
WFP	World Food Programme
WHO	World Health Organization
WMO	World Meteorological Organization

INDEX